The Yeast Handbook

More information about this series at http://www.springer.com/series/5527

Graham G. Stewart

Brewing and Distilling Yeasts

Springer

Graham G. Stewart
International Centre for Brewing and Distilling
Heriot Watt University
Edinburgh, Scotland

GGStewart Associates
Cardiff, Wales, United Kingdom

The Yeast Handbook
ISBN 978-3-319-69124-4 ISBN 978-3-319-69126-8 (eBook)
https://doi.org/10.1007/978-3-319-69126-8

Library of Congress Control Number: 2017955013

Printed on acid-free paper

This Springer imprint is published by Springer Nature
The registered company is Springer International Publishing AG
The registered company address is: Gewerbestrasse 11, 6330 Cham, Switzerland

Foreword

Brewing and distilling are ancient processes. As Primo Levi puts it in his classic book of autobiographical short stories, *The Periodic Table*: "Distilling is a slow philosophic occupation; it gives you time to think. You are aware of repeating a ritual consecrated by the centuries ...". Levi was considering distillation from the point of view of a chemist in the narrow context of the simple separation of X from Y as an analytical technique. Distilling is unquestionably more complex when used in the production of, for example, Scotch whisky, and there is plenty of "time to think" in the even slower "philosophic occupation" of employing fermentation in the production of alcoholic beverages.

Graham Stewart certainly gives us much to ponder upon in a scholarly work which brings out not only the many similarities between the two types of yeast used in the production of spirits and beers but also their differences. A glance at the contents page of the substantial volume demonstrates its scope. Taxonomy, cell structure and metabolism are considered in depth, and due space is given to practical aspects of yeast cropping, handling and storage. Flavour production, yeast stress in fermentation and vitality/viability are thoroughly and thoughtfully discussed. Bioethanol production, genetic manipulation, killer yeasts and non-*Saccharomyces* ethanol producers also receive due attention.

A particular strength of the book is the historical account it gives of the most significant features of the development of knowledge regarding yeast and fermentation. The formidable accomplishments in the application of science in the field are clearly demonstrated. Such an account will be useful to those engaged in the fermentation industries who seek an understanding of the development of the processes and practical methodology of their profession. It may also be read with advantage by historians who consider both the chemical and biological sciences.

I have known Graham for more than forty years, from a time, which now seems so distant, when no major brewing company worth its salt would be without a research laboratory and European Brewing Convention (EBC) Congresses were attended by the great and the good of the world's brewing industry, regularly attracting over 1500 delegates. Only Graham could have written this authoritative and detailed account of brewing and distilling yeast. He brings the perspective and experience of a venerable informed insider who has made so many important contributions of his own to our knowledge of the subject.

May 2017 Raymond G Anderson

Preface

Anne Anstruther

Inge Russell

When I was conducting research for my PhD degree, and then beginning to write the thesis, it was emphasized to me, more than once, by Rod Brunt, my PhD supervisor, that, if at all possible, PhD theses should "tell a story"! They should start with "Once upon a time . . ." and finish with ". . .and they lived happily ever after". Dr. Brunt also expressed the view (as time has gone by that I agree with) that a basic difference between pure and applied microbiology research does not really exist; there is only good and bad research! This book is certainly not a PhD thesis (indeed, its contents refer to many PhD theses—a number of which I have had the privilege to supervise or examine). Nevertheless, I hope it tells an informative story of essentially good research that is focused on yeast and is applicable to both the brewing and distilling processes.

I am extremely grateful to Anne Anstruther for her support, patience, encouragement and guidance throughout the course of this project. I am also indebted to

Inge Russell for critically reading all the chapters in this book. Lastly, and by no means least, I owe my wife, Olga, considerable gratitude for her understanding throughout my scientific career, particularly when I was attempting to commit pen to paper! I am a great advocate of the phrase "I hate to write but love to have written!" (Parker D (1925) A certain lady. The New Yorker 1(2):15–16).

Edinburgh, Scotland
August 2017

Graham G. Stewart

Contents

Chapter 1
Introduction

1.1 Yeast Is a Eukaryote and a Single-Cell Fungus

Another book on yeast with an emphasis on brewing and distilling—why? Many review publications (books, monographs, articles, etc.) have considered brewer's and distiller's yeast systems, but these have usually been in separate documents. This book is a personal appreciation of yeast which has also been elaborated in greater detail in a number of other personal publications (e.g. Stewart 1977, 2009, 2010, 2015a, 2016a; Stewart and Russell 2009; Hill and Stewart 2009; Russell and Stewart 2014; Sammartino and Stewart 2015; Walker and Stewart 2016; Stewart et al. 2016). The primary objective of this book is to compare and contrast the characteristics of brewing and distilling yeast cultures in a single document because the two types of yeast have much to learn from each other but there are many differences! Although the central objective of both types of yeast culture is to produce ethanol, carbon dioxide and glycerol plus other metabolites, primarily through the glycolytic Embden-Meyerhof pathway (EMP) (Chap. 6), there are many differences between the two types of yeast. Indeed, as will be discussed in detail later, there are also many differences within these two yeast types (e.g. between ale and lager brewing strains, distilling yeast for potable and industrial ethanol purposes) and the products (beer, whisk(e)y, gin, vodka, rum, saké, neutral and fuel alcohol, etc.) of various manufactures of brewer's and distiller's yeast. Similarities and differences will be discussed, together with the implications of these differences in the production of beer and distilled products (potable and industrial). Although wine is not within the subject area to be considered here, where appropriate, examples from the uses of yeast in the production of this alcoholic beverage will be discussed.

This text considers 17 chapters and they consider a plethora of subjects relevant to brewing and distilling yeast strains. In each chapter, every effort has been made for them to be "stand alone" regarding the subjects considered. However, some duplication and overlap between chapters has occurred (e.g. Chaps. 6 and 9; 11 and

G.G. Stewart, *Brewing and Distilling Yeasts*, The Yeast Handbook,
https://doi.org/10.1007/978-3-319-69126-8_1

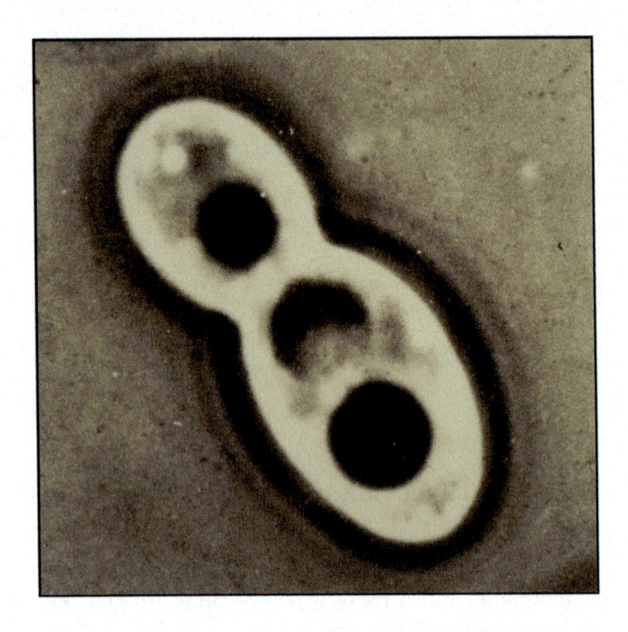

Fig. 1.1 Phase contrast micrograph of a budding cell of *Saccharomyces cerevisiae*

12; 8 and 14; 15 and 17) and every effort has been made to identify convergences in the text of relevant chapters.

Yeasts are unicellular fungi (Fig. 1.1), and they are economically the most important grouping of microorganisms employed on this planet (Kurtzman et al. 2011). However, this chapter is only an introduction to a very extensive story which this book attempts to relate. The industrial and economic uses of yeast are summarized in Table 1.1 (Stewart 2016a). However, this text will focus on yeast's fermentation properties in order to produce potable and industrial ethanol employing a wide range of substrates: starches, cellulose, hemicellulose, sucrose, inulin, lactose and other fermentable and potentially fermentable substrates (e.g. lignocellulose) (details later).

Although the focus here is on fermentation alcohol produced by yeast from a variety of substrates, it should be emphasized that yeast is a eukaryote (Fell 2012). Organisms (mammals, plants, protozoa, fungi, etc.), whose cells contain a nucleus and other structures (organelles), enclosed within membranes, are also eukaryotes (Fig. 1.2). The nucleus and its membrane-bound structure set eukaryotic cells apart (e.g. animals, plants, algae and fungi) (Adl et al. 2012; Youngson 2006) from prokaryotic cells (bacteria, cyanobacteria, etc.) (Whitman et al. 1998). Prokaryotes are unicellular organisms that lack a membrane-bound organelle. Further details regarding the structure of alcohol producing yeast cells will be discussed later (Chap. 5). Suffice to say, the predominant yeast genus employed in the production of fermentation alcohol (but not the only genus) is *Saccharomyces* (Chap. 9). Other yeast genera are used in specialized alcohol fermentation systems, for example, *Schizosaccharomyces*, *Pichia*, *Pachysolen* and *Kluyveromyces* (Castro and Roberto 2014) (Chap. 17). It is also worthy of note that some bacteria (not many) (e.g. *Zymomonas mobilis*) also produce ethanol (Rogers et al. 1982; Barrow et al. 1984).

Table 1.1 Industrial and economic uses of yeast

Product	Function
Potable ethanol	Beer, cider, wine, spirits (whisk(e)y, gin, vodka, rum, liquors, etc.)
Industrial ethanol	Fuel, pharmaceuticals, sterilants, solvent
Baker's yeast	Biomass (animal feed) and carbon dioxide
Yeast extracts	Cell walls, membranes, mannans, glucans, vitamins, food flavourings
Heterologous proteins and peptides	Plethora of medical applications

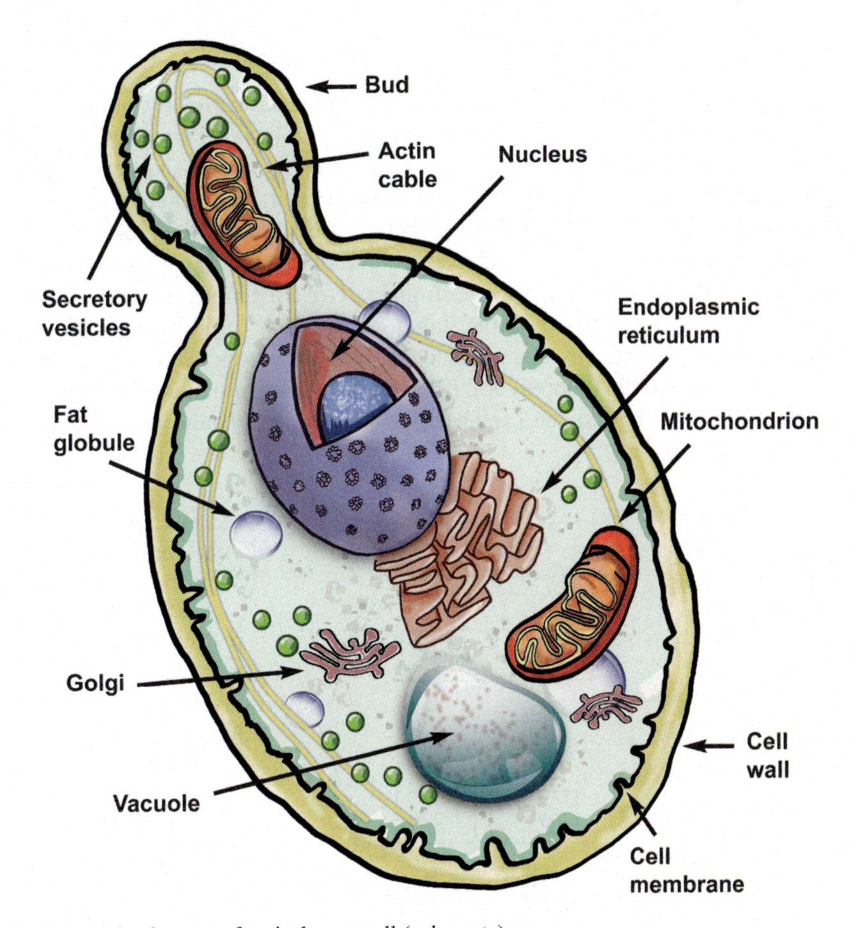

Fig. 1.2 Major features of typical yeast cell (eukaryote)

1.2 The Brewing and Distilling Processes

Brewing was one of the earliest processes to be undertaken on a commercial scale, and, of necessity, it became one of the first to develop from a craft into a technology. In many instances, this trend has recently turned full circle with the introduction of craft breweries (and distilleries) worldwide (Oliver 2012; Lyons 2018) (Chap. 12). Beer production can be divided into five distinct unit processes (and a number of subprocesses) (Fig. 1.3):

- *Malting* is the germination of steeped barley or other appropriate cereals (e.g. wheat and sorghum) and controlled drying (or kilning) of the germinated cereal which will preserve the cereal's enzyme activity (mainly amylases, proteinases, glucanases, etc.).
- *Mashing* is the extraction of the ground (milled) malted barley (or other malted cereals) with water and followed by the enzyme hydrolysis of starch and proteins. The extracted material is then separated (with a lauter tun or mash filter) from the insoluble material (spent grains) to produce sweet wort (Becher et al. 2017).
- *Adjuncts*—starch from unmalted cereals (corn, rice, wheat, barley, sorghum)—is often used as part of the grist to supplement the starch that comes from the malt. This starch is hydrolysed by enzymes from malt (or exogenous enzymes) and the application of heat to gelatinize the starch. Also, syrups (produced as a result of starch hydrolysis—usually in a separate facility—or sucrose [from cane or beet]) are added to the kettle (copper) boil (Stewart 2006). The use of adjuncts also has an effect on wort sugar spectra and beer flavour stability (details later) (Chaps. 7, 11 and 16).

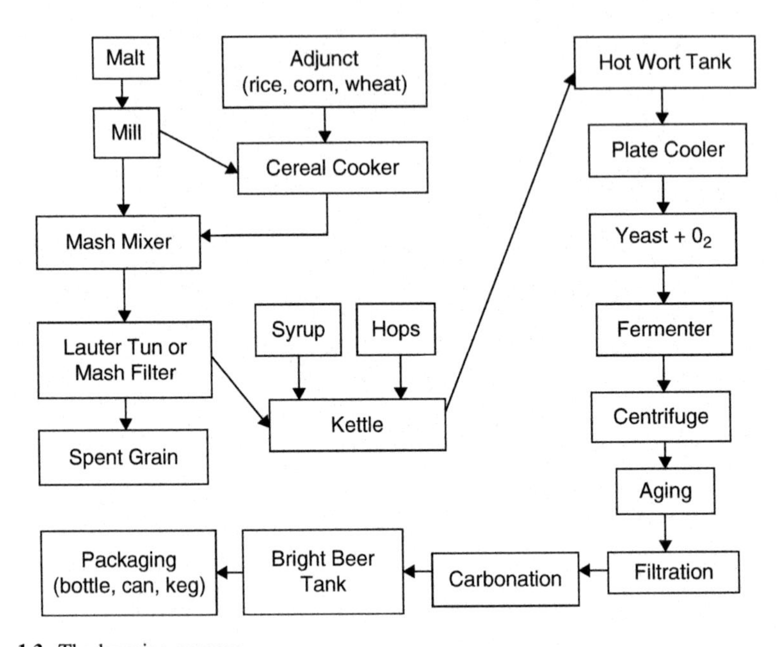

Fig. 1.3 The brewing process

- *Wort boiling* (in a kettle or copper) sterilizes, precipitates protein and slightly concentrates the sweet wort. The inclusion of hops and hop extracts and, as already discussed, sometimes syrups contain fermentable sugars (adjuncts), which are also added to this boiling wort (Leiper and Miedl 2006; Miedl-Appelbee 2018).
- *Fermentation, maturation, dilution and filtration*—details later. However, it is important to emphasize that brewer's yeast cultures are cropped during the closing stages of fermentation and recycled into a subsequent fermentation (Russell 2016; Hung et al. 2017), which is not the case in the production of whisk(e)y and most other distilled products (details in Chap. 13) (Stewart 2014, 2018; Taylor 2018).
- *Packaging* (meaning: kegging, casking, bottling and canning) is beyond the scope of this book. However, there are comprehensive publications that consider this process (e.g. Klimovitz and Ockert 2014; Partridge 2018).

The use of adjuncts (unmalted cereals) (Stewart 2016a, b, c, 2018) in brewing is permitted in most countries. However, there are a few countries that do not permit the use of these non-malted carbohydrate sources in brewing. Germany is the pre-eminent country that strictly controls the production of its beer. The German Purity Law (Reinheitsgebot) dates from 1516 (Narziss 1986) and states that only barley malt, hops and water (yeast was included later) can be employed in the brewing process and the use of adjuncts is forbidden! The 500th anniversary of the proclamation of the Purity Law Decree was celebrated in 2016 by an event arranged by the German Brewers' Association in Ingolstadt, and the keynote speech was given by Angela Merkel—Germany's Chancellor (Winkelmann 2016).

It is interesting to note that the Purity Law is as strict as the Scotch Whisky Regulations (formerly the Scotch Whisky Act)—details later—which dates from 1909 and was last revised in 2009 (Gray 2013). Two further European countries have adopted the Brewing Purity Law—Norway and Greece (Dornbusch and Karl-Ullrich 2011). However, brewing companies from many countries all have malt beers in their product portfolio.

Brewing fermentation and yeast management are diverse processes that can be quite complicated (Stewart 2015b). Major differences include two distinct yeast species—*Saccharomyces cerevisiae* and *Saccharomyces pastorianus*—for ales and lagers, respectively (Chap. 3), together with their different wort metabolic characteristics (Chaps. 6 and 7), fermentation and maturation temperature preferences and sedimentation (flocculation) (Chap. 13) properties (Russell 2016).

The production processes for Scotch and other whiskies have many features in common with brewing, but there are important differences. Basically, hydrolysed cereal starch from barley malt, maize (corn) and/or wheat is fermented to ethanol, CO_2 and a plethora of other metabolites (many that possess flavouring properties). The fermented wort (wash) is distilled, and the distillate matured in charred oak casks (barrels) for a minimum of 3 years to produce Scotch whisky (often considerably longer) (Russell and Stewart 2014).

The primary differences between brewing and distilling production processes are:

- Wort employed for distilling (unlike brewing) is unhopped and not boiled. Distiller's wort, produced exclusively from unmalted corn or wheat (grain wort), is not filtered, whereas malt wort is filtered with a lauter tun or mash filter in order to remove spent grains and solid protein/phenolic complexes—similar to brewing.
- The unboiled distiller's wort is not sterile and contains enzyme activity (amylases and proteinases)—details of the implications to the process (particularly fermentation) will be discussed later (Chaps. 5 and 7)
- Only strains of *Saccharomyces cerevisiae* (ale type) are employed for the fermentation of whisky worts. This also sometimes includes a portion of spent brewer's yeast, and, as will be discussed in detail later, "special" strains of some yeast that can metabolize small dextrins are also employed by some (not all) distillers (Russell and Stewart 2014) (Chaps. 7 and 17).
- Distiller's fermentations are conducted at higher temperatures (28–32 °C for potable alcohol and even higher for industrial alcohol) than brewing and in different geometry fermenters. Fermentation temperatures for the production of industrial fermentation alcohol are 32–36 °C (Chap. 9).
- Following the fermentation of distiller's wort, the yeast is not cropped and, consequently, only used once. The yeast plus fermented wort goes directly into a still [batch for malt wort (Fig. 1.4) and continuous for grain wort (Fig. 1.5)]. The process implications of yeast management, distiller's and brewer's wort, are discussed in detail later (Chap. 12).

Fig. 1.4 A typical pot still employed for the batch distillation of malt whisky

Fig. 1.5 A continuous (Coffey) still

Although Scotch whisky is an important distilled product, it is certainly not the only one! Potable spirits such as vodka, gin, rum and brandy cannot be overlooked, and appropriate fermentation procedures and other aspects are discussed (Chap. 15). Also, industrial fermentation ethanol production is increasingly important (blended into gasoline) as a liquid fuel, particularly in the United States, Brazil and other countries, which are currently involved (details in Chap. 9). This extensive development has resulted in considerable advances with yeasts that metabolize substrates such as hydrolysed starch (glucose, maltose and maltotriose) and cellulose (pentoses and hexoses), inulin, sucrose and lactose.

References

Adl M, Sina M, Adl A, Lane CE, Lukeš J, Bass D, Bowser SS, Brown MW, Burki F, Dunthorn M, Hampl V, Heiss A, Hoppenrath M, Lara E, Le Gall L, Lynn DH, McManus H, Mitchell EA, Mozley-Stanridge SE, Parfrey LW, Pawlowski J, Rueckert S, Shadwick L, Schoch CL, Smirnov A, Spiegel FW (2012) The revised classification of eukaryotes. J Eukaryot Microbiol 59:429–493

Barrow KD, Collins JD, Leight A, Rogers PL, Warr RS (1984) Sorbitol production by *Zymomonas mobilis*. Appl Microbiol Biotechnol 20:225–232

Becher T, Ziller K, Wasmuht K, Gehrig K (2017) A novel mash filtration process (Part 1). Brauwelt Int 35:191–194

Castro RC, Roberto IC (2014) Selection of a thermotolerant *Kluyveromyces marxianus* strain with potential application for cellulosic ethanol production by simultaneous saccharification and fermentation. Appl Biochem Biotechnol 172:1553–1564

Dornbusch H, Karl-Ullrich H (2011) Reinheitsgebot. In: Oliver G (ed) The Oxford companion to beer. Oxford University Press, Oxford, New York, pp 692–693

Fell JW (2012) Yeasts in marine environments. In: Jones EBG, Pang K-L (eds) Marine fungi and fungal-like organisms. De Gruyter, Springer, Berlin, pp 91–102

Gray AS (2013) The Scotch Whisky Industry review, 36th edn. Sutherlands, Edinburgh

Hill AE, Stewart GG (2009) A brief overview of brewer's yeast. Brew Dist Int 5:13–15

Hung J, Turgeon Z, Dahabieh M (2017) A bright future for brewer's yeast. Brauwelt Int 35:195–197

Klimovitz R, Ockert K (2014) Beer packaging, 2nd edn. Master Brewers Association of Americas, Minneapolis, MN

Kurtzman CP, Fell JW, Boekhout T, Robert V (2011) Methods for isolation, phenotypic characterization and maintenance of yeasts. In: The yeasts, a taxonomic study, 5th edn. Elsevier, Boston, pp 87–110

Leiper KH, Miedl M (2006) Brewhouse technology. In: Priest FG, Stewart GG (eds) Handbook of brewing, 2nd edn. Taylor & Francis, Boca Raton, FL, pp 383–446

Lyons TP (2018) Craft brewing. In: Stewart GG, Anstruther A, Russell I (eds) Handbook of brewing, 3rd edn. Taylor & Francis, Boca Raton, FL

Miedl-Appelbee M (2018) Brewhouse technology In: Stewart GG, Anstruther A, Russell I (eds) Handbook of brewing, 3rd edn. Taylor & Francis, Boca Raton, FL

Narziss L (1986) Centenary review technological factors of flavour stability. J Inst Brew 92:346–353

Oliver G (2012) Craft brewing. In: The Oxford companion to beer. Oxford University Press, Oxford, pp 270–273

Partridge M (2018) Packaging: a historical perspective and packaging technology. In: Stewart GG, Anstruther A, Russell I (eds) Handbook of brewing, 3rd edn. Taylor & Francis, Boca Raton, FL

Rogers P, Lee K, Shotnichi M, Tribe D (1982) Microbial reactions: ethanol production by *Zymomonas mobilis*. Springer, New York, pp 37–84

Russell I (2016) Yeast. In: Bamforth CW (ed) Brewing materials and processes. Academic Press, Elsevier, Boston, MA, pp 77–96

Russell I, Stewart GG (2014) Whisky: technology, production and marketing, 2nd edn. Academic Press, Elsevier, Boston, MA

Sammartino M, Stewart GG (2015) A lifelong relationship with yeast. MBAA Tech Quart 52:146–151
Stewart GG (1977) Fermentation – yesterday, today and tomorrow. MBAA Tech Quart 14:1–15
Stewart GG (2006) Studies on the uptake and metabolism of wort sugars during brewing fermentations. MBAA Tech Quart 43:265–269
Stewart GG (2009) Forty years of brewing research. J Inst Brewing 115:3–29
Stewart GG (2010) A love affair with yeast. MBAA Tech Quart 47:4–11
Stewart GG (2014) The concept of nature-nurture applied to brewer's yeast and wort fermentations. MBAA Tech Quart 51:69–80
Stewart GG (2015a) Seduced by yeast. J Am Soc Brew Chem 73:1–21
Stewart GG (2015b) Yeast quality assessment, management and culture maintenance. In: Hill AE (ed) Brewing microbiology: managing microbes, ensuring quality and valorising waste. Elsevier, Woodhead, Oxford, pp 11–29
Stewart GG (2016a) Industrial uses of yeast – brewing and distilling. New Food 19:20–24
Stewart GG (2016b) Saccharomyces species in the production of beer. Beverages 2:34
Stewart GG (2016c) Adjuncts. In: Bamforth CW (ed) Brewing materials and processes. Academic Press, London, pp 27–46.
Stewart GG (2018) Fermentation. In: Stewart GG, Anstruther A, Russell I (eds) Handbook of brewing, 3rd edn. Taylor & Francis, Boca Raton, FL
Stewart GG, Russell I (2009) An introduction to brewing science and technology. In: Brewer's yeast, 2nd edn. The Institute of Brewing and Distilling
Stewart GG, Marshall DL, Speers A (2016) Brewing fundamentals – fermentation. MBAA Tech Quart 53:2–22
Taylor D (2018) Aging, dilution and filtration. In: Stewart GG, Anstruther A, Russell I (eds) Handbook of brewing, 3rd edn. Taylor & Francis, Boca Raton, FL
Walker GM, Stewart GG (2016) *Saccharomyces cerevisiae* in the production of fermented beverages. Beverages 2:30
Whitman WB, Coleman DC, Wiebe WJ (1998) Prokaryotes: the unseen majority. Proc Natl Acad Sci USA 95:6578–6583
Winkelmann L (2016) 500 years of beer purity law. Brauwelt Int 34:188–189
Youngson RM (2006) Collins dictionary of human biology. Harper-Collins, Glasgow

Chapter 2
History of Brewing and Distilling Yeast

It is worth repeating in order to emphasize that yeasts are unicellular fungi (Fig. 2.1). They are classified into three groups: ascosporogenous yeasts, basidiomyces yeasts and imperfect yeasts (Kurtzman et al. 2011). *Saccharomyces* is the representative genus of ascosporogenous yeasts and historically (and currently) the most familiar microorganism to humans. Yeast cells were first microscopically observed by a Dutchman, Antonie van Leeuwenhoek (the father of microbiology) (Fig. 2.2). He was born and raised in Delft and worked there as a draper and founded his own business in 1654. He developed an interest in lens making and used handcrafted microscopes. He was the first to observe and describe single-celled organisms, which he originally referred to as "animalcules" and they are now referred to as microorganisms. As well as yeast cells (Fig. 2.3) from wine jars, he was the first to record microscopic observations of bacteria, spermatozoa, small blood vessels and muscle fibres. Leeuwenhoek did not author any scientific papers or books—his observations came to light through his correspondence with the Royal Society in London, which published his letters. In 1680, he was elected a Fellow of the Royal Society in London (Dobell 1932).

With improvements in microscopes early in the nineteenth century, yeasts were seen to be living organisms (Barnett 2007). Although some famous scientists [e.g. Lavoisier (Fig. 2.4), one of the founders of modern chemistry who described the phenomenon of alcoholic fermentation as "one of the most extraordinary in chemistry"] ridiculed this notion and their influence delayed the development of microbiology, in the 1850s and 1860s, yeasts were established as microbes and responsible for alcoholic fermentation, and this led to studies in the role of bacteria in lactic and other fermentations, as well as bacterial pathogenicity. At this time, there were difficulties in distinguishing between the activities of microbes and of extracellular enzymes. Between 1884 and 1894, Emil Fischer's study of sugar utilization by yeasts generated an understanding of enzymatic specificity and the nature of enzyme-substrate complexes—details later.

The *Saccharomyces* genus was first described by Meyen (1838) when he designated beer yeast as *Saccharomyces cerevisiae* in 1838, from the observations of

G.G. Stewart, *Brewing and Distilling Yeasts*, The Yeast Handbook,
https://doi.org/10.1007/978-3-319-69126-8_2

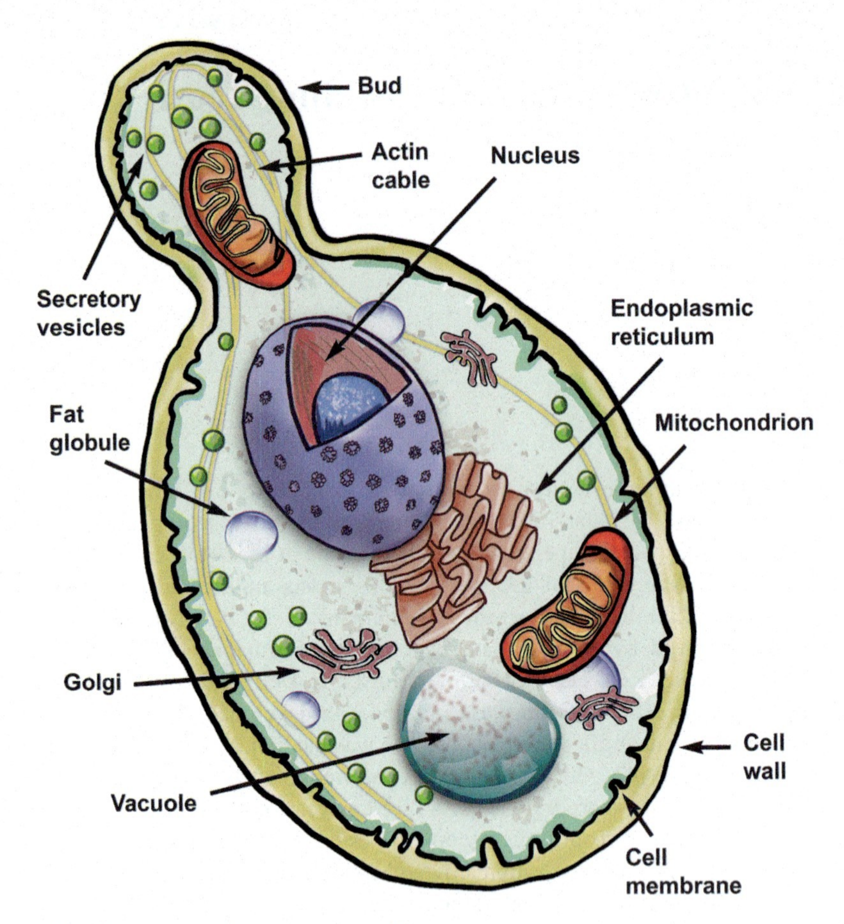

Fig. 2.1 Main features of a typical yeast cell

ascospores and their germination. This name is derived from the Greek words sákcharon (sugar) and mykes (fungus). The number of *Saccharomyces* species has changed over the years according to the criteria used to delimit species, and nine species (maybe ten species—details later) are now accepted in the genus *Saccharomyces* (Fig. 2.5) (Kurtzman 1994; Kurtzman et al. 2011).

As already discussed in Chap. 1, the vegetative cells of the *Saccharomyces* species are round, oval or cylindrical and reproduce by multilateral budding (Fig. 2.6). They may form pseudohyphae or chains (Fig. 2.7) where the daughter cell does not separate from the mother cell (Hough 1959) and fails to form septate hyphae. Chain formation is not uncommon in ale yeast strains of *S. cerevisiae* but has not been reported in lager strains of *S. pastorianus* (Stewart 1975; Botstein et al. 1997). Yeasts are predominantly diploid (with a haploid phase) but can be of higher ploidy which is often (but not always) the case with brewing and distilling yeast strains (Stewart and Russell 1986). Asci, which are persistent and usually transformed by direct change from the vegetative cells, may contain one to four

Fig. 2.2 Antonie van Leeuwenhoek

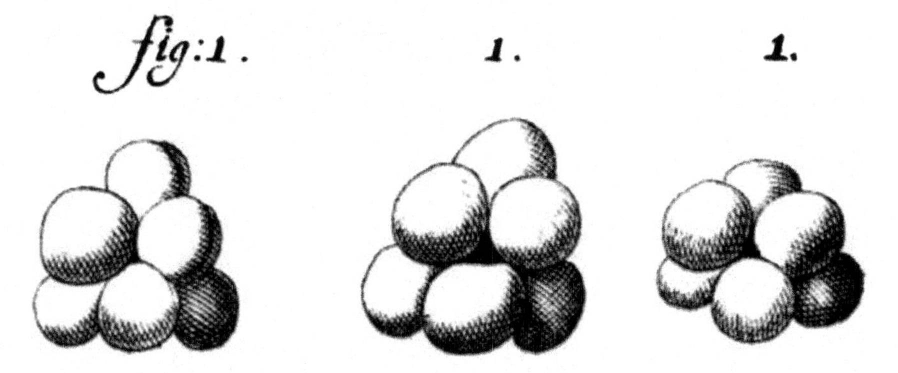

Fig. 2.3 Drawings of yeast cells observed by van Leeuwenhoek

ascospores (Bilinski et al. 1986, 1987). The ascospores are round or slightly oval (Fig. 2.8) with smooth walls.

There are ample photomicrographs of yeast published in the literature—many of excellent quality. A foremost yeast cytologist was Carl Robinow (Fig. 2.9) who produced a large number of microphotographs of yeast cells that have been employed by many publications to illustrate the author's prose (Robinow 1975, 1980). Professor Robinow was a distinguished academic in the Department of Microbiology and Immunology at the University of Western Ontario, London, Ontario, from 1949 until his death in 2006 (Hyams and Johnson 2007). He was a friend, neighbour and close confidante for over 25 years!

Fig. 2.4 Antoine-Laurent Lavoisier

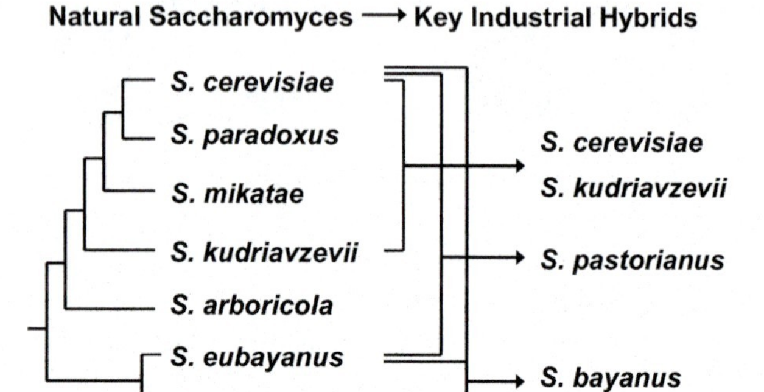

Fig. 2.5 Taxonomic relationships in the genus *Saccharomyces*

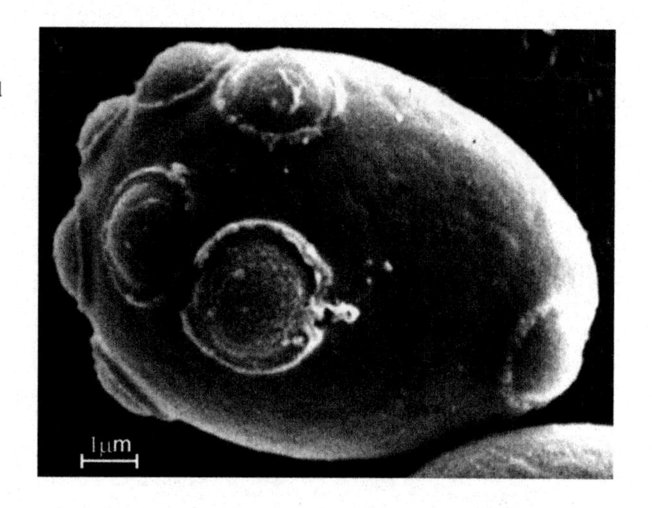

Fig. 2.6 Electron micrograph of a yeast cell with multiple bud scars and a birth scar. Bar represents 1 μm. Photograph courtesy of the late C.F. Robinow, University of Western Ontario, London, Ontario, Canada

The initial study of yeast genetics was conducted in the Carlsberg Laboratory, Copenhagen (details of the history of this laboratory will be discussed later in this chapter and in Chap. 16). Wingë (Wingë and Laustsen 1937, 1939) first described the haploid-diploid life cycle in yeast (Fig. 2.10). *Saccharomyces* species can alternate between the haploid (a single set of chromosomes in the nucleus) and diploid (two sets of chromosomes in the nucleus) states. This yeast genus contains 16 chromosomes (Zörgö et al. 2012)—one set for haploid cells and two sets for diploid cells, etc. The haploid cells display two mating types (sexes) designated MAT"**a**" and MAT"**α**", which are manifested by the extracellular production of MAT"**a**" or MAT"**α**" protein mating factors (pheromone) (Nickoloff et al. 1986). Details of these pheromones will be discussed later (Chap. 16). When "**a**" haploids are mixed with "**α**" haploids, mating takes place, and diploid (**a/α**) zygotes are formed that contain the genetic composition of both haploid partners (Fig. 2.10). In order to achieve zygote formation (hybridization), the ascus wall should be removed with specific lytic enzymes (glucanases) (Russell et al. 1973). The four spores from each ascus can be isolated by the use of a needle attached to a micromanipulator, a microscope and, these days, a closed circuit camera and TV screen (Fig. 2.11). Each spore can be induced to germinate, and these germinated vegetative cultures can be tested for their fermentation, flocculation and other characteristics and subsequently employed for further hybridization studies. It should be emphasized that haploid and diploid cultures can exist stably as vegetative cultures and undergo cell division via mitosis and budding (Stewart 1985). However, it has already been discussed that brewing (and other industrial) strains

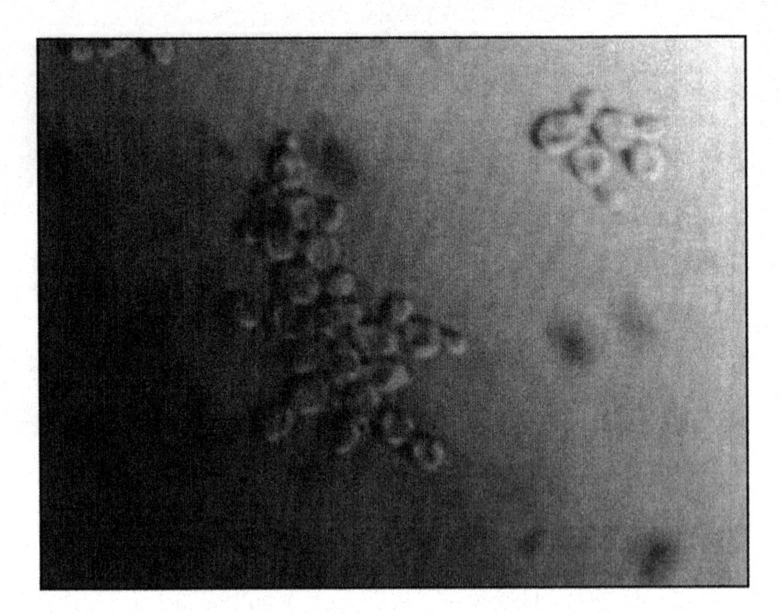

Fig. 2.7 Yeast pseudohyphae (chain formation)

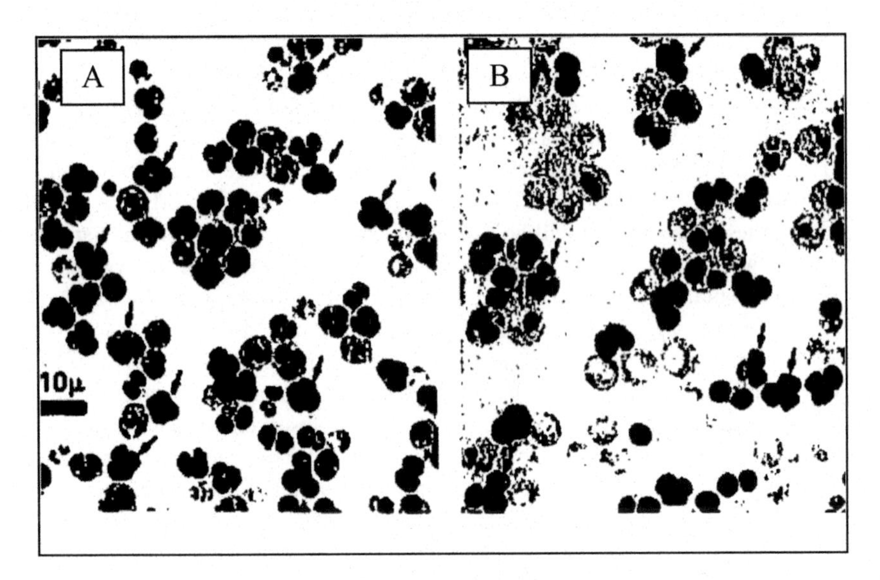

Fig. 2.8 Sporulating yeast cell; (a) wet mount preparation and (b) stained preparation

are not readily amenable to hybridization techniques because they are not usually (not always) either haploid or diploid but rather aneuploid or polyploid. Consequently, such strains possess little or no mating ability and poor sporulation, and the spores that do form have poor viability. It has been shown that it is possible to increase the sporulation ability of both brewing and distilling yeast strains by

Fig. 2.9 Carl Robinow

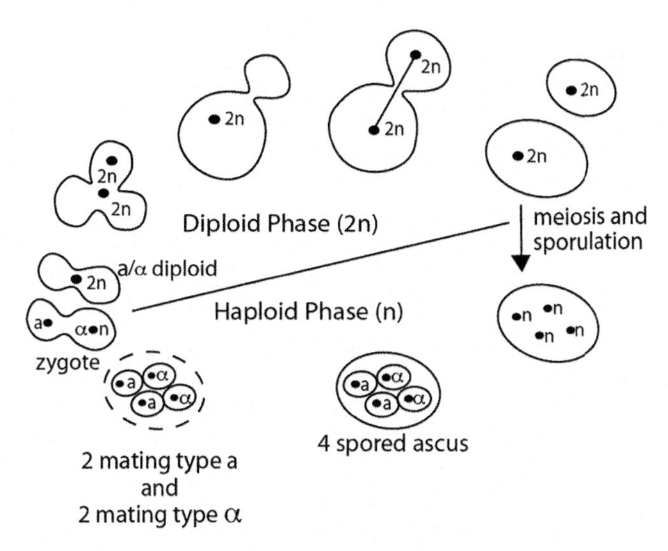

Fig. 2.10 Haploid/diploid life cycle of *Saccharomyces cerevisiae*

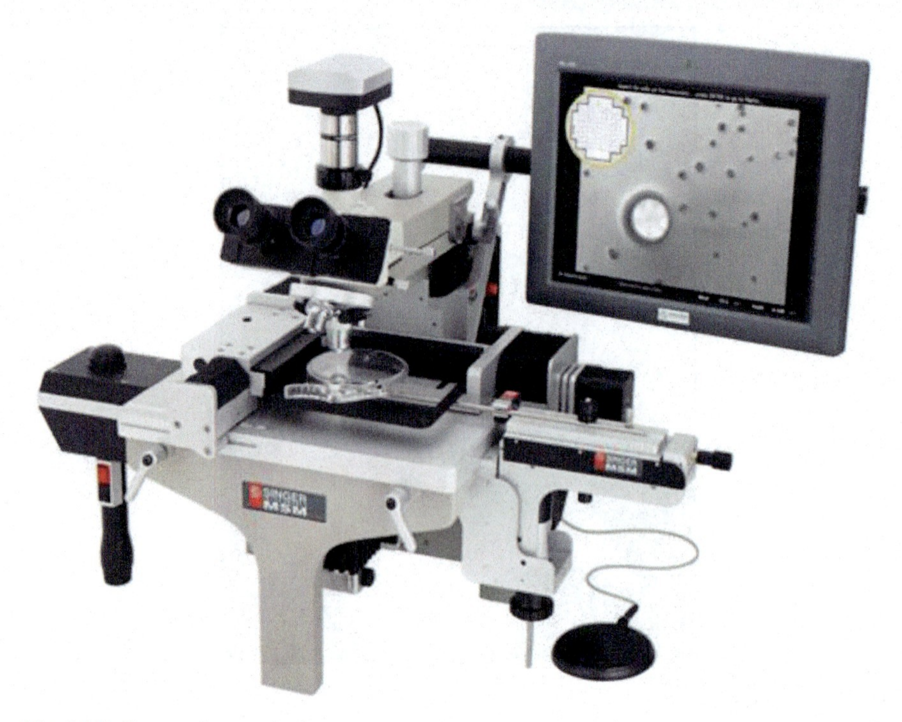

Fig. 2.11 Singer micromanipulator

manipulation of the presporulation environment, the sporulation medium and the incubation conditions (Bilinski et al. 1986). Hybridization studies with spores induced by these modified conditions have produced a number of hybrids, some of which possess improved wort fermentation efficiencies (Bilinski et al. 1987). A number of these hybrid strains have been studied for their potential to produce industrial (fuel) fermentation alcohol from starch hydrolysates, cellulose, lactose and/or sucrose substrates (details in Chap. 9). Also, some hybrids possess enhanced temperature, osmotic pressure and ethanol tolerance (Bilinski et al. 1987). However, brewing trials with these hybrids [containing *DEX* (glucoamylase) genes] produced undrinkable beers (for most consumers) because they possessed a number of distinct off flavours such as 4-vinylguaiacol [cloves, due to the presence of the phenolic off-flavour gene (POF)] as a result of the decarboxylation of ferulic acid (Fig. 2.12), a phenolic acid obtained from malt (Russell et al. 1983), but did exhibit enhanced overall wort fermentation characteristics due to the utilization of dextrin material (Fig. 2.13) (Erratt and Stewart 1981; Panchall et al. 1984).

Hybridization (cross-breeding) as a manipulation technique fell into disfavour when new biotechnological manipulation techniques, predominantly recombinant DNA (Stewart 1983) and to a lesser extent spheroplast fusion (Stewart et al. 1982), were thought to be the solution for the development of novel brewing and distilling yeast strains. However, for reasons that will be discussed later, hybridization, mass

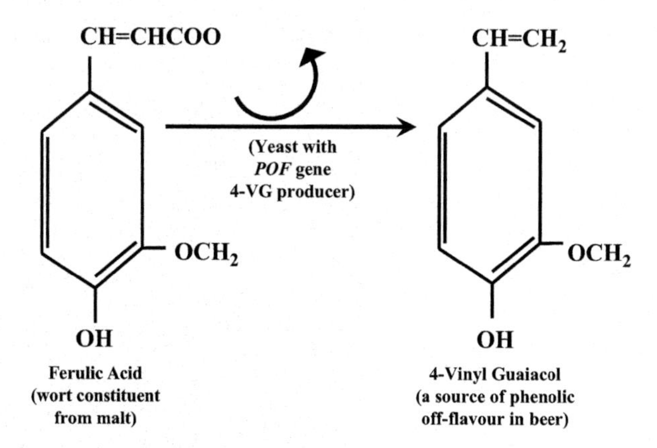

Fig. 2.12 Enzymatic decarboxylation of ferulic acid to 4-vinyl guaiacol (4-VG) by *Saccharomyces* spp.

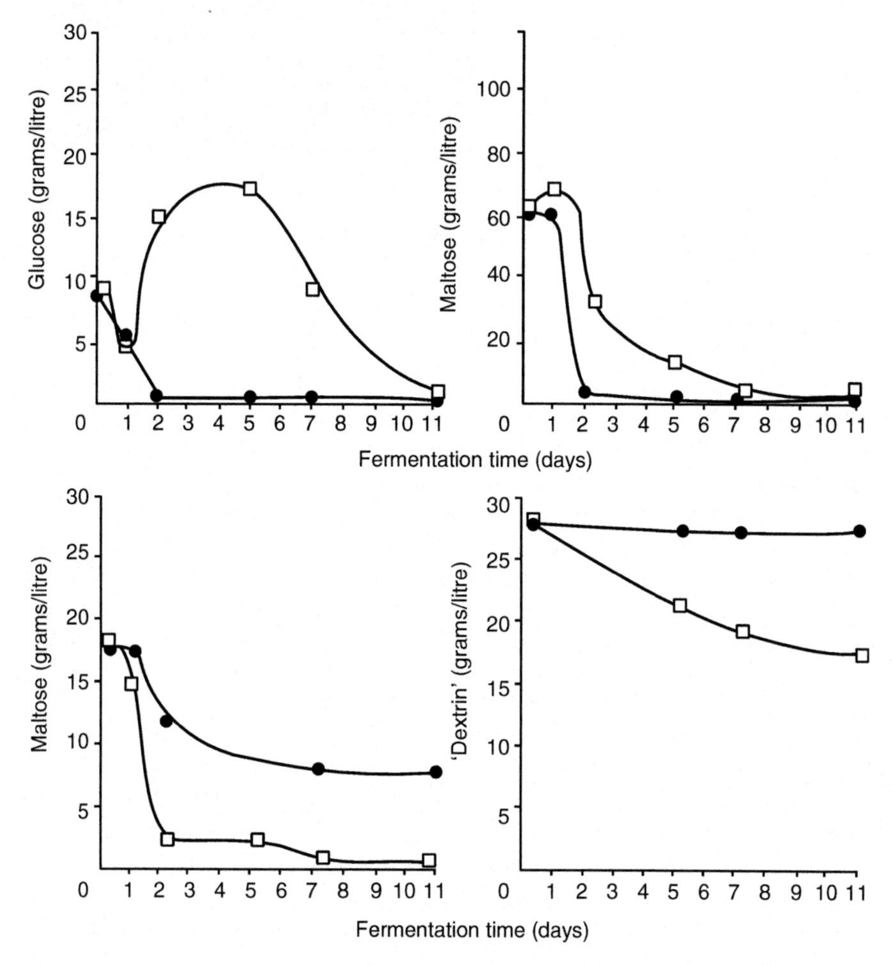

Fig. 2.13 Wort fermentation patterns of a brewing ale yeast strain (filled circle) and a diploid yeast strain (open square) containing *DEX* genes

Fig. 2.14 Louis Pasteur

and rare mating techniques are becoming popular again (Sanchez et al. 2014). Details of currently available genetic manipulation techniques will be discussed in detail later (Chap. 16). The reasons why genetically manipulated brewing and distilling yeast strains have not yet extensively been employed industrially will also be considered (Chap. 16).

It has already been stated that yeast is most closely associated with humankind and that *Saccharomyces cerevisiae* has long been used for brewing, distilling (for both potable and industrial alcohol), wine making, baking bread and confectionary and yeast extracts for food and flavouring (Pyke 1958) (Table 1.1). Yeast has also been used for therapeutic purposes (de Oliveira Leite et al. 2013). It is by far the most studied and best understood species of the yeast domain (Stewart 2014a, b) and is an important model system for basic research into the biology and molecular biology of eukaryotic cells (Botstein et al. 1997). Indeed, the ability to rationally manipulate all aspects of its gene expression by in vitro genetic techniques offers *S. cerevisiae* a unique place amongst eukaryotes and confirms its model "eukaryote" designation (Chap. 16) (Replansky et al. 2008; Rose 1980).

A number of scientists played important roles in developing ideas regarding the genus *Saccharomyces*. Meyen (1838) and Reese (Schwann 1837) have already been mentioned here, and, of course, Pasteur's pioneering work cannot be overlooked (Anderson 1995). Louis Pasteur (1822–1895) (Fig. 2.14) was a famous French polymath scientist (chemist and microbiologist) renowned for his discoveries on the principles of vaccination, microbial fermentation and pasteurization. When he died in 1895, Pasteur was the most famous scientist in the world! He had been bestowed with academic, national and international honours in France and by many other countries. France honoured his passing with a state funeral at Notre Dame Cathedral, complete with full military honours! (What scientist would receive such treatment today?) (Anderson 1995). He did not receive a Nobel Prize for his studies because this prize was not initiated until 1901, 6 years after his death, and Nobel Prizes are not awarded posthumously (Abrams 2001).

Many biographies of Pasteur have been published (e.g. Debré 2000) and most of them all but ignore his important work on beer. For example, a comprehensive biography (Feinstein 2008) of Pasteur only devotes three pages to his brewing studies in a 400 page text! Although his studies on fermentation did not reach the great heights as did the demonstration of his anthrax vaccine in public when all the sheep he vaccinated lived and all the unvaccinated sheep died! This feat dramatically advanced his career and elevated him to the status of a hero—the great scientist protecting French citizens against all the ravages of mad dogs' serious infections.

The most comprehensive discussion of Pasteur's research on brewing fermentation was a plenary lecture entitled: "Louis Pasteur (1822–1895): An assessment of his impact on the brewing industry". This lecture commemorated the centenary of Pasteur's death and was delivered by Ray Anderson (who has written the Foreword to this book) at the European Brewery Convention Congress held in Brussels in 1995 (Anderson 1995). The commentary in this chapter, of Pasteur's work, is a précis (with permission) of Anderson's lecture. Pasteur was a chemist who spent 10 years working on the crystal structure of various substances, and tartaric acid had special significance for him because it came from wine and led to his first interest in life processes. He then began working on the fermentation of French sugar beet juice to alcohol. As part of these studies, he observed yeast budding under the microscope and confirmed that amyl alcohol, as well as ethanol, was a major fermentation by-product. Most importantly, Pasteur renounced the dominant chemical theory of fermentation, which was supported by Justus Liebig and others (Hein 1961). However, Pasteur's basic position was that fermentation is a biological process, not chemical and not novel or obscure. A number of scientists, including some associated with the brewing industry, had reached similar conclusions a few years before Pasteur entered the field (Cagniard-Latour 1837; Schwann 1837). What Pasteur did, by virtue of his experimental and debating skills, was to convince the scientific world of the true nature of yeast as a living organism and the causative agent of alcoholic fermentation. He also extended the "germ theory of fermentation" and showed that each fermentation was caused by specific organisms during lactic, acetic, butyric or alcoholic fermentation (Debré 2000).

Pasteur's extensive fermentation studies were on wine, and his research culminated in a book entitled "Etudes sur le Vin" (Pasteur 1876) published in 1876, [which was translated into English in 1879 as "Studies on Fermentation" (Faulkner and Robb 1879)]. It reviewed familiar diseases and problems with wine—they were called "turned", "acid", "ropy" and "oily" wine, and each condition was associated with a microscopic organism. This eventually led to heat treatment as a reliable and practical way of preserving wine. Pasteurization—a term almost immediately

Fig. 2.15 Emile Duclaux

adopted—became a reality, and Pasteur patented the process in April 1865 (Feinstein 2008).

Pasteur subsequently turned his attention to diseases affecting silk worms, another commercially important French industry. Two separate diseases were found to be working in synergy here. His close colleague Emile Duclaux (Fig. 2.15) considered this study to be "a beautiful example of scientific investigation" (Duclaux 1899). Following a period of serious ill health, Pasteur continued to devise experiments with considerable care, but their actual execution was left to colleagues for the rest of his life!

When Pasteur's health was partially recovered (he never returned to full health), he turned his attention to beer, which occurred in 1871 in the aftermath of the Franco-Prussian War. Pasteur at no time forgave Germany for their diabolical treatment of France during this conflict, and he returned the Medicinae Doctor (MD) (Doctor of Medicine or Teacher of Medicine) that he had received in 1868 from the University of Bonn and refused permission for his publication "Etudes sur le Biere" to be translated into German. When he was dying in 1895, he declined the Prussian Order of Merit offered by the Kaiser.

Pasteur examined samples of beer with a microscope and found in spoilt beer, yeast and "the presence of disease" (his terminology). He identified a similar situation in a number of French breweries (large and small) and in beer from Whitbread's Brewery in London and Younger's Brewery in Edinburgh (Philliskirk 2012).

Fig 2.16 Letter from Pasteur to Henry Younger 1884

Pasteur visited Edinburgh in 1884 as part of Edinburgh University's tercentenary celebrations. He took an interest in the city's famous breweries, and he was able to advise on their microbiological situation. While in Edinburgh, Pasteur stayed with Henry Younger of Younger's Brewery, and Younger introduced him to his business partner Alexander Low Bruce. During their discussions, Alexander Bruce decided he would like to found a new Chair of Public Health within Edinburgh University. Although Bruce died shortly thereafter, a sum of £15,000 was offered to the University to endow the "Bruce and John Usher Chair of Public Health"—it was the first chair of its kind in Britain (Keir 1951).

Many of the beers examined by Pasteur were rife with infection, and he identified the contaminating organisms in them (Fig. 2.16—a letter from Pasteur to Henry Younger in 1887). At this point in time, he did not advise pasteurization.

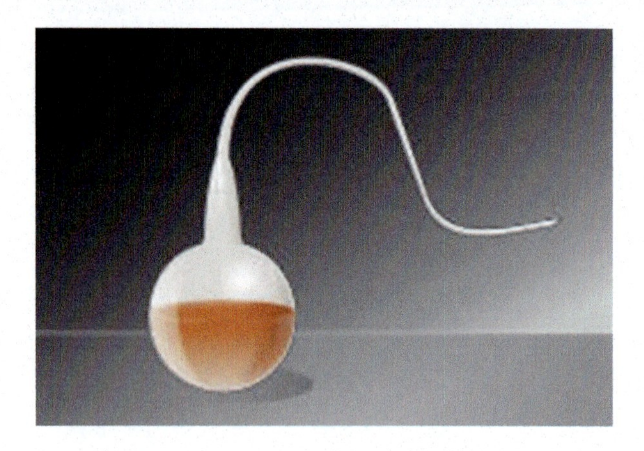

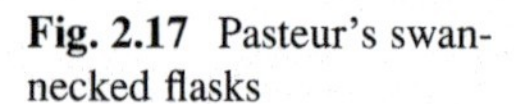

Fig. 2.17 Pasteur's swan-necked flasks

In *Etudes sur le Biere* he noted: "The process of pasteurisation is less successful in the case of beer than that of wine for the delicacy of flavour which distinguishes beer because it is more easily affected by heat than wine." (Pasteur 1873). Much later, experience with the pasteurization of beer has shown this not to be the case under appropriate operating conditions (Philliskirk 2012).

Pasteur did not like the taste of beer—he preferred wine! He was surprised that his colleagues could tell one beer from another. His objective in studying beer was to make French beer better than what could be produced in Germany. In terms of Pasteur's research on beer, a major objective was to recommend a procedure for preserving beer, and he correctly advocated prevention rather than cure. Nevertheless, in *Etudes sur le Biere* (Pasteur 1873) only the first two and the last chapter were actually concerned with beer. The rest of the publication was devoted to a discussion of spontaneous generation and the nature of fermentation. He was able to show that a boiled medium could remain uninfected for years, even when exposed to air, so long as dust and germs were prevented from entering it. This he achieved by trapping the dust and germs in his famous swan-necked flasks (Fig. 2.17). This was one of the last, and most important, experiments disproving the theory of spontaneous generation (Pasteur 1879). He concluded: "Never will the doctrine of spontaneous generation recover from the mortal blow of this simple experiment. There is no known circumstance in which it can be confirmed that microscopic beings came into the world without germs, without parents similar to themselves".

In his later years, Pasteur had become a legend! Nevertheless, he was a complex man: difficult, brilliant, vain, unforgiving and very patriotic that could develop into chauvinism (Debré 2000)! Unfortunately, he has also been accused of selective treatment of results (Anderson 1993, 1995; Altman 1995), but the evidence in support of this accusation is unproven.

Although showing that fermentation was the result of the action of living microorganisms was a breakthrough, it did not explain the biochemical nature of the fermentation process or prove that it is caused by the microorganisms that appear to always be present in fermentation. Many scientists, including Pasteur, had unsuccessfully attempted to extract the fermentation enzymes from yeast. Success

Fig. 2.18 Eduard Buechner (Edward Buchner)

came in 1897 when the German chemist Eduard Buchner (Buechner) ground yeast with alumina, extracted a juice from this mixture and found to his surprise that this "dead" liquid would ferment a sugar solution, producing carbon dioxide and alcohol much like whole living yeasts. Buchner's results are considered by many to mark the birth of biochemistry as a research science (Fig. 2.18). These unorganized ferments (the word ferment is derived from the Latin verb ferveo, which means "to boil") behaved like whole yeast cells. It was then understood that fermentation is caused by enzymes that are produced by microorganisms and signifies the birth of biochemistry (Kohler 1971, 1972). In 1907, Buchner was awarded the Nobel Prize in Chemistry: "for his biochemical research and his discovery of cell-free fermentation". Further biographical details of both Pasteur and Buchner are in Chap. 6.

Detailed research studies on brewer's yeast have been carried out in the Carlsberg Laboratory, which was established in 1875 in Copenhagen, Denmark, by the founder of the Carlsberg Brewery, J.C. Jacobsen, who administered the laboratory personally until the inauguration of the Carlsberg Foundation on September 25, 1876. It was Pasteur's reputation that encouraged Jacobsen to found the Carlsberg Laboratory. At this time, it was placed under the management of a board of trustees elected by the Royal Danish Academy of Sciences and Letters. In April 1972, the research laboratory's management was transferred to the Carlsberg Brewery, and it became the Carlsberg Research Centre [located in a new building (Fig. 2.19)]—with current designation and venue in Copenhagen. I have been privileged to be invited to present plenary lectures all over the world. The

Fig. 2.19 The Carlsberg Research Centre

one I presented at the new Carlsberg Research Centre in 1975 (considering our research at Labatt's on "yeast flocculation"), to commemorate the opening of this building and the establishment of the Centre, was personally memorable (Fig. 2.20)!

This institution has been (and still is) of considerable service to industry, particularly the brewing industry. J.C. Jacobsen (Holter and Møller 1976) defined the aims of the Carlsberg Foundation as:

- To continue and extend the activities of the chemical and physiological Carlsberg Laboratory in conformity with the institute's aims
- To promote various branches of science, together with mathematics, philosophy, history and philosophy.

Over the years, although the Carlsberg Laboratory/Foundation was a relatively small research institute, it has published a large number of profound concepts that are still important today. The laboratory/foundation published its own peer-reviewed journal—"Comptes rendus des travaux du Laboratoire Carlsberg"—which was published from 1878 to 1972 (Holter and Møller 1976). Important concepts developed in the laboratory/foundation included:

- Nitrogen determination (Johan Kjeldahl)
- Concept of pH (Søren Sørensen)
- Bacterial morphology (Johannes Schmidt)
- Concept of colloidal solutions (Raj Luderstrøm-Laing)
- Isolation of pure yeast strains and their propagation (Emil Christian Hansen)

Fig. 2.20 Presenting a lecture considering "yeast flocculation" to mark the opening of the Carlsberg Research Centre in Copenhagen in 1975

- Yeast life cycle (Øjvind Wingë)
- Electron microscopy and molecular biology of plants (Diter von Wettstein)
- Yeast molecular biology (Morton Keilland-Brandt) (Kielland-Brandt et al. 1983).

As this book is about brewer's and distiller's yeast, only the significant contributions of this research centre to our knowledge of yeast will be discussed further beginning with Emil Christian Hansen (Fig. 2.21). He joined the Physiology Department of the Carlsberg Laboratory in 1877 and on January 1, 1879, became its head. His first thesis (report) was entitled: "About organisms in beer and beer wort" for which he received a doctoral degree in April 1879. In 1883, he announced his system of isolating pure lager yeast cultures, and the Carlsberg Brewery in Copenhagen conducted the first experiments with pure lager yeast strains (details in Holter and Møller 1976).

Hansen had the notion that the main reason for the frequent occurrence of aberrant brews resulting in unsaleable beer was not only bacterial infection [this was assumed by Pasteur (1876)] but also contamination by wild yeasts. He devised a method by which he could isolate a single yeast cell and propagate it into a pure

Fig. 2.21 Emil Christian Hansen with W.E. Jansen (coppersmith) in the background

strain culture. J.C. Jacobsen (the Carlsberg Breweries Head Brewer) initially had his doubts about Hansen's ideas but soon concurred with him when confronted with Hansen's results, and he watched with great interest Hansen's attempts to separate a pure and usable yeast strain from the mixed culture that was being employed in the brewery.

Hansen successfully cultured sufficient quantities of the pure strain yeast (Carlsberg's bottom yeast no.1—it is still called to this today) to use in the brewery's fermenters. This yeast strain was used for the first time on a production scale on November 12, 1883, at Gamle Carlsberg. These fermentation trials were successful, and Jacobsen's doubters were appeased, and very quickly he used his influence to rapidly communicate these developments widely. He sent reports to a number of notable lager producing breweries (in Europe and the United States) offering them this pure yeast strain for their experimentation. This grouping of lager yeasts was designated "*Saccharomyces carlsbergensis*" and is now termed "*Saccharomyces pastorianus*" (further details regarding this species are in Chap. 3).

Hansen soon found that it was too tedious and inconvenient for his laboratory to regularly furnish the Carlsberg Brewery with enough pure culture and it would be easier to obtain a specific apparatus for this purpose. With the assistance of a coppersmith, W.E. Jansen, Hansen started to construct such an apparatus (Fig. 2.22). By the beginning of 1886, the apparatus was working effectively in the Copenhagen Carlsberg Brewery. Jansen then began to market the apparatus, and Heineken in the Netherlands was one of the first breweries to purchase this equipment, followed shortly thereafter by the Joseph Schlitz Brewery, Milwaukee in the United States (Stewart 2017).

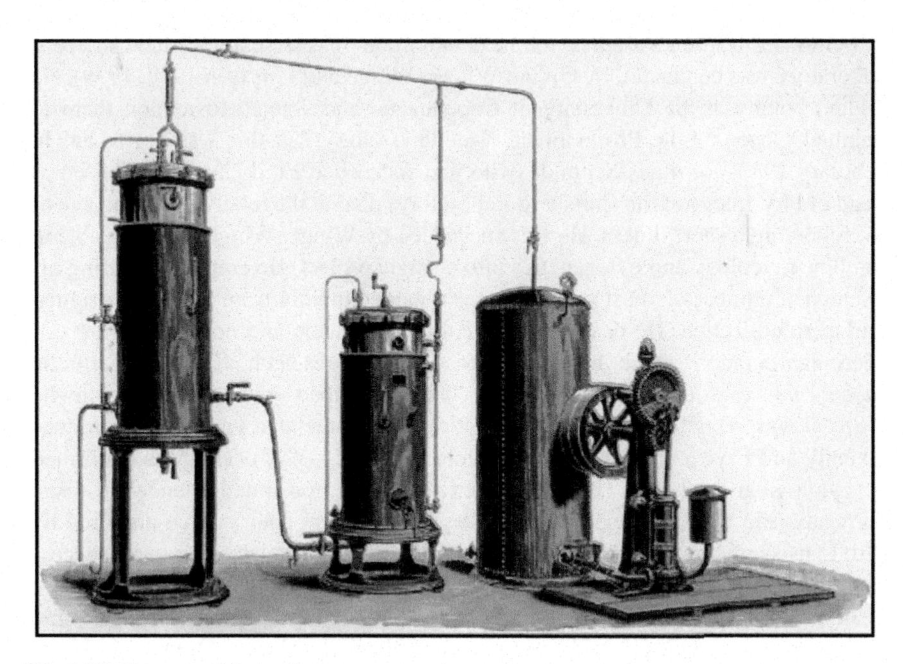

Fig. 2.22 Hansen and Jansen's yeast propagation apparatus

As a result of Hansen and Jansen's work, the practice of employing a pure yeast strain for lager production was soon adopted by breweries all over the world, particularly in the United States. Ale-producing regions (mainly in parts of the United Kingdom), however, met this "radical innovation" with skepticism, and this method was merely regarded as a means of reducing infection by wild yeasts and bacteria! The British brewing scientist Horace Brown (his biographical details are discussed later) obtained a Hansen pure yeast culture apparatus and installed it in Worthington's Burton-on-Trent Brewery in order to culture ale yeast cultures. After 9 years of use, it was discontinued because "its benefits were marginal" (Brown 1916).

In 1959, it was reported (Hough 1959) that of 39 ale yeast cultures in use commercially in Britain, 12 contained a single strain, 16 had two major strains and the rest still contained three or more yeast strains. These ale strains are currently taxonomically classified as part of the species "*Saccharomyces cerevisiae*" (Kurtzman and Fell 2011). Also, in 1971, we reported that the major brewing company in Canada (at the time) used a top cropping ale culture consisting of two strains of *S. cerevisiae* that exhibited co-flocculating properties where two strains were non-flocculent alone but flocculent when mixed together (Stewart and Garrison 1972)—further details of co-flocculation are in Chap. 13. The matter of the establishment and management of culture collections together with greater details regarding yeast propagation are discussed in Chap. 4.

Following Hansen's studies, the next important yeast research in the Carlsberg Laboratory was conducted by Øjvind Wingë (Wingë and Laustsen 1937, 1939). He studied botany at the University of Copenhagen and wanted to remain there to eventually become the Professor of Plant Physiology, but this was not to be! In February 1933, Johannes Schmidt (who had succeeded Emil Christian Hansen as Head of Physiology at the Carlsberg Laboratory) died at the relatively young age of 56, following a short illness. He was succeeded by Wingë. Wingë was now able to combine mycology and cytogenetics into a single project. He employed, during his research, a number of yeast stock cultures obtained mainly from Hansen's original Carlsberg collection. He designed new culture chambers and novel techniques of micromanipulation which he applied to his yeast research. The basics of the *Saccharomyces* life cycle had already been described (review by Herskowitz 1988). It was Wingë who demonstrated, for the first time, that yeasts can reproduce sexually and have a regular haploid-diploid cycle (Fig. 2.10). This was confirmed by cytological methods. The first genetic work demonstrated Mendelian yeast segregation in 1937 (Wingë and Laustsen 1937). This paper was co-authored by Otto Laustsen whose skill with a micromanipulator owed much to his early success with yeast.

In 1938, again in collaboration with Otto Laustsen, Wingë described the first case of artificial species hybridization in yeast (details in Chap. 16). From 1939, all his work was devoted to yeast genetics, including a paper on cytoplasmic inheritance (further details of yeast respiratory deficiency later) (Chap. 14), and the inheritance of yeast fermentation ability. These latter studies showed various types of gene interaction in the control of a yeast's fermentation ability (Chap. 16) (Wingë and Laustsen 1939).

Wingë's discovery of sexual reproduction in yeast opened new possibilities for producing improved strains through cell crossing and zygote formation. These methods, which Wingë called "breeding under the microscope" (Fig. 2.11), resulted in the production of new strains of both baker's and distiller's yeast but, as already briefly discussed (further details later), have not been as successful in producing improved and different strains of brewer's yeast, which is not the situation with baker's yeast (Wingë and Laustsen 1939; Pyke 1958). However, recent hybridization studies (Sanchez et al. 2014) with brewer's yeast strains may be focusing on a change in emphasis—details later (Chap. 16).

This chapter would be incomplete if we did not also discuss, in some detail, the contributions of the British brewing scientist Horace Brown (Fig. 2.23) to developments in brewer's yeast. In 1866, he started work in the Worthington Brewery, Burton-on-Trent. Although largely self-taught, Horace Brown was a true polymath, who left his mark on virtually all areas of science as applied to brewing, in a career that lasted over 50 years. His research work considered many aspects of the brewing process—barley germination, beer microbiology and contamination, wort composition, oxygen and fermentation, beer haze formation and beer analysis. His focus was upon practical brewing problems by employing and developing fundamental scientific principles.

Fig. 2.23 Horace Tabberer Brown

Brown's research on brewing microbiology and yeast fermentation relied on Pasteur's original work. He identified contaminating bacteria and wild yeasts in wort fermentations and characterized their effects on brewer's yeast activity in wort and beer flavour, stability and drinkability. His work on brewer's yeast physiology focused on the role of oxygen and confirmed its importance for wort fermentations. He also continued Pasteur's studies on yeast acid washing as a means to eliminate contaminating bacteria (Brown 1916). Recent studies on the oxygenation of wort at yeast pitching and yeast acid washing, particularly during high-gravity brewing (HGB) conditions, will be discussed in detail later (Chap. 11).

In 1916, the London Section of the Institute of Brewing presented Horace Brown with a portrait (Fig. 2.23) "as an expression of the affection, esteem and homage of the Members of the Institute". Unfortunately, the portrait was destroyed during the 1941 London bombing. In reply to the 1916 presentation, Brown gave a lecture entitled: "Reminiscences of Fifty Years' Experience of the Application of Scientific Methods to Brewing Practice". Brown began his presentation with the following statement: "The origin of this paper is the result of a retrospect and a process of mental stocktaking, supported by the fact that I have been closely connected, for half-a-century, with one of the most interesting industries in the world" (Brown 1916).

At the end of his wide-ranging lecture, which was published as a 91-page paper in the *Journal of the Institute of Brewing* (Brown 1916) and today is freely available to the public from the journal's website, Brown expressed a number of personal sentiments that are, in part, still relevant today. Brown stated that he was deeply disappointed that more had not been achieved to exploit science in brewing. He blamed two factors—the lack of education of brewers and the short-termed vision of brewing directors when it came to technological change (Brown 1916). He concluded: "Those who are really the responsible heads of manufacturing businesses seldom have any desire to know anything of the inner meaning of their processes or the scientific principles which underlie them". One has to wonder if Brown overstated the problem that prevailed in the brewing industry at the beginning of the last century. It has already been discussed in this text that a great deal of fundamental research in brewing and distilling occurred during the second half of the nineteenth century. One cannot help speculating whether the industry has come "full circle" and that its interest in learning more about basic brewing and distilling's scientific principles (particularly in North America and the United Kingdom) is waning!

Most of the period of Brown's research coincided with the industrial revolution. Consequently, brewing, and to a lesser extent distilling, was being converted from a cottage craft into a major industry and with this conversion came the need for greater scientific understanding of the process. At that time, the brewing industry was a centre of much scientific research, and it is interesting to note that four Fellows of the Royal Society (including Horace Brown) were employed by breweries in the United Kingdom (Stewart and Russell 1986)—currently there are none! This was occurring in other countries as well. For example, in Denmark, J.C. Jacobsen founded the Carlsberg Laboratory/Foundation in 1876, which became, and still is, the home of a number of distinguished scientists (Holter and Møller 1976), and similar trends occurred in Germany and Czechoslovakia (as it was named then). Today, the focus of brewing yeast research is in Germany, Belgium, Finland, Japan and China.

Since the notable studies by Horace Brown and Øjvind Wingë, research on brewer's yeast (distiller's yeast will be discussed separately) continued apace throughout the twentieth century in brewing research laboratories, brewing research institutes (e.g. the Carlsberg Research Centre, Campden BRI and the VTT in Helsinki, Finland) and universities (e.g. Heriot Watt University; University of California, Davis; Berlin University of Technology; Technical University of Munich in Weihenstephan; University of Louvain; University of Abertay; and latterly the University of Nottingham). The relevant results published by these centres will be discussed, as appropriate, in the ensuing chapters of this text.

In addition, the notable reputation that yeast (particularly the genus *Saccharomyces*) enjoys as an experimental eukaryote (Rose 1980) cannot be overlooked even though this book's principal focus is on brewing and distilling yeasts. Over 25 years ago, Botstein and Fink (1988) speculated that yeast would be the ideal experimental organism for modern biology. It has been argued that the combination of recombinant DNA technology and classical biochemistry, enzymology and protein structure and function has created a revolution that has presented biologists and zymologists access to an array of novel methods for connecting proteins (structural and enzymes) and genes with their multiple roles in the biology and applications of an organism (further details in Chap. 16).

In the intervening 25 years since 1988, Botstein and Fink (2011) consider that yeast has graduated from a position as the premier model for eukaryotic cell biology to become the pioneer organism that has facilitated the establishment of entirely new fields of study called "functional genomics", "systems biology" and "synthetic genome engineering" (Pretorius 2016). These new fields look beyond the function of individual genes and proteins, focusing on how they interact and work together to determine the properties of living cells and organisms. Also, research on yeast as a model organism has assisted consideration of cancer diagnosis and highlighted genetics and metabolic simulation between yeast and cancer cells and other human maladies (Mager and Winderickx 2005; Guaragnella et al. 2014). Diseases induced by human mitochondrial disorders may initially be considered by many to be unrelated to the influence of yeast mitochondria on brewing fermentations (Stewart 2014a, b). Fundamental research on brewer's yeast mitochondrial function has been extrapolated to an understanding of mammalian mitochondrial disorders (Reinhardt et al. 2013; Seo et al. 2010) (Chap. 14). In the context of this book, results from functional genomics can be applied to both brewing and distilling fermentation (Giaever et al. 2002; Conant and Wolfe 2008). Also, results with brewing and distilling yeast mitochondria are being applied to mammalian mitochondrial disorders (Reinhardt et al. 2013; Seo et al. 2010; Stewart 2014b) (Chap. 14).

In the context of this book's focus, results from functional genomics can be applied to brewing, distilling and wine fermentations (Giaever et al. 2002; Conant and Wolfe 2008; Pretorius 2016)—details in Chap. 16.

References

Abrams I (2001) Reflections on the first century of the Nobel Peace Prize. Det Norske Nobelinstitutts Skriftserie (The Norwegian Nobel Institute Series 1 - No. 5). Peace and Change 26:525–549

Altman LK (1995) The doctor's world; Revisionist history sees Pasteur as liar who stole rival's ideas. New York Times, 16 May 1995

Anderson C (1993) Pasteur notebooks reveal deception. Science 259:1117

Anderson RG (1995) Louis Pasteur (1822–1895): an assessment of his impact on the brewing industry. In: Proceedings of the 25th European brewery convention congress. Brussels, pp 13–23

Barnet J (2007) Beginnings of microbiology and biochemistry: the contribution of yeast research. MBAA Tech Quart 44:256–263

Bilinski CA, Russell I, Stewart GG (1986) Analysis of sporulation in brewer's yeast: induction of tetrad formation. J Inst Brew 92:594–598

Bilinski CA, Russell I, Stewart GG (1987) Physiological requirements for induction of sporulation in lager yeast. J Inst Brew 93 p216-219

Botstein D, Fink GR (1988) Yeast: an experimental organism for modern biology. Science 240:1439–1449

Botstein D, Fink GR (2011) Yeast: an experimental organism for 21st century biology. Genetics 189:695–704

Botstein D, Chervitz SA, Cherry JM (1997) Yeast as a model organism. Science 277:1259–1260

Brown HT (1916) Reminiscences of fifty years' experience of the application of scientific method to brewing practice. J Inst Brew:22: 267–22: 354

Cagniard-Latour C (1837) Memoire sur la fermentation vineuse. C R Lab Carlsberg 4:905–906

Conant GC, Wolfe KH (2008) Turning a hobby into a job: how duplicated genes find new functions. Nat Rev Genet 9:938–950

de Oliveira Leite AM, Miguel MA, Peixoto RS, Rosado AS, Silva JT, Paschoalin VM (2013) Microbiological, technological and therapeutic properties of kefir: a natural probiotic beverage. Braz J Microbiol 44:341–349

Debré P (2000) Louis Pasteur. The Johns Hopkins University Press, Baltimore. 978-0-8018-6529-9

Dobell C (1932) Anthony van Leeuwenhoek and his "Little Animals". Staples Press, London

Duclaux E (1899) Louis Pasteur: the history of a mind. Translated by Smith EF, Hedges F. Saunders, Philadelphia, Penn

Erratt J, Stewart GG (1981) Fermentation studies using *Saccharomyces diastaticus* yeast strains. Devel Ind Microbiol 22:577–586

Faulkner F, Robb CD (1879) Studies on fermentation: the diseases of beer, their causes, and the means of preventing them. Macmillan

Feinstein S (2008) Louis Pasteur - the father of microbiology. Series: Inventors who changed the world. Enslow Publishers

Giaever G, Chu AM, Ni L, Connelly C, Riles L, Véronneau S, Dow S, Lucau-Danila A, Anderson K, André B, Arkin AP, Astromoff A, El-Bakkoury M, Bangham R, Benito R, Brachat S, Campanaro S, Curtiss M, Davis K, Deutschbauer A, Entian KD, Flaherty P, Foury F, Garfinkel DJ, Gerstein M, Gotte D, Güldener U, Hegemann JH, Hempel S, Herman Z, Jaramillo DF, Kelly DE, Kelly SL, Kötter P, LaBonte D, Lamb DC, Lan N, Liang H, Liao H, Liu L, Luo C, Lussier M, Mao R, Menard P, Ooi SL, Revuelta JL, Roberts CJ, Rose M, Ross-Macdonald P, Scherens B, Schimmack G, Shafer B, Shoemaker DD, Sookhai-Mahadeo S, Storms RK, Strathern JN, Valle G, Voet M, Volckaert G, Wang CY, Ward TR, Wilhelmy J, Winzeler EA, Yang Y, Yen G, Youngman E, Yu K, Bussey H, Boeke JD, Snyder M, Philippsen P, Davis RW, Johnston M (2002) Functional profiling of the *Saccharomyces cerevisiae* genome. Nature 418(6896):387–391

Guaragnella N, Palermo V, Galli A, Moro L, Mazzoni C, Giannattasio S (2014) The expanding role of yeast in cancer research and diagnosis: insights into the function of the oncosuppressors. FEMS Yeast Res 14:2–16

Hein GE (1961) The Liebig-Pasteur controversy: vitality without vitalism. J Chem Educ 38:614

Herskowitz H (1988) Life cycle of the budding yeast *Saccharomyces cerevisiae*. Microbiol Rev 52:536–553

Holter H, Møller KM (eds) (1976) The Carlsberg Laboratory 1876–1976. Rhodos International Science and Art Publishers

Hough JS (1959) Flocculation characteristics of strains present in some typical British pitching yeasts. J Inst Brew 65:479–482

Hyams J, Johnson B (2007) Carl F Robinow (1909–2006): an appreciation. Fungal Genet Biol 44:1215–1218

Keir D (1951) The younger centuries: the story of William Younger & Co Ltd 1749 to 1949. William Younger, Edinburgh

Kielland-Brandt MC, Nilsson-Tillgren T, Petersen JGL, Holmberg S, Gjermansen C. (1983) Approaches to the genetic analysis and breeding of brewer's yeast. In: JFT Spencer (ed) Yeast genetics, Chap. 3. Springer, New York, pp 421–425

Kohler R (1971) The background to Eduard Buchner's discovery of cell-free fermentation. J Hist Biol 4:35–61

Kohler R (1972) The reception of Eduard Buchner's discovery of cell-free fermentation. J Hist Biol 5:327–353

Kurtzman CP (1994) Molecular taxonomy of the yeasts. Yeast 10:1727–1740

Kurtzman GP, Fell JW (2011) The yeast – a taxonomic study, 5th edn. Elsevier, Amsterdam

Kurtzman CP, Fell JW, Boekhout T (2011) The yeasts - a taxonomic study, 5th edn. Elsevier, Amsterdam

Mager WH, Winderickx J (2005) Yeast as a model for medical and medicinal research. Trends Pharmacol Sci 26:265–273

Meyen J (1838) Jahresbericht uber die Resultate der Arbeiten im Felde der physiologischen Botanik von dem Jahre 1837. Wiegmann Arch. Naturgesch. 4:1–186

Nickoloff JA, Chen EY, Heffron F (1986) A 24-base-pair DNA sequence from the *MAT* locus stimulates intergenic recombination in yeast. Proc Natl Acad Sci USA 83:7831–7835

Panchall CJ, Russell I, Sills AM, Stewart GG (1984) Genetic manipulation of brewing and related yeast strains. Food Technol 38:99–106

Pasteur L (1873) Manufacture of beer and yeast. US Patent 141,072, 9 May 1873

Pasteur L (1876) Etudes sur la Biere. Gauthier-villars, Paris (French)

Pasteur L (1879) Studies on fermentation. Macmillan, London

Philliskirk G (2012) Louis Pasteur. In: Oliver G (ed) The Oxford companion to beer. Oxford University Press, Oxford, pp 642–643

Pretorius I (2016) Conducting wine symphonies with the aid of yeast genomics. Beverages 2(4):36

Pyke M (1958) The technology of yeast. In: Cook AH (ed) The chemistry and biology of yeasts. Academic Press, New York, pp 535–586

Reinhardt K, Dowling DK, Morrow EH (2013) Medicine, mitochondrial replacement, evolution and the clinic. Science 341:1345–1346

Replansky T, Koufopanou V, Greig D, Bell G (2008) *Saccharomyces* sensu stricto as a model for evolution and ecology. Trends Ecol Evol 23:494–501

Robinow CF (1975) The propagation of yeasts for the light microscope. Methods Cell Biol XI:1–22

Robinow CF (1980) The view through the microscope. In: Stewart GG, Russell I (eds) Current developments in yeast research. Pergamon Press, Toronto, pp 623–634

Rose AH (1980) *Saccharomyces cerevisiae* as a model eukaryote. In: Stewart GG, Russell I (eds) Current developments in yeast research. Pergamon Press, Toronto, pp 645–652

Russell I, Garrison IF, Stewart GG (1973) Studies on the formation of spheroplasts from stationary phase cells of *Saccharomyces cerevisiae*. J Inst Brew 79:48–54

Russell I, Hancock IF, Stewart GG (1983) Construction of dextrin fermentable yeast strains that do not produce phenolic off-flavors in beer. J Am Soc Brew Chem 41:45–51

Sanchez A, Steensels J, Snoek T, Meersman E (2014) Improving industrial yeast strains: exploiting natural and artificial diversity. FEMS Microbiol Rev 38:947–995

Schwann T (1837) Vorläufige Mittheilung, bettreffend Versuche über die Weingährung und Fäulniss. Ann Phys Chem 41:184–193

Seo AY, Joseph AM, Dutta D, Hwang JC, Aris JP, Leeuwenburgh C (2010) New insights into the role of mitochondria in aging: mitochondrial dynamics and more. J Cell Sci 123:2533–2542

Stewart GG (1975) Yeast flocculation – practical implications and experimental findings. Brew Dig 50:42–62

Stewart GG (1983) Genetically engineered yeasts. Food Eng 55:114–118

Stewart GG (1985) New developments in ethanol fermentation. J Am Soc Brew Chem 43:61–65

Stewart GG (2014a) *Saccharomyces cerevisiae*. In: Batt CA (ed) Encyclopedia of food microbiology. Elsevier, Boston, pp 297–315

Stewart GG (2014b) Yeast mitochondria. MBAA Tech Quart 51:3–11

Stewart GG (2017) Brewer's yeast propagation – the basic principles. MBAA Tech Quart 54:125–131

Stewart GG, Garrison IF (1972) Some observations on co-flocculation in *Saccharomyces cerevisiae*. Proc Am Soc Brew Chem 118–131

Stewart GG, Russell I (1986) One hundred years of yeast research and development in the brewing industry. J Inst Brew 92:537–558

Stewart GG, Russell I, Panchal CJ (1982) The genetics of alcohol metabolism in yeast. Brew Dist Int 12(23–25):36–40

Wingë Ø, Laustsen O. (1937) On two types of spore germination, and on genetic segregations in *Saccharomyces*, demonstrated through single-spore cultures. C R Trav du Lab Carlsberg, Série Physiologique 22:99–117
Wingë Ø, Laustsen O (1939) On 14 new yeast types, produced by hybridization. C R Trav du Lab Carlsberg, Série Physiologique 22:337–353
Zörgö E, Chwialkowska K, Gjuvsland AB, Garré E, Sunnerhagen P, Liti G, Blomberg A, Omholt SW, Warringer J (2012) Life history shapes trait heredity by accumulation of loss-of-function alleles in yeast. Mol Biol Evol 29:1781–1789

Chapter 3
Taxonomy of Brewing and Distilling Yeasts and Methods of Identification

It has already been discussed that brewing and distilling yeast strains (with a few exceptions—details later, Chap. 17) belong to the genus *Saccharomyces* (Chap. 1). There are basically three different types of beer: lager, ale and stout (a form of dark beer). In reality, stout is a type of ale. The commercial worldwide production volume of ale has always been much lower than that of lager, and over the years this difference has grown wider globally, approximately 10% of total beer production from a total of 1.96 billion hectolitres/pa (Barth 2014). However, the difference between ale and lager beer volumes has narrowed slightly during the past decade in the United States and is currently 4.6% ale, largely due to the increasing growth of the "craft brewing" sector (Oliver 2012), which predominantly (but not exclusively) produces ales. It is also forecast that the global lager market will grow at approximately 4.2% per annum during the period 2017–2020.

Reasons for the differences between ale, lager and distilling yeast strains have intrigued many students of *Saccharomyces* yeasts, including this author, for over the past 40 years (e.g. Stewart 1977, 1988, 2009; Stewart and Russell 1993, 2009; Hill and Stewart 2009; Stewart et al. 2016). Considerable research by many industrial and academic institutions has been conducted (Stewart 2009), and a number of typical differences between ale, lager and distiller's yeast strains have been established and accepted (Table 3.1). There are several differences in the production of ale and lager beers, one of the main ones being the characteristics of the ale and lager yeast strains employed and the fermentation temperatures. Considerable research by many breweries and related institutions on this topic has been conducted (Gallone et al. 2016; Pulvirenti et al. 2000; Rainieri et al. 2003; Libkind et al. 2011)—details later.

In 1996, the genome of a haploid *Saccharomyces cerevisiae* strain (S288c) was the first eukaryote genome to be completely sequenced (Goffeau et al. 1996). Since then, the availability of a wide range of whole genome sequences has enabled a detailed comparison of industrial yeast strains, and it has been concluded that brewing strains are interspecies hybrids (Libkind et al. 2011; Krogerus et al. 2015; Gallone et al. 2016; Bing et al. 2014). The nomenclature arguments in terms of the terminology of brewing and distilling yeast strains and their exact

G.G. Stewart, *Brewing and Distilling Yeasts*, The Yeast Handbook,
https://doi.org/10.1007/978-3-319-69126-8_3

Table 3.1 Differences between ale and lager yeast strains

Ale yeast	Lager yeast
Saccharomyces cerevisiae (ale type)	*Saccharomyces carlsbergensis*
Saccharomyces cerevisiae (ale and distillers yeast)	*Saccharomyces uvarum* (*carlsbergensis*)
	Saccharomyces cerevisiae (lager type)
	Saccharomyces pastorianus (current taxonomic name)
Fermentation temperature (18–25 °C)	Fermentation temperature (8–15 °C)
Cells can grow at 37 °C and higher	Cells cannot grow above 34 °C
Cells cannot ferment the disaccharide melibiose (galactose-glucose)	Ferments melibiose
Strains with distinctive colonial morphology on wort-gelatin medium	Strains do not have a distinctive morphology on wort-gelatin medium
"Top" fermentation	"Bottom" fermentation

origin have been ongoing for a long time. With the advent of molecular biology-based methodologies, particularly gene sequencing of ale and lager brewing strains has shown that they are interspecies hybrids with homologous relationships to one another and also to *Saccharomyces bayanus*, a yeast species used in wine fermentations and identified as a wild (contaminating) yeast during brewing and distilling fermentations. The gene homology (Pedersen 1995) between *S. pastorianus* and *S. bayanus* strains is high at 72%, whereas the homology between *S. pastorianus* and *S. cerevisiae* is much lower at 50%. It is interesting to note that gene homology between humans and chimpanzees is 96% (Britten 2002)!

A landmark paper (Libkind et al. 2011) has reported the isolation of a new yeast species in Patagonia, South America. Named *Saccharomyces eubayanus*, this yeast is reported to be an early ancestor of lager yeast strains and to have endowed this yeast with the capacity to ferment at cold temperatures <10 °C. Lager brewers ferment wort at relatively cool temperatures (8–18 °C) and then condition the (immature) green beer under refrigeration conditions. *S. pastorianus* [the correct taxonomic name for the lager species that brewers used to refer to as *Saccharomyces carlsbergensis* (Gibson et al. 2013a, b)] can tolerate lower temperatures than is the situation with ale-producing yeasts—*Saccharomyces cerevisiae* (Table 3.1). This so-called cryotolerant yeast is a hybrid of the ale yeast *S. cerevisiae* and a yeast species that, until recently, has evaded conclusive identification (Hebly et al. 2015). When the *S. eubayanus* genome was sequenced, it was found to be distinct from any previously described species but was a 99.5% match with the missing piece of the hybrid lager yeast *S. pastorianus*—the part of the hybrid not accounted for by the well-studied warm-fermenting ale yeast *S. cerevisiae* (Libkind et al. 2011). Nguyen and Gaillardin (2005) also studied the hybridization history of lager *Saccharomyces* strain's mosaic genomes and patterns of introgression between *Saccharomyces bayanus*, *Saccharomyces uvarum*, and *S. cerevisiae*. The same novel species named *S. eubayanus* by Libkind et al. (2011) was identified by Nguyen and Gaillardin (2005) and called *Saccharomyces* lager strains. The proposal that *S. eubayanus* has come exclusively from Patagonia has been questioned in a 2014 publication by Bing et al. (2014) who successfully isolated this yeast species from the bark and rotten wood of different oak and other deciduous trees collected in the Tibetan Plateau, including high-altitude areas in west China provinces. The results reported in the Bing et al. (2014) publication strongly suggest that the Tibetan population of *S. eubayanus* is also the direct donor of the non-*S. cerevisiae* subgenome of lager yeasts. As Europe and Asia are located on the same landmass, it would have been much easier for Tibetan *S. eubayanus* strains to make their way to Europe through the Eurasian continental bridge. Consequently, there would have been plenty of opportunities for Tibetan *S. eubayanus* strains to colonize Europe before they were domesticated for lager brewing in central Europe in the 1400s. However, there are probably a number of sources of *S. eubayanus*—Patagonia, Tibet and others yet to be identified from different regions of the globe. In addition, Rainieri et al. (2006) have studied pure and mixed genetic lines of *S. bayanus* and *S. pastorianus* and have identified their contribution to the lager brewing strain genome (Stewart 2016; Hill and Stewart 2008; Stewart et al. 1988).

In support of the geographical diversity theory of *Saccharomyces* species that are used in alcoholic beverage production, it should be noted that there are a number of such species that are considered to be subspecies of *S. bayanus* (Guan et al. 2013). This species of yeast can be further reclassified into different strains in the same manner. There are currently thousands of different *S. cerevisiae* strains, many of which are used for a plethora of industrial purposes (Barnett 2003; Bothast and Schlicher 2005; Nevoigt 2008). The basics of *Saccharomyces* strain hybridization are common between domesticated yeasts used in alcohol production (potable and industrial). The yeasts used in the whisk(e)y industry are mainly *S. cerevisiae* (Walker and Hill 2016), although various secondary species are also used—details later (Chaps. 9 and 17). Baker's yeast is usually *S. cerevisiae* (Dequin 2001; Randez-Gil et al. 2003), types of brewer's yeast have already been discussed and rum is fermented primarily with *S. cerevisiae* and the fission yeast *Schizosaccharomyces pombe* (together with various wild yeasts—details in Chap. 17) (Pauley and Maskell 2016). The wine industry mostly uses *S. cerevisiae* and/or *S. bayanus*, together with a plethora of various wild yeasts (e.g. *Kloeckera*, *Saccharomycodes*, *Schizosaccharomyces*, *Hansenula*, *Candida*, *Pichia* and *Torulopsis*) and numerous bacteria (Jarboe et al. 2007; Pretorius 2016). A discussion of these species is beyond the scope of this publication. Also, beyond the scope of this chapter (there was a brief reference in Chap. 2) is the developing application of *S. cerevisiae* as a model for medical applications (Chap. 14) (Menacho-Marquez and Murguia 2007; Stewart 2014a).

Experiments by Piotrowski et al. (2012) examined how selective pressures such as temperature, osmotic pressure and ethanol can lead to different genomic outcomes owing to interspecific hybridization as yeast strains evolve over time. This study concerned the suggestion that the evolution of yeast species may be due to users (e.g. brewer's or distiller's) of yeast unknowingly applying selective pressures on the yeasts for adaptation to the environment (Boulton 2011). The phenomenon of nature-nurture effects on brewer's and distiller's yeast strains (Stewart et al. 1975; Stewart 2014b) will be discussed later (Chap. 4).

Recently, researchers from the University of Manchester and the National Collection of Yeast Cultures (NCYC) have discovered a new species of *Saccharomyces* sp. (Naseeb et al. 2017). This species was found at altitude, growing 1000+ metres above sea level on an oak tree in Saint Auban, in the foothills of the French Alps. In order to survive at this altitude, the yeast has developed an ability to tolerate colder conditions than most characterized strains of *Saccharomyces* yeasts (Stewart et al. 2016). This will increase the number of species in the *Saccharomyces* genus from nine to ten.

This novel yeast species is of interest to brewers, as they rely on lager yeasts to thrive under cold conditions, and it also creates the opportunity for novel yeast hybrids with improved biotechnology properties to be applied to fermentation conditions. Most of the isolates were found to be known species, but two were confirmed, through DNA sequencing and genetic testing, to belong to a new yeast species. It is a cold-tolerant species which can also ferment maltose efficiently. The new yeast species has been formally recognized using recognized scientific criteria and named *Saccharomyces jurei* in memory of the yeast researcher Jure Piskur (Fig. 3.1) who prematurely died of cancer at the young age of 54 in 2014!

A large number of studies have been conducted in order to characterize different types of industrial yeast strains:

- There are numerous differences in strains' flocculation and sedimentation characteristics (Verstrepen et al. 2003; Vidgren and Londesborough 2011; Stewart et al. 2016) which will be discussed in a later chapter (Chap. 13).
- There are subtle differences in the uptake of wort sugars, particularly maltose and maltotriose (Zheng et al. 1994a, b; Day et al. 2002a, b; Dietvorst et al. 2005; Gibson et al. 2013b; Stewart 2006; Vidgren et al. 2009) and wort amino acids and peptides (Jones and Pierce 1964; Lekkas et al. 2005, 2007, 2009; Stewart et al. 2013). Lager strains utilize maltotriose more efficiently than ale and distiller's strains, with less residual maltotriose remaining at the end of a lager wort fermentation (Zheng et al. 1994b; Day et al. 2002b; Salema-Oom et al. 2005; Vidgren et al. 2009). This is one reason why lager strains metabolize high-gravity worts more efficiently than ale strains during static fermentation conditions (Stewart and Murray 2012; Stewart 2014a, b). Another reason for static fermentation differences is variations in yeast flocculation characteristics (Hill and Stewart 2009; Stewart and Russell 2009) (details in Chaps. 7 and 13).
- Ale and distiller's cultures are more amenable to drying whereas lager cultures are not easily dried (Finn and Stewart 2002). Although the exact reason(s) for this difference is not immediately apparent, it is thought that there is a relationship between intracellular trehalose levels (Bolat 2008; Eleutherio et al. 2015; Stewart 2014a, b), the ability to dry yeast cells and cell viability and vitality (Gadd et al. 1987; D'Amore et al. 1991).
- Lager strains, under normal wort fermentation conditions, produce considerably more sulphur dioxide than ale/distilling strains (Donalies and Stahl 2002). This difference is thought to be due to divergent sulphur metabolism pathways and

Fig. 3.1 Jure Piskur

the lower fermentation temperature during lager strain fermentations (Stewart and Russell 2009; Samp and Sedin 2017).

- Lager strains possess the *FSY1* gene, which encodes for a specific fructose/$H^{(+)}$ symporter in the wild-type lager. This gene is not present in ale yeast strains (Gonçalves et al. 2000; Anjos et al. 2013; Rodrigues-Pousada et al. 2004)—details later.
- Ale/lager differences between diacetyl and the metabolism of other vicinal diketones (VDK) are also apparent with most strains [there are some exceptions (Stewart and Russell 2009)]. Generally, lager strains produce more α-acetolactate and subsequently diacetyl than ale strains (Kielland-Brandt et al. 1989; Kristoffer and Gibson 2013). However, under similar environmental conditions (temperature, wort composition, gravity, etc.), the rate of subsequent removal of diacetyl late in the fermentation is similar for both ale and lager yeast strains (Kristoffer and Gibson 2013).
- It has already been discussed that lager strains exhibit a poorer upper temperature tolerance than *S. cerevisiae* (Gibson et al. 2013a). Ale/distiller's strains can grow at 37 °C and above, whereas lager strains will not grow at 37 °C. Nutrient agar plates containing lager yeast, and incubated at 37 °C immediately after inoculation (the plates must *not* be allowed to remain at room temperature at all!), will not show any growth. Yeast growth on these plates indicates the presence of contamination with either an ale yeast (which can occur in a multiproduct brewery) or a wild yeast. This is a useful check for yeast purity in a lager pitching yeast (Stewart and Russell 2009; Bokulich and Bamforth 2013; Russell 2016) but is not a rapid test. Studies (Naseeb et al. 2017) on a recently isolated and identified yeast species, *Saccharomyces jurei*, that can tolerate colder conditions and metabolize maltose have already been discussed in this chapter.

As detailed in Table 3.1, a traditional (and classic) difference between ale/distiller's and lager yeast strains *S. cerevisiae* and *S. pastorianus*, respectively, is that lager strains are able to metabolize the disaccharide melibiose (glucose-galactose). The ability to metabolize this sugar depends on the presence of the enzyme melibiase (α-galactosidase), which is secreted into the periplasmic space of the yeast cell, and hydrolyses melibiose into its constituent monosaccharides—glucose and galactose. These resulting monosaccharides are subsequently taken up and metabolized by the lager yeast culture. The production of α-galactosidase (melibiase) is possible because of the presence in the lager strain genome of one or more *MEL* genes—a polymeric series of genes (Turakainen et al. 1993). Alternatively, ale yeast strains are unable to metabolize melibiose because they cannot produce melibiase owing to the absence of an active *MEL* gene (no α-galactosidase). This inability to utilize melibiose by distiller's strains has industrial relevance in the production of bioethanol from sugar beet because this root crop contains significant amounts of raffinose [approx. 3% (w/w) dry weight]—a trisaccharide, two thirds of which is melibiose. It is worthy of note that cane sugar does not contain significant amounts of raffinose.

Today, differentiation of ale, distiller's and lager yeast strains can be achieved by employing a number of novel methods employing molecular biology techniques (Botstein and Fink 1988)—these will be reviewed later (Chap. 16). One of the most basic and suitable methods available in the 1960s for the purpose of strain and species differentiation that still has relevance today is the giant morphology plate method (Richards 1967). This method involves inoculating the yeast culture onto solid media using a large petri dish and examining the colonial morphology that develops after incubating under standard humid conditions at approximately 18 °C for *at least* 3 weeks. It has been found that gelatin, as the solidifying matrix, enhances the distinctive colonial morphological features to a much greater extent than agar. It is usual to find that malt-wort, corn steep liquor, fruit and vegetable infusions induce greater colony differentiation than complete media based on nutrients from animal sources or defined synthetic media (not sure why this is the case). This method has, however, one major shortcoming (as well as the prolonged incubation period necessary at below room temperature). It gives no information on the value of a yeast strain for brewing or distilling purposes. To quote Cook (1969), who stated in his address to the European Brewery Convention Congress in 1969: "It is important to realize that this procedure (i.e., the giant colony procedure) is rather like taking photographs of those in this hall. The photographs would enable us to identify individuals elsewhere but tells us nothing of their performance as maltsters, brewers and scientists". However, although strains of *S. cerevisiae* (ale) exhibit characteristic colonial morphologies when grown on wort-gelatin media (Fig. 3.2), *S. pastorianus* (lager) strains exhibit similar, uncharacteristic and somewhat uniform morphology, strain to strain (Fig. 3.3). The application of this giant colony morphologies technique during our study of co-flocculation with ale strains will be discussed later (Chap. 13).

The application of DNA fingerprinting techniques (Chambers et al. 2014) and other novel molecular biology methods to identify yeast strains has superseded identification methods such as giant colony culturing plating methods, but they suffer from a number of impediments. They also do not tell us anything about a strain's fermentation growth and sedimentation characteristics. Also, for identification purposes, the DNA fingerprint of a particular yeast strain must be held on file (similar to fingerprints in forensic science—Cole 2001)—details in Chap. 17.

Fig. 3.2 Giant colony morphologies of ale yeast strains

Fig. 3.3 Giant colony morphologies of lager yeast strains

References

Anjos J, Rodrigues de Sousa H, Roca C, Cássio F, Luttik M (2013) Fsy1, the sole hexose-proton transporter characterized in *Saccharomyces* yeasts, exhibits a variable fructose:H+ stoichiometry. Biochim Biophys Acta 1828:201–207

Barnett JA (2003) Beginnings of microbiology and biochemistry: the contribution of yeast research. Microbiology 149:557–567

Barth R (2014) The chemistry of beer: the science in the suds. Wiley, Chichester

Bing J, Han PJ, Liu WQ, Wang QM, Bai FY (2014) Evidence for a Far East Asian origin of lager beer yeast. Curr Biol 24:R380–R381

Bokulich NA, Bamforth CW (2013) The microbiology of malting and brewing. Microbiol Mol Biol Rev 77:157–172

Bolat I (2008) The importance of trehalose in brewing yeast survival. Innov Roman Food Biotechnol 2:1–10

Bothast RJ, Schlicher MA (2005) Biotechnological processes for conversion of corn into ethanol. Appl Microbiol Biotechnol 67:19–25

Botstein D, Fink GR (1988) Yeast: an experimental organism for modern biology. Science 240:1439–1443

Boulton C (2011) Yeast handling. Brew Dist Inst 7:7–10
Britten RJ (2002) Divergence between samples of chimpanzee and human DNA sequences is 5% counting indels. Proc Natl Acad Sci USA 99:13633–13635
Chambers GK, Curtis C, Millar CD, Huynen L, Lambert DM (2014) DNA fingerprinting in zoology: past, present, future. Invest Genet 5:3
Cole S (2001) Suspect identities: a history of fingerprinting and criminal identification. Harvard University Press, Cambridge, MA, pp 60–61
Cook AH (1969) Yeast performance – its significance and assessment. In: Proceedings of 12th congress European brewing convention, pp 225–240
D'Amore T, Crumplen R, Stewart GG (1991) The involvement of trehalose in yeast stress tolerance. J Ind Microbiol 7:191–196
Day RE, Higgins VJ, Rogers PJ, Dawes IW (2002a) Characterization of the putative maltose transporters encoded by *YDL247w* and *YJR160c*. Yeast 19:1015–1027
Day RE, Rogers PJ, Dawes IW, Higgins VJ (2002b) Molecular analysis of maltotriose transport and utilization by *Saccharomyces cerevisiae*. Appl Environ Microbiol 68:5326–5335
Dequin S (2001) The potential of genetic engineering for improving brewing, wine-making and baking yeasts. Appl Microbiol Biotechnol 56:577–588
Dietvorst J, Londesborough J, Steensma HY (2005) Maltotriose utilization in lager yeast strains: MTT1 encodes a maltotriose transporter. Yeast 22:775–788
Donalies UE, Stahl U (2002) Increasing sulphite formation in *Saccharomyces cerevisiae* by overexpression of *MET14* and *SSU1*. Yeast 19:475–488
Eleutherio E, Panek A, De Mesquita JF, Trevisol E, Magalhães R (2015) Revisiting yeast trehalose metabolism. Curr Genet 61:263–274
Finn D, Stewart GG (2002) Fermentation characteristics of dried brewer's yeast, the effect of drying on flocculation and fermentation. J Am Soc Brew Chem 108:424–433
Gadd GM, Chalmers A, Reed RH (1987) The role of trehalose in dehydration resistance of *Saccharomyces cerevisiae*. FEMS Microbiol Lett 48:249–254
Gallone B, Steensels J, Prahl T, Soriaga L, Saels V, Herrera-Malaver B, Merlevede A, Roncoroni M, Voordeckers K, Miraglia L, Teiling C, Steffy B, Taylor M, Schwartz A, Richardson T, Christopher White C, Baele G, Maere S, Verstrepen KJ (2016) Domestication and divergence of *Saccharomyces cerevisiae* beer yeasts. Cell 166:1397–1410
Gibson BR, Londesborough J, Rautio J (2013a) Transcription of α-glucoside transport and metabolism genes in the hybrid brewing yeast *Saccharomyces pastorianus* with respect to gene provenance and fermentation targets. J Inst Brew 119:23–31
Gibson BR, Storgårds E, Krogerus K, Vidgren V (2013b) Comparative physiology and fermentation performance of Saaz and Frohberg lager yeast strains and the parental species *Saccharomyces eubayanus*. Yeast 30:255–266
Goffeau A, Barrell BG, Bussey H, Davis RW, Dujon B, Feldmann H, Galibert F, Hoheisel JD, Jacq C, Johnston M, Louis EJ, Mewes HW, Murakami Y, Philippsen P, Tettelin H, Oliver SG (1996) Life with 6000 genes. Science 274:563–567
Gonçalves P, Rodrigues de Sousa H, Spencer-Martins I (2000) *FSY1*, a novel gene encoding a specific fructose/H(+) symporter in the type strain of *Saccharomyces carlsbergensis*. J Bacteriol 182:5628–5630
Guan Y, Dunham MJ, Troyanskaya OG, Caudy AA (2013) Comparative gene expression between two yeast species. BMC Genomics 14:33
Hebly M, Brickwedde A, Bolat I, Driessen M, de Hulster EA, van den Broek M, Pronk JT, Geertman JM, Daran JM, Daran-Lapujade P (2015) *S. cerevisiae* × *S. eubayanus* interspecific hybrid, the best of both worlds and beyond. FEMS Yeast Res 15:fov005
Hill A, Stewart GG (2008) A brief overview of brewer's yeast. Brew Dist Int 5:13–15
Hill AG, Stewart GG (2009) A brief overview of brewer's yeast. Brew Dist Int 5:13–15
Jarboe LR, Grabar TB, Yomano LP, Shanmugan KT, Ingram LO (2007) Development of ethanologenic bacteria. Adv Biochem Eng Biotechnol 108:237–261
Jones M, Pierce J (1964) Absorption of amino acids from wort by yeasts. J Inst Brew 70:307–315

Kielland-Brandt MC, Gjermansen C, Nilsson-Tillgren T, Holmberg S (1989) Proceedings of the 22nd congress of the European Brewery Convention, Zurich. IRL Press, Oxford, pp 37–45
Kristoffer K, Gibson BR (2013) Diacetyl and its control during brewery fermentation. J Inst Brew 119:86–97
Krogerus K, Magalhães F, Vidgren V, Gibson B (2015) New lager yeast strains generated by interspecific hybridization. J Ind Microbiol Biotechnol 42:769–778
Lekkas C, Stewart GG, Hill A, Taidi B, Hodgson J (2005) The importance of free amino nitrogen in wort and beer. MBAA Tech Quart 42:113–116
Lekkas C, Stewart GG, Hill A, Taidi B, Hodgson J (2007) Elucidation of the role of nitrogenous wort components in yeast fermentation. J Inst Brew 113:183–191
Lekkas C, Stewart GG, Hill A, Taidi B, Hodgson J (2009) The role of small peptides in brewing fermentations. J Inst Brew 115:134–138
Libkind D, Hittinger CT, Valério E, Gonçalves C, Dover J, Johnston M, Gonçalves P, Sampaio JP (2011) Microbe domestication and the identification of the wild genetic stock of lager brewing yeast. Proc Natl Acad Sci USA 108:14539–14544
Menacho-Marquez M, Murguia JR (2007) Yeast on drugs: *Saccharomyces cerevisiae* as a tool for anticancer drug research. Clin Transl Oncol 9:221–228
Naseeb S, James SA, Alsammar H, Michaels CJ, Gini B, Nueno-Palop C, Bond CJ, McGhie H, Roberts IN, Delneri D (2017) *Saccharomyces jurei* sp. isolation and genetic identification of a novel yeast species from *Quercus robur*. Int J Syst Evol Microbiol 67:2046–2052
Nevoigt E (2008) Progress in metabolic engineering of *Saccharomyces cerevisiae*. Microbiol Mol Biol Rev 72:379–412
Nguyen HV, Gaillardin C (2005) Evolutionary relationships between the former species *Saccharomyces uvarum* and the hybrids *Saccharomyces bayanus* and *Saccharomyces pastorianus*; reinstatement of *Saccharomyces uvarum* (Beijerinck) as a distinct species. FEMS Yeast Res 5:471–483
Oliver G (2012) "Märzenbier" by Dorst Hornbusch. In: Oliver G (ed) The Oxford companion to beer. pp 271–273
Pauley M, Maskell D (2016) The role of *Saccharomyces cerevisiae* in the production of gin and vodka. Beverages 3:13
Pedersen MB (1995) Recent views and methods for the classification of yeasts. Cerevisia – Belg J Brew Biotechnol 20:28–33
Piotrowski JS, Nagarajan S, Kroll E, Stanbery A, Chiotti KE, Kruckeberg AL, Dunn B, Sherlock G, Rosenzweig F (2012) Different selective pressures lead to different genomic outcomes as newly-formed hybrid yeasts evolve. BMC Evol Biol 12:46–52
Pretorius IS (2016) Conducting wine symphonics with the aid of yeast genomics. Beverages 2:36–64
Pulvirenti A, Nguyen H-V, Caggia C, Giudici P, Rainieri S, Zambonelli C (2000) *Saccharomyces uvarum*, a proper species within *Saccharomyces* sensu stricto. FEMS Microbiol Lett 192:191–196
Rainieri S, Zambonelli C, Kaneko Y (2003) *Saccharomyces sensu stricto*: systematics, genetic diversity and evolution. J Biosci Bioeng 96:1-9
Rainieri S, Kodama Y, Kaneko Y, Mikata K, Nakao Y, Ashikari T (2006) Pure and mixed genetic lines of *Saccharomyces bayanus* and *Saccharomyces pastorianus* and their contribution to the lager brewing strain genome. Appl Environ Microbiol 72:3968–3974
Randez-Gil FJ, Aguilera J, Codón A, Rincón AM, Estruch F, Prieto JA (2003) Baker's yeast: challenges and future prospects. In: de Winde JH (ed) Functional genetics of industrial yeasts. Springer, Berlin, pp 58–97
Richards M (1967) The use of giant-colony morphology for the differentiation of brewing yeasts. J Inst Brew 73:162–166
Rodrigues-Pousada CA, Nevitt T, Menezes R, Azevedo D, Pereira J, Amaral C (2004) Yeast actuation proteins and stress response: an overview. FEBS Lett 567(1):80–85
Russell I (2016) Yeast. In: Bamforth CW (ed) Brewing materials and processes: a practical approach to beer excellence. Academic Press, Amsterdam, pp 77–96

Salema-Oom M, Pinto VV, Gonçalves P, Spencer-Martins I (2005) Maltotriose utilization by industrial *Saccharomyces* strains: characterization of a new member of the α-glucoside transporter family. Appl Environ Microbiol 71:5044–5049

Samp EJ, Sedin D (2017) Important aspects of controlling sulfur dioxide in brewing. MBAA Tech Quart 54:60–71

Stewart GG (1977) Fermentation – yesterday, today and tomorrow. MBAA Tech Quart 14:1–15

Stewart GG (1988) Twenty-five years of yeast research. Devel Ind Microbiol 29:1–21

Stewart GG (2006) Studies on the uptake and metabolism of wort sugars during brewing fermentation. MBAA Tech Quart 43:265–269

Stewart GG (2009) Forty years of brewing research. J Inst Brew 115:3–29

Stewart GG (2014a) The concept of nature-nurture applied to brewer's yeast and wort fermentations. MBAA Tech Quart 51:69–80

Stewart GG (2014b) Yeast mitochondria – their influence on brewer's yeast fermentation and medical research. MBAA Tech Quart 72:6–11

Stewart GG (2016) *Saccharomyces* species in the production of beer. Beverages 2:34

Stewart GG, Murray JP (2012) Brewing intensification – successes and failures. MBAA Tech Quart 49:111–120

Stewart GG, Russell I (1993) Fermentation – the "black box" of the brewing process. MBAA Tech Quart 30:159–168

Stewart GG, Russell I (2009) An introduction to brewing science and technology. Brewer's yeast, 2nd edn. The Institute of Brewing and Distilling, London

Stewart GG, Russell I, Goring T (1975) Nature-nurture anomalies – further studies on yeast flocculation. Proc Am Soc Brew Chem 33:137–147

Stewart GG, D'Amore T, Panchal CJ, Russell I (1988) Factors that influence the ethanol tolerance of brewer's yeast strains during high gravity wort fermentations. MBAA Tech Quart 25:47–53

Stewart GG, Hill A, Lekkas C (2013) Wort FAN – its characteristics and importance during fermentation. J Am Soc Brew Chem 71:179–185

Stewart GG, Maskell DL, Spears A (2016) Brewing fundamentals – fermentation. MBAA Tech Quart 53:2–22

Turakainen H, Naumov G, Naumova E, Korhola M (1993) Physical mapping of the *MEL* gene family in *Saccharomyces cerevisiae*. Curr Genet 24:461–464

Verstrepen KJ, Derdelinckx G, Verachtert H, Delvaux FR (2003) Yeast flocculation: what brewers should know. Appl Microbiol Biotechnol 61:197–205

Vidgren V, Londesborough J (2011) Yeast flocculation and sedimentation in brewing. J Inst Brew 117:475–487

Vidgren V, Huuskonen A, Virtanen H, Ruohonen L, Londesborough J (2009) Improved fermentation performance of a lager yeast after repair of its *AGT1* maltose and maltotriose transporter genes. Appl Environ Microbiol 75:2333–2345

Walker G, Hill AE (2016) *Saccharomyces cerevisiae* in the production of whisk(e)y. Beverages 2:38–53

Zheng X, D'Amore T, Russell I, Stewart GG (1994a) Factors influencing maltotriose utilization during brewery fermentations. J Am Soc Brew Chem 52:41–47

Zheng X, D'Amore T, Russell I, Stewart GG (1994b) Transport kinetics of maltotriose in strains of *Saccharomyces*. J Ind Microbiol 13:159–166

Chapter 4
Yeast Culture Collections, Strain Maintenance and Propagation

4.1 Introduction

Microbial culture collections (including yeast cultures) focus on collecting, maintaining and distributing microbial strains amongst microbiologists, brewers and distillers. Such collections are a means of preserving microbial diversity. This chapter considers the early history of relevant culture collections, which has already been preliminarily discussed in Chap. 2 with a consideration of the early studies of Emil Christian Hansen working at the Carlsberg Laboratory in Copenhagen. Also, in this chapter, the international connections between yeast culture collections and their relationship to the World Federation for Culture Collections (WFCC), the International Congress of Culture Collections (ICCC) and the International Union of Microbiology Societies (IUMS) will be discussed.

The importance of selecting optimal yeast strains for research and/or industrial applications is often underestimated. For example, utilizing a strain's background that already provides the desired stress tolerance or nutrient utilization profile can often eliminate costly strain optimization later. Yeast culture collections can provide not only yeast strains themselves but also data and curator expertise to assist in the selection of an optimal yeast strain (Boundy-Mills 2012). Some yeast collections are renowned for the broad range of cultures and services that they supply, while "boutique" collections can provide a broad selection of strains possessing certain properties, categories, characterization data and assistance in selecting strains.

Yeast cultures stored in a collection (or bank) are usually (but not always) purified yeast cultures. Once yeast was in pure form in the Carlsberg Laboratories (Holter and Moller 1976), other laboratories started maintaining a bank of strains and supplying cultures to brewers and some (not many) distillers. In 1885, the scientific station for brewing in Munich reported distributing 107 yeast cultures to breweries (Rubin 2012). Today, it is possible for brewers, both home and professional brewers, to obtain yeast strains from culture collections and have them shipped worldwide in various forms.

G.G. Stewart, *Brewing and Distilling Yeasts*, The Yeast Handbook,
https://doi.org/10.1007/978-3-319-69126-8_4

A typical example of a culture collection is held at the United Kingdom's Centre for Agriculture and Bioscience International (CABI). This centre manages a collection of over 28,000 viable strains including the National Collection of Fungus Cultures. This culture collection contains over 6000 species (many of them yeasts) isolated from environmental and agricultural systems worldwide, making it one of the most significant collections in the world! In addition, the collection contains a burgeoning number of strains developed as a result of genetic manipulation. Another UK yeast culture collection is the National Collection of Yeast Cultures (NCYC) (based at the University of East Anglia in Norwich), which is part of the United Kingdom National Culture Collection (UKNCC).

Another yeast culture collection, worthy of mention, is the Herman J. Phaff Yeast Culture Collection at the University of California, Davis. The study of yeast at UC Davis has a notable scientific history with contributions from many remarkable scientists that began before the establishment of Davis as a separate campus of the University of California. In the 1890s, F.T. Bioletti, a cellar foreman in the UC College of Agriculture's Central Experiment Station, experimented with selected yeasts for wine fermentation. William Cruess worked with Bioletti and pure yeast cultures were used during wine fermentations. Some of the first pure yeast cultures from Californian wineries are still maintained in the Phaff collections. Emil Mrak (along with William Cruess) became an active contributor of pure yeast cultures to the collection during his academic years prior to his service as the second chancellor of UC Davis commencing in 1959.

Herman Phaff (Fig. 4.1) consolidated and expanded the Davis yeast collection from the study of direct application to yeast properties as a fundamental resource for yeast taxonomy, ecology, physiology and subsequently molecular biology investigations. In 1996, the collection was officially dedicated and named the Herman J. Phaff Yeast Culture Collection, recognizing Phaff's outstanding contributions over 50 years of yeast research. He also founded the Yeast Newsletter which is currently edited by Marc-André Lachance (one of Phaff's PhD students), now a Microbiology Professor in the University of Western Ontario, London, Ontario. Phaff was certainly one of the most renowned yeast microbiologists of the twentieth century!

In 1980, at the Fifth International Yeast Symposium held in London, Ontario, Canada, a colloquium was held, honouring Herman Phaff, and organized by Sally Anne Meyer and Arnie Demain (both of them were Phaff's former PhD students) (Stewart and Russell 1980). Phaff's research record was discussed and acclaimed! It was also emphasized during the colloquium that his interests had shifted away from the solution of problems relating to the food and beverage industries to the acquisition of a better understanding of the basic phenomena underlying these problems. Since 2001, the yeast collection at Davis has been curated by Kyria Boundy-Mills and currently contains over 6000 yeast strains (Boundy-Mills 2012).

It has already been discussed in this chapter, and Chap. 2, that pure yeast cultures were first introduced into the brewing industry by Emil C. Hansen (Fig. 2.21). He employed dilution techniques to isolate single cells from brewing yeast cultures, assessed them individually and then selected the specific strain that produced the

Fig. 4.1 Herman Phaff

desired brewing properties. As a result of their introduction in 1883 into the Copenhagen Carlsberg brewery on a production scale, the benefits of pure yeast cultures rapidly became clear. By 1892, many brewing companies such as Heineken, Schlitz, Pabst, Anheuser Busch and 50 or so smaller breweries were using pure lager cultures and employing propagation plants based on the Hansen model (Fig. 2.22) (Stewart 2017).

4.2 Preservation of Stock Yeast Cultures

Prior to actual propagation procedures, it is imperative that no change in the character and properties of the yeast strain occurs. Many yeast strains are difficult to maintain in a stable state and long-term preservation by lyophilization (freeze drying), which has proven useful for some mycelial fungi and bacteria (Kirsop and Doyle 1991; Tan et al. 2007), has been found to give poor wort fermentation results with many brewing yeast strains (Russell and Stewart 1981).

Storage studies have been conducted with a number of ale and lager brewing and also distiller's yeast strains. The Labatt studies were over a 2-year storage period. A number of wort fermentation tests were conducted (Russell and Stewart 1981). These tests included wort fermentation rate and wort sugar uptake efficiency, flocculation tests, sporulation ability, formation of respiratory-deficient (RD) colonies and ease of revival. The results were compared to the characteristics of the unstored control culture.

Low-temperature storage in liquid nitrogen (−196 °C) (Wellman and Stewart 1973) appears to be the storage method of choice if cost and availability of the appropriate equipment is not a significant factor. Since the above studies were largely conducted in the 1970s, −70 °C refrigerators have become commercially available. A study in the early 1980s used a −70 °C refrigerator and showed that cultures stored in it had the same low death rate and were easy to revitalize, equivalent to −196 °C storage (Panchal et al. 1986). Also, the degree of flocculation, wort fermentation and sporulation ability and proportion of mutants present [details of respiratory-deficient (RD) mutants are in Chap. 14] were all unaffected by this storage method. Storage at 4 °C on nutrient agar slopes, subcultured every 6 months, was the next method of preference to low-temperature storage. Today, many breweries store their strains (or contract store them) at −70 °C. Routine subculturing of cultures on solid media every 6 months is a less desirable but cost effective method. Lyophilization of brewer's yeast cultures should be avoided (Stewart and Russell 2009).

4.3 Propagation of Yeast Cultures

The propagation of a pure yeast culture has been an essential part of the brewing process (particularly lager yeast) for over 100 years. However, unlike the production of other alcoholic beverages (e.g. distilled potable spirits, wine and saké), where a yeast culture is only used for one cycle (also erroneously called a generation), a brewing yeast culture is used for a number of cycles. Nevertheless, the yeast propagation process is still not fully understood (Nielsen 2005, 2010). The reason for this situation is unclear and this reflects yeast's facultative nature (Crabtree 1928). Brewer's yeast cultures do not live forever and must be replaced on a regular basis with a fresh yeast culture (details in Chap. 11).

The desire is to use yeast with good fermenting and appropriate sedimentation characteristics but also with the ability to grow rapidly while maintaining these properties during propagation (Nielsen 2005). The problem with yeast growth is that catabolite repression—the Crabtree effect—limits the yeast's ability to take up oxygen during yeast propagation in wort (Crabtree 1928). Furthermore, in recent years, there has been a growing awareness that certain factors during propagation may stress the yeast and, consequently, influence yeast viability, vitality and beer quality (Stewart and Russell 2009). Stress factors during propagation are closely related to the aeration (oxygenation) methods employed (details in Chap. 11).

Although both the Crabtree and Pasteur effects will be discussed in Chap. 6, their definition is worth detailing here:

- *Pasteur effect* (Pasteur et al. 1879)—inhibition of alcoholic fermentation in the presence of oxygen. Activation of glycolysis by anaerobiosis
- *Crabtree effect* (Crabtree 1928)—instantaneous aerobic ethanol formation following transition from carbon limitation to carbon excess

Catabolic repression severely limits the yeast's aerobic growth in a concentrated sugar solution, such as wort, and this makes it difficult for a brewer to develop biomass. The brewing industry has chosen to live with this handicap, because optimizing yeast growth in a substrate other than wort could jeopardize yeast growth during fermentation and result in beer with poor flavour and overall quality. The brewing industry has focussed on strict sanitary standards to avoid infection (unlike the propagation of baker's and distiller's yeast biomass—details later) and also to minimize yeast stress during the propagation process (e.g. maximum viability, vitality, intracellular glycogen levels and minimum proteinase A secretion during the propagation process) in order that negative effects on fermentation and beer quality are prevented. The question to be answered is how to control the process by ensuring optimal air (oxygen) supply and also minimizing any stress that the yeast may encounter during its growth cycle during the propagation phase.

The difference in biomass formation is remarkable comparing the production of baker's yeast with brewer's yeast propagation. Table 4.1 shows that if yeast is fed with a dilute substrate such as molasses (as is the situation with distillers and baker's strains), supplemented with nutrients and vitamins, a typical yield factor would be 54%—meaning that every 100 g of sugar consumed would develop into 54 g of yeast dry matter. This requires an oxygen supply of 74%—meaning 740 mg of oxygen would be absorbed for each gram of yeast dry matter produced. When yeast is propagated in wort containing a much more concentrated fermentable sugar level (usually a maximum gravity of 12°Plato), the Crabtree effect severely inhibits yeast growth to the point that the yield factor is only 5–10%. However, the culture will be acclimatized to the wort environment—particularly the ability to metabolize maltose, maltotriose, wort amino acids (except proline), ammonium ions and di- and tripeptides (Stewart and Russell 2009; Stewart et al. 2013; Stewart 2016).

Today the sanitary problems in relation to yeast propagation have been largely solved, and the focus on the propagation process has latterly been on minimizing yeast stress factors. These factors include (1) shear and turbulence forces (Chlup et al. 2008), (2) oxidative stress (D'Amore et al. 1991; Stewart and Russell 2009), (3) carbon dioxide toxicity and (4) alcohol toxicity (further details of various aspects of yeast stress are in Chap. 11). The first three stress categories are mainly linked to aggregation in the propagation plant, while the fourth occurs when the process is prolonged. As well as acknowledging that maximum aeration is not optimum oxygenation, it is now accepted that maximum cell numbers will not produce the most vital yeast culture. With typical lager strains, the most vital yeast is obtained with approximately 100×10^6 cells per mL (Nielsen 2005, 2010). The essential parameters that are important during brewer's yeast propagation are wort composition and gravity (wort over 12°Plato should not be employed for brewer's yeast propagation), aeration (oxygenation) intensity, yield factor, sterility, doubling time, gas production and foam formation. Although yeast propagation was pioneered with lager strains and the value of ale strain propagation was questioned, with the advent of HG wort fermentation and its use in ale brewing, propagation has become applicable for ale brewing procedures as well (Stewart 2017).

Table 4.1 Biomass yield depending on the propagation procedure and the substrate employed[a]

	Yield factor (g yeast dry solids/g carbohydrate)	Oxygen consumption/g yeast dry solid
Molasses, nutrients and vitamins	0.54	740
10°Plato wort	0.1	120

[a]Nielsen (2005)

References

Boundy-Mills K (2012) Yeast culture collections of the world: meeting the needs of industrial researchers. J Ind Microbiol Biotechnol 39:673–680

Chlup PH, Bernard D, Stewart GG (2008) Disc stack centrifuge operating parameters and their impact on yeast physiology. J Inst Brew 114:45–61

Crabtree HG (1928) The carbohydrate metabolism of certain pathological overgrowths. Biochem J 22:1289–1298

D'Amore T, Crumplen R, Stewart GG (1991) The involvement of trehalose in yeast stress tolerance. J Ind Microbiol 7:191–196

Holter H, Moller KM (1976) The Carlsberg Laboratory 1876/1976. Rhodes International Science and Art, Copenhagen

Kirsop BE, Doyle A (eds) (1991) Maintenance of microorganisms and cultured cells, 2nd edn. Academic Press, Elsevier, Boston, MA

Nielsen O (2005) Control of the yeast propagation process - how to optimize oxygen supply and minimize stress. MBAA Tech Quart 42:128–132

Nielsen O (2010) Status of the yeast propagation process and some aspects of propagation for re-fermentation. Cerevisia 35:71–74

Panchal CJ, Bilinski CA, Russell I, Stewart GG (1986) Yeast stability in the brewing and industrial fermentation industries. Crit Rev Biotechnol 4:253–262

Pasteur L, Faulkner F, Robb DC (1879) Studies on fermentation. MacMillan, London, pp 235–336

Rubin J (2012) White labs. In: Oliver G (ed) The Oxford companion to beer. Oxford University Press, New York, NY, p 843

Russell I, Stewart GG (1981) Liquid nitrogen storage in yeast cultures compared to more traditional storage methods. J Am Soc Brew Chem 38:19–24

Stewart GG (2016) *Saccharomyces* species in the production of beer. Beverages 2:34

Stewart GG (2017) Brewer's yeast propagation – the basic principles. MBAA Tech Quart 54:125–131

Stewart GG, Hill A, Lekkas C (2013) Wort FAN – Its characteristics and importance during fermentation. J Amer Soc Brew Chem 71:179–185

Stewart GG, Russell I (1980) Current developments in yeast research. Pergamon, Toronto, pp 635–644

Stewart GG, Russell I (2009) An introduction to brewing science and technology. Brewers yeast, 2nd edn. The Institute of Brewing and Distilling, London

Tan CS, van Ingen CW, Stalpers JA (2007) Freeze-drying fungi using a shelf-freeze drier. Methods Mol Biol 368:119–125

Wellman A, Stewart GG (1973) Storage of brewing yeasts by liquid nitrogen refrigeration. Appl Microbiol 26:577–583

Chapter 5
The Structure and Function of the Yeast Cell Wall, Plasma Membrane and Periplasm

5.1 Introduction

This chapter largely considers the yeast species *Saccharomyces cerevisiae* and its related interbreeding species, all of which are involved in the alcohol-producing industries. There are other yeast species, for example, the fission yeast *Schizosaccharomyces pombe*, which are only distantly related to *S. cerevisiae* but which have equally important features, regarding alcohol fermentation, will be discussed in Chap. 17. Also, a number of other yeast genera and specific species of *Kluyveromyces* (Lane and Morrissey 2010) (e.g. *K. marxianus* and *K. lactis*) will also be discussed in Chap. 17 together with their overall metabolism regarding the production of fermentation ethanol compared to *S. cerevisiae* and related yeast species in Chaps. 6 and 7. Also, bacteria that produce ethanol (*Zymomonas mobilis*) will receive attention in Chap. 17.

It has already been discussed (Chaps. 1 and 3) that yeast is a eukaryote that consists of a number of membrane-bound organelles—nucleus, vacuoles, mitochondria, Golgi bodies, endoplasmic reticulum, etc. (Fig. 1.2). Also, the exterior of each yeast cell consists of a distinct wall and plasma membrane with a space (the periplasm), in between the two (Arnold 1991-details in Chap. 13).

The cell wall can be enzymatically removed from the cell using a mixture of glucanases and proteinases (Kuo and Yamamoto 1975) to produce a spheroplast (Fig. 5.1). The integrity of each spheroplast (protoplast) can be preserved by maintaining these structures in a high osmotic pressure environment (e.g. 2 M sorbitol). Yeast spheroplasts can be employed to produce novel strains (spheroplast fusion). Their use in the fusion technique will be discussed in the chapter discussing yeast genetic manipulation (Chap. 16) (van Solingen and van der Plaat 1977; Jones et al. 1987). Spheroplasts can be regenerated into whole dividing cells by incubating them in a nutrient growth medium with an even higher osmotic pressure (3 M sorbitol) medium (Svoboda 1966).

G.G. Stewart, *Brewing and Distilling Yeasts*, The Yeast Handbook,
https://doi.org/10.1007/978-3-319-69126-8_5

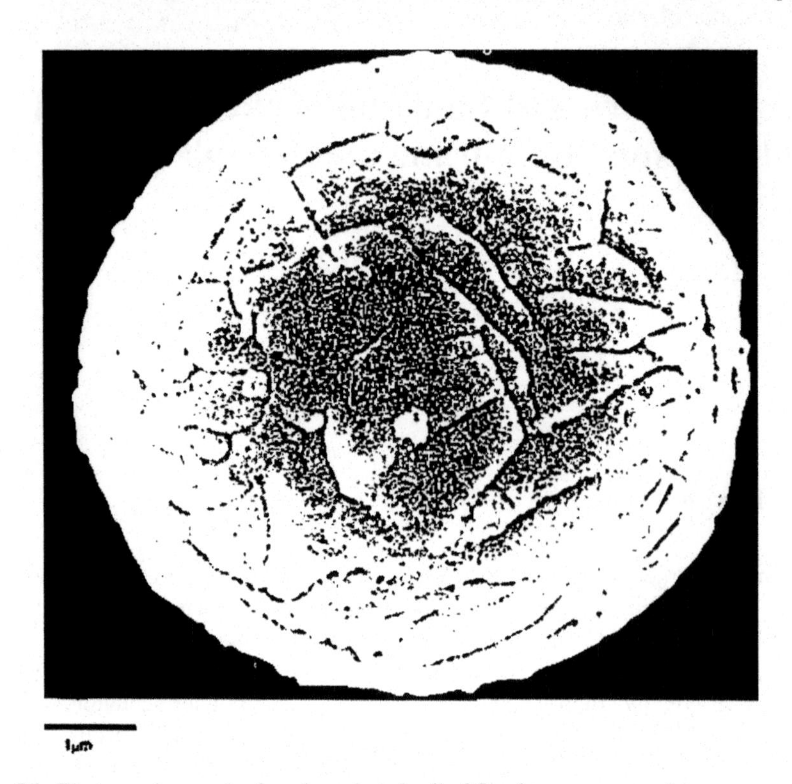

Fig. 5.1 Electron micrograph of a spheroplasted cell of *Saccharomyces cerevisiae*

The yeast cell wall is a dynamic organelle that determines the cell shape and integrity of the organism during growth and cell division. It must provide the cell with mechanical strength in order to withstand changes in osmotic pressure imposed by the environment (Gibson et al. 2007; Dague et al. 2010). It must also protect the cell from mechanical stresses imposed upon a culture such as centrifugation (Chlup et al. 2007) and stress during storage between fermentations (Stoupis et al. 2002).

The basic chemical composition and structural aspects of the *S. cerevisiae* cell have been known for some time. The two major components are β-glucans (formed by 1,3-β- and 1,6-linkages) and mannoproteins (proteins highly N- or O-glycosylated mannose residues linked by 1,2,-1,3-, 1,4- and 1,6-α-linkages), which represent, respectively, about 50–60% and 40–50% of the cell wall mass. A third component is chitin, which is manufactured by 1,4-β-*N*-acetylglucosamine. Although chitin is present in low amounts in cell residues (1–3% of the cell wall mass), compared to glucan and mannan, it is an essential component of cell wall composition—the bud and birth scars (Figs. 5.2 and 5.3) (Cabib et al. 2001).

The basic chemistry of the *S. cerevisiae* cell wall is well understood, but few studies have reported its physical and mechanical properties because of a lack of tools that are able to measure these properties at the cellular level (Hinterdorfer and

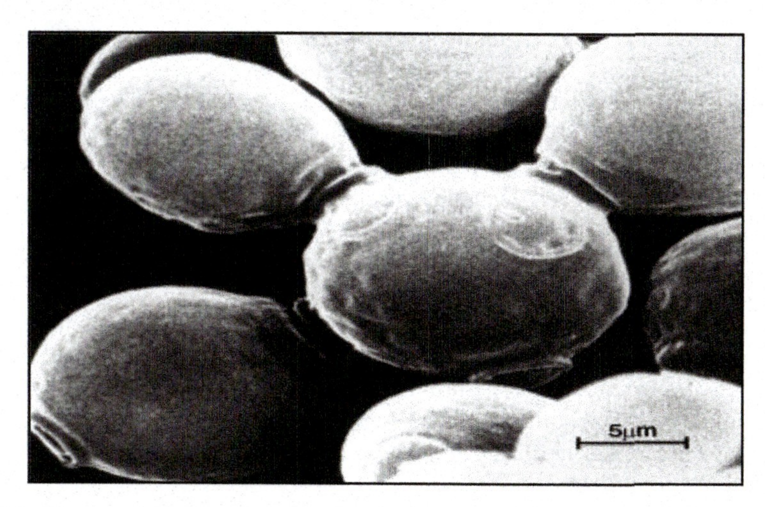

Fig. 5.2 Electron micrograph of budding yeast cells showing chitin rings

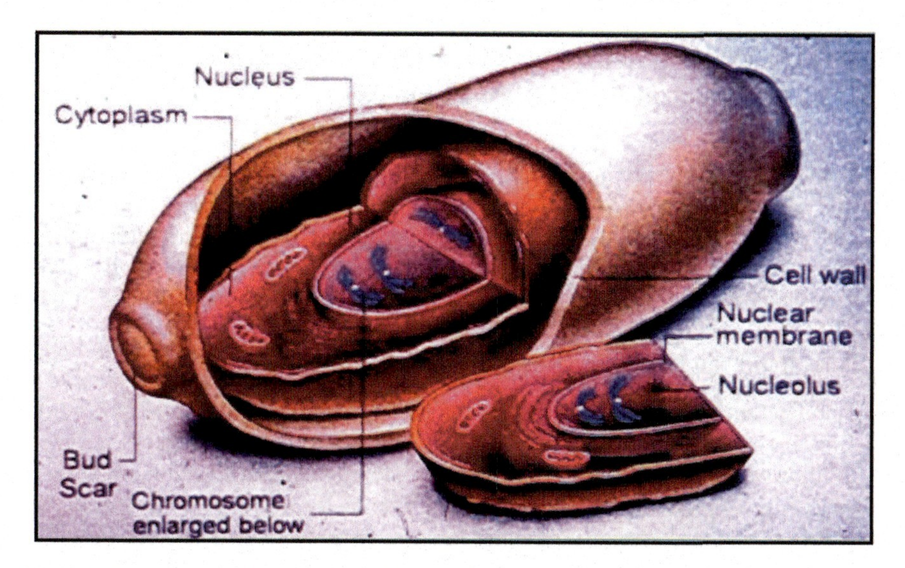

Fig. 5.3 Anatomy and shape of a typical yeast cell

Dufrêne 2006). However, during the past decade or so, the complexities of the cell wall architecture have emerged as a result of detailed genetic, molecular and biochemical studies, which have led to the discovery of several interconnections amongst the various wall components to form macromolecular complexes. The precise assembly processes and proteins involved in these interconnections have not yet been fully resolved. However, a modular yeast cell wall model has been proposed to account for its molecular organization (Smits et al. 1999). This model proposes that the nonreducing ends of the 1,3-β-glucan insoluble fibres may function as attachment sites for 1,6-β-glucan and chitin chains. At the outside (external) of the network are 1,6-β-glucans onto which mannoproteins can be connected, whereas chitin chains are attached at the inside of the 1,3-β-glucan network (Klis et al. 2006). The different modules are interconnected by non-covalent interactions in the β-glucan-chitin layer and by covalent cross-links in the mannoprotein layer, including disulphide bonds between mannoproteins (Orlean 1997), as well as through other mannoprotein-glucan links that are not yet fully characterized (Klis et al. 2006). The molecular organization of the yeast cell wall is dynamic, not static, and must evolve, depending on the growth conditions, morphological development and as a response to changing and challenging environmental conditions. Cell wall damage can be induced by a number of parameters, for example, wall-perturbing drugs such as calcofluor white, caffeine or sodium dodecyl sulphate (SDS), enzymes such as zymolyase, mutations in cell wall-related genes and mechanical agitation (Stoupis et al. 2002; Heinisch 2005; Chlup and Stewart 2011).

5.2 Cell Wall Structure and Function in *Saccharomyces cerevisiae* and Related Yeast Species

Brewers and brewing scientists/microbiologists consider that a primary function, from a brewery perspective (not the only one), of the cell wall of brewer's yeast strains is flocculation (Vidgren and Londesborough 2011) and the structure of mannan in relation to this function. However, the structure and the applications of yeast mannan other than flocculation are discussed later in this chapter. Flocculation is discussed in detail in Chap. 13. In addition to flocculation, the yeast cell wall has four major functions (Klis et al. 2006):

- Stabilization of internal osmotic conditions. The osmolarity of the cytoplasm of *S. cerevisiae* and other yeasts is usually higher than outside of the cell. In order to limit water influx, which would perturb the intercellular reaction conditions and cause excessive swelling of the cell eventually leading to rupture of the plasma membrane, yeasts (and other fungi) construct a sturdy and elastic wall. Extension of the wall creates a counteracting pressure by the wall, which prevents water influx. This parameter plays an important role during the fermentation of high-gravity worts, which will be discussed in Chap. 11.
- Protection against physical stress. The cell wall is not only involved in maintaining osmotic stability, but it also functions as a physically protective coat. The combination of considerable mechanical strength and high elasticity allows the wall to redistribute physical stresses and offers efficient protection against mechanical damage (Morris et al. 1986). In the context of the brewing process, physical stress, for example, the centrifugation of yeast at the end of fermentation without appropriate controls, can adversely affect yeast quality and beer stability (flavour, physical and foam) (Chlup and Stewart 2011). Further

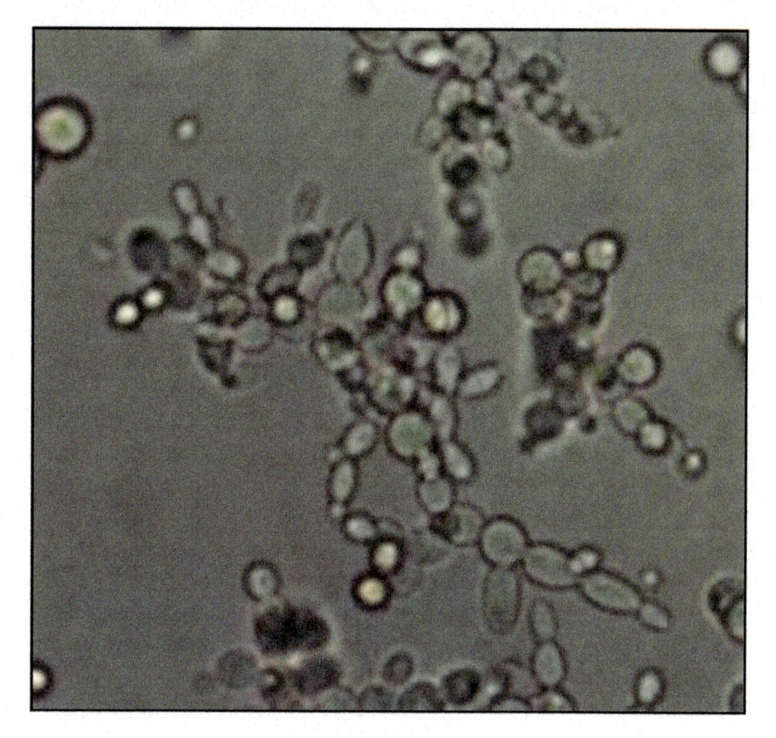

Fig. 5.4 Pseudohyphae and chain formation in yeast cells

details of the effects of centrifugation on yeast during cell cropping are discussed in Chap. 13.

- Maintenance of cell shape is a precondition for morphogenesis. Yeast cells will usually grow as oval cells (Fig. 5.3) or in a more elongated form during nitrogen limitation. Some ale strains will grow pseudohyphally [also called chain formation—Reynolds and Fink 2001; Stewart et al. 2016] (Fig. 5.4) (further details in Chap. 13).
- The stress-bearing polysaccharides of the brewer's and distiller's yeast cell wall function as a scaffold for an external layer of glycoproteins (Osumi 1998). Collectively, these glycoproteins, and particularly their N-linked carbohydrate side chains, limit the permeability of the cell wall for macromolecules, thus shielding the skeletal polysaccharides from attack by foreign proteins (De Nobel et al. 1990). Conversely, they may also limit the escape of soluble intermediates into the medium during cell wall construction. The limited permeability of the external protein layer may also permit the creation of a microenvironment in the inner region of the wall adjacent to the plasma membrane, particularly in cell colonies. The high degree of glycosylation in the cell wall proteins and the presence of negatively charged phosphate groups in their carbohydrate side chains probably also contribute to water retention (Lyons and Hough 1970).

As well as the above collective functions, specific purposes of individual proteins are important. The external protein layer of the cell wall may consist of at least 70 different glycoproteins, and the composition of this protein layer may vary depending on the growth conditions (De Groot et al. 2005; Kitagaki et al. 1997; Shimoi et al. 1998; Zlotnik et al. 1984). Cell wall proteins allow the cells to flocculate (see Chap. 13) (Kobayashi et al. 1998), recognize mating partners (see Chap. 16), form a biofilm by attaching to the inner surface of dispensing lines (Mamvura et al. 2016) (details in Chap. 13) and grow pseudohyphally and invasively (this also applies to pathogenic yeasts such as *Candida albicans*) (Kaur et al. 2005). They also assist the cells in retaining iron and facilitate the uptake of sterols, which are required for growth under anaerobic conditions (Abramova et al. 2001; Alimardani et al. 2004; Cappellaro et al. 1991; Kobayashi et al. 1998; Protchenko et al. 2001; Reynolds and Fink 2001). In addition, cell wall proteins also strongly affect cell hydrophobicity, which is important during wort fermentations (Chap. 13). Also, some cell wall proteins may be specifically involved in cell wall synthesis and repair.

Cell wall construction is a tightly controlled function. The polysaccharide composition, together with the structure and thickness of the cell wall, varies considerably depending on the environmental conditions (Aguilar-Uscanga and Francois 2003). Cell wall synthesis is strictly coordinated during the cell cycle. The majority of the cell wall protein-encoding genes are cell cycle regulated (Smits et al. 1999; Bühler et al. 2005). Oxygen levels in the growth medium also affect the wall's protein composition (Stewart et al. 2016).

S. cerevisiae exerts a considerable amount of metabolic energy during cell wall construction. Depending on the growth conditions, its mass in terms of dry weight may account for about 10–25% of the total cell mass (Klis et al. 2006). The cell wall consists of an inner layer of load-bearing polysaccharides, acting as a scaffold for a protective outer layer of mannoproteins that extends into the medium (Fig. 5.5) (Klis et al. 2002; Ovalle et al. 1998; Orlean 1997; Yin et al. 2005). The major load-bearing polysaccharide is a branched 1,3-β-glucan (Fleet 1991; Kwiatkowski et al. 2009). Due to the presence of side chains, the 1,3-β-glucan molecules can only locally associate through hydrogen bonds, resulting in the formation of a three-dimensional network. This network is highly elastic and is considerably extended under normal osmotic conditions. When cells are transferred to hypertonic solutions, they rapidly shrink and may lose up to 60% of their original volume, which corresponds to an estimated surface loss of approximately 40–50% (Morris et al. 1986). Shrinkage is fully reversible, as seen when the cells are transferred to the original medium. This elasticity of the cell wall reflects the structure of the individual 1,3-β-glucan molecules, which have a flexible and helical shape.

In addition, the 1,4-α-glucan content of the yeast cell wall can vary from as little as 1% to 29% (Sedmak 2006) of the dry weight, depending upon the nutritional status of the cells, the method of isolation, the environmental conditions and the phase in the growth cycle that the cells were harvested (Vink et al. 2004). Industrially produced brewer's yeast cells, as described in a patent application by

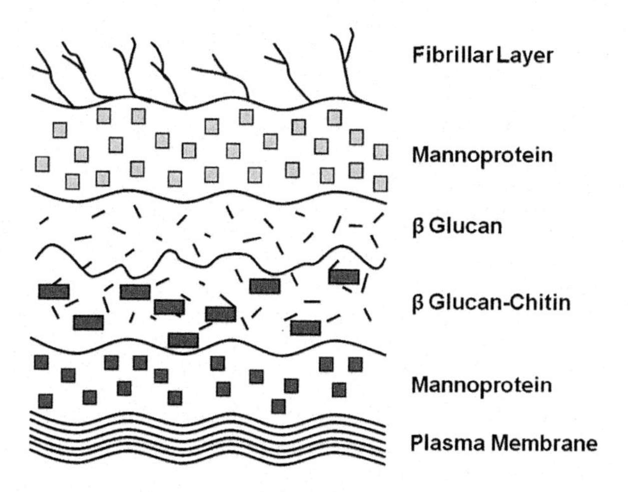

Fig. 5.5 Schematic diagram of the architecture of the yeast cell wall. Components such as monoprotein are found distributed throughout the entire wall and therefore the layered arrangement shows zones of enrichment. It is a matrix, not a sandwich!

Sedmak (2006), contained glucans at a 28.9% (dry weight concentration) and included 12.4% (dry weight) of 1,4-α-glucan.

The majority of cell wall proteins (CWPs) are modified by **g**lycosyl-**p**hosphatidyl-**i**nositol (GPI) and are indirectly linked to the 1,3-β-glucan network. Many proteins of eukaryotic cells (including *Saccharomyces* spp) are anchored to membranes by covalent linkage to glycosyl-phosphatidyl-inositol (GPI) (Brown and Waneck 1992). These proteins lack a transmembrane domain, have no cytoplasmic tail and are, therefore, located on the extracellular side of the plasma membrane. A number of GPI-anchored proteins have been identified such as adhesion molecules, activation antigens and other miscellaneous glycoproteins.

To prove the presence of a putative GPI-CWP in the *Saccharomyces* cell wall and to determine how it is linked to the cell wall, the following evidence is discussed:

- Immunofluorescence studies or immunogold labelling should show its presence in intact cells and SDS-extracted cell walls (Schreuder et al. 1993; Cappellaro et al. 1994; Ram et al. 1998; Rodriguez-Pena et al. 2000).
- Deletion of the 30–40 amino acids should result in secretion of the protein into the suspending medium (Wojciechowicz et al. 1993; Shimoi et al. 1995).
- Extending a secretory reporter protein with the putative GPI-CWP should direct the reporter protein into the wall (Van Berkel et al. 1994; Hamada et al. 1999).
- The protein should be releasable from isolated walls by cleaning the phosphodiester bridge in the GPI remnant using either H7-pyridine or a phosphodiesterase (Kapteyn et al. 1997; De Groot et al. 2004).
- The proteins should be released from isolated walls by either a 1,6-β-glucanase (resulting in protein-bound 1,6-β-glucan epitopes) or, in some special cases, by a chitinase (resulting in protein-bound 1,6-β-glucan and chitin epitopes) (Kapteyn et al. 1997).

As well as the GPI-CWPs, a smaller group of proteins are directly linked to the 1,3-β-glucan network through a linkage (not completely characterized) that is sensitive to mild alkali treatment. These proteins are called ASL (*a*lkali-*s*ensitive *l*inkage) CWPs and include the family of Pir-CWPs (Pir, *p*roteins with *i*nternal *r*epeats) (De Groot et al. 2007). Whereas the GPI-CWPs are found in the outer layer of the wall, the Pir-CWPs appear to be uniformly distributed throughout the inner skeletal layer, which is consistent with their being directly connected to 1,3-β-glucan macromolecules (Kapteyn et al. 1999). In contrast to the other three Pir-CWPs, Cislp has a single repeat sequence. It has been shown (Castillo et al. 2003) that this particular sequence is required for binding to 1,3-β-glucan.

Three of the four Pir-CWPs contain multiple internal repeats. Also, some of GPI-CWPs, such as the flocculins, Aga1p, Tix1p, Tir4p and Dan4p, contain multiple repeats (Verstrepen et al. 2005) that are predominantly found in the cell surface proteins. The number of repeats often varies between yeast strains, and this is due to a comparatively high frequency of recombination events in the repeat domain. This variability affects cell surface properties, such as hydrophobicity and flocculation competence in brewer's yeast strains (details in Chap. 13), also virulence in pathogenic yeasts such as *Candida albicans* (Weig et al. 2004; Kaur et al. 2005).

It has also been claimed that many cytoplasmic proteins, such as the glycolytic enzymes (details in Chap. 6), may also be intrinsic components of the cell wall and that they arrive at the cell surface through a nonclassical protein export pathway (Cleves et al. 1996; Molina et al. 2000).

5.3 Dynamics of the Cell Wall

Approximately 1200 genes exhibit a cell wall-related phenotype when deleted. This supports the concept that cell wall construction is an integral part of the overall cell physiology (De Groot et al. 2001). It is consequently not surprising that the cell wall may vary in composition and thickness, depending on the composition of the growth medium, growth temperature, external pH and oxygen levels (Aguilar-Uscanga and Francois 2003). This sensitivity is reflected in the resistance of an intact cell to cell wall-degrading enzymes (Russell et al. 1973). Much less is known about possible corresponding changes at the molecular level in terms of the length and branching degree of the structural polysaccharides, the degree of glycosylation and the phosphorylation of cell wall proteins. It is not known whether a yeast cell favours the use of specific CWP-polysaccharide complexes in specific phases of the cell cycle or during morphogenetic processes such as mating, pseudohyphal growth, biofilm formation or when entering the stationary growth phase and during sporulation (Kollár et al. 1997). It has been shown that the walls of exponentially growing cells cultured in rich media contain at least 20 different cell wall proteins (Cappellaro et al. 1998; Yin et al. 2005). There are many predicted cell wall proteins (Caro et al. 1997), and this raises the question of whether the usage of specific cell wall proteins may depend on the growth conditions. There is evidence of this from global transcript analysis (Boer et al. 2003; Boorsma et al. 2004; Garcia et al. 2004; Tai et al. 2005; Ter Linde and Steensma 2003) and from studies focusing on particular genes (Abramova et al. 2001; Smits et al. 2006). Klis et al. (2006) have discussed three relevant examples: exponentially growing cells compared to stationary phase cells, aerobically grown cells compared to anaerobically grown cells (Chap. 6) and the changes that occur in the cell wall in response to stress during the cell cycle (Chap. 11).

The walls of cells that grow exponentially in a rich medium differ significantly from cells that are on the verge of entering stationary phase (Chap. 13). The latter cultures possess cells that are more resistant to glucanases (Russell et al. 1973; De Nobel et al. 1990) usually during stationary growth phase (Russell and Stewart 1979). The number of disulphide bridges in the walls of these cells increases about sixfold. This is accompanied by a radical change in the protein profiles obtained by gel filtration of proteolytically released CWP fragments (Shimoi et al. 1998). Sed1p, which is a minor protein in exponentially growing cells, becomes the most abundant CWP in stationary phase cells. Global transcript analyses of cells in stationary phase are consistent with these results, showing that the SED1 transcript level increases more than tenfold (Gasch et al. 2000). Sed1p has seven potential N-glycosylation sites and is heavily N-glycosylated, which limits cell wall permeability. It also contains four cysteine residues. This may largely explain the observed phenotypes in stationary phase cells (Shimoi et al. 1998). Indeed, deletion of SED1 results in a strongly reduced resistance of the intact cells to zymolyase. SED1 is also strongly upregulated during cell wall stress, heat stress, oxidative stress and hyperosmotic stress (Chap. 11). This suggests that Sed1p has a general stress-protective function (Gasch et al. 2000; Boorsma et al. 2004; Hagen et al. 2004). Other genes that are strongly upregulated in cells entering stationary phase include the GPI-CWP-encoding gene SP11 and also GSC2, which codes for the alternative catalytic subunit of the 1,3-β-glucan synthase complex (Gasch et al. 2000).

Kitagaki et al. (1997) have shown that the gel filtration profiles of proteolytically released CWP fragments of the same strain, from shaken and non-shaken cultures, differed sharply and that Tir1p was enriched in the cell walls of the non-shaken cells. Stagnant cells were introduced to low-oxygen levels, and this was responsible for the differences in the gel filtration profiles of CWPs. Several research groups have extended and agreed with Kitagaki's publication. It has also been shown that under anaerobic conditions the following GPI-CWP-encoding genes: *DAN1*, *DAN4*, *MAC1*, *FLO11*, *TIR1*, *TIR2*, *TIR3* and *TIR4* are strongly upregulated (Ter Linde et al. 1999; Abramova et al. 2001; Cohen et al. 2001; Tai et al. 2005). When either *TIR1*, *TIR3* or *TIR4* is deleted, the cells are not able to grow under nitrogen limitation demonstrating that these proteins play a vital role during these growth conditions (Abramova et al. 2001). However, the mechanisms of this process are still not fully understood. The appearance of a new group of GPI-CWP-encoding genes, expressed under anaerobic conditions, is accompanied by a reduction in the transcript level of the GPI-CWP-encoding genes *CWP1* and *CWP2*, which are expressed under aerobic conditions. The transcript levels of the *PAU1*, *PAU3*, *PAU4*, *PAU5* and *PAU6* genes, which encode secretory proteins of unknown function and destination, are upregulated under anaerobic conditions (Rachidi et al. 2000; Tai et al. 2005).

5.4 Cell Wall Construction During the Cell Cycle

The construction of the *Saccharomyces* cell wall is coordinated with the cell cycle. For example, chitin is laid down at three locations:

- Dispersed in the cell wall of the growing daughter cell after cytokinesis has occurred (G1 cells)
- As a ring at the presumptive bud site (late G1 cells)
- In the primary septum (M/G1 cells) (Cabib and Durán 2005) (Fig. 5.6).

Cells also switch between isotropic (uniformity in all orientations) and apical (the last common ancestor of an entire group) during the cell cycle. Young daughter cells grow mainly isotropically, requiring the insertion of new cell wall macromolecules into the existing polymer network. During this phase of the cell cycle, the Pir-CWPs are highly expressed and are suited for coping with the difficulties related to isotropic growth. It is during this phase of the cell cycle that chitin synthase produces chitin, which is also laid down in the lateral walls. After the cell has committed itself to cell division, a site is selected for the formation of a new bud. This site is delineated by a chitin ring (Fig. 5.2). The ring will determine the diameter of the neck between mother and bud cells. The chitin ring is anchored in the existing glucan network by attachment to 1,3-β-glucan chains (Cabib 2004). The cell wall at the bud site is weakened, and a tiny bud emerges. Initially, the bud becomes apical. However, when the bud becomes larger, the contribution of isotropic growth increases and finally takes over from apical growth. Full independence of the bud from the mother cell begins with the formation of a primary septum, consisting of CHI-produced chitin growing from the chitin ring. The CH11-produced chitin is largely unbound to other cell wall polymers, probably allowing for crystallization, whereas the remainder is linked to 1,3-β-glucan to anchor it in place. The primary septum becomes covered on both sides with a secondary septum. Secretion of a chitinase by the daughter cell dissolves the primary septum and, in collaboration with other cell wall-degrading enzymes, allows for the release of the daughter cell, leaving the mother cell with a prominent bud scar and the daughter cell with a much less conspicuous birth scar (Figs. 5.2 and 5.3) (Cabib 2004).

Cell separation may go amiss, probably explaining why the cell expresses, amongst other enzymes, a specific chitin synthase (CHSI) during this phase of the cell cycle. The daughter cell will grow further until it has achieved the critical size for the next round of cell division. This results in the formation of a type

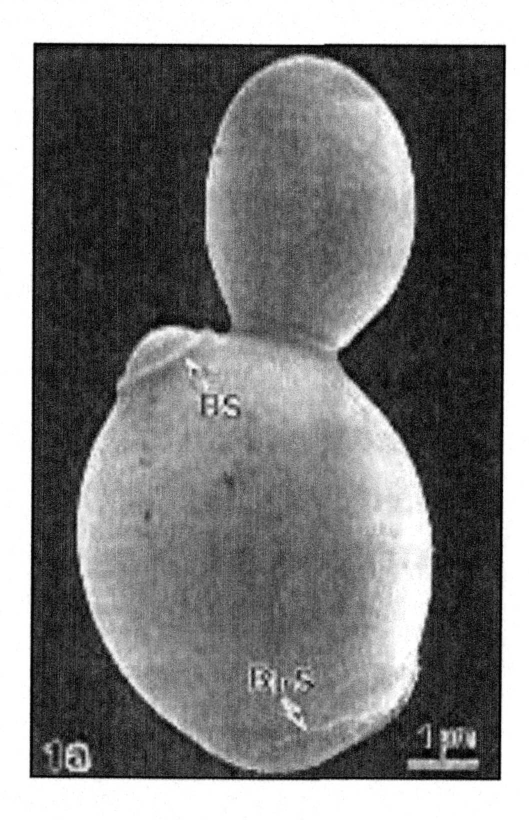

Fig. 5.6 Yeast bud and birth scars with chitin layers

pseudohyphae or chain formation that has already been briefly mentioned in this chapter (Fig. 5.4) (Stewart et al. 2016), this will be discussed further in Chap. 13.

The cell does not only employ a specific chitin synthase during various phases of the cell cycle; it also uses specific cell wall proteins during the cell cycle. Microarray (also known as DNA chip technology) studies (Hirasawa et al. 2007) offer a glimpse of the interrelationships between cell wall synthesis and the cell cycle. Indeed, they correspond well between the actual protein levels and the expected protein locations (Rodriguez-Pena et al. 2000; Baladrón et al. 2002; Smits et al. 2006). Using promoter swap experiments, Smits et al. (2006) have shown that incorporation of the GPI-CWPs Cwp2p and Tiplp in the walls of medium-sized buds and mother cells, respectively, is indeed fully determined by the timing of transcription during the cell cycle. However, GPI-CWP Cwp1p does not follow this role. Although CWP1 is transcribed at about the same time as CWP2, Cwplp is incorporated into the cell wall at a later stage and is found exclusively in the birth scar (Caro et al. 1998; Smits et al. 2006).

5.5 Yeast Mannan

Mannan is a polymer of the monosaccharide mannose. It has already been discussed that, as well as glucose (glucan), it is a major component of the *Saccharomyces* cell wall. This type of mannan has a α(1–6) linked backbone and α-(1–2) and α-(1–3) linked branches (Lessage and Bussey 2006). From a brewing perspective, its major function is its involvement in flocculation (details in Chap. 13). However, mannan-oligosaccharides (MOS) are widely used in nutrition as a natural additive for humans and many farm animals. MOS has been shown to improve gastrointestinal health, thus improving wellbeing, energy levels and performance. Most MOS products have been extensively reviewed (e.g. Biggs et al. 2007; Terre et al. 2007).

The initial interest in MOS to protect gastrointestinal health originated from studies conducted in the late 1980s. At that time the ability of mannose, the pure version of the complex sugar in MOS, to inhibit *Salmonella* infections was identified. Binding to mannan reduces the risk of pathogen colonization in the intestinal tract (Oyofo et al. 1989). Different forms of mannose-type sugars interact differently with type-1-fimbriae. The mannan present in the cell wall of *S. cerevisiae* is particularly effective at binding pathogens (White et al. 2002). The effect of yeast MOS has been found to be positive, particularly with fish, poultry, pigs and calves (Dimitroglou et al. 2009; Parks et al. 2005; Rodriguez-Estrada et al. 2009, Le Dividich et al. 2009).

5.6 Enzymes Involved in Cell Wall Construction

Cell surface protein enzymes that are involved in cell wall construction can be divided into:

- Synthetic enzymes
- Remodelling enzymes
- Assembly enzymes
- Degrading enzymes

1,3-β-glucan is synthesized as a linear polymer (Baladrón et al. 2002), which during or after extrusion through the plasma membrane is remodelled, resulting in branching and possibly further extension and coupling to existing 1,3-β-glucan molecules.

Chitin is a linear polymer that may become attached through its reducing end to 1,3-β-glucan and 1,6-β-glucan. The three synthases CSI, CSH and CSIII, including their respective catalytic subunits Chs1p, Chs2p and Chs3p, are located in the plasma membrane and extrude the growing chain through the membrane. Chitin synthesis and its role in various key events of the cell cycle and during cell wall stress have been extensively studied (Bulik et al. 2003). The assembly enzymes involved in coupling chitin to 1,3-β-glucan and 1,6-β-glucan are uncharacterized.

However, evidence is emerging that the Crh family (Crh1p, Crr1p and Utr2p) may be involved. The evidence conferring a role for the GPI protein Crh1p in interconnecting chitin to β-glucan is as follows:

- Crh1p shows similarity to 1,3-/1,4-β-glucanases (Coutinho and Henrissat 1999).
- Mutagenesis of amino acids that are predicted to have a catalytic function results in loss of protein function (Rodriguez-Pena et al. 2002).
- Loss of function causes a decrease of alkali-insoluble glucan in the wall (Rodriguez-Pena et al. 2000), as the attachment of chitin to 1,3-β-glucan results in 1,3-β-glucan becoming less soluble in alkali (Hartland et al. 1994; Mol and Wessels 1987). This observation is consistent with a role for Crh1p in attaching chitin to 1,3-β-glucan.
- The temporal expression of CRH1 during the cell cycle, which is characterized by two maxima, coincides with the temporary location of Crh1p at the presumptive bud site, when a chitin ring is synthesized and, in a later phase of the cell cycle at the bud neck, when the primary system is being synthesized in the bud scar of the mother cell (Fig. 5.6) (Spellman et al. 1998).
- When cells are confronted with cell wall stress, the chitin content of the walls is enhanced and also the CWP-GPIr→1,6-β-glucan←chitin complex seems to be strongly increased (Kapteyn et al. 1997; Bulik et al. 2003). This is accompanied by strong upregulation of CRH1 (Boorsma et al. 2004; Garcia et al. 2004; Lagorce et al. 2003). It seems likely that Crr1p and Utr2p may have a similar function.

1,6-β-glucan is a highly branched polymer. Synthesis seems to occur at the level of the plasma membrane, but several endoplasmic reticulum (ER) and Golgi proteins are in some unknown way involved in 1,6-β-glucan synthesis (Montijn et al. 1999). It is unknown whether this polymer is synthesized as a linear molecule, and subsequently becomes branched, or is formed by repetitively attaching a preformed branched oligosaccharide to a growing chain. An in vitro method for the synthesis of 1,6-β-glucan has been developed that will help identify the genes directly involved in its synthesis (Vink et al. 2004).

GPI-proteins destined for the cell wall arrive at the cell surface with their GPI anchors intact, whereas mature GPI-CWPs possess truncated lipidless GPI anchors that have been cleaved between the first mannose residue and glucosamine (Kollár et al. 1997; Fujii et al. 1999). The identification of the enzymes that are involved in releasing GPI-CWPs membrane-bound intermediates is unclear, but the two-member protein family, consisting of GPI-PMPs Dfg5p and DCW1p, are likely candidates. The double deletant is non-viable, even in an osmotically stabilized growth medium, and it exhibits homology to bacterial endomannases (Kitagaki et al. 2002, 2004).

5.7 Yeast Plasma Membrane and Structure

Similar to all eukaryotes, the plasma membrane of yeast (specifically *Saccharomyces* spp.) forms the barrier between the cytoplasm and the cell wall.

The plasma membrane consists principally of lipids and proteins in approximately equal proportions. As a result of the large number of functions performed by the membrane, it must be assumed that most of the membrane proteins are functional and not structural (Rank and Robertson 1983). Membrane proteins are responsible for regulating solute transport in and out of the cell (details in Chap. 6). They include the enzymes that catalyse cell wall synthesis, some receptors and the ATPase responsible for maintaining the plasma membrane protein-motive force (Van der Rest et al. 1995). The membrane lipid biosynthetic enzymes are located in the yeast's mitochondria (details in Chap. 14).

The yeast plasma membrane has several distinct roles and these are:

- To present a barrier to the free diffusion of solutes
- To catalyse specific change reactions—details later
- To store energy in the form of transmembrane ions and solute gradients
- To regulate the rate of energy dissipation
- To provide sites for binding specific molecules to invoke catabolic signalling pathways
- To provide an organized support matrix for the site of enzyme pathways involved in the biosynthesis of the cell components (Hazel and Williams 1990; Ram et al. 1998)

The lipid components of the membrane are amphipathic and confer the plasma membrane with hydrophobic properties. The principal membrane lipids are phospholipids and sterols. Phospholipids are phosphatidylinositol, phosphatidylserine, phosphatidylcholine and phosphatidylethanolamine. The major sterol is ergosterol, and trace quantities such as lanosterol and esterified zymosterol are also present.

5.8 Periplasm

The periplasm is not a continuous space because of interrupting invaginations into the plasma membrane and irregularities in the inner surface of the cell wall. Indeed, the periplasm is not an organelle as such. Nevertheless, it is the location where a number of important yeast enzymes are located and active.

Arnold (1991) has defined periplasmic enzymes as "those enzymes which may be assayed in intact cells without disruption of the plasma membrane". The most notable periplasmic enzymes are invertase (Fig. 5.7) and acid phosphatase. Acid phosphatase hydrolyses phosphate esters in the medium. However, it may have an additional role along with other periplasmic enzymes via dephosphorylation. Other periplasmic enzymes include a variety of binding proteins and the enzyme melibiase (α-galactosidase) (details in Chap. 3) for lager yeast strains.

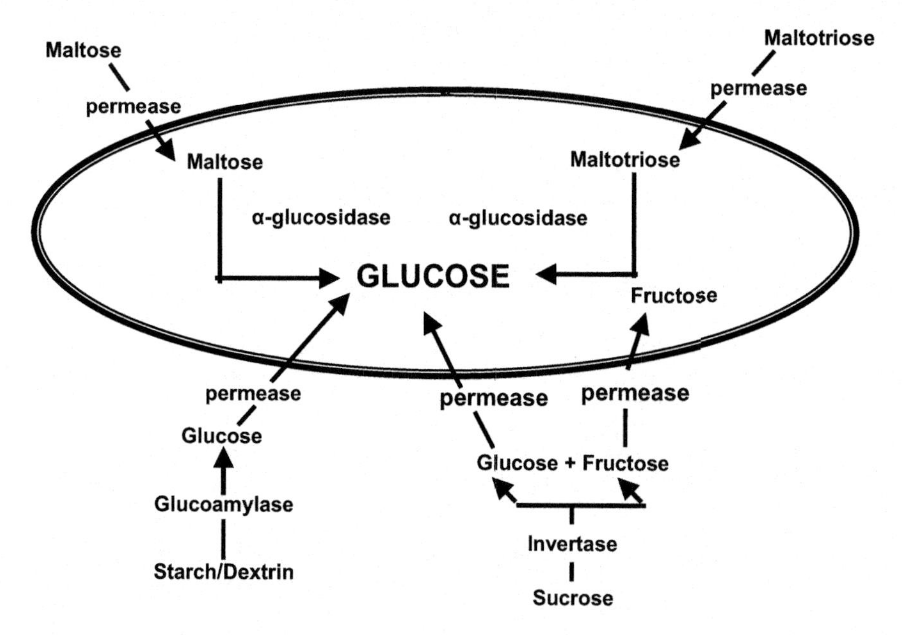

Fig. 5.7 Hydrolysis and uptake of sugars by a yeast cell

Invertase (β-fructosidase) catalyses the hydrolysis of sucrose into glucose and fructose (Fig. 5.7), and this enzyme exists in either an extracellular or an intracellular form (Lahiri et al. 2012). The glucose and fructose are then taken up into the yeast cell. The external yeast invertase is predominantly in the periplasmic space and contains approximately 50% carbohydrate (glucose moieties), 5% mannose and 3% glucosamine, whereas the intracellular invertase is not glycosylated (Workman and Way 1983). It has been established that in derepressed cells (Ali and Haq 2007) most of the invertase is external, whereas in the fully repressed state, all of the invertase is intracellular (Vu and Lee 2008).

5.9 Summary of the *Saccharomyces* Cell Wall Biosynthesis, Plasma Membrane and Periplasm Structure and Function

Knowledge of the structure and function of the *Saccharomyces* cell wall is an expanding field. These studies offer potential for understanding cellular processes related to morphogenesis, for elucidating the regulatory networks that assist in the coordination of cellular processes with cell cycle progression and for practical applications such as the mechanism of flocculation and biofilm formation during the brewing process (see Chap. 13). Genetic engineering of cell wall surface properties by the introduction of new proteins has also attracted active interest (Kondo and Ueda 2004) (details in Chaps. 13 and 16). These findings offer valuable predictions for how their construction might be regulated and how their cell surface properties might be regulated and re-engineered (De Groot et al. 2005).

References

Abramova N, Sertil O, Mehta S, Lowry CV (2001) Reciprocal regulation of anaerobic and aerobic cell wall mannoprotein gene expression in *Saccharomyces cerevisiae*. J Bacteriol 183:2881–2887

Aguilar-Uscanga B, Francois JM (2003) A study of the yeast cell wall composition and structure in response to growth conditions and mode of cultivation. Lett Appl Microbiol 37:268–274

Ali S, Haq I (2007) Kinetics of improved extracellular β-D-fructofuranosidase fructohydrolase production by a derepressed *Saccharomyces cerevisiae*. Lett Appl Microbiol 45:160–167

Alimardani P, Regnacq M, Moreau-Vauzelle C, Rossignol T, Blondin B, Bergès T (2004) SUT1-promoted sterol uptake involves the ABC transporter Aus1 and the mannoprotein Dan1 whose synergistic action is sufficient for this process. Biochem J 381:195–202

Arnold WN (1991) Periplasmic space. In: Rose AH, Harrison JS (eds) The yeasts, vol 4. Academic Press, London, pp 279–293

Baladrón V, Ufano S, Duenas EV, Martín-Cuadrado AB, del Rey F, Vázquez de Aldana CR (2002) Eng1p, an endo-1,3-β-glucanase localized at the daughter side of the septum, is involved in cell separation in *Saccharomyces cerevisiae*. Eukaryot Cell 1:774–786

Biggs P, Parsons C, Fahey G (2007) The effects of several oligosaccharides on growth performance, nutrient digestibilities, and cecal microbial populations in young chicks. Poultry Sci 86:2327–2336

Boer VM, de Winde JH, Pronk JT, Piper MD (2003) The genome-wide transcriptional responses of *Saccharomyces cerevisiae* grown on glucose in aerobic chemostat cultures limited for carbon, nitrogen, phosphorus, or sulfur. J Biol Chem 278:3265–3274

Boorsma A, de Nobel H, ter Riet B, Bargmann B, Brul S, Hellingwerf KJ, Klis FM (2004) Characterization of the transcriptional response to cell wall stress in *Saccharomyces cerevisiae*. Yeast 21:413–427

Brown D, Waneck GL (1992) Glycosyl-phosphatidylinositol-anchored membrane proteins. J Am Soc Nephrol 3:895–906

Bühler M, Mohn F, Stalder L, Mühlemann O (2005) Transcriptional silencing of nonsense codon - containing Ig minigenes. Mol Cell 18:307–317

Bulik DA, Olczak M, Lucero HA, Osmond BC, Robbins PW, Specht CA (2003) Chitin synthesis in *Saccharomyces cerevisiae* in response to supplementation of growth medium with glucosamine and cell wall stress. Eukaryot Cell 2:886–900

Cabib E (2004) The septation apparatus, a chitin-requiring machine in budding yeast. Arch Biochem Biophys 426:201–207

Cabib E, Durán A (2005) Synthase III-dependent chitin is bound to different acceptors depending on location on the cell wall of budding yeast. J Biol Chem 280:9170–9179

Cabib E, Roh D-H, Schmidt M, Crotti LB, Varma A (2001) The yeast cell wall and septum as paradigms of cell growth and morphogenesis. J Biol Chem 276:19679–19682

Cappellaro C, Hauser K, Mrsa V, Watzele M, Watzele G, Gruber C, Tanner W (1991) *Saccharomyces cerevisiae* a-agglutinin and α-agglutinin: characterization of their molecular interaction. EMBO J 10:4081–4088

Cappellaro C, Baldermann C, Rachel R, Tanner W (1994) Mating type-specific cell-cell recognition of *Saccharomyces cerevisiae*: cell wall attachment and active sites of a- and a-agglutinin. EMBO J 13:4737–4744

Cappellaro C, Mrsa V, Tanner W (1998) New potential cell wall glucanases of *Saccharomyces cerevisiae* and their involvement in mating. J Bacteriol 18:5030–5037

Caro LHP, Tettelin H, Vossen JH, Ram AF, van den Ende H, Klis FM (1997) In silicio identification of glycosyl-phosphatidylinositol-anchored plasma-membrane and cell wall proteins of *Saccharomyces cerevisiae*. Yeast 13:1477–1489

Caro LHP, Smits GJ, Van Egmond P, Chapman JW, Klis FM (1998) Transcription of multiple cell wall protein-encoding genes in *Saccharomyces cerevisiae* is differentially regulated during the cell cycle. FEMS Microbiol Lett 161:345–349

Castillo L, Martinez AI, Garcera A, Elorza MV, Valentín E, Sentandreu R (2003) Functional analysis of the cysteine residues and the repetitive sequence of *Saccharomyces cerevisiae* Pir4/Cis3: the repetitive sequence is needed for binding to the cell wall β-1,3-glucan. Yeast 20:973–983

Chlup PH, Stewart GG (2011) Centrifuges in brewing. MBAA Tech Quart 48:13–19

Chlup PH, Wang T, Lee EG, Stewart GG (2007) Assessment of the physiological status of yeast during high- and low-gravity wort fermentations determined by flow cytometry. MBAA Tech Quart 44:286–295

Cleves AE, Cooper DNW, Barondes SH, Kelly RB (1996) A new pathway for protein export in *Saccharomyces cerevisiae*. J Cell Biol 133:1017–1026

Cohen BD, Sertil O, Abramova NE, Davies KJA, Lowry CV (2001) Induction and repression of DAN1 and the family of anaerobic mannoprotein genes in *Saccharomyces cerevisiae* occurs through a complex array of regulatory sites. Nucleic Acids Res 29:799–808

Coutinho PM, Henrissat B (1999) Carbohydrate-active enzymes: an integrated database approach. In: Gilbert HJ, Davies G, Henrissat B, Svensson B (eds) Recent advances in carbohydrate bioengineering. Royal Society of Chemistry, Cambridge, pp 3–12

Dague E, Bitar R, Ranchon H, Durand F, Yken HM, François JM (2010) An atomic force microscopy analysis of yeast mutants defective in cell wall architecture. Yeast 27:673–684

De Groot PW, Ruiz C, Vázquez de Aldana CR, Duenas E, Cid VJ, Del Rey F, Rodríquez-Peña JM, Pérez P, Andel A, Caubín J, Arroyo J, García JC, Gil C, Molina M, García LJ, Nombela C, Klis FM (2001) A genomic approach for the identification and classification of genes involved in cell wall formation and its regulation in *Saccharomyces cerevisiae*. Comp Funct Genom 2:124–142

De Groot PWJ, De Boer AD, Cunningham J. De Groot PW, de Boer AD, Cunningham J, Dekker HL, de Jong L, Hellingwerf KJ, de Koster C, Klis FM (2004) Proteomic analysis of *Candida albicans* cell walls reveals covalently bound carbohydrate-active enzymes and adhesins. Eukaryot Cell 3:955–965

De Groot PWJ, Ram AF, Klis FM (2005) Features and functions of covalently linked proteins in fungal cell walls. Fungal Genet Biol 42:657–675

De Groot MJ, Daran-Lapujade P, Van Breukelen B (2007) Quantitative proteomics and transcriptomics of anaerobic and aerobic yeast cultures reveals post-transcriptional regulation of key cellular processes. Microbiology 153:3864–3878

De Nobel JG, Klis FM, Priem J, Munnik T, Van den Ende H (1990) The glucanase-soluble mannoproteins limit cell wall porosity in *Saccharomyces cerevisiae*. Yeast 6:491–499

Dimitroglou A, Merrifield DL, Moate R, Davies SJ, Spring P, Sweetman J, Bradley G (2009) Dietary mannan oligosaccharide supplementation modulates intestinal microbial ecology and improves gut morphology of rainbow trout, *Oncorhynchus mykiss* (Walbaum). J Anim Sci 87:3226–3234

Fleet GH (1991) Cell walls. In: Rose AH, Harrison JS (eds) The yeast. Academic Press, New York, pp 199–277

Fujii T, Shimoi H, Iimura Y (1999) Structure of the glucan-binding sugar chain of Tip1p, a cell wall protein of *Saccharomyces cerevisiae*. Biochim Biophys Acta 1427:133–144

Garcia R, Bermejo C, Grau C, Pérez R, Rodríguez-Peña JM, Francois J, Nombela C, Arroyo J (2004) The global transcriptional response to transient cell wall damage in *Saccharomyces cerevisiae* and its regulation by the cell integrity signalling pathway. J Biol Chem 279:15183–15195

Gasch AP, Spellman PT, Kao CM, Carmel-Harel O, Eisen MB, Storz G, Botstein D, Brown PO (2000) Genomic expression programs in the response of yeast cells to environmental changes. Mol Biol Cell 11:4241–4257

Gibson BR, Lawrence SJ, Leclaire JP, Powell CD, Smart KA (2007) Yeast responses to stresses associated with industrial brewery handling. FEMS Microbiol Rev 31:535–569

Hagen I, Ecker M, Lagorce A, Francois JM, Sestak S, Rachel R, Grossmann G, Hauser NC, Hoheisel JD, Tanner W, Strahl S (2004) Sed1p and Srl1p are required to compensate for cell wall instability in *Saccharomyces cerevisiae* mutants defective in multiple GPI-anchored mannoproteins. Mol Microbiol 52:1413–1425

Hamada K, Terashima H, Arisawa M, Yabuki N, Kitada K (1999) Amino acid residues in the omega-minus region participate in cellular localization of yeast glycosylphosphatidylinositol-attached proteins. J Bacteriol 181:3886–3889

Hartland RP, Vermeulen CA, Klis FM, Sietsma JH, Wessels JG (1994) The linkage of (1–3)-β-glucan to chitin during cell wall assembly in *Saccharomyces cerevisiae*. Yeast 10:1591–1599

Hazel JR, Williams EE (1990) The role of alterations in membrane lipid composition in enabling physiological adaptation of organisms to their physical environment. Prog Lipid Res 29:167–227

Heinisch JJ (2005) Baker's yeast as a tool for the development of antifungal kinase inhibitors—targeting protein kinase C and the cell integrity pathway. Biochim Biophys Acta 1754:171–182

Hinterdorfer P, Dufrêne YF (2006) Detection and localization of single molecular recognition events using atomic force microscopy. Nat Methods 3:347–355

Hirasawa T, Yoshikawa K, Nakakura Y, Nagahisa K, Furusawa C, Katakura Y, Shimizu H, and Shioya S (2007) Identification of target genes conferring ethanol stress tolerance to *Saccharomyces cerevisiae* based on DNA microarray data analysis. J Biotechnol 131:34–44

Jones RM, D'Amore A, Russell I, Stewart GG (1987) The use of spheroplast fusion to improve yeast osmotolerance. J Am Soc Brew Chem 48:0058

Kapteyn JC, Ram AFJ, Groos EMJ, Kollar R, Montijn RC, Van Den Ende H, Llobell A, Cabib E, Klis FM (1997) Altered extent of cross-linking of β 1,6-glucosylated mannoproteins to chitin in *Saccharomyces cerevisiae* mutants with reduced cell wall β 1,3-glucan content. J Bacteriol 179:6279–6284

Kapteyn J, Van Egmond P, Sievi E, Ende VD, Makarow M, Klis FM (1999) The contribution of the O-glycosylated protein Pir2p/Hsp150 to the construction of the yeast cell wall in wild-type cells and beta 1,6-glucan-de¢cient mutants. Mol Microbiol 31:1835–1844

Kaur R, Domergue R, Zupancic ML, Cormack BP (2005) A yeast by any other name: *Candida glabrata* and its interaction with the host. Curr Opin Microbiol 8:378–384

Kitagaki H, Shimoi H, Itoh K (1997) Identification and analysis of a static culture-specific cell wall protein, Tir1p/Srp1p in *Saccharomyces cerevisiae*. Eur J Biochem 249:343–349

Kitagaki H, Wu H, Shimoi H, Ito K (2002) Two homologous genes, *DCW1 (YKL046c)* and *DFG5*, are essential for cell growth and encode glycosylphosphatidylinositol (GPI)-anchored membrane proteins required for cell wall biogenesis in *Saccharomyces cerevisiae*. Mol Microbiol 46:1011–1022

Kitagaki H, Ito K, Shimoi H (2004) A temperature-sensitive dcw1 mutant of *Saccharomyces cerevisiae* is cell cycle arrested with small buds which have aberrant cell walls. Eukaryot Cell 3:1297–1306

Klis FM, Mol P, Hellingwerf K, Brul S (2002) Dynamics of cell wall structure in *Saccharomyces cerevisiae*. FEMS Microbiol Rev 26:239–256

Klis FM, Boorsma A, De Groot PWJ (2006) Cell wall construction in *Saccharomyces cerevisiae*. Yeast 23:185–202

Kobayashi O, Hayashi N, Kuroki R, Sone H (1998) Region of *FLO1* proteins responsible for sugar recognition. J Bacteriol 180:6503–6510

Kollár R, Reinhold BB, Petráková E, Yeh HJ, Ashwell G, Drgonová J, Kapteyn JC, Klis FM, Cabib E (1997) Architecture of the yeast cell wall. β(1–6)-glucan interconnects mannoprotein, β(1–3)-glucan, and chitin. J Biol Chem 272:17762–17775

Kondo A, Ueda M (2004) Yeast cell-surface display – applications of molecular display. Appl Microbiol Biotechnol 64:28–40

Kuo S, Yamamoto S (1975) Preparation and growth of yeast protoplasts. Methods Cell Biol 11:169–183

Kwiatkowski S, Thielen U, Glenny P, Moran C (2009) A study of *Saccharomyces cerevisiae* cell wall glucans. Inst Brew Distil 115:151–158

Lagorce A, Hauser NC, Labourdette D, Rodriguez C, Martin-Yken H, Arroyo J, Hoheisel JD, François J (2003) Genome-wide analysis of the response to cell wall mutations in the yeast *Saccharomyces cerevisiae*. J Biol Chem 278:20345–20357

Lahiri S, Basu A, Sengupta S, Banerjee S, Dutta T, Soren D, Chattopadhyay K, Ghosh AK (2012) Purification and characterization of a trehalase-invertase enzyme with dual activity from *Candida utilis*. Arch Biochem Biophys 522:90–99

Lane MM, Morrissey JP (2010) *Kluyveromyces marxianus*: a yeast emerging from its sister's shadow. Fungal Biol Rev 24:17–26

Le Dividich J, Martel-Kennes Y, Coupel A (2009) Bio-Mos in diets for sows: effects on piglet performance. J Rech Porcine 41:249–250

Lessage G, Bussey H (2006). Cell wall assembly in *Saccharomyces cerevisiae*. Microbiol Mol Biol Rev 70:317–343

Lyons TP, Hough JS (1970) Flocculation of brewer's yeast. J Inst Brew 76:564–571

Mamvura TA, Paterson AE, Fanucchi D (2016) The impact of pipe geometry variations on hygiene and success of orbital welding of brewing industry equipment. J Inst Brew 123:81–97

Mol PC, Wessels JGH (1987) Linkages between glucosaminoglycan and glucan determine alkali-insolubility of the glucan in walls of *Saccharomyces cerevisiae*. FEMS Microbiol Lett 41:95–99

Molina M, Gil C, Pla J, Arroyo J, Nombela C (2000) Protein localisation approaches for understanding yeast cell wall biogenesis. Microsc Res Technol 51:601–612

Montijn RC, Vink E, Muller WH, Verkleij AJ, Van Den Ende H, Henrissat B, Klis FM (1999) Localization of synthesis of β1,6-glucan in *Saccharomyces cerevisiae*. J Bacteriol 181:7414–74120

Morris GJ, Winters L, Coulson GE, Clarke KJ (1986) Effect of osmotic stress on the ultrastructure and viability of the yeast *Saccharomyces cerevisiae*. J Gen Microbiol 132:2023–2034

Orlean P (1997) Biogenesis of yeast wall and surface components. In: Pringle JR, Broach JR, Jones EW (eds) Molecular and cellular biology of the yeast *Saccharomyces*, vol. 3 Cell cycle and cell biology. Cold Spring Harbor Laboratory Press, pp 229–362

Osumi M (1998) The ultrastructure of yeast: cell wall structure and formation. Micron 29:207–233

Ovalle R, Lim ST, Holder B, Jue CK, Moore CW, Lipke PN (1998) A spheroplast rate assay for determination of cell wall integrity in yeast. Yeast 14:1159–1166

Oyofo BA, DeLoach JR, Corrier DE, Norman JO, Zipren RL, Mollenhauer HH (1989) Effect of carbohydrates on *Salmonella typhimurium* colonization in broiler chickens. Avian Dis 531–534

Parks CW, Grimes JL, Ferket PR (2005) Effects of virginiamycin and a mannanoligosaccharide-virginiamycin shuttle program on the growth and performance of large white female turkeys. Poultry Sci 84:1967–1973

Protchenko O, Ferea T, Rashford J, Tiedeman J, Brown PO, Botstein D, Philpott CC (2001) Three cell wall mannoproteins facilitate the uptake of iron in *Saccharomyces cerevisiae*. J Biol Chem 276:49244–49250

Rachidi N, Martinez MJ, Barre P, Blondin B (2000) *Saccharomyces cerevisiae* PAU genes are induced by anaerobiosis. Mol Microbiol 35:1421–1430

Ram AFJ, Kapteyn JC, Montijn RC, Caro LH, Douwes JE, Baginsky W, Mazur P, van den Ende H, Klis FM (1998) Loss of the plasma membrane-bound protein Gas1p in *Saccharomyces cerevisiae* results in the release of β1,3-glucan into the medium and induces a compensation mechanism to ensure cell wall integrity. J Bacteriol 180:1418–1424

Rank GH, Robertson AJ (1983) Protein and lipid composition of the yeast plasma membrane. In: Spencer JFT, Spencer DM, Smith ARW (eds) Yeast genetics, fundamental and applied aspects. Springer, Berlin, pp 225–241

Reynolds TB, Fink GR (2001) Bakers' yeast, a model for fungal biofilm formation. Science 291:878–881

Rodriguez-Estrada U, Satoh S, Haga Y, Fushimi H, Sweetman J (2009) Effects of single and combined supplementation of *Enterococcus faecalis*, mannan oligosaccharide and polyhydroxybutyrate acid on growth performance and immune status of rainbow trout, *Oncorhynchus mykiss*. Agric Int 15:607–617

Rodriguez-Pena JM, Cid VJ, Arroyo J, Nombela C (2000) A novel family of cell wall-related proteins regulated differently during the yeast life cycle. Mol Cell Biol 20:3245–3255

Rodriguez-Pena JM, Rodriguez C, Alvarez ARodriguez-Peña JM, Rodriguez C, Alvarez A, Nombela C, Arroyo J (2002) Mechanisms for targeting of the *Saccharomyces cerevisiae* GPI-anchored cell wall protein Crh2p to polarised growth sites. J Cell Sci 115:2549–2558

Russell I, Stewart GG (1979) Spheroplast fusion of brewer's yeast strains. J Inst Brew 85:95–98

Russell I, Garrison IF, Stewart GG (1973) Studies in the formation of spheroplasts from stationary phase cells of *Saccharomyces cerevisiae*. J Inst Brew 79:48–54

Schreuder MP, Brekelmans S, Van den Ende H, Klis FM (1993) Targeting of a heterologous protein to the cell wall of *Saccharomyces cerevisiae*. Yeast 9:399–409

Sedmak, JJ (2006) Production of β-glucans and mannans. US Patent Application US 2006/ 0263415 A1

Shimoi H, Iimura Y, Obata T (1995) Molecular cloning of CWP1: a gene encoding a *Saccharomyces cerevisiae* cell wall protein solubilized with *Rarobacter faecitabidus* protease I. J Biochem 118:302–311

Shimoi H, Kitagaki H, Ohmori H, Iimura Y, Ito K (1998) Sed1p is a major cell wall protein of *Saccharomyces cerevisiae* in the stationary phase and is involved in lytic enzyme resistance. J Bacteriol 180:3381–3387

Smits GJ, Kapteyn JC, van den Ende H, Klis FM (1999) Cell wall dynamics in yeast. Curr Opin Microbiol 2:348–352

Smits GJ, Schenkman LR, Brul S, Pringle JR, Klis FM. (2006) Role of cell-cycle regulated expression in the localized incorporation of cell-wall proteins in yeast. Mol Biol Cell 17:3267–3280

Spellman PT, Sherlock G, Zhang MQ, Iyer VR, Anders K, Eisen MB, Brown PO, Botstein D, Futcher B (1998) Comprehensive identification of cell cycle-regulated genes of the yeast *Saccharomyces cerevisiae* by microarray hybridization. Mol Biol Cell 9:3273–3297

Stewart GG, Marshall DL, Speers A (2016) Brewing fundamentals – fermentation. Tech Quart Master Brew Assoc Am 53:2–22

Stoupis T, Stewart GG, Stafford RA (2002) Mechanical agitation and rheological considerations of ale yeast slurry. J Am Soc Brew Chem 60:58–62

Svoboda A (1966) Regeneration of yeast protoplasts in agar gels. Exp Cell Res 64:640–642

Tai SL, Boer VM, Daran-Lapujade P, Walsh MC, de Winde JH, Daran JM, Pronk JT (2005) Two-dimensional transcriptome analysis in chemostat cultures. Combinatorial effects of oxygen availability and macronutrient limitation in *Saccharomyces cerevisiae*. J Biol Chem 280:437–447

Ter Linde JJM, Steensma HY (2003) Transcriptional regulation of YML083c under aerobic and anaerobic conditions. Yeast 20:439–454

Ter Linde JJ, Liang H, Davis RW, Steensma HY, van Dijken JP, Pronk JT (1999) Genome-wide transcriptional analysis of aerobic and anaerobic chemostat cultures of *Saccharomyces cerevisiae*. J Bacteriol 181:7409–7413

Terre M, Calvo MA, Adelantado C, Kocher A, Bacha A (2007) Effects of mannan oligosaccharides on performance and microorganism fecal counts of calves following an enhanced-growth feeding program. Anim Feed Sci Technol 137:115–125

Van Berkel MAA, Caro LHP, Montijn RC, Klis FM (1994) Glucosylation of chimeric proteins in the cell wall of *Saccharomyces cerevisiae*. FEBS Lett 349:135–138

Van der Rest ME, Kamminga AH, Nakano A, Anraku Y, Poolman B, Konings WN (1995) The plasma membrane of *Saccharomyces cerevisiae*: structure, function and biogenesis. Microbiol Rev 59:304–322

Van Solingen P, van der Plaat JB (1977) Fusion of yeast protoplasts. J Bacteriol 130:946–947

Verstrepen KJ, Jansen A, Lewitter F, Fink GR (2005) Intragenic tandem repeats generate functional variability. Nat Genet 37:986–990

Vidgren V, Londesborough J (2011) Yeast flocculation and sedimentation in brewing. J Inst Brew 117:475–487

Vink E, Rodriguez-Suarez RJ, Gerard-Vincent M, Ribas JC, de Nobel H, van den Ende H, Durán A, Klis FM, Bussey H (2004) An *in vitro* assay for (1–6)-β-D-glucan synthesis in *Saccharomyces cerevisiae*. Yeast 21:1121–1131

Vu TK, Lee VV (2008) Biochemical studies on the immobilization of the enzyme invertase (EC.3.2.1.26) in alginate gel and its kinetics. Asean Food J 15:73–78

Weig M, Jansch L, Gross U, De Koster CG, Klis FM, De Groot PW (2004) Systematic identification in silico of covalently bound cell wall proteins and analysis of protein polysaccharide linkages of the human pathogen *Candida glabrata*. Microbiology 150:3129–3144

White LA, Newman MC, Cromwell GL, Lindemann MD (2002) Brewer's dried yeast as a source of mannan oligosaccharides for weanling pigs. J Anim Sci 80:2619–2628

Wojciechowicz D, CF L, Kurjan J, Lipke PN (1993) Cell surface anchorage and ligand-binding domains of the *Saccharomyces cerevisiae* cell adhesion protein α-agglutinin, a member of the immunoglobulin superfamily. Mol Cell Biol 13:2554–2563

Workman E, Way DF (1983) Purification and properties of the β-fructofuranosidase from *Kluyveromyces fragilis*. FEBS Lett 160:16e 20

Yin QY, de Groot PWJ, Dekker HL, de Jong L, Klis FM, de Koster CG (2005) Comprehensive proteomic analysis of *Saccharomyces cerevisiae* cell walls: identification of proteins covalently attached via glycosylphosphatidylinositol remnants or mild alkali-sensitive linkages. J Biol Chem 280:20894–20901

Zlotnik H, Fernandez MP, Bowers B, Cabib E (1984) *Saccharomyces cerevisiae* mannoproteins form an external cell wall layer that determines wall porosity. J Bacteriol 159:1018–1026

Chapter 6
Energy Metabolism by the Yeast Cell

6.1 Introduction

Most (not all) genera and species of yeast can ferment sugars to ethanol anaerobically (van Dijken and Scheffers 1986). It is discussed in a number of chapters of this text (e.g. Chaps. 5, 7, 9 and 15) that yeasts are renowned for this ability, especially on an industrial scale for the production of potable and industrial ethanol. This is why research on yeast alcohol fermentation has (and still does) received extensive financial support from the public and private sectors of many countries. Although many aspects of yeast, as they apply to brewing and distilling, are discussed in this text, the question of ethanol metabolism and related areas is the most important in this context. As a consequence, this aspect receives specific and extensive attention here. This chapter includes a review of the history (encompassing the relevant research scientists who published their studies on the glycolytic metabolism with yeasts, other eukaryotes and a few bacteria). Indeed, studies on glycolysis (particularly with yeast) rival its significance, scientifically and economically, with Fleming's (Fig. 6.1) discovery of penicillin (McIntyre 2007) followed by Florey (Fig. 6.2) and Chain's (Fig. 6.3) subsequent studies to develop this first antibiotic into a viable product for initial medical use by the Allies during D-Day in 1944 and thereafter (Bud 2007). Fleming, Florey and Chain shared the 1945 Nobel Prize for Physiology or Medicine as recognition of their discovery and development of penicillin production on an industrial scale. Although the research on penicillin involved relatively few notable scientists (originally located in St. Mary's Hospital in London, Oxford and, subsequently, in the United States), studies on glycolysis included a much larger number of scientists working in many research laboratories in both Europe and the United States.

G.G. Stewart, *Brewing and Distilling Yeasts*, The Yeast Handbook,
https://doi.org/10.1007/978-3-319-69126-8_6

Fig. 6.1 Alexander Fleming

6.2 Glycolysis

The elucidation of the glycolytic, and related, pathways from glucose into pyruvate and subsequently ethanol (or lactic acid) involves a number of enzyme-catalysed stages (10–12 steps). The glycolytic step elucidation was a complex and diverse biochemical/enzymatic research programme that involved a number of excellent scientists. Because of the importance of this pathway, particularly in the context of this text, its development (including the senior research scientists involved) will be discussed in some detail. Although this is not intended to be a biochemistry textbook [there are many that fulfil this role (e.g. Berg et al. 2012; Cox and Nelson 2013)], full details of carbohydrate metabolism by yeast is relevant here.

It has already been discussed (Chap. 2) that during the mid-nineteenth century, Louis Pasteur (Fig. 2.13) carried out extensive physiological fermentation studies with intact living yeast cells (Barnett 2000). Later, in 1897, Eduard Buchner (Buchner 1897a, b) (Fig. 6.4) achieved fermentation by cell-free extracts, making it practicable to study the biochemistry of fermentation in vitro (Barnett and

Fig. 6.2 Howard Florey

Fig. 6.3 Ernst Chain

Fig. 6.4 Eduard Buchner

Fig. 6.5 Summary of glycolysis

$$C_6H_{12}O_6 + 2\ ADP + 2\ P_i + 2\ NAD^+ \rightarrow$$

$$2\ CH_3COCOO^- + 2\ ATP + 2\ NADH + 2\ H_2O + 2H^+$$

Lichtenthaler 2001). During the twentieth century, this research was central for generating major advances in biochemistry, with massive economic applications!

Buchner concluded that:

> ... the initiation of fermentation does not require so complicated a structure as the living cell. The agent responsible for the fermenting activity of the extracted juice is a dissolved substance, no doubt a protein, this will be called zymase.

Buchner's findings opened the way to elucidating in yeast (and other organisms) the principal reactions of glycolysis.

Once the wort sugars (sucrose, glucose, fructose, maltose and maltotriose) are inside the yeast cell (Fig. 5.7 and details in Chap. 7), they are converted via the glycolytic (also known as the Embden-Meyerhof-Parnas, EMP, glycolysis—details later) pathway into pyruvate and subsequently ethanol. The summaries in Figs. 6.5 and 6.15 show the basic steps in the glycolytic pathway and where energy-rich adenosine triphosphate (ATP) is broken down and synthesized. This conversion to pyruvate generates a net of 2 ATP molecules, which supplies the yeast cell with energy. The enzyme cofactor NAD^+ (nicotinamide adenine dinucleotide), a

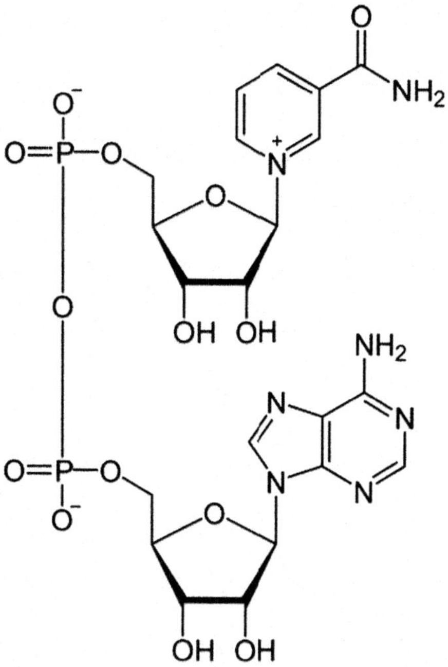

Fig. 6.6 Nicotinamide adenine dinucleotide

cofactor for dehydrogenase enzymes controlling oxidative reactions during catabolism, was characterized (Fig. 6.6). Reduced NAD^+ (or $NADH_2$) is formed when electrons are transferred to NAD^+ as hydrogen ions $[H^-]$:

$$NAD^+ + [2H] \rightarrow NADH + H^+ \ (\text{or } NADH_2)$$

When yeast cells are respiring in an aerobic environment, the Krebs cycle [also known as the Tricarboxylic Acid (TCA) Cycle or the Citric Acid Cycle] oxidative phosphorylation (also called the electron transfer chain) occurs—details later.

As a result of biochemical studies with yeast, and later with mammalian muscle cells, the glycolytic pathway was elucidated during the first half of the twentieth century (although its detailed regulation was subsequently understood—details later). When pyruvate, the end-product of glycolysis, is metabolized anaerobically, yeast converts it to ethanol, while muscle converts it to lactic acid (Meyerhof 1930). Meyerhof explains: "… a muscle resting in nitrogen (anaerobic conditions) produces lactic acid steadily—in the presence of oxygen no lactic acid accumulates". There have been many parallels and interconnections regarding research on the two kinds of eukaryotic cells—yeast cells and muscle cells. By 1940, the complete glycolytic pathway had been elucidated, largely by some remarkable, outstanding biochemists. Six were Nobel Prize winners!

The following are brief biographical notes on some of these outstanding biochemists who elucidated the glycolytic pathway (Barnett 2003):

Fig. 6.7 Gustav Embden

- *Carl Ferdinand Cori* (1896–1984) was born in Prague (then within the Austro-Hungarian Empire) and studied medicine in both Budapest and Prague. In 1922, he emigrated to the United States. He was invited to work in Buffalo, New York, and moved to Washington University Medical School in 1933. Carl and his wife, Gerty Cori, jointly received the Nobel Prize for Physiology or Medicine in 1947 for their studies on the EMP pathway in the liver (Cori and Cori 1929; Cori 1983; Randle 1986).
- *Gerty Cori*, like her husband Carl, was born in Prague and, similar to her husband, studied medicine and emigrated to the United States with him and worked closely with him thereafter. She was the third woman to receive a Nobel Prize in science—the others were the mother and daughter, Marie Curie (Reid 1978) and Irène Joliot-Curie (Byers and Moszkowski 1956) for their studies on isotopes.
- *Gustav Embden* (1874–1933) (Fig. 6.7) studied medicine at the universities of Freiburg-im-Breisgau, Munich and Strasbourg and later worked with Paul Ehrlich in Frankfurt. Embden became a professor and subsequently rector of Bonn University. Working with muscle, he made significant research contributions on glycolysis (Cori 1983). His research on glycolysis was marked by the glycolytic pathway being named the Embden-Meyerhof-Parnas (EMP) pathway in his honour.
- *Arthur Harden* (1865–1940) (Fig. 6.8) studied chemistry at Manchester University, was awarded a doctorate at the University of Erlangen in Germany and subsequently returned to Manchester. He was a polymath, who published numerous papers in many subject areas. He shared the Nobel Prize for Chemistry in 1929 (with Hans

Fig. 6.8 Arthur Harden

Karl August Simon von Euler-Chelpin) for their research on fermentation (Ihde 1971). In an obituary published in Notices of Fellows of the Royal Society, Hopkins and Martin (1942) wrote:

> Harden's outstanding qualities as an investigator were his clarity of mind, precision of observation, and capacity to analyse dispassionately the results of an experiment and define their significance. He mistrusted the use of his imagination beyond a few paces in advance of the facts. If he had exercised less restraint, he might have gone further!

- *Otto Fritz Meyerhof* (1882–1951) (Fig. 6.9) qualified in medicine at Heidelberg University and was actively interested in philosophy for much of his life. In 1918, Meyerhof chose muscle as a tissue for his experimental work (Nachmansohn 1972). With the Nazis in power, Meyerhof left Germany and worked in Paris from 1938 to 1940. When the Germans occupied Paris, he left for the United States in 1940 and became a biochemistry professor at the University of Pennsylvania (Meyerhof 1930). He was welcomed there, having shared the 1922 Nobel Prize in Physiology or Medicine with a colleague A.V. Hill (Peters 1952).
- *Carl Neuberg* (1877–1956) (Fig. 6.10) is one of the founders of modern biochemistry but had unfortunately enjoyed a less illustrious scientific career than Meyerhof. He was the founding editor of *Biochemische Zeitschrift* and edited 278 volumes of this notable journal during the next 30 years (Neuberg 1918). He moved to the Netherlands and then Palestine and finally emigrated to the United States. Like

Fig. 6.9 Otto Fritz Meyerhof

Fig. 6.10 Carl Neuberg

Fig. 6.11 Otto Warburg

many others, his career reflected the political upheavals of his time (Gottschalk 1956; Nord 1958).

- *Jakub Karol Parnas* (1884–1949) also had a life much affected by the political geography of the twentieth century. He was born in a part of the Austro-Hungarian Empire that is now part of the Ukraine. Parnas held professorships in Strasbourg and Warsaw, and then he was head of the Biological and Medical Chemistry Institute in Moscow (Korzybski 1974).
- *Otto Heinrich Warburg* (1883–1970) (Fig. 6.11) in this author's opinion was one of the most outstanding biochemists of all time! He studied for a doctorate under Emil Fischer in Berlin. As well as an enormous research output of over 500 publications, mostly on basic cell metabolism (Warburg 1926), Warburg was responsible for advances in biochemical methodology, particularly the Warburg manometer (Fig. 6.12). This was developed for measuring rates of gas exchange, and it was the standard equipment in many research laboratories from the 1930s until the early 1970s (Korzybski 1974). The gas phase in the manometer vessel was achieved by constant shaking of the vessels in a temperature-controlled water bath. Warburg was also responsible for important developments in spectrophotometry and he received the Nobel Prize for Physiology or Medicine in 1931. Warburg displayed a very different attitude to research to that of Harden, which is illustrated by the following comment by him:

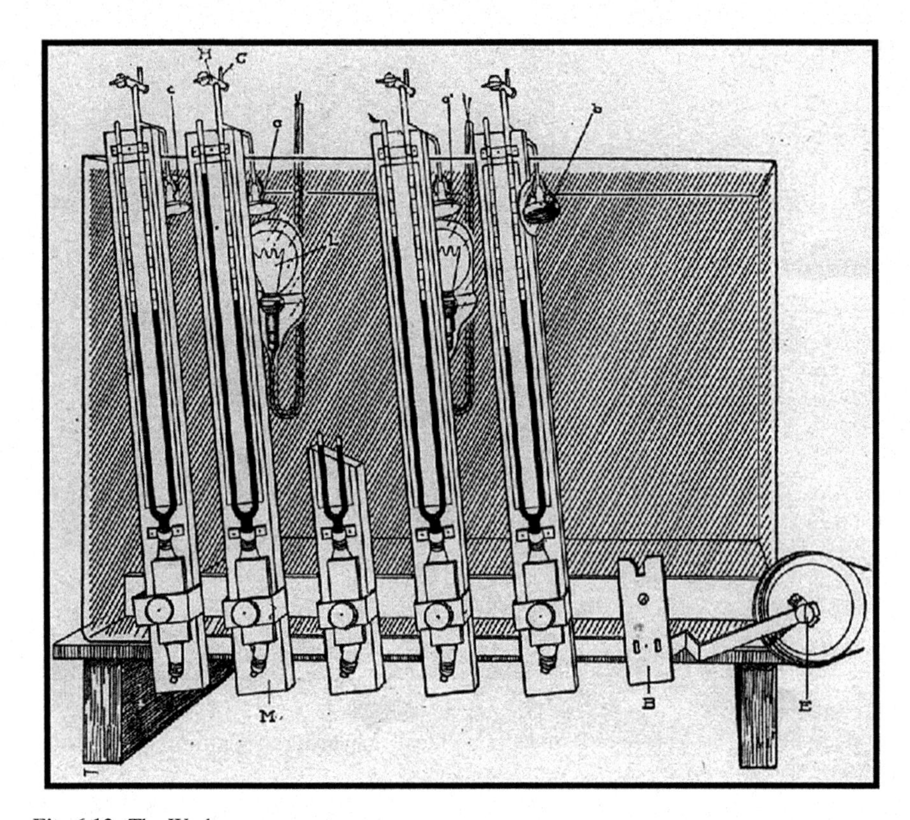

Fig. 6.12 The Warburg manometer

> . . . I learned that a scientist must have the courage to attack the great unsolved problems of his time, and solutions usually have to be forced by carrying out innumerable experiments without much critical hesitation

As well as having considerable empathy with this approach, I also spent a great deal of time during my PhD studies employing a Warburg respirometer (manometer) to measure oxygen uptake rates with baker's yeast cultures that had been treated with monofluoroacetic acid (HFA) (Brunt and Stewart 1967; Stewart and Brunt 1968). The oxygen uptake rates of these yeast cultures had been significantly affected as a result of HFA's synthesis to monofluorocitric acid (lethal synthesis) (Fig. 6.13) (Peters 1952). HFA has also been shown to be a convulsant in rats because of its effects on this mammal's central nervous system following its conversion to fluorocitric acid (Stewart et al. 1969). Indeed, Prof. Sir Rudolf Peters (Fig. 6.14) was the external examiner of my PhD thesis in 1968 (Stewart 1968).

The overall progress of glycolysis is:

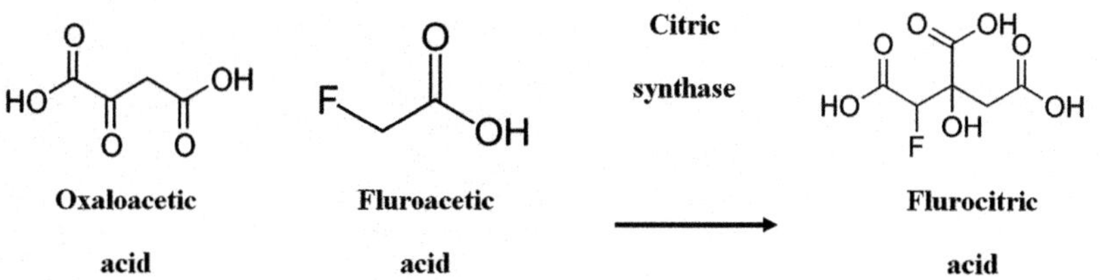

Oxaloacetic acid + Fluroacetic acid —Citric synthase→ Flurocitric acid

Fig. 6.13 Lethal synthesis of fluorocitric acid. Fluorocitric acid causes inhibition of aconitase and consequently inhibits the TCA Cycle

$$\text{Glucose} + 2\text{NAD}^+ + 2\text{ADP} + 2\text{Pi} \rightarrow 2\ \text{Pyruvate} + 2\text{NADH} + 2\text{H}^+ + 2\text{ATP} + 2\text{H}_2\text{O}$$

If glycolysis were to continue indefinitely, all of the NAD^+ would be metabolized and it would cease. To allow glycolysis to continue, yeast (and other tissues such as mammalian muscle) must be able to oxidize NADH back to NAD^+. How this occurs depends on which external electron acceptor is available—details later.

The glycolytic pathway can be divided into two phases:

1. The preparatory phase—in which ATP is consumed and, consequently, is also known as the investment phase
2. The pay-off phase—in which ATP is produced

As already discussed, during glycolysis, the metabolism of one molecule of glucose consumes two molecules of ATP. Most cells, including many yeast species (e.g. *Saccharomyces* spp.), will carry out further reactions to "repay" the used NAD^+ and produce a final product of either ethanol or lactic acid plus carbon dioxide—details later.

Cells in the presence of oxygen perform aerobic respiration and synthesize considerably more ATP. These aerobic reactions use pyruvate and NADH + H^+ obtained from glycolysis. Eukaryotic aerobic respiration (including *Saccharomyces*) produces approximately 34 additional molecules of ATP for each molecule of glucose taken into the cell. However, most of these are produced by a considerably different series of enzymatic reactions at the substrate-level phosphorylation that occurs during glycolysis—details later. The lower-energy production, with glucose, of anaerobic respiration relative to aerobic respiration results in much greater flux through the pathway under hypoxic (low-oxygen) conditions, unless alternative sources of aerobically metabolized substrates such as fatty acids and sterols are found—details of glucose aerobic metabolism (Krebs cycle) can be found later in this chapter. The first step in glycolysis is the phosphorylation of glucose by a family of enzymes called hexokinases to form glucose-6-phosphate (G6P) (Fig. 6.15). The reaction consumes ATP and it acts to maintain the glucose concentration low, promoting continuous transport of glucose into the cell through plasma membrane transporters. In addition, glucose is blocked from leaking out of the cell because the cell lacks transporters for glucose-6-phosphate (G6P), and free

Fig. 6.14 Rudolf Peters

diffusion out of the cell is prevented due to the charge characteristic of G6P. Glucose may alternatively be formed from the phosphorolysis or hydrolysis of intracellular glycogen (details later).

G6P is then rearranged into fructose 6-phosphate (F6P) by glucose phosphate isomerase. Fructose can also enter the glycolytic pathway by phosphorylation at this point in the pathway (Fig. 6.15). The energy expenditure of another ATP molecule in this step can be justified in two ways. The glycolytic process (up to this point in the process) is now irreversible and the energy supplied destabilizes the molecule. The reaction catalysed by phosphofructokinase (PFK-1) is coupled to the hydrolysis of ATP (an energetically favourable step); it is, in essence, irreversible, and a different pathway must be used to conduct the reverse conversion during gluconeogenesis. This makes the reaction a key regulatory point. This is also the rate-limiting step of glycolysis.

The second phosphorylation event is necessary to allow for the formation of two charged groups (rather than only one) in the subsequent glycolytic step, ensuring the prevention of free diffusion of substrates out of the cell. The same reaction can also be catalysed by a pyrophosphate-dependent phosphofructokinase (PFP or PPi-PFK), which is found in yeast and most bacteria but not in mammals. This enzyme uses pyrophosphate (PPi) as a phosphate donor instead of ATP. It is a reversible reaction, increasing the flexibility of glycolytic metabolism.

Destabilizing fructose 1,6-diphosphate (F1,6BP) permits the hexose ring to be split by aldolase into two triose sugars—dihydroxyacetone phosphate and glyceraldehyde 3-phosphate, an aldose. Electrons are delocalized in the carbon-carbon

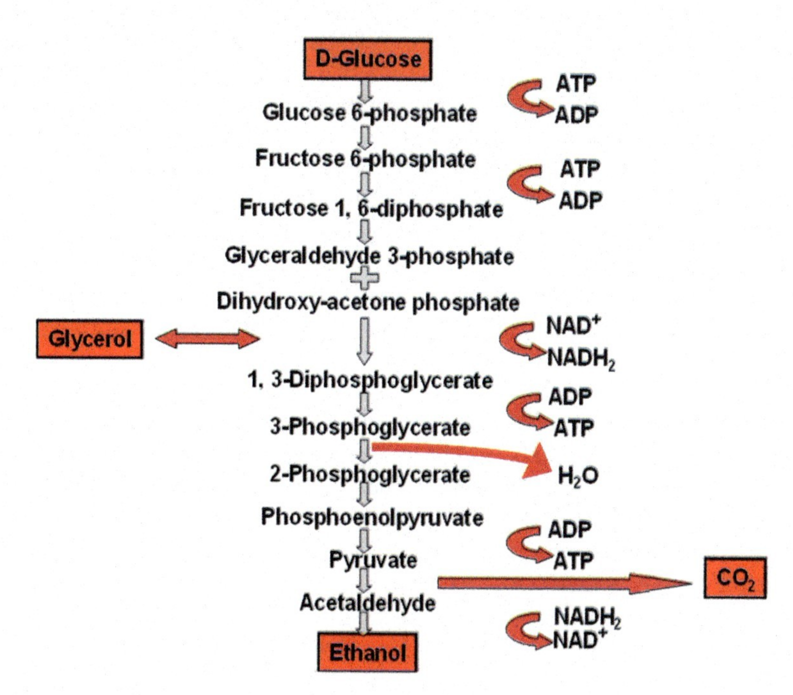

Fig. 6.15 The Embden-Meyerhof-Parnas glycolytic pathway

bond cleavage associated with the alcohol group. Triosephosphate isomerase rapidly interconverts dihydroxyacetone phosphate with D-glyceraldehyde 3-phosphate (GADP) that proceeds further into glycolysis. This is advantageous, because it directs dihydroxyacetone phosphate along the same pathway as glyceraldehyde 3-phosphate, simplifying its regulation.

It has already been stated that the second half of glycolysis is known as "the pay-off" phase. This is characterized by a net gain of the energy-rich ATP and NADH. Since glucose leads to two triose sugars in "the preparatory phase", each reaction in the pay-off phase occurs twice per glucose molecule. This yields 2 NADH molecules and 4 ATP molecules, leading to a net gain of 2 NADH molecules and 2 ATP molecules from the glycolytic pathway per glucose molecule.

The aldehyde groups of the triose sugars are oxidized and inorganic phosphate is added to them, forming 1,3-diphosphoglycerate. The hydrogen is used to reduce two molecules of NAD^+, a hydrogen carrier, to produce $NADH + H^+$ for each triose. The hydrogen atom balance and the charge balance are both maintained because the phosphate (Pi) group actually exists in the form of a hydrogen phosphate anion (HPO_4^{++}) (Lane et al. 2009), which dissociates to contribute the extra H^+ ion and gives a net charge of −3 on both sides of the equation.

The next step is the enzymatic transfer of a phosphate group from 1,3-diphosphoglycerate to ADP by phosphoglycerate kinase forming ATP from ADP and 3-phosphoglycerate. At this stage, glycolysis has reached the metabolic energy break-even point: two molecules of ATP have been consumed and two new molecules synthesized. This step, one of the two substrate-level phosphorylation steps, requires ADP. However, where the cell has plenty of ATP (and little ADP), this reaction does not occur. Because ATP decays relatively rapidly when it is not metabolized, this is an important regulatory point in the glycolytic pathway. ADP actually exists as $ADPMg^-$ and ATP as $ATPMg^{++}$. This balances the charges at −5 on both sides of the reaction. The phosphoglycerate mutase then isomerizes 3-phosphoglycerate into 2-phosphoglycerate. Enolase subsequently forms phosphoenolpyruvate from 2-phosphoglycerate. A final substrate-level phosphorylation now forms a molecule of pyruvate and a molecule of ATP by means of the enzyme pyruvate kinase.

With the formation of pyruvate, glycolysis is nearing completion. However, if glycolysis is to continue, yeast (and other organisms) must be able to oxidize the NADH to NAD. In yeast, this occurs in a process called ethanol fermentation, whereby pyruvate is converted first to the carbonyl acetaldehyde and carbon dioxide and then to ethanol (Fig. 6.15). Also, in some bacteria and muscle cells, pyruvate is not converted to ethanol but to lactate in a process called lactic acid fermentation:

$$\text{Pyruvate} + \text{NADH} + \text{H}^+ \rightarrow \text{Lactate} + \text{NAD}^+$$

Lactic acid and ethanol formation can occur in the absence of oxygen. This anaerobic fermentation allows many single-cell organisms to use glycolysis as their only energy source. Lactic acid formed in muscles (particularly during exercise) results in a decrease in tissue pH and cramps can result.

The glycolytic pathway is subject to regulation by four enzymes (hexokinase, glucokinase, phosphofructokinase and pyruvate kinase). The flux through the glycolytic pathway is adjusted in response to conditions both inside and outside the yeast cell. The internal factors that regulate glycolysis do so primarily to provide ATP in adequate concentrations for the cell's requirements. The external factors relate to the quantity of glucose in the culture medium.

Phosphofructokinase (PFK-1) is an important control point in the glycolytic pathway, since it is one of the irreversible steps and has allosteric effectors: AMP and fructose 2,6-diphosphate (F2,6DP). F2,6DP is a potent activator of PFK-1 that is synthesized when F6P is phosphorylated by a second phosphofructokinase (PFK-2). ATP competes with AMP for the allosteric effector site on the PFK enzyme. The ATP concentrations in yeast cells are much higher than those of AMP. However, the concentration of ATP does not change more than about 10% under physiological conditions, whereas a 10% decrease in ATP results in a sixfold increase in AMP (Voet and Voet 2011). Citrate inhibits PFK when tested in vitro in yeast, resulting in an inhibitory effect on ATP formation (Stewart and Brunt 1968). However, it is doubtful that this is a meaningful effect in vivo, because citrate in the cytosol is utilized mainly for conversion to acetyl-CoA for fatty acid synthesis (Berovic and Legisa 2007) (Fig. 6.22).

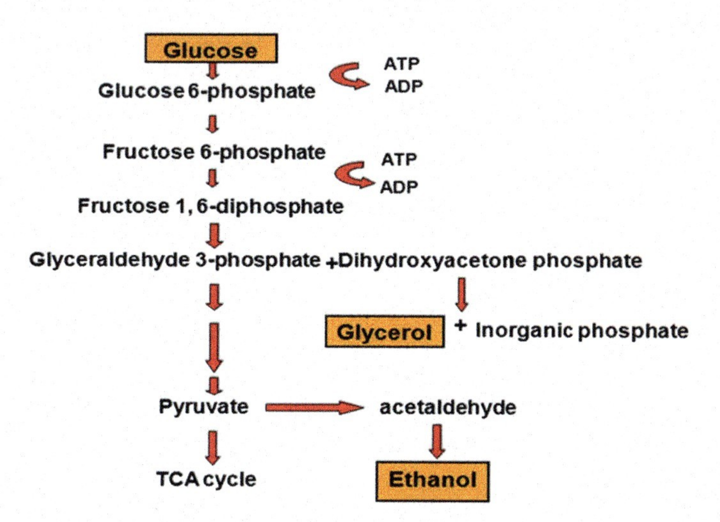

Fig. 6.16 Summary of the formation of glycerol and ethanol

Pyruvate kinase catalyses the final stage of the glycolytic pathway, in which pyruvate and ATP are formed. Pyruvate kinase catalyses the transfer of a phosphate group from phosphoenolpyruvate (PEP) to ADP, yielding one molecule of pyruvate and one molecule of ATP. Many of the metabolites in the glycolytic pathway are also used for the anabolic synthesis of complex molecules—carbohydrates, amino acids, peptides and proteins, nucleotides, etc. As a consequence, flux through this pathway is critical to maintain a supply of carbon skeletons for biosynthesis.

The following metabolic pathways are all reliant upon glycolysis as a source of metabolites:

- Pentose phosphate pathway (Fig. 6.17), which begins with the dehydrogenation of glucose-6-phosphate (Keller et al. 2014). This pathway produces various pentose (five-carbon) sugars and NADPH for the synthesis of fatty acids and cholesterol—details later.
- Glycogen (Fig. 6.18) synthesis also starts with glucose-6-phosphate—details later.
- Glycerol is produced from the glycolytic intermediate glyceraldehyde-3-phosphate (Fig 6.16) (Cronwright et al. 2002).
- Various post-glycolytic pathways:
  - Fatty acid synthesis
  - Sterol synthesis
  - The citric acid (TCA Cycle) (Fig. 6.22) which in turn leads to:

    Energy (ATP, etc.)
    Amino acid synthesis
    Nucleotide synthesis
    Tetrapyrrole

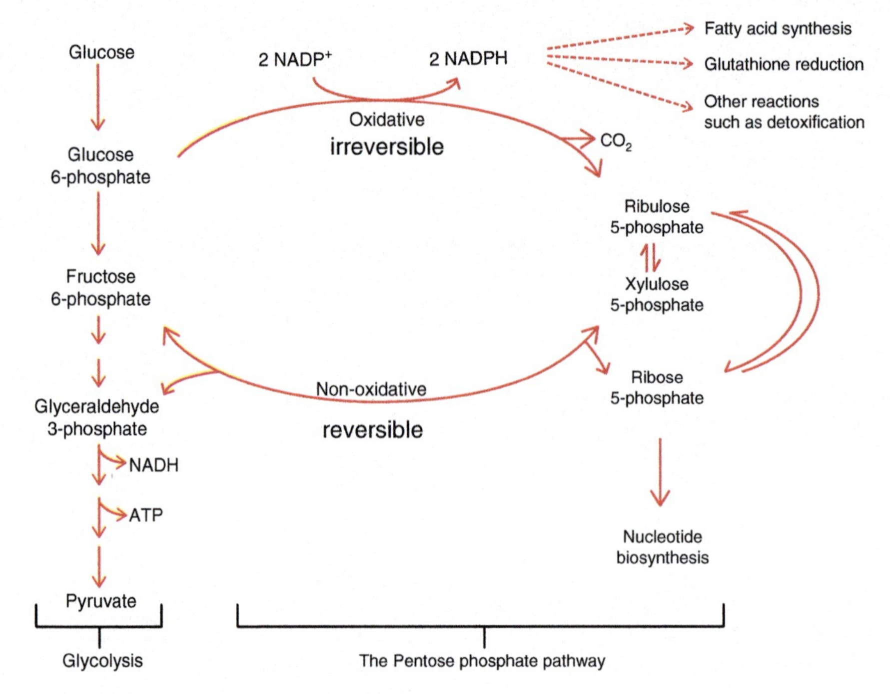

Fig. 6.17 The pentose phosphate pathway

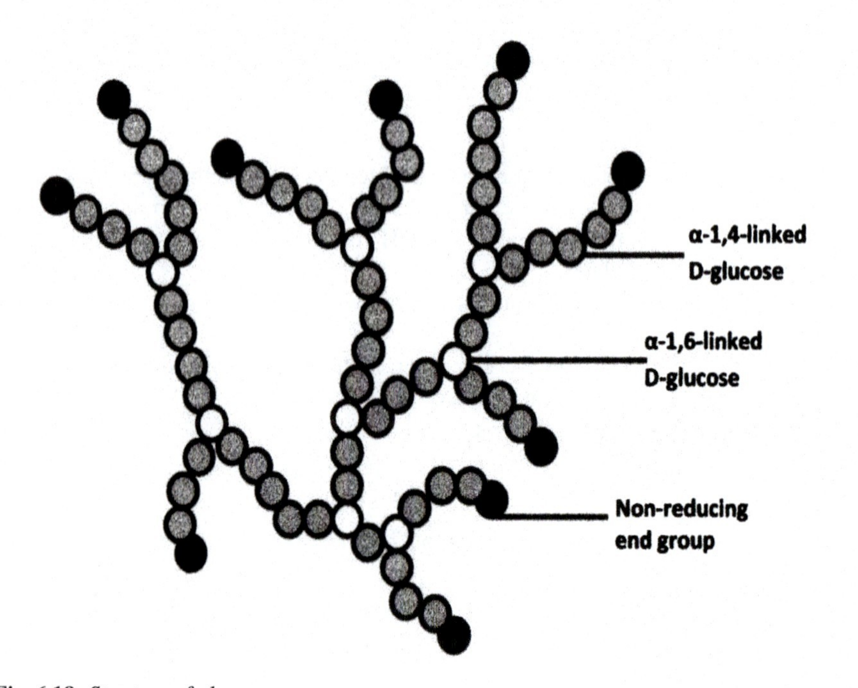

Fig. 6.18 Structure of glycogen

It has already been discussed in this chapter that NAD^+ is the oxidizing agent during glycolysis, as it is in most energy yielding reaction (e.g. oxidation of fatty acids and during the Citric Acid Cycle). The NADH produced is primarily used to transfer electrons to O_2 in order to produce water. However, as already discussed, when oxygen is not available, compounds such as lactate or ethanol are produced. NADH is rarely used for synthetic purposes, the notable exception being gluconeogenesis (a metabolic pathway that results in the generation of glucose from certain non-carbohydrate carbon substrates, e.g. glycogen, fatty acids, proteins, lipids, etc.).

The source of the NADPH is twofold:

- Oxidative decarboxylation by $NADP^+$ is linked to the malic enzyme, pyruvate, and CO_2 and NADPH are formed.
- NADPH is also formed by the pentose phosphate pathway (Fig. 6.17), which converts glucose into ribose, which can then be used during the synthesis of nucleotides and nucleic acids, or it can be catabolized (a set of metabolic reactions that degrade molecules into smaller units that are oxidized to release energy) to pyruvate.

6.3 Glycerol

Glycerol is a major by-product of glycolysis by *S. cerevisiae* (Kutyna et al. 2012; Vijaikishore and Karanth 1987) (Fig. 6.16). It is of major importance to the beer, wine and spirit production industries. In a beer, which contains approximately 5% (v/v) ethanol, the glycerol concentration will be approximately 0.2% (v/v) and approximately 0.5% (v/v) in wine (Stewart 2014). Although fermented spirit wash will contain glycerol, the distillation process ensures that the final spirit does not contain measurable quantities of this alcohol (Agarwal 1990).

Glycerol is formed by the reduction of dihydroxyacetone phosphate (DHAP) to glycerol, concomitant with NADH oxidation by NAD^+-dependent glycerol dehydrogenase (Fig. 6.16) (Bisping and Rehm 1986; Bisping et al. 1990; Cronwright et al. 2002). Glycerol 3-phosphate is then dephosphorylated to glycerol by glycerol 3-phosphatase. Under anaerobic and glucose-repressing growth conditions, yeast can quickly adapt to a preferred carbon and energy source—this is usually achieved through inhibition of enzyme synthesis involving in the catabolism of carbon sources. It is widely believed that yeast cells produce glycerol to assist and maintain a cytosolic redox state conducive to sustain glycolytic catabolism (Nordstrom 1966). The inability of a mutant defective in glycerol production to grow under anaerobic conditions confirms this observation (Costenoble et al. 2000).

Glycerol can be produced either by microbial fermentation or by chemical synthesis (usually by the soap industry). Glycerol was first produced on a large scale using the sulphite-steered yeast process during World War I, when demands for glycerol in explosive manufacture exceeded the supply from the soap industry (Prescott and Dunn 1959). However, the wartime process technology could never adapt to the peacetime competition that came from the chemically synthesized process that developed after World War II as yields of glycerol from sugar by fermentation were low and recovery by distillation was inefficient.

Glycerol production by yeast fermentation has been known since the investigations of Pasteur (1858). In *S. cerevisiae*, glycerol is a fermentation by-product of sugar to ethanol metabolism in a redox-neutral process. Substantial overproduction of glycerol from monosaccharides can be obtained during yeast fermentation by the following routes:

- Forming a complex of acetaldehyde with the bisulphite ion that limits ethanol production and promotes reoxidation of glycolytically formed NADH by glycerol synthesis (Enjalbert et al. 2000)
- Growing yeast at pH values around 7 or above (Deshpande et al. 2011)
- By using an osmotolerant yeast (e.g. *Zygosaccharomyces rouxii*) (Pribylova et al. 2007; Agarwal 1990)

Overproduction of glycerol by *S. cerevisiae* from monosaccharides can be achieved by combining acetaldehyde with the bisulphite ion (known as the steering agent) or by growing the yeast at a pH of 7 or above. The first method, based on the trapping of acetaldehyde by bisulphite ions, yields the following reaction:

$$\text{Hexose} + \text{bisulphite} \rightarrow \text{acetaldehyde} - \text{bisulphite} + CO_2 + H_2O + \text{glycerol}$$

In some instances, the fermentation process using the steering agent has been adapted to improve the efficiency of glycerol production (Cocking and Lilly 1919; Cronwright et al. 2002; Kalle and Naik 1985; Bisping and Rehm 1986; Vijaikishore and Karanth 1987; Bisping et al. 1990; Hecker et al. 1990; Petrovska et al. 1999).

The second method, operated at pH 7 and above, is based on the following reaction:

$$2\ \text{hexose} \rightarrow 2\ \text{glycerol} + \text{ethanol} + \text{acetic acid} + CO_2 + 2H_2O$$

It has already been discussed that the sulphite-steered yeast process was used for commercial glycerol production during World War I in Germany. Six thousand tons of beet sugar (sucrose) was fermented by *S. cerevisiae* monthly to yield 1000 tons of "dynamite" glycerol (Bernhauer 1957). The maximum conversion efficiency of glycerol production for this process was 20% of the metabolized sugar. The process was also used to produce glycerol during World War II in Germany, Japan, the former Soviet Union, Poland and Brazil. However, by 1960, production of glycerol by sugar fermentation was found to be uneconomic when compared to the chemical synthesis process from petrochemical feedstocks (Freeman and Donald 1957). The Chinese Academy of Science also carried out research on sulphite-steered yeast processes from 1967 to 1971. Although the conversion efficiency of glycerol produced, based on the concentration of total sugar used, was approximately 27%, this was still regarded to be too low for commercial development. Indeed, today adaptive evolution is being applied to strains *S. cerevisiae* in order to generate enhanced glycerol production (Kutyna et al. 2012).

6.4 Osmotolerant Yeasts

Nickerson and Carroll (1945) first reported on the production of glycerol by osmotolerant yeasts. They noted significant glycerol production by an osmotolerant yeast species, *Zygosaccharomyces acidifaciens*, which did not require a steering agent. This observation stimulated a comprehensive investigation of osmotolerant or osmophilic yeasts. The objective was to identify yeast(s) that could produce glycerol without the need of a steering agent. Also, as well as glycerol, osmotolerant yeasts were found to produce other polyols including erythritol, D-arabitol and mannitol (Spencer and Sallons 1956).

Most osmotolerant yeast species considered for glycerol production belong to the genera: *Candida*, *Debaryomyces*, *Hansenula*, *Pichia*, *Saccharomyces*, *Schizosaccharomyces*, *Torulaspora* and *Zygosaccharomyces* (Petrovska et al. 1999). The main advantages of osmotolerant yeasts for glycerol production, compared to the processes based on alkali and sulphite, are the following:

- Aerobic rather than anaerobic or oxygen-limited conditions are required for cell growth and fermentation.
- No steering agents or osmotic solutes are needed.
- Considerably higher sugar concentrations can be used with an improved glycerol production rate and yield.
- A much simpler process technology is used with less contamination.

There have been significant improvements in the microbial production of glycerol by strain manipulation and selection when compared to the processes that were employed in the first half of the last century (Petrovska et al. 1999). The understanding of glycerol metabolism in yeast species such as *S. cerevisiae* has made significant progress over the past 20 years, but knowledge in this area has only been applied to the manipulation of wine yeasts with regard to glycerol production (Prior et al. 1999). The challenge for other biotechnology disciplines will be to apply the fundamental knowledge on glycerol metabolism in order to manipulate yeast strains appropriately for economic glycerol production by microbial fermentation.

6.5 Pentose Phosphate Pathway

The pentose phosphate pathway (also called the phosphogluconate pathway and the hexose monophosphate shunt) has already been mentioned in this chapter. It is a metabolic pathway parallel to glycolysis that generates NADPH and pentoses, as well as ribose 5-phosphate, a precursor for the synthesis of nucleotides (Fig. 6.17). While it does not involve oxidation of glucose, its primary role is as an anabolic pathway (a metabolic pathway that produce molecules from smaller units) rather than a catabolic pathway (a metabolic pathway that breaks down molecules into smaller units).

There are two distinct phases in the pathway. The first is the oxidative phase, in which NADPH is generated, and the second is the non-oxidative phase, in which there is the synthesis of 5-carbon sugars. In yeast, the pentose phosphate pathway takes place in the cytosol (Kruger and von Schaewen 2003). Similar to glycolysis, the pentose phosphate pathway appears to have an ancient evolutionary origin (Keller et al. 2014). It is suggested that the origins of this pathway could date back to the prebiotic world.

The primary results of the pathway are as follows:

- The generation of reducing equivalents, in the form of NADPH, used in reductive biosynthetic reactions within cells (e.g. fatty acid synthesis)
- Production of ribose 5-phosphate (R5P), used in the synthesis of nucleotides and nucleic acids
- Production of erythrose 4-phosphate (E4P), which is used in the synthesis of aromatic amino acids (details in Chap. 7)

One of the uses of NADPH in the cell is to prevent oxidative stress. It reduces glutathione via glutathione reductase, which converts reactive H_2O_2 into H_2O by glutathione peroxidase. If this enzyme is absent, the H_2O_2 will be converted to hydroxyl free radicals by Fenton chemistry (a solution of H_2O_2 with ferrous iron as a catalyst) is used to oxidize contaminants.

$$Fe^{++} + H_2O_2 \rightarrow Fe^{+++} + HO^{\bullet} + OH^{-}$$
$$Fe^{+++} + H_2O_2 \rightarrow Fe^{++} + HOO^{\bullet} + OH^{+}$$

The pentose phosphate pathway can be divided into two phases:

- The oxidative phase, where two molecules of $NADP^+$ are reduced to NADPH, utilizing the energy from the conversion of glucose 6-phosphate into ribulose 5-phosphate. The overall reaction is:

$$\text{Glucose 6} - \text{phosphate} + 2\text{NADP}^+ + \text{H}_2\text{O}$$
$$\rightarrow \text{ribulose} - 5 - \text{phosphate} + 2\text{NADPH} + 2\text{H}^+ + \text{CO}_2$$

- The non-oxidative phase, where the net reaction is:

$$3 \text{ ribulose} - 5 - \text{phosphate} \rightarrow \text{ribose} - 5 - \text{phosphate} + 2 \text{ xylulose} - 5 - \text{phosphate} \rightarrow 2 \text{ fructose} - 6 - \text{phosphate} + \text{glyceraldehyde} - 3 - \text{phosphate}$$

Glucose-6-phosphate dehydrogenase is the rate-controlling enzyme of this pathway. It is allosterically stimulated (the regulation of a protein by binding an effector molecule at a site other than the enzyme's active site) by $NADP^+$ and strongly inhibited by NADPH (Voet and Voet 2011). The ratio of $NADPH/NADP^+$ is normally approximately 3:1 in the yeast cytoplasm, and this makes the cytoplasm a highly reducing environment. An NADPH-utilizing pathway from $NADP^+$ stimulates glucose-6-phosphate dehydrogenase to produce more NADPH. This reaction is also inhibited by acetyl-CoA.

6.6 Glycogen: Structure, Biosynthesis and Metabolism

Glycogen is the major intracellular carbohydrate in yeast cells, along with the disaccharide trehalose. However, they have different metabolic functions. Glycogen is a multibranched polysaccharide of glucose molecules that serves as a source of energy in yeast (other fungi) insects and animals (Quain and Tubb 1982). The polysaccharide structure represents the main intracellular storage form of glucose. It is an analogue of starch and has a similar structure to amylopectin (a starch component), but is more extensively branched and compact than starch (Fig. 6.18). Glycogen forms an energy reserve that can be rapidly mobilized to meet a sudden requirement for glucose but is less compact than the energy reserves of triglycerides (lipids). It is a branched biopolymer consisting of linear chains of glucose residues with further chains of glucose residues with branched chains every 8 to 12 glucose units or so. The glucose units are linked together linearly by $\alpha(1 \rightarrow 4)$ glycosidic bonds from one glucose to the next. Branches are linked to the chains from which they are branched by $\alpha(1 \rightarrow 6)$ glycosidic bonds between the first glucose of the new branch and a glucose on the stem chain (Bernhauer 1957).

Glycogen synthesis, unlike its breakdown, requires an input of energy. Energy for glycogen synthesis comes from uridine triphosphate (UTP), which reacts with glucose-1-phosphate, forming UDP-glucose, in a reaction catalysed by UTP-glucose-1-phosphate uridylyltransferase. Glycogen is synthesized from monomers of UDP-glucose initially by the protein glycogenin, which has two tyrosine anchors for the reducing end of glycogen. After about eight glucose molecules have been added to a tyrosine residue, the enzyme glycogen synthase progressively lengthens the glycogen chain using UDP-glucose, adding $\alpha(1 \rightarrow 4)$-bonded glucose. The glycogen branching enzyme catalyses the transfer of a terminal fragment of six or seven glucose residues from a nonreducing end to the C-6 hydroxyl group of a glucose residue deeper into the interior of the glycogen

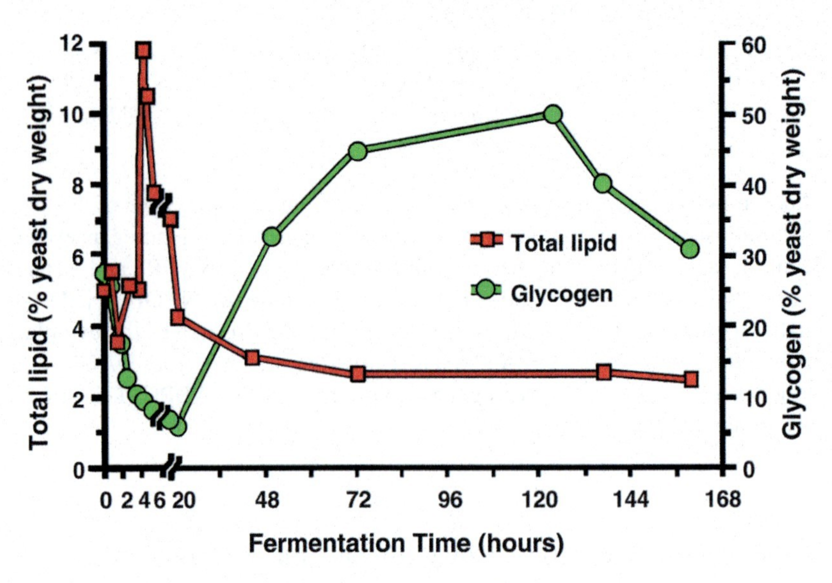

Fig. 6.19 Intracellular concentration of glycogen and lipids in a lager yeast culture during fermentation of a 15°Plato wort

molecule. The branching enzyme can act upon a branch having at least 11 residues, and the enzyme may transfer to the glucose chain or adjacent glucose chains (Rothman and Cabib 1969).

One of the factors that will affect brewer's wort fermentation rate is the condition under which the yeast culture is stored between fermentations. Of particular importance in this regard is the influence of temperature during these storage conditions on the cell's intracellular glycogen level. It has already been discussed that glycogen serves as a source of biochemical energy during the fermentation lag phase, when the energy demand is intense for the synthesis of cell membrane compounds such as sterols and unsaturated fatty acids (Fig. 6.19) (Quain et al. 1981). Consequently, it is important that appropriate levels of cell glycogen are maintained in the course of storage, so that in the initial stages of fermentation, the yeast culture is able to synthesize sterols and unsaturated fatty acids and trehalose. Trehalose is a nonreducing disaccharide (Fig. 6.20) that plays a protective role against stresses such as osmoregulation, nutrient depletion and starvation. It improves cell resistance to high and low temperatures, elevated ethanol concentrations, etc. (D'Amore et al. 1991) (further details of trehalose glycogen and stress tolerance in yeast are in Chap. 11). A novel method to measure glycogen, neutral lipid and trehalose in yeast has recently been described (Chan et al. 2016).

It has been proposed that there are two pools of glycogen—cell wall-bound glycogen and intracellular glycogen; it is the latter that correlates with a yeast culture's fermentation efficiency (Enjalbert et al. 2000; Deshpande et al. 2011). The cell wall-bound glycogen has been found to be approximately three times higher than the intracellular glycogen. Using a synthetic medium, cell wall-bound

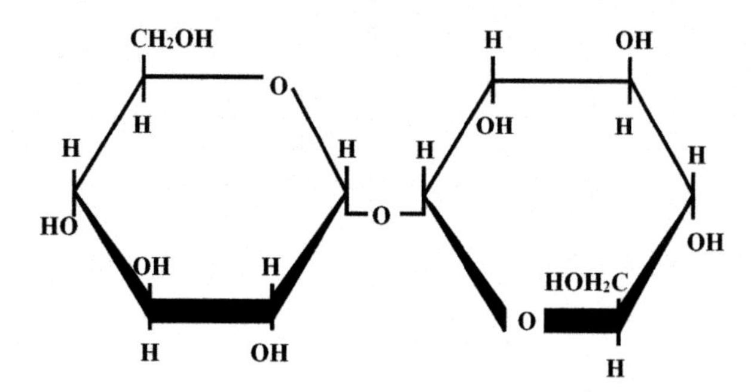

Fig. 6.20 Structure of trehalose

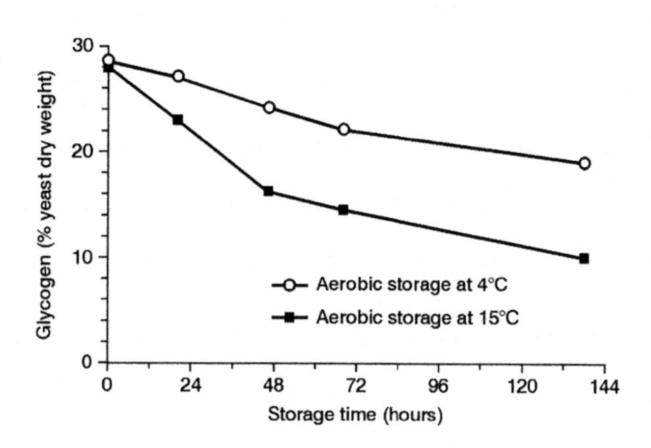

Fig. 6.21 The effect of storage temperatures on the intracellular glycogen concentration of yeast cultures

glycogen was reduced by 85% during the first 6–8 h after pitching, when almost 50% of the medium's sugar content was utilized. The cell wall-bound glycogen correlated with fermentation performance. Cells grown in an 8% (w/v) sugar content medium and rich in cell wall-bound glycogen, when subsequently pitched into a 1% (w/v) sugar medium, showed enhancement in ethanol content by 21%. Also, its depletion affected overall fermentation performance. Low glycogen levels in pitching yeast cultures result from unsatisfactory yeast management between fermentations (Stewart 2015a, b)—such as high temperatures, semi-aerobic storage conditions, improper centrifugation procedures, etc. (Chlup and Stewart 2011). It has also been proposed that the degradation of glycogen provides the energy for the viability of respiratory-deficient (RD) mutants in glucose-starved media (Deshpande et al. 2011)—a situation that is still unclear and that requires further clarification!

Yeast storage temperature and oxygen level (Fig. 6.21) has a direct influence on the rate and extent of glycogen utilization, as might be expected, considering the effect that temperature has on cellular metabolic rates in general. Although strain dependent, of particular interest is the fact that within 48 h, the yeast stored semi-

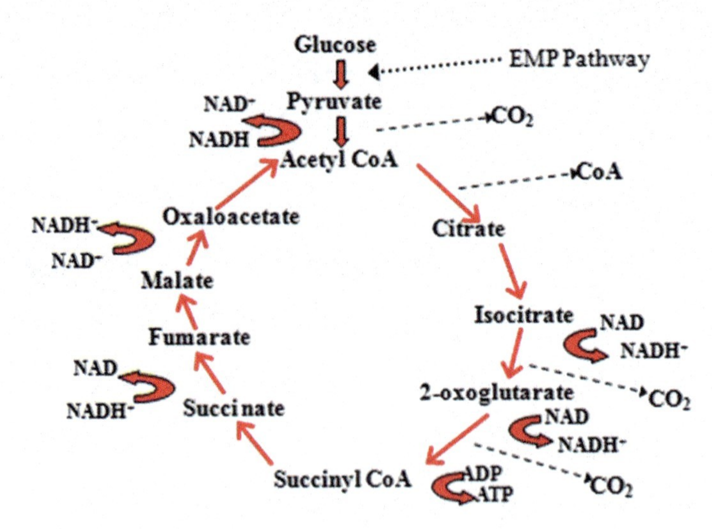

Fig. 6.22 Details of the Tricarboxylic Acid (TCA) Cycle. ~36 molecules of ATP formed for every molecule of glucose metabolized aerobically

aerobically at 15 °C has only 15% of the original glycogen concentration remaining. Glycogen reductions to this extent will have a profound effect on wort fermentation, whereas, after storage for 44 h at 4 °C, there was 26% of the original glycogen concentration remaining (Stewart 2015a, b) (Fig. 6.21).

6.7 Citric Acid Cycle or Tricarboxylic Acid Cycle

The Citric Acid Cycle—also known as the Tricarboxylic Acid (TCA) Cycle and/or the Krebs cycle (Lowenstein 1969)—is basically a series of chemical reactions used by aerobic organisms (including yeast) to generate energy through the oxidation of acetyl-CoA derived from carbohydrates, fats and proteins (Fig. 6.22) into carbon dioxide and chemical energy. In addition, the cycle provides precursors for certain amino acids as well as the reducing agent NADH, which has already been described in this chapter and which is used in numerous other biochemical reactions. Its central importance to many biochemical pathways suggests that it was one of the earliest established components of cellular metabolism (Lane et al. 2009).

The name of this metabolic pathway is derived from citric acid (a type of tricarboxylic acid) that is consumed and then regenerated by this sequence of reactions that complete the cycle. In addition, the cycle consumes acetate (in the form of acetyl-CoA) and water, reduces NAD^+ to NADH and produces carbon dioxide as a waste product. The NADH generated by the TCA Cycle is fed into the oxidative phosphorylation (electron transport) pathway. The net result of these two

Fig. 6.23 Albert Szent-Györgyi

closely linked pathways is the oxidation of nutrients to form usable chemical energy in the form of ATP.

In eukaryotic cells (including yeast), the TCA Cycle occurs in the matrix of the mitochondrion (details in Chap. 14). In prokaryotic cells, such as bacteria (bacteria lack mitochondria), the TCA reaction sequence is performed in the cytoplasm, with the proton gradient for ATP production being across the plasma membrane rather than the inner membrane of the eukaryotic mitochondrion.

The major components and reactions of the TCA Cycle were established in the 1930s by a number of notable scientists. The premier initial research on this cycle was conducted by Albert Szent-Györgyi and Hans Adolf Krebs. In recognition for their accomplishments, they each received Nobel Prizes (details below).

- *Albert Szent-Györgyi* (Fig. 6.23) was born in Budapest in 1893. He began his studies at the Semmelweis University in 1911 in his uncle's anatomy laboratory. He served as an army medic during World War I and was invalided out of the army and completed his MD in 1917. He began full-time research in Pozsony and ended up at the University of Groningen, where his studies focussed on the chemistry of cellular respiration which led to a position as a Rockefeller Foundation Fellow at Cambridge University. He received his PhD from Fitzwilliam College, Cambridge, in 1927 for work isolating an organic acid which he called "hexuronic acid". In 1937, he received the Nobel Prize in Physiology or Medicine "for his discoveries in connection with the biological combustion process with special reference to vitamin C and the catalysis of

Fig. 6.24 Hans Krebs

fumaric acid". In 1947, Szent-Györgyi established the Institute for Muscle Research at the Marine Biological Laboratory in Woods Hole, Massachusetts. He continued research in Woods Hole for the rest of his life. During the mid-1950s, Szent-Györgyi began using electron microscopy to study muscles at the subunit level. In the late 1950s, he developed a research interest in cancer and applied ideas related to quantum mechanics that were associated with the biochemistry of cancer. Late in his life, Szent-Györgyi began to pursue free radicals as a potential cause of cancer. He came to see cancer as being ultimately an electronic problem at the molecular level. Szent-Györgyi died in Woods Hole, Massachusetts, on October 22, 1986. During over 60 years, his research he proved to be a real polymath with interests in the biological and physical sciences as well as medicine!

- *Hans Krebs* (Fig. 6.24) was a German-born British physician and biochemist (Leigh 2009). Although he is best known for elucidating the urea and citric acid cycles, the latter was of greater importance in the context of this text because it applies to a broad spectrum of biological systems including yeast. He was born in Hildesheim, Germany, in 1900 to Georg Krebs, an ear, nose and throat surgeon, and Alma Krebs. He was the middle of three children. He followed in his father's footsteps and studied medicine and completed his medical degree in 1923 from the Third Medical Clinic in the University of Berlin. He then studied in the Department of Chemistry at the Pathological Institute of the Charité Hospital, Berlin, for training in chemistry and biochemistry and in 1925 earned an MD degree from the University of Hamburg (Quayle 1982). In 1926, he joined Otto Warburg (biography already discussed in this chapter) as a research assistant at the Kaiser Wilhelm Institute for Biology in Berlin-Dahlem. For the next 6 years, he worked on various aspects of urea synthesis. By 1932 he had developed the basic chemical reactions of the urea cycle (Holmes 1980), which established his scientific reputation.

Krebs life as a reputed German scientist came to an abrupt halt because of his Jewish ancestry. He was dismissed from his position at the Kaiser Wilhelm Institute in April 1933, but Frederick Gowland Hopkins at Cambridge University recruited Krebs to work there in the Department of Biochemistry. Before the end of 1933, he had settled in Cambridge with financial support from the Rockefeller Foundation. He was fortunate to be allowed to take his equipment and research samples to Cambridge, as they proved to be pivotal to later discoveries. This was particularly true of Warburg's manometer (Fig. 6.12) for the measurement of oxygen consumption of thin slices of tissue—which was the basis for his research (Wilson and Walker 2010). In 1935, the University of Sheffield offered him the post of Lecturer in Pharmacology and he worked there for 19 years. In 1938, the university established a Department of Biochemistry with Krebs as its first Head, and eventually Professor in 1945. In 1944, the British Medical Research Council established the MRC Unit for Cell Metabolism Research at Sheffield with Krebs as its Director. Krebs moved, with his MRC unit, to the University of Oxford in 1954 as Whitley Professor of Biochemistry, a post he held until his retirement in 1967.

While at the University of Sheffield, Krebs and a research assistant William Johnson investigated cellular separation by which oxygen was consumed to produce energy from glucose breakdown. They investigated possible chemical reactions and came up with numerous hypothetical pathways. Krebs used the manometer to test the hypotheses. After 4 months of experimental work, he established the sequence of the chemical cycle, which is called the "Citric Acid Cycle" also subsequently called the Tricarboxylic Acid Cycle or the Krebs cycle—details later (Krebs and Johnson 1937).

Krebs received a large number of honours and awards. He became a naturalized British subject in 1939. He was elected a Fellow of the Royal Society in 1947 and received the Nobel Prize in Physiology or Medicine for the "discovery of the Citric Acid Cycle" in 1953, which he shared with Fritz Lipmann. He was knighted in 1958. He received honorary doctorates from 21 universities!

The TCA Cycle unifies carbohydrate, fat and protein metabolism. The reactions of the cycle are carried out by eight enzymes that completely oxidize acetate, in the form of acetyl-CoA (Fig. 6.22), into two molecules each of carbon dioxide and water. It has already been discussed in this chapter that the primary source of acetyl-CoA is from the breakdown of sugars by glycolysis, which yields pyruvate that, in turn, is decarboxylated by pyruvate dehydrogenase to acetyl-CoA:

$$\underset{\text{pyruvate}}{CH_3COCOOH} + CoASH + NAD^+ \rightarrow \underset{\text{acetyl-CoA}}{CH_3COSCoA} + NADH + CO_2$$

Acetyl-CoA is the starting point of the TCA Cycle. It may also be obtained from the oxidation of fatty acids. The cycle can be outlined as follows:

- The Citric Acid Cycle begins with the transfer of a two-carbon acetyl group from acetyl-CoA to the four-carbon acceptor compound (oxaloacetate) to form a six-carbon compound (citrate).
- The citrate then goes through a series of chemical transformations, losing two carboxyl groups as CO_2. The carbons lost as CO_2 originate from what was oxaloacetate, not directly from acetyl-CoA. The carbons donated by acetyl-CoA become part of the oxaloacetate carbon backbone after the first turn of the Citric Acid Cycle. Loss of the acetyl-CoA-donated carbons as CO_2 requires several revolutions of the Citric Acid Cycle. However, because of the role of the Citric Acid Cycle in anabolism, they might not be lost, since many TCA Cycle intermediates are also used as precursors for the biosynthesis of other molecules.
- Most of the energy made available by the oxidative steps of the cycle is transferred as energy-rich electrons to NAD^+, forming NADH. For each acetyl group that enters the Citric Acid Cycle, three molecules of NADH are produced.
- Electrons are also transferred to the electron acceptor Q, forming QH_2.
- At the end of each cycle, the four-carbon oxaloacetate has been regenerated, and the cycle continues.

The theoretical maximum yield of ATP molecules through the oxidation of one molecule of glucose in glycolysis, Citric Acid Cycle, and oxidative phosphorylation is 38 (assuming 3 molar equivalents of ATP per equivalent NADH and 2 ATP per $FADH_2$). However, as will be discussed later, this proposed yield of ATP is now thought to be an overstatement! In yeast, and other eukaryotes, two equivalents of NADH are generated in glycolysis, which takes place in the cytoplasm. Transport of these two equivalents into the mitochondria consumes two equivalents of ATP, thus reducing the net production of ATP to 36. Furthermore, inefficiencies in oxidative phosphorylation due to leakage of protons across the mitochondrial membrane and slippage of the ATP synthase/proton pump commonly reduce the ATP yield from NADH and $FADH_2$ to less than the theoretical maximum yield. The observed yields are, therefore, closer to ~2.5 ATP per NADH and ~1.5 ATP per $FADH_2$, further reducing the total net production of ATP to approximately 30. A re-assessment of the total ATP yield with newly revised proton-to-ATP ratios provides an estimate of 29.85 ATP per glucose molecule.

Several catabolic pathways converge on the TCA Cycle. Most of these reactions add intermediates to the TCA Cycle and are therefore known as anaplerotic reactions, from the Greek meaning "to fill up". These increase the amount of acetyl-CoA that the cycle is able to carry, increasing the mitochondrion's capability to carry out respiration if this is otherwise a limiting factor. Processes that remove intermediates from the cycle are termed "cataplerotic" reactions.

The total energy gained from the complete breakdown of one (six-carbon) molecule of glucose by glycolysis, the formation of two acetyl-CoA molecules, their catabolism in the Citric Acid Cycle and oxidative phosphorylation equals about 38 ATP molecules (or less), in yeast and other eukaryotes.

6.8 Conclusions

Ethanol production by brewer's and distiller's yeast strains (*Saccharomyces* spp.) is the major focus of this chapter—not the complete oxidation of glucose (and related sugars) (although it is discussed) under aerobic conditions. This metabolic pathway (the EMP pathway) is, however, an integral part of the TCA Cycle but the energetics are different. The overall EMP pathway in yeast is:

$$\underset{\text{glucose}}{C_6H_{12}O_6} \rightarrow \underset{\text{ethanol}}{2C_2H_5OH} + \underset{\text{carbon dioxide}}{2CO_2}$$

Only two ADP molecules are converted to two ATP molecules, instead of a larger number of ATP molecules being formed for each glucose molecule being completely metabolized under aerobic conditions.

References

Agarwal GP (1990) Glycerol. Adv Biochem Eng Biotechnol 41:95–128

Barnett JA (2000) A history of research on yeasts 2: Louis Pasteur and his contemporaries, 1850–1880. Yeast 16:755–771

Barnett JA (2003) A history of research on yeasts 5: the fermentation pathway. Yeast 20:509–543

Barnett JA, Lichtenthaler FW (2001) A history of research on yeasts 3: Emil Fischer, Eduard Buchner and their contemporaries, 1880–1900. Yeast 18:363–388

Berg JM, Tymoczko JL, Stryer L (2012) Biochemistry, 7th edn. W.H. Freeman, Basingstoke

Bernhauer K (1957) Glycerol. In: Bernhauer K (ed) Ullmanns Encyklopadie der technischen Chemie, vol 8. Urban and Scwarzenberg, München, pp 800–832

Berovic M, Legisa M (2007) Citric acid production. Biotechnol Annu Rev 13:303–343

Bisping B, Rehm HJ (1986) Glycerol production by cells of *Saccharomyces cerevisiae* immobilized in sintered glass. Appl Microbiol Biotechnol 23:174–179

Bisping B, Baumann U, Rehm HJ (1990) Production of glycerol by immobilized *Pichia farinosa.* Appl Microbiol Biotechnol 32:380–386

Brunt RV, Stewart GG (1967) The effect of monofluoroacetic acid upon the glucose metabolism of *Saccharomyces cerevisiae*. Biochem Pharmacol 16:1539–1545

Buchner E (1897a) Alkoholische Ġahrung ohne Hefezellen (Vorĭaufige Mittheilung). Berichte der Deutschen Chemischen Gesellschaft 30:117–124

Buchner E (1897b) Alkoholische Ğarung ohne Hefezellen (Zweite Mittheilung). Berichte der Deutschen Chemischen Gesellschaft 30:1110–1113

Bud R (2007) Penicillin: triumph and tragedy. Oxford University Press, Oxford

Byers N, Moszkowski SAM, (1956) (via 1956 Nature obituary), Irène Joliot-Curie contributions and bibliography. Nature 177:964

Chan LL-Y, Kury A, Wilkinson A, Berkes C, Pirani A (2016) Measuring glycogen, neutral lipid and trehalose contents using fluorescence-based image cytometry. MBAA Tech Quart 53:108–113

Chlup PH, Stewart GG (2011) Centrifuges in brewing. MBAA Tech Quart 48:46–50

Cocking AT, Lilly CM (1919) Improvements in the production of glycerine by fermentation. Br Patent 1919: 164,034

Cori CF (1983) Embden and the glycolytic pathway. Trends Biochem Sci 8:257–259

Cori CF, Cori GT (1929) Glycogen formation in the liver from D- and L-lactic acid. J Biol Chem 81:389–403

Costenoble R, Valadi H, Gustafsson L, Niklasson C, Franzen CJ (2000) Microaerobic glycerol formation in *Saccharomyces cerevisiae*. Yeast 16:1483–1495

Cox M, Nelson DL (2013) Lehninger principles of biochemistry, 6th edn. isbn-13: 978-1464109621

Cronwright GR, Rohwer JM, Prior BA (2002) Metabolic control analysis of glycerol synthesis in *Saccharomyces cerevisiae*. Appl Environ Microbiol 68:4448–4456

D'Amore T, Crumplen R, Stewart GG (1991) The involvement of trehalose in yeast stress tolerance. J Ind Microbiol 7:191–195

Deshpande PS, Santch SN, Arvindekar AU (2011) Study of two pools of glycogen in *Saccharomyces cerevisiae* and their role in fermentation performance. J Inst Brew 117:113–119

Enjalbert B, Passou JL, Vincent O, Francois J (2000) Mitochondrial respiratory mutants of *Saccharomyces cerevisiae* accumulate glycogen and readily mobilize it in a glucose-depleted medium. Microbiology 146:2685–2694

Freeman GG, Donald GMS (1957) Fermentation processes leading to glycerol. The influence of certain variables on glycerol formation in the presence of sulfites. Appl Microbiol 5:197–210

Gottschalk A (1956) Prof. Carl Neuberg. Nature 178(4536):722–723

Hecker D, Bisping B, Rehm H-J (1990) Continuous glycerol production by the sulphite process with immobilized cells of *Saccharomyces cerevisiae*. Appl Microbiol Biotechnol 32:627–632

Holmes FL (1980) Hans Krebs and the discovery of the ornithine cycle. Fed Proc 39:216–225

Hopkins FG, Martin CJ (1942) Arthur Harden. Obit Not Fellow R Soc 4:3–14

Ihde D (1971) Hermeneutic phenomenology: the philosophy of Paul Ricoeur. Northwestern University Press, 198p

Kalle GP, Naik SC (1985) Continuous fed-batch vacuum fermentation system for glycerol from molasses by the sulfite process. J Ferment Technol 63:411–414

Keller MA, Turchyn AV, Ralser M (2014) Non-enzymatic glycolysis and pentose phosphate pathway-like reactions in a plausible Archean ocean. Mol Syst Biol 10:725

Korzybski TW (1974) Parnas, Jakub Karol. In: Gillispie CC (ed) Dictionary of scientific biography, vol 10. Scribner, New York, pp 326–327

Krebs HA, Johnson WA (1937) Metabolism of ketonic acids in animal tissues. Biochem J 31:645–660

Kruger NJ, von Schaewen A (2003) The oxidative pentose phosphate pathway: structure and organisation. Curr Opin Plant Biol 6:236–246

Kutyna DR, Varela C, Stanley GA, Borneman AR, Henschke PA, Chambers PJ (2012) Adaptive evolution of *Saccharomyces cerevisiae* to generate strains with enhanced glycerol production. Appl Microbiol Biotechnol 93:1175–1184

Lane AN, Fan TW-M, Higashi RM (2009) Metabolic acidosis and the importance of balanced equations. Metabolomics 5:163–165

Leigh FW (2009) Sir Hans Adolf Krebs (1900-81), pioneer of modern medicine, architect of intermediary metabolism. J Med Biog 17:149–154

Lowenstein JM (1969) Methods in enzymology, vol 13: Citric acid cycle. Academic Press, Boston

McIntyre N (2007) Sir Alexander Fleming. J Med Biogr 15:234

Meyerhof O (1930) The chemistry of muscular contraction. Lancet 219: vol. II:1415–1422

Nachmansohn D (1972) Biochemistry as part of my life. Annu Rev Biochem 41:1–28

Neuberg C (1918) Neuberg C Uberführung der Fructose-diphosphoršaure in Fructose-monophosphoršaure. Biochemische Zeitschrift 88:432–436

Nickerson WJ, Carroll WR (1945) On the metabolism of *Zygosaccharomyces*. Arch Biochem 7:257–271

Nord FF (1958) Carl Neuberg, 1877–1956. Adv Carbohydr Chem 13:1–7

Nordstrom K (1966) Yeast growth and glycerol formation, carbon and redox balances. Acta Chem Scand 20:6–15

Pasteur L (1858) On the formation of glycerol during alcoholic fermentation. Journal fur Praktishe Chemie.73:506 Les Comptes rendus de l'Academie des Sciences XLVI(18) 857

Peters RA (1952) Croonian Lecture: Lethal synthesis. Proc R Soc (Lond) B 139:143–170

Petrovska B, Winkelhausen E, Kuzmanova S (1999) Glycerol production by yeasts under osmotic and sulfite stress. Can J Microbiol 45:695–699

Prescott SC, Dunn CG (eds) (1959) Industrial microbiology. McGraw Hill, New York

Pribylova L, Straub ML, Sychrova H, de Montigny J (2007) Characterisation of *Zygosaccharomyces rouxii* centromeres and construction of first *Z. rouxii* centromeric vectors. Chromosome Res 15:439–445

Prior BA, Baccari C, Mortimer RK (1999) Selective breeding of *Saccharomyces cerevisiae* to increase glycerol levels in wine. Int Sci Vigne Vin 33:57–65

Quain DE, Tubb RS (1982) The importance of glycogen on brewing yeasts. MBAA Tech Quart 19:29–33

Quain DE, Thurston PA, Tubb RS (1981) The structural and storage carbohydrates of *Saccharomyces cerevisiae* changes during fermentation of wort and a role for glycogen metabolism in lipid synthesis. J Inst Brew 87:108–111

Quayle JR (1982) Obituary. Sir Hans Krebs, 1900-1981. J Gen Microbiol 128:2215–2220

Randle P (1986) Carl Ferdinand Cori. Biogr Mem Fellows R Soc 32:67–95

Reid RW (1978) Marie Curie. New American Library, New York

Rothman LB, Cabib E (1969) Regulation of glycogen synthesis in the intact yeast cell. Biochemistry 8:3332–3341

Spencer JFT, Sallons HR (1956) Production of polyhydric alcohols by osmophilic yeasts. Can J Microbiol 2:72–79

Stewart GG (1968) The influence of fluoroacetic acid on the control of yeast metabolism. PhD thesis, Bath University

Stewart GG (2014) Brewing intensification. American Society for Brewing Chemists, St. Paul, MN

Stewart GG (2015a) Seduced by yeast. J Am Soc Brew Chem 73:1–21

Stewart GG (2015b) Yeast quality assessment, management and culture maintenance. In: Hill AE (ed) Brewing microbiology: managing microbes, ensuring quality and valorising waste. Elsevier, Oxford, UK, pp 11–29

Stewart GG, Brunt RV (1968) The effect of monofluoroacetic acid upon the carbohydrate metabolism of *Saccharomyces cerevisiae*. Biochem Pharmacol 17:2349–2354

Stewart GG, Abbs ET, Roberts DJ (1969) Biochemical effects of fluoroacetate administration in rat brain, heart and blood. Biochem Pharmacol 19:1861–1866

van Dijken JP, Scheffers WA (1986) Redox balances in the metabolism of sugars by yeasts. FEMS Microbiol Rev 32:199–224

Vijaikishore P, Karanth NG (1987) Glycerol production by fermentation: a fed-batch approach. Biotechnol Bioeng 30:153–333

Voet D, Voet JG (2011) Biochemistry, 4th edn. New York: Wiley

Warburg O (1926) Über den Stoffwechsel der Tumoren. Springer, Berlin

Wilson K, Walker J (2010) Principles and techniques of biochemistry and molecular biology, 7th edn. Cambridge University Press, Cambridge, UK

Chapter 7
Yeast Nutrition

7.1 Background

The major raw materials employed for fermentation by yeast in the production of most beers and many potable and industrial spirits are water, barley, wheat, corn (maize), rice, sorghum, oats, sugar and its derivatives (from cane and beet) and hops for beer (Roberts and Wilson 2006). Also, barley, wheat and sorghum are usually malted (germinated and normally kiln-dried). Raw materials and the process used to produce industrial (fuel) alcohol by fermentation are discussed in Chap. 9. With the exception of sucrose (a glucose-fructose disaccharide), a sugar obtained from either cane or beet, various forms of starch are the principal raw materials, and these have to be hydrolysed (usually by a spectrum of amylases—Fig. 7.1) to fermentable sugars. In this context, the predominant fermentable sugars are glucose, fructose, maltose and maltotriose (Fig. 7.5) (Stewart 2006) together with minority sugars such as melibiose, raffinose and cellobiose (Fig. 7.17) that can be metabolized to ethanol by specialized yeast species (details later).

Although this publication endeavours to focus on brewer's and distiller's yeast fermentation, the formation mechanisms whereby starch and sucrose (and lactose) are enzymatically hydrolysed to fermentable sugars cannot be overlooked. The source of starch hydrolytic enzymes (amylases) is usually malted (germinated) barley, wheat or sorghum. Alternatively, amylases can be exogenously added to the starch. These enzymes are usually of bacterial and/or fungal origin (Stewart 1987). This use of exogenous enzymes is prohibited in the production of Scotch whisky for both malt and grain spirit (Scotch Whisky Regulations 2009; Gray 2013; Murray 2014) and for some other whiskies, for example, Japanese whisky (Fukuyo and Myojo 2014). However, exogenous amylases are permitted in the mashing of unmalted barley in the production of Irish whiskey (Irish Whiskey Act—1980) (Quinn 2014) and Canadian rye whisky (Canadian Federal Food and Drugs Act—1993) (Lyons 2014).

G.G. Stewart, *Brewing and Distilling Yeasts*, The Yeast Handbook,
https://doi.org/10.1007/978-3-319-69126-8_7

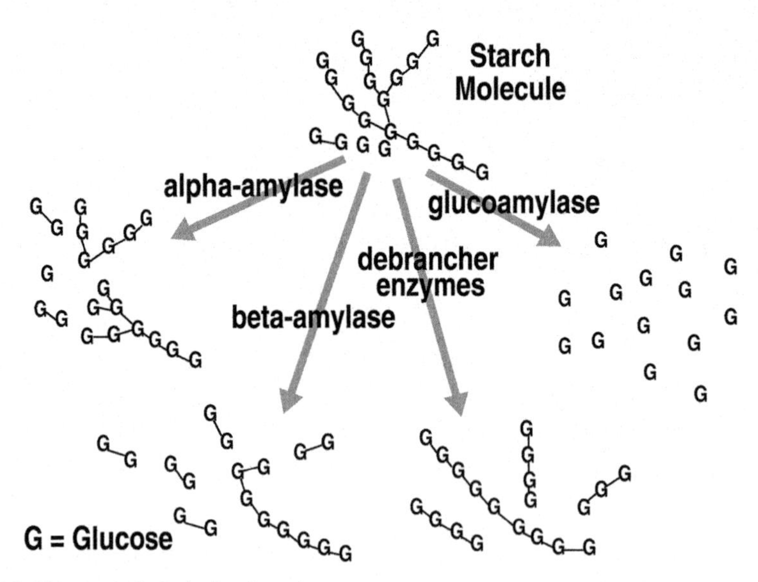

Fig. 7.1 Enzymatic hydrolysis of starch

In brewing, the major raw materials employed are both malted and unmalted cereals together with a number of flexible fermentable materials (sugars and syrups). However, there are raw materials that are essential in conventional brewing procedures. Indeed, in some countries, their use is dictated by appropriate brewing production regulations such as the German Purity Law (Reinheitsgebot of 1516), which only permits malted cereals, water and hops—use of yeast was approved much later (Narziss 1984; Winkelmann 2016). Unmalted cereals (called adjuncts) are not within this category as they are not essential for beer production (Stewart 2016). However, most (but not all) countries employ adjuncts in the brewing of some beers (Hertrich 2013). Adjuncts include corn (maize), rice, sorghum, cassava, barley, wheat, sugars and syrups (hydrolysed starch and sucrose).

The unfermented liquid medium used for fermentation, prepared from cereals (malted and unmalted) in both brewing and whisk(e)y (and other spirit production), is termed wort (named after Wort, a town located in Baden Wüttenderg, and sometimes for whisky-wash). The production procedures for both types of product are basically similar (Kapral 2008). However, hops are used in brewing, and the wort is boiled, whereas it is not boiled in spirit production procedures and consequently is not sterile and contains active enzymes and microbial contaminants. This latter production difference will influence the properties and composition of the two types of wort, particularly because the unboiled wort will still possess amylase and proteinase (and other enzymes) activity (wort sugar and amino acid spectrum will be discussed later). Also, as it is not sterile, it will contain a number of bacteria and wild yeasts. Whereas, boiled brewer's wort is sterile.

Compared to other unfermented media [e.g. grape must, mash (an unfiltered, unboiled cereal fermentation medium), various fermentable syrups, etc.] used in the production of fermentation alcohol (both industrial and potable), wort is by far the most complex! Therefore, when yeast is pitched (inoculated) into wort, it is introduced into a very complex environment that consists of simple sugars, dextrins, amino acids, peptides, proteins, vitamins, ions, nucleic acids and other constituents too numerous to discuss (Stewart and Russell 2009). One of the major advances in brewing science and, to a lesser extent, distilling science, during the past 40 years or so, has been the elucidation of the mechanisms by which the yeast cell utilizes, in an orderly manner, the above plethora of wort nutrients. Wort sugars (Stewart 2006) and amino acids (Jones and Pierce 1964; Lekkas et al. 2007; Stewart et al. 2013) (details later) are removed in a distinct order (or priority) from the wort at various points during the fermentation cycle.

7.2 Wort Sugar Uptake

Wort contains the sugars: sucrose, fructose, glucose, maltose and maltotriose, together with dextrin material that is usually (not always) non-fermentable. A typical percentage sugar spectrum of brewer's wort is shown in Table 7.1. Under normal conditions, brewing yeasts are capable of utilizing sucrose, glucose, fructose, maltose and maltotriose (and others—see below) in this approximate sequence (or priority). Some degree of overlap does occur, however, leaving maltotetraose (G4) and the larger dextrins unfermented (Fig. 7.2). Because the wort (or wash), used for distilling, is unsterile and contains active amylases, the sugar uptake spectrum during a spirit fermentation is rather different—this will be discussed later in this chapter.

Saccharomyces cerevisiae strains, including ale brewing and whisky distilling strains, have the ability to take up and ferment a wide range of sugars, for example, sucrose, glucose, fructose, galactose, mannose, trehalose, maltose, maltotriose and raffinose (in part). Also, *S. pastorianus* (lager yeast—details in Chap. 3) is able to utilize the disaccharide melibiose (Fig. 7.3). In addition, *S. diastaticus* (a subspecies of *S. cerevisiae*) is also able to utilize dextrins (partially hydrolysed starch material) because it produces glucoamylase as an extracellular enzyme (Erratt and Stewart 1978). Recently, the characterization of a novel species of *Saccharomyces* has been described (*Saccharomyces jurei*). This species exhibits efficient maltose fermentation at cold temperatures (>10 °C) (Naseeb et al. 2017). These studies have been discussed further in Chap. 3.

The initial step in the utilization of a sugar by yeast (or any eukaryotic cell) is usually either its passage intact across the cell (plasma) membrane (details of membrane structure in Chap. 5) or its hydrolysis outside the cell membrane, followed by entry into the cell, of some or all, of its hydrolysis products (Fig. 7.4). Maltose and maltotriose (Fig. 7.5) (together with glucose and fructose) are examples of sugars that pass intact across the cell membrane. Sucrose,

Table 7.1 Typical wort sugar spectrum

	Percentage composition
Glucose	10–15
Fructose	1–2
Sucrose	1–2
Maltose	50–60
Maltotriose	15–20
Dextrins	10–20

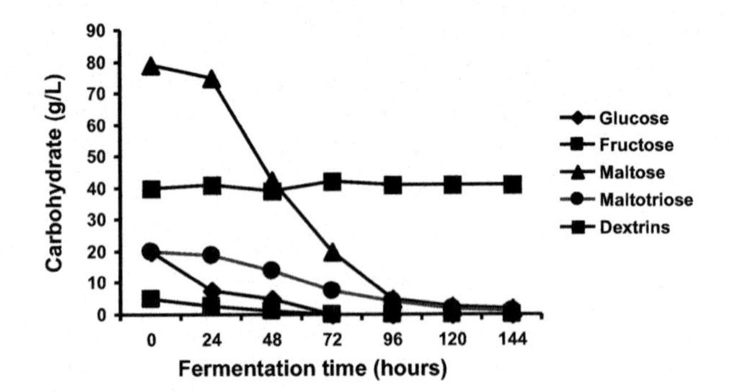

Fig. 7.2 Order of the uptake of wort sugars by brewer's yeast cultures

melibiose (*S. pastorianus*) and dextrins (*S. diastaticus*) are hydrolysed by the periplasmic/extracellular enzymes [invertase (β-fructosidase), melibiase (α-galactosidase) and glucoamylase (amyloglucosidase) for dextrins] which are located in the plasma layer, and the hydrolysis products are taken into the cell. An important metabolic difference between the uptake of glucose and fructose compared to the uptake of maltose and maltotriose is that metabolic energy (ATP) is required for maltose and maltotriose (Fig. 7.6) uptake (active transport), whereas glucose and fructose are taken up passively (minimal energy is required) (Bisson et al. 1993). The fermentation implications of the metabolic difference between the uptake of glucose and fructose compared to maltose and maltotriose will be discussed later in this chapter.

It has already been discussed that maltose and maltotriose are the major sugars in most brewer's worts, spirit mash and wheat dough (Stewart and Russell 2009), and, as a consequence, for brewing, distilling and baking, the yeast's ability to use these two sugars is vital and depends upon the correct genetic complement, which is a diverse and complex system (details later). Competition occurs between the transporters for maltose and maltotriose, with maltose being the preferred substrate (Fig. 7.7). This results in maltose being used by a yeast culture first, because sugar transport is the rate-limiting step (Day et al. 2002; Han et al. 1995).

Maltose fermentation by brewing, distilling and baking yeast strains requires at least one of five unlinked polymeric (*MAL*) loci located in the telomeric regions (a region of repetitive nucleotide sequences at the end of each chromosome) of the different chromosomes (*MAL1-MAL4* and *MAL6*) (Table 7.2). The genes for

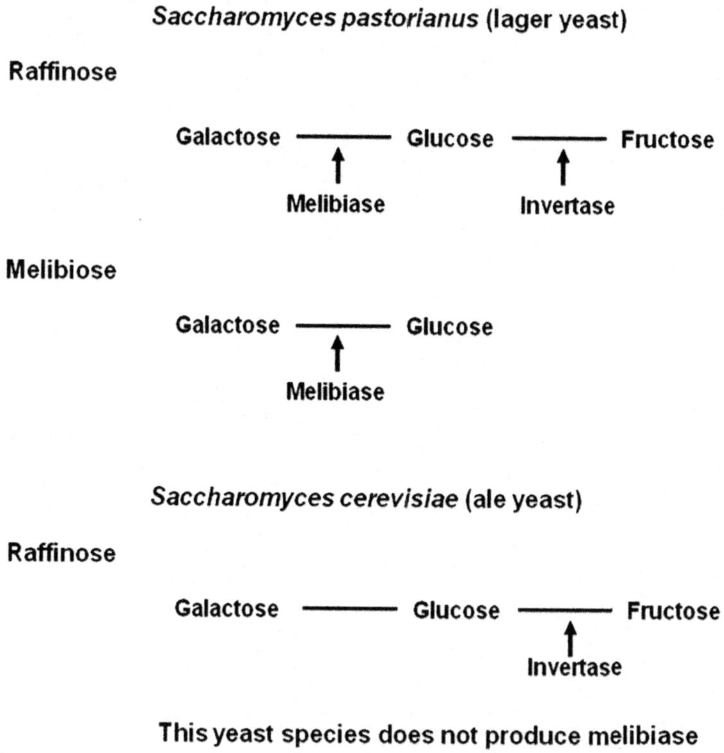

Saccharomyces pastorianus (lager yeast)

Raffinose

Galactose — Glucose — Fructose
Melibiase Invertase

Melibiose

Galactose — Glucose
Melibiase

Saccharomyces cerevisiae (ale yeast)

Raffinose

Galactose — Glucose — Fructose
Invertase

This yeast species does not produce melibiase

Fig. 7.3 Utilization of the sugars raffinose and melibiose by lager (*Saccharomyces pastorianus*) and ale (*Saccharomyces cerevisiae*) yeast strains—ale yeast strains do not possess melibiase activity

maltose and maltotriose metabolism are located in the *MAL* loci (Kodama et al. 1995; Salema-Oom et al. 2005). Each *MAL* locus consists of three genes encoding: (1) the structural gene for α-glucosidase (maltase), (2) maltose permease and (3) an activator protein needed for regulation of the expression of both the α-glucosidase and the permease genes.

The expression of the maltose and the maltose transporter is also regulated by maltose induction and glucose repression. When glucose concentrations are high (usually greater than 10 g/L), the *MAL* genes are repressed, and only when 40–50% of the glucose has been taken up from the wort will the uptake of maltose and maltotriose commence (Fig. 7.2). The question of glucose repression of the uptake of maltose and maltotriose will be discussed in detail later (Adams 2004).

Brewer's yeast strains possess independent uptake mechanisms (maltose and maltotriose permeases) in order to transport the two sugars across the plasma membrane into the cell. Four different types of transporters have been identified, *Malx1*, *Mtt1*, *Mphx* and *Agt1*, and they differ in their distribution and in their substrate range (Alves et al. 2008; Day et al. 2002; Han et al. 1995; Jespersen

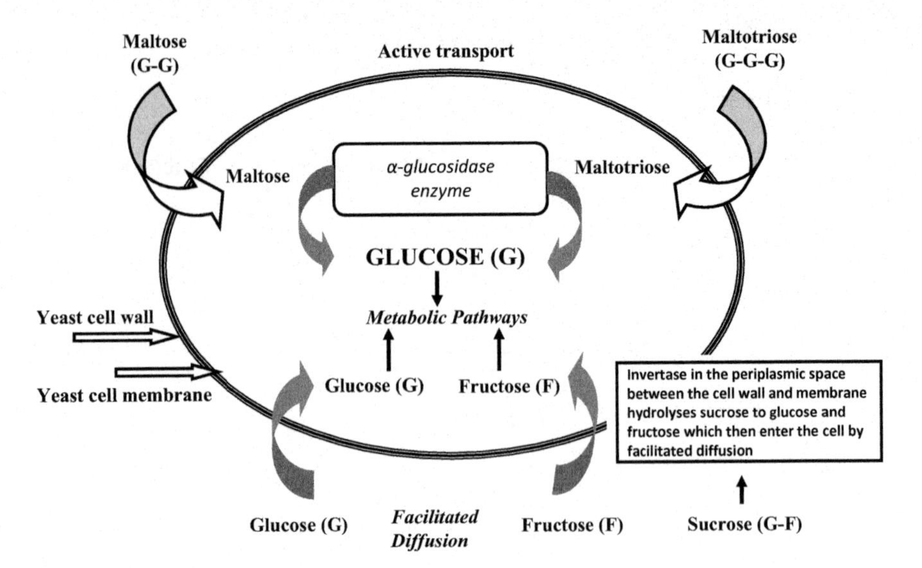

Fig. 7.4 Uptake of wort sugars by a yeast culture

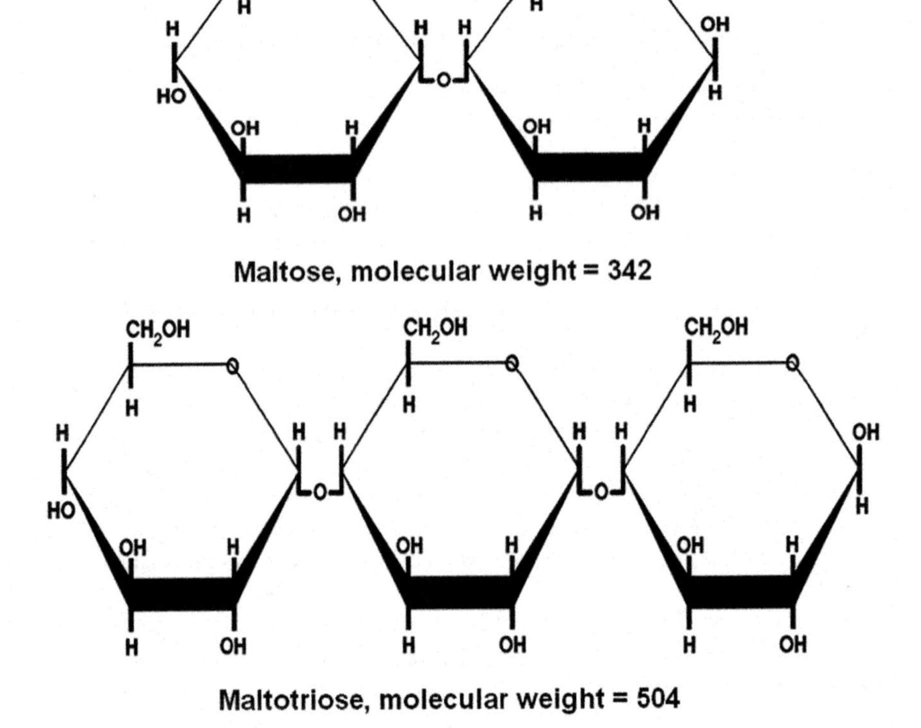

Fig. 7.5 Structure of maltose and maltotriose

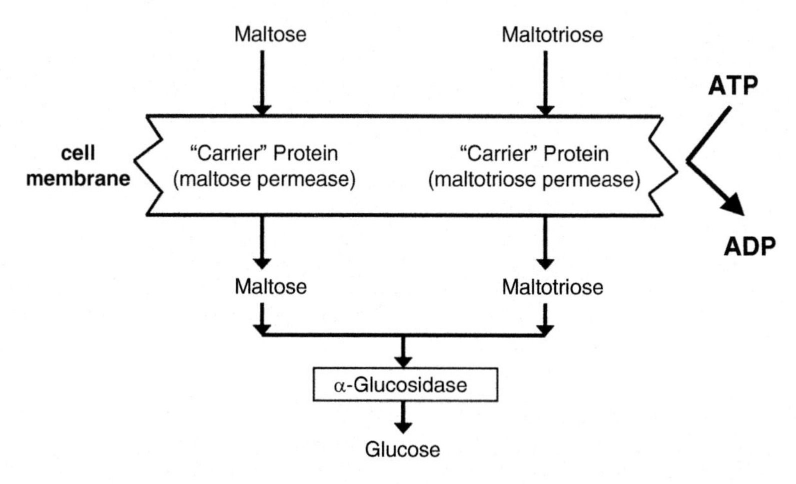

Fig. 7.6 Uptake and metabolism of maltose and maltotriose by yeast cells

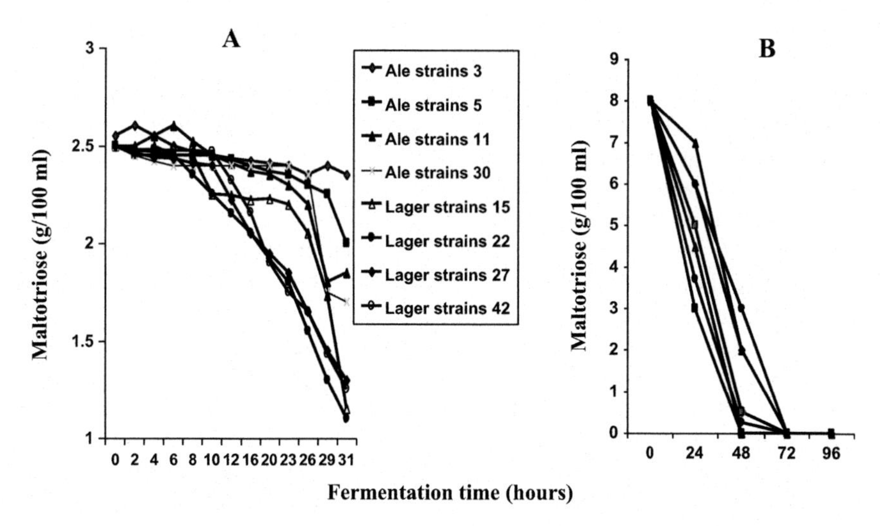

Fig. 7.7 Maltotriose (**a**) and maltose (**b**) uptake profiles from a 16°Plato wort by ale and lager yeast strains

et al. 1999; Kodama et al. 1995; Salema-Oom et al. 2005). Almost all brewing strains of the *S. cerevisiae* species examined contained the *AGT1* gene and an increased copy number of *MALx1* permeases. The *MAL* transporter genes are generally regarded as specific for maltose (Kodama et al. 1995), but activity towards the uptake of maltotriose has also been claimed (Stewart et al. 1995). The *Agt1* transporter can transport trehalose, turanose, α-methylglucoside and sucrose, in addition to maltose and maltotriose (Han et al. 1995). Once inside the cell, all of these sugars are hydrolysed to glucose units by the α-glucosidase system.

Table 7.2 Characteristics of maltose (*MAL*) gene loci in breweries and distiller's yeast strains[a]

Loci	Chromosome location
MAL1	VII
MAL2	III
MAL3	VIII
MAL4	XI
MAL6	II/VIII

Three genes: (1) maltose permease, (2), α-glucosidase, (3) MAL activator

[a]Oda Y and Tonomura K (1996) Detection of maltose fermentation genes in the baking yeast strains. Lett. Appl. Microbiol. 23:266–268

This glucose is then metabolized through the glycolytic pathway to pyruvate and, subsequently, acetaldehyde and ethanol (details in Chap. 6).

Maltose transport by ale and lager yeast strains has been compared and contrasted in detail. *S. pastorianus* (lager) yeast strains utilize maltotriose more efficiently than *S. cerevisiae* strains (Zheng et al. 1994a, b; Crumplen et al. 1996), and it has been found that glucose inhibited maltose transport more strongly with ale strains than with lager strains. Rautio and Londesborough (2003) found that trehalose and sucrose strongly inhibited maltose transport with an ale strain but only weakly inhibited maltose transport with a lager yeast strain. Taken together, these results suggest that the dominant maltose transporters of the ale strains studied had broader specificity than those of lager strains and were probably *Agt1* proteins.

Hybridization studies have shown that all of the ale and lager strains studied contained *AGT1* and several *MAL* genes (Day et al. 2002). This discrepancy has been resolved, in part, by the finding that the *AGT1* genes contained stop codons rendering these genes defective (Vidgren et al. 2007). The same defect has been identified in other lager stains but not in ale yeast strains (Vidgren et al. 2009). Glucose was coded to repress expression of the *Agt1* gene in ale strains (Vidgren et al. 2013). Gibson et al. (2013) have also examined transcription of the α-glucoside transport and metabolism genes in *S. pastorianus*.

A strong temperature dependence for maltose transport has been found for ale yeasts and a markedly smaller temperature dependence for the transport of this sugar observed with lager strains (Vidgren et al. 2009). The more rapid wort fermentation of lager yeast strains at low temperatures may be the result of different maltose transporters, but this possibility is, as yet, unproven. This may also explain why lager fermentations are conducted at lower temperatures (8–18 °C) than ale fermentations (21–26 °C) (Boulton and Quain 2001). This temperature effect for maltose and maltotriose uptake with ale and lager yeast strains has already been discussed in this chapter and in Chap. 3. It has also been reported that in the ale and lager strains studied, the strains employed different types of maltose and maltotriose transporters (Vidgren et al. 2009; Cousseau et al. 2012). In the ale strains, the *Agt1* transporter was dominant, whereas in the lager strains the *Malx1* and *Mtt1* type was found to be dominant.

A number of ale and lager yeast strains have been employed in order to explore further the mechanisms of both maltose and maltotriose uptake from wort. A 16°Plato all-malt wort was fermented in a 30 L static fermentation vessel. Under these conditions, lager strains utilized maltotriose more efficiently than ale strains (Fig. 7.7), whereas maltose utilization efficiency was not dependent on the type of brewing strain employed (Fig. 7.7) (Zheng et al. 1994a). This supports the proposal that maltotriose and maltose possess independent, but closely linked, uptake (permease) systems (Zheng et al. 1994b). In addition, this consistent difference observed, with a number of ale and lager strains, supports the observation (Zheng et al. 1994a) that ale strains appear to have greater difficulty completely fermenting wort (particularly high-gravity wort) than lager strains (Stewart 2009) because of impediments with the uptake of wort maltotriose.

As a consequence of these latter results, the possible contribution of the different maltotriose transporters present in the genome of lager strains explains (at least in part) the superior overall wort fermentation performance of this hybrid yeast. The results indicate that both the *MTY1* and *AGT1* genes from the *S. eubayanus* subgenome (part of the lager yeasts—*S. pastorianus*) encode for functional maltotriose transporters that permit the fermentation of this sugar by the yeast, despite the apparent differences in the kinetics of maltotriose-H^+ symport activity. The presence of two maltotriose transporters in the *S. eubayanus* subgenome not only highlights the importance of sugar transport for efficient maltotriose utilization by industrial yeasts (brewing, distilling, etc.), but these genomes can be used in breeding and/or selection programs aimed at increasing yeast fitness for efficient fermentation of brewer's wort (particularly high-gravity wort—details later in Chap. 11) (Cousseau et al. 2012). Indeed, it would be interesting to characterize *AGT1*- and *MALx1*-related genes found in other species of the *Saccharomyces* genus, for example, those found in the genomes of *S. paradoxus* and *S. mikatae* species.

The *MAL* gene cassettes have been investigated further. Using hybridization techniques (Sherman 1991), a strain with two *MAL2* and two *MAL4* gene copies was constructed (Stewart 1981). The wort fermentation rate was compared with a strain containing only one copy of *MAL2*. As expected, the overall wort fermentation rate with the strain containing multiple *MAL* genes was considerably faster than the strain containing a single copy of *MAL2* (Fig. 7.8). The principal reason for this faster fermentation rate was due to an increased rate of maltose uptake and subsequent metabolism compared with the yeast strain containing the single *MAL2* copy.

Strains of *S. pastorianus* usually ferment maltose and maltotriose rapidly due to efficient transport of these α-glucosides into the cell. It is thought that this is the rate-limiting step in the fermentation of these sugars from wort. A novel maltotriose transporter gene in the *S. eubayanus* subgenome of a production lager brewing strain (Weihenstephan 34/70) has been reported that permits efficient maltotriose fermentation by yeast cells. The characterization of maltotriose transporters from yeast being used for the production of some lager beers opens new opportunities to enhance yeast fitness for the fermentation of brewer's wort, where the historical

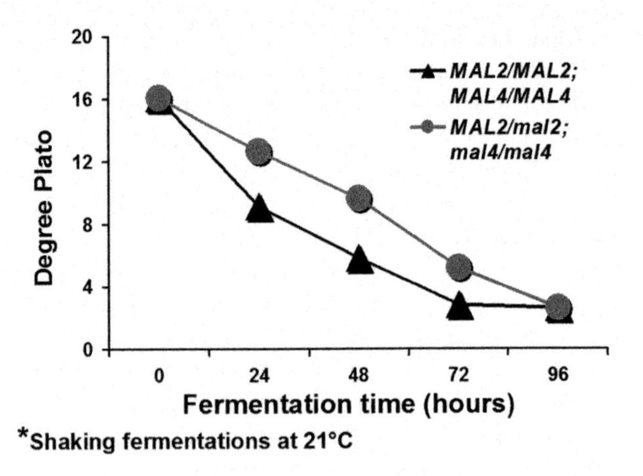

Fig. 7.8 Wort fermentation of sugars with multiple *MAL* gene copies

strain currently exhibits incomplete or sluggish maltotriose fermentation (Cousseau et al. 2012). The yeast's growth and subsequent fermentation of high-gravity wort are particularly relevant in this regard (Stewart 2014).

It has already been briefly discussed that maltose transport capacity is regulated by the glucose repression and maltose induction of the genes encoding maltose transporters (Ernandes et al. 1992, 1993). There is also glucose-triggered catabolite inactivation of the existing maltose transporter proteins. In addition, there is a third factor, namely, the cellular lipid composition, in particular the sterol content of the plasma membrane (Guimarães et al. 2006). What happens to maltose transport capacity when cropped brewer's yeast is pitched into an aerated wort will depend upon the interaction of the above three factors, as well as the particular yeast strain employed and the wort sugar composition and gravity. The maltose transport capacity of cropped yeast is relatively low, partly because of a sterol deficiency. The sterol and unsaturated fatty acid content of the culture will rise rapidly during the first few hours after pitching (Murray et al. 1984) (Fig. 6.19). The presence of oxygen is critical for the synthesis of both sterols and unsaturated fatty acids (Kirsop 1974). In wort containing high concentrations of glucose, maltose transport capacity falls during the first 20 h after pitching, probably as a consequence of glucose-induced catabolite inactivation (Rautio and Londesborough 2003). Expression of the maltose transporter genes is also activated during the first 24 h of fermentation (James et al. 2003), and maltose transport activity recovers and reaches a peak at approximately the same time as the concentration of yeast in suspension achieves its maximum (Stewart 2006). At this stage in a fermentation cycle, maltose transporter genes are well expressed, glucose has been partially or wholly consumed (bringing an end to catabolite inactivation) and the plasma membrane contains adequate amounts of sterols and unsaturated fatty acids (UFA). Thereafter during the fermentation cycle, the specific activity of maltose transport declines steadily to the low value found in cropped (harvested) yeast. During the second half of fermentation, the remaining maltose and maltotriose must be transported into the yeast cell by transporters that are operating suboptimally because the plasma membrane does not contain sufficient sterols. In particular, the daughter cells formed in the final cell division will probably be deficient in sterols because, although their membranes may have received limited sterols from the mother cells, any size increase after separation of the daughter bud occurs in an anaerobic environment, where no new sterol can be synthesized due to oxygen deficiency (Kirsop 1974; Murray et al. 1984).

The uptake of wort sugars is a very important consideration regarding the overall metabolism of wort as a whole. A particularly critical aspect is during high-gravity brewing and distilling, which will be considered in Chap. 11. The repressing effects of glucose on the uptake of maltose and maltotriose during wort fermentation have already been discussed in this chapter but not in detail. This repression is a major factor that influences wort fermentation and the overall rate and extent. The uptake of maltose only commences when approximately 50% (this is both yeast strain and wort composition dependent) (Stewart et al. 1988) of the wort glucose (and fructose) has been taken up by the yeast cells (Fig. 7.2). In other words, in most strains of *S. cerevisiae* and *S. pastorianus*, maltose utilization is subject to control by carbon catabolite repression (Russell et al. 1986). In a similar manner, the presence of glucose represses the production of glucoamylase by *S. diastaticus*, thereby inhibiting the hydrolysis of wort dextrins and starch (Erratt and Stewart 1978). Repression of this nature has a negative effect on the overall fermentation rate (Fig. 7.9).

Studies have been conducted in which glucose was added to fermenting wort when the yeast culture was actively metabolizing maltose. The added glucose caused inhibition (repression) of maltose uptake. Once this added glucose had been taken up by the yeast culture, the metabolism of maltose recommenced

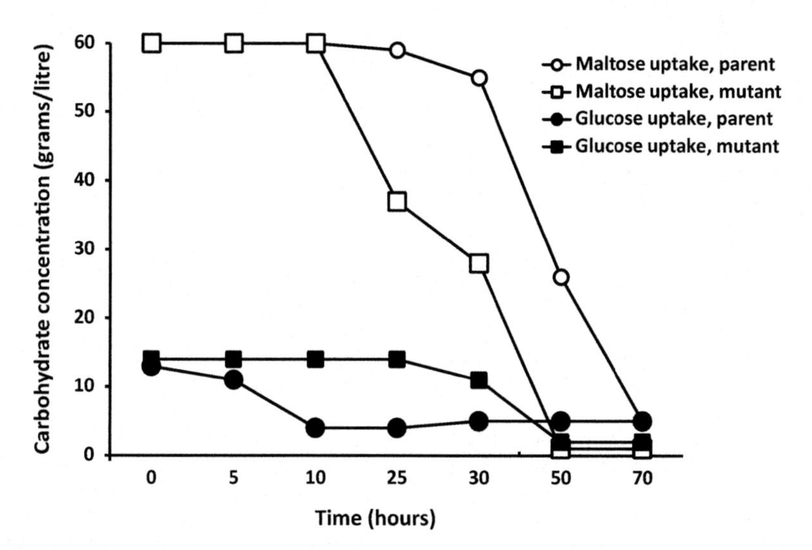

Fig. 7.9 Fermentation of a 12°Plato wort by a *Saccharomyces diastaticus* haploid strain 1479 and its derepressed dextrin mutant

(D'Amore et al. 1989; Ernandes et al. 1992, 1993) (Fig. 7.10). Batistote et al. (2006) have shown that modified patterns of maltose and glucose utilization by brewing and wine yeasts inducing strong effects on biomass production were influenced by the structural complexity of the nitrogen source employed. The results suggest that

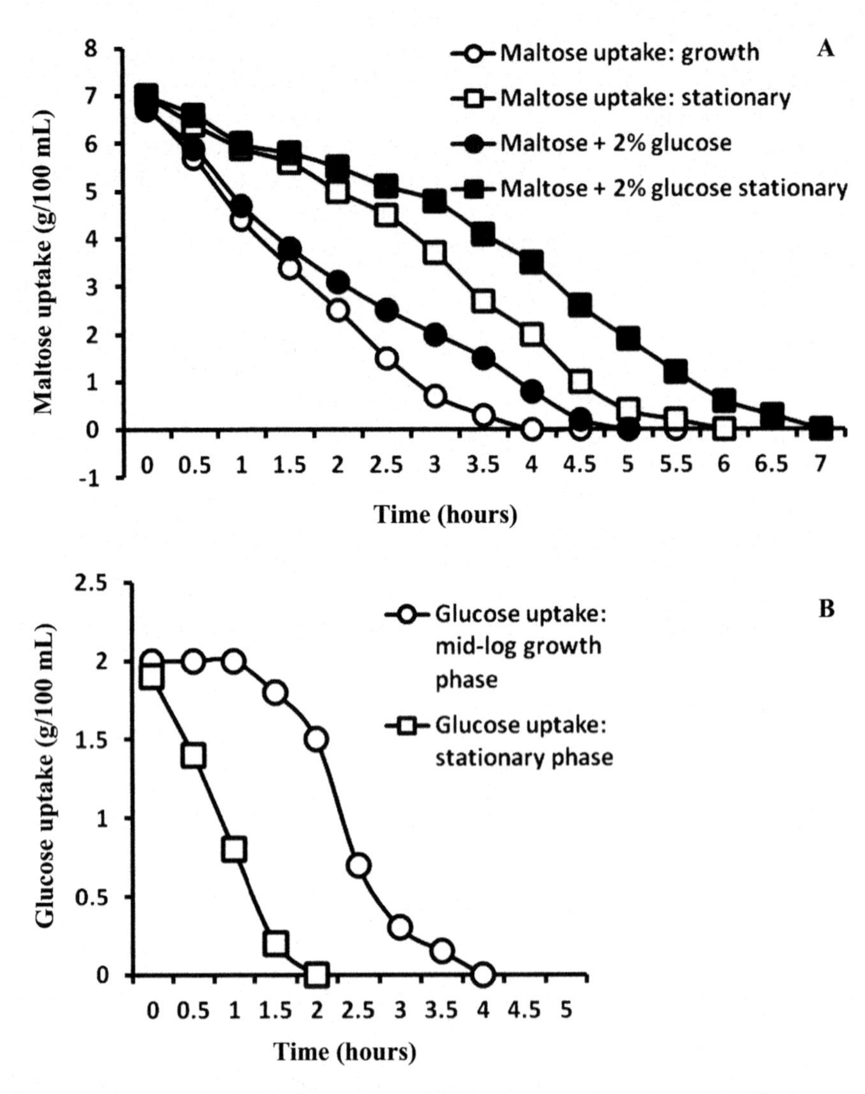

Fig. 7.10 Sugar uptake during fermentation of 7% maltose and 7% maltose plus 2% glucose in PY medium by maltose pre-grown strain 3021, at 30 °C, and harvested at the mid-log [circles (open, filled)] and stationary [(open square, filled square)] growth phases. Maltose uptake: 7% maltose (open circle), (open square); 7% maltose plus 2% glucose (filled circle), (filled square). Glucose uptake: mid-log (open circle) and stationary (open square) growth phases. Fermentation at 21 °C, pH 5.4, inoculum 20 mg cells/mL. Each data point represents the average of three separate experiments

with a medium containing a complex nitrogen source such as peptone, *Saccharomyces* is not subject to the same control mechanisms as those involved in the utilization of a simpler nitrogen source such as Difco Yeast Nitrogen Base (Wickerham 1971) and that there is a mutual interaction between the carbon and nitrogen sources, including the mechanisms involved in sugar and nitrogen catabolite repression (Cruz et al. 2003). These factors may play an important role in the induction/repression processes for nitrogen and sugar utilization in yeasts. Further details of nitrogen and sugar utilization in both brewer's and distiller's yeast are discussed later in this chapter.

In an attempt to overcome this repression, the glucose analogue, 2-deoxy-glucose (2-DOG) (Fig. 7.11), has been successfully employed for the selective isolation of spontaneous mutants of yeasts (Jones et al. 1986) and other fungi. These 2-DOG mutants were derepressed for the production of carbohydrate-hydrolysing enzymes when employing this non-metabolizable glucose analogue. Derepressed mutants of brewing and other industrial yeast strains have been isolated that were able to metabolize wort maltose and maltotriose in the presence of glucose (Fig. 7.15) (Novak et al. 1990a, b) (Fig. 7.12). Fermentation and ethanol formation rates in 12°Plato wort were also shown to have been increased in the 2-DOG mutants when compared to the parental strains (Novak et al. 1991). In addition, 2-DOG starch mutants of *S. diastaticus* have been isolated that exhibited increased fermentation ability in brewer's and distiller's wort, cassava and corn mash (Jones et al. 1986)—further details regarding *S. diastaticus* and other diastatic yeasts will be discussed later here and in Chap. 17 (Whitney et al. 1985).

All of the above studies with 2-DOG mutants were conducted with *S. cerevisiae* ale and with distilling yeast strains (Fig. 7.13). Our research group was unable to isolate any 2-DOG mutants from the range of *S. pastorianus* lager strains screened. This is the major reason why large-scale lager trials were not conducted with 2-DOG mutants in Canada that, by the late 1980s, was predominantly a lager beer-producing country. Since these studies, a major Spanish brewing company,

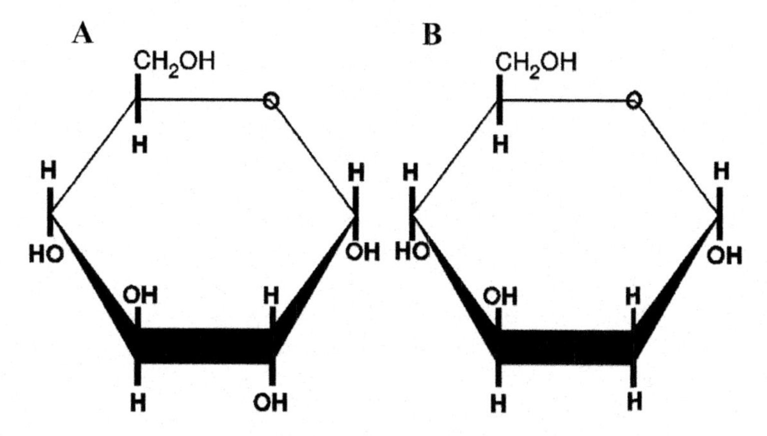

Fig. 7.11 Structure of (**a**) D-glucose and (**b**) 2-deoxy-D-glucose

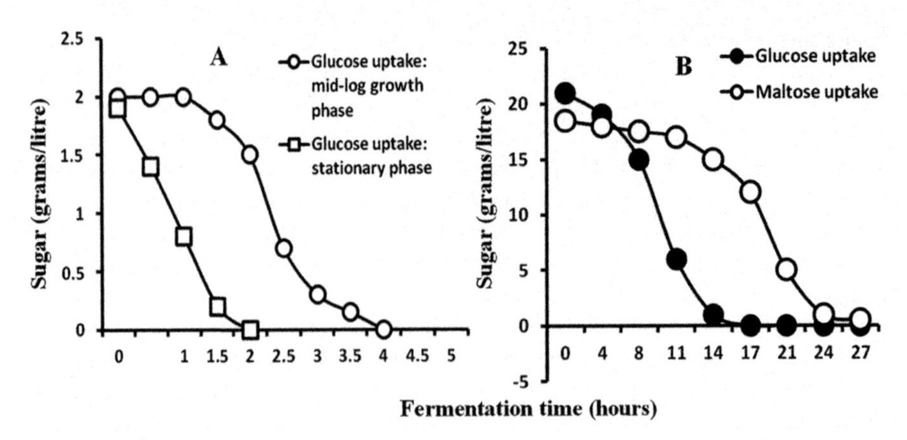

Fig. 7.12 Sugar uptake during fermentation of a 20 g/L glucose/20 g/L maltose medium. The strains employed were (**a**) *Saccharomyces cerevisiae* 1190 and (**b**) its 2-DOG-resistant mutant 1620. Glucose uptake (filled circle) and maltose uptake (open circle)

with university collaborators, has re-examined the use of 2-DOG mutants to ferment a 25°Plato wort, this time with a lager yeast strain (Piña et al. 2007). Stable 2-DOG mutants of this lager yeast strain were isolated. The fermentation characteristics of these 2-DOG lager yeast mutants were assessed in 25°Plato wort employing 2 L static fermenters at 13 °C. Improved fermentation capacity, where wort glucose did not repress maltose uptake, was achieved but with changes in the beer flavour profile. However (similar to the situation with the Canadian trials), the increased wort fermentation rate was insufficient to introduce their 2-DOG lager yeast mutants into production brewing.

In addition to the use of 2-DOG to isolate derepressed mutants in order to select yeast strains capable of taking up glucose and maltose simultaneously, strains expressing deregulation of hexose transporters (*HXTI*) have been induced. A diploid strain of *S. cerevisiae* was exposed to UV light for 50 s after growth on yeast nitrogen base media containing 2% maltose and either 0.3% 5-thioglucose (5-TG) or 0.04% 2-DOG. The viable cells (the mutants) acquired the ability to utilize glucose simultaneously with maltose and possibly also sucrose and galactose. Northern blot analysis (measurement of gene expression by detection of RNA) showed that the mutant strain expressed only the HXT6/7 gene, irrespective of the glucose concentration in the medium, indicating a deregulation of the induction/repression pathways modulating HXT gene expression. Interestingly, maltose-grown cells of the mutant displayed inverse diauxy in a glucose/maltose mixture, preferring maltose uptake to glucose. The conclusion to be drawn from this study is that the glucose transport step was probably uncoupled from downstream regulation, because it seemed to be unable to sense the abundant concentrations of glucose. Consequently, maltose uptake was preferred to glucose (Salema-Oom et al. 2011).

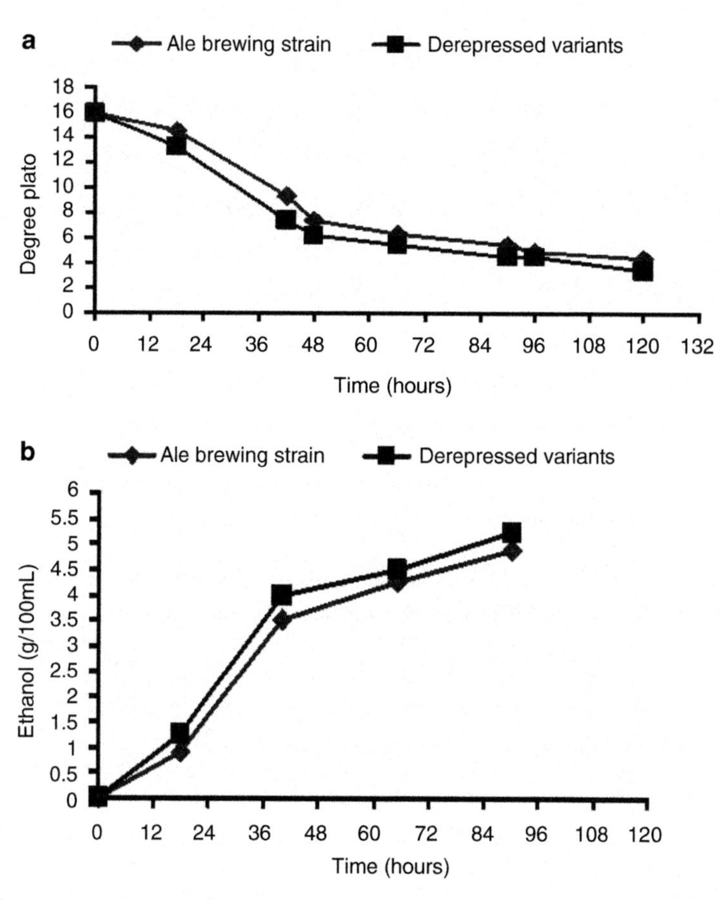

Fig. 7.13 Degree Plato reduction (**a**) and ethanol formation (**b**) by an ale brewing strain and its 2-DOG derepressed variant

The uptake of sugars from distiller's (whisky) wort is more variable and complex than from brewer's wort (Fig. 7.14). It has already been discussed that distiller's wort is not boiled (unlike brewer's wort) and therefore contains active enzymes, particularly amylases and proteinases, as well as contaminating microorganisms coming from the raw materials (barley malt and unmalted cereals) and the distillery equipment (mash mixers, lauter tuns, mash filters, fermenters, etc.). As a consequence, during mashing and fermentation, fermentable sugars are continuously being produced by the amylases from the wort dextrins and starch, which are taken up by the yeast and to a lesser extent by contaminating bacteria such as *Lactobacilli* (and other bacterial species) (Bathgate 2016). This means that the sugar uptake patterns in a distillery fermentation are more variable, complex and dynamic than in a typical brewing wort fermentation! The sugar uptake profiles in a distilling fermentation are more poorly characterized compared to brewing, and this is an area that requires further attention (Russell and Stewart 2014) (Fig. 7.14).

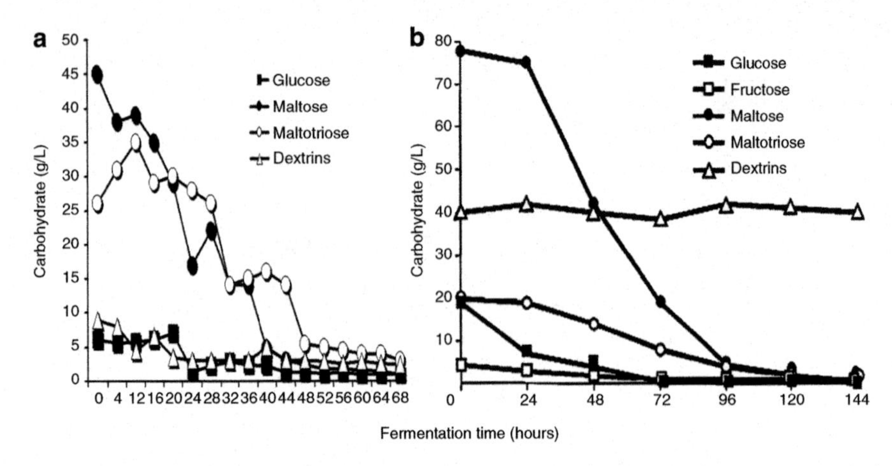

Fig. 7.14 Order of sugar uptake during a typical whisky all-malt fermentation (**a**) and a typical 12°Plato beer (**b**) fermentation

The approximate sugar composition of a typical unfermented malt distilling wort at pitching is as follows: sucrose 2%, fructose 1%, glucose 10%, maltose 50%, maltotriose 15%, maltotetraose 10% and other dextrins 10% (Table 1.1). As discussed above, the dextrin levels in unboiled wort continue to decrease during fermentation because of the ongoing amylolytic enzyme activity in order to hydrolyse dextrins into smaller fermentable molecules. The sugars present are not utilized simultaneously. Glucose is immediately available to the yeast and, as in brewing, is taken up first by the yeast because all the required transport systems for its uptake and subsequent metabolism are immediately present and available (Bisson and Fraenkel 1984). Maltose and maltotriose are present in the largest amounts but, as would be expected, are not utilized initially by the yeast culture. During fermentation, smaller molecular weight fermentable sugars are continuously being produced from the larger dextrins as a result of amylolytic enzyme activity. In addition, glucose is produced from the maltose and maltotriose that are already present in the wort as one of the fermentable sugars produced during mashing. As a result, the sugar uptake patterns in the distilling fermentations are very different from those observed with brewing wort fermentations where, as already described, the order of sugar uptake progresses in an orderly and consistent manner (Stewart 2006) (Fig 7.14).

It has already been discussed that brewer's wort contains unfermentable dextrins (partially hydrolysed starch). These dextrins remain in the finished beer and consequently contribute to the beer's mouth feel and contribute to its calorific value (Erratt and Stewart 1978). In order to produce a low-calorie beer, dextrins must be reduced. There are a number of techniques to reduce dextrins in beer (Owades and Koch 1989). One such method would be to employ a yeast strain that possesses diastatic ability to metabolize some (or all) of the wort dextrins. It has already been discussed that there is a grouping of yeast strains—*S. diastaticus*—that are taxonomically closely related to ale brewer's yeast strains (Stewart 2014). These strains contain the genetic ability to produce an extracellular enzyme (glucoamylase) that can hydrolyse the dextrins to glucose, which will be metabolized by the yeast during wort fermentation. These genes have been identified as *STA1/DEX1*, *STA2/DEX2* and *STA3/DEX3* (Erratt and Stewart 1980; Tamaki 1978; Lyness et al. 1993). A yeast strain incorporating these genes was constructed using classical hybridization techniques (Sherman 1991) and its fermentation characteristics assessed by Erratt and Stewart (1978) employing wort as the fermentation medium. This amylolytic yeast exhibited a faster fermentation rate and a lower final wort degree Plato than the control yeast, which did not contain genes for producing amylolytic enzymes and therefore could not metabolize wort dextrins (Erratt and Stewart 1981; Adams 2004) (Fig. 7.15).

In this author's opinion, these yeast genes should be denoted as *DEX* genes because the gene product (glucoamylase) only hydrolyses dextrins, not starch. To denote these genes as *STA* genes is incorrect. However, it has recently been reported (Ogata et al. 2017) that a glucoamylase (secreting diploid yeast) has been constructed using mating techniques that will be useful for producing novel pints of beer owing to its differing fermentation pattern and the concentrations of ethanol and flavour compounds produced as a result to wort fermentation. Unfortunately, glucoamylase gene in this strain has been incorrectly denoted as *STA1*!

The extracellular glucoamylase produced by this group of yeast was thermotolerant, probably because it was heavily glycosylated (a mannoprotein) (Erratt and Stewart 1980; Lyness et al. 1993). As a consequence of this, the glucoamylase was not inactivated during typical brewery tunnel pasteurization conditions (12 PU) (1 PU ≡ 1 min of treatment at 60 °C) of the low-dextrin beer, and the resultant beer became progressively sweeter in the finished package (bottle, can and keg) over time. Also, as already described, this enzyme did not possess

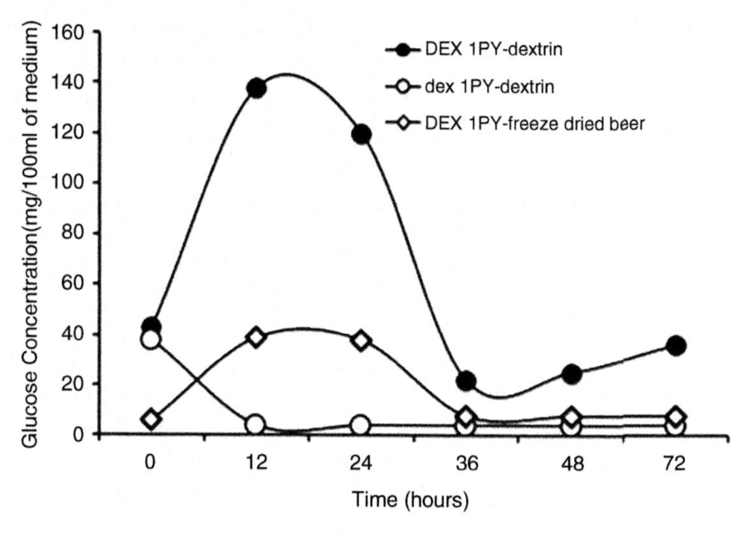

Fig. 7.15 Glucose concentration during growth of a *DEX 1* marked yeast

de-branching activity (was unable to hydrolyse the α(1–6) branch linkages of dextrin and starch); only the α(1–4) linkages were attacked (Erratt and Stewart 1980, 1981; Latorre-García et al. 2008). A similar low-dextrin beer was produced with a glucoamylase-producing yeast (where the gene was introduced using cloning rather than traditional hybridization techniques), and this beer was produced on a semi-production scale. It was called "Nutfield Lyte" (Fig. 7.16) and was part of a collaboration project between the Brewing Research Foundation of Nutfield, Surrey, UK (now Campden BRI, Brewing Division), and Heriot-Watt University in Edinburgh, Scotland (Hammond 1995). This particular strain was approved for use on a production scale by the UK's Novel Food Products and Processes (Baxter 1995). However, it is not currently being employed on a production scale for brewing low carbohydrate beer.

Employing protoplast (spheroplast) fusion (Panchal et al. 1984; Jones et al. 1987) as a genetic manipulation technique (details in Chaps. 16 and 17), a yeast strain with the ability to utilize partially dextrinized (hydrolysed) cassava starch at higher fermentation temperatures (>40 °C) has been produced and successfully trialled (Russell et al. 1986). However, brewing trials with this particular strain produced beer (although low carbohydrate) with an undrinkable flavour character!

Fig. 7.16 Nutfield Lyte

Nevertheless, this yeast strain's potential for producing industrial fermentation alcohol was recognized, and it consequently received patent protection, which has now expired. Further details of spheroplast fusion can be found in Chaps. 9 and 14 (Stewart et al. 1988).

Kilonzo et al. (2008), Latorre-García et al. (2008) and Wang and Xiu (2010) have all developed recombinant *S. cerevisiae* strains with glucoamylase production ability. Low-calorie beer has been produced with these strains following incorporation of a number of the *STA* (*DEX*) genes into the yeast genome, and the negative off-flavour production characteristics have been reduced [particularly 4-vinyl guaiacol formation from ferulic acid as a result of its decarboxylation (Fig. 2.12)] (Russell et al. 1983). Cellobiose is a reducing disaccharide with two β-glucose molecules linked by β (1–4) bonds (Fig. 7.17). It is obtained by the enzyme acid or acidic hydrolysis of cellulose, from cellulose-rich materials such as cotton, jute or paper. *Saccharomyces* spp. cannot take up cellobiose, but other yeasts, such as *Schwanniomyces* spp., can metabolize this disaccharide to ethanol (Sills and Stewart 1985). Further details of this process are in Chap. 9.

7.3 Free Amino Nitrogen (FAN) Metabolism

It is well documented that nitrogen is an essential element for yeast growth and metabolism (Pugh et al. 1997; Stewart 2012). It has been known for a long time that active yeast growth is critical for efficient wort fermentation (Thorne 1949). Active yeast growth involves the uptake of nitrogen, and as yeast multiplication ceases, nitrogen utilization also begins to decelerate. However, not all wort nitrogen can be utilized by the yeast for growth and to conduct a plethora of metabolic activities (Lekkas et al. 2007).

Free amino nitrogen (FAN) is the grouping of wort nitrogenous compounds available for consumption by yeast. FAN is the sum of the individual wort amino acids, ammonium ions and small peptides (di-, tripeptides). FAN is an important general measure of yeast nutrients, which constitutes the yeast assimilable nitrogen during brewing and distilling fermentations (Pickerell 1986; Lekkas et al. 2009; Stewart et al. 2013).

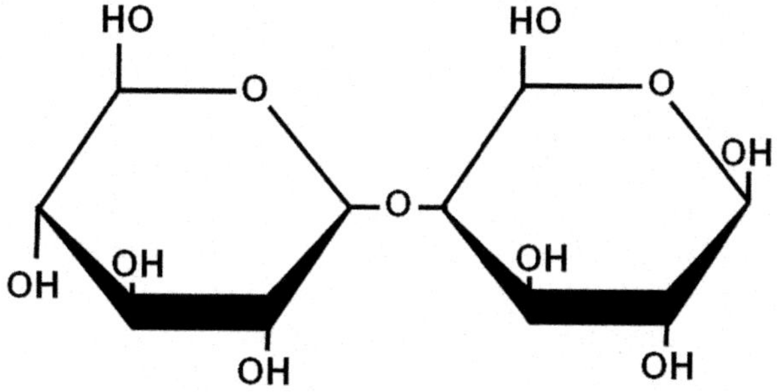

Fig. 7.17 Structure of cellobiose

Research in this field during the last 35 years (and longer) has confirmed that, even if wort carbohydrate utilization is complete, a similar quality of fermented wort (beer or spirit) is not guaranteed to be produced. This confirms that the fermentable sugar content alone is not a good indicator of yeast performance (Inoue et al. 1995). Indeed (as will be described in greater detail later—Chap. 8), the uptake of sugar is stimulated by higher nitrogen levels in high-gravity wort (Stewart and Murray 2011). Consequently, FAN is regarded as a better index for the prediction of healthy yeast growth, viability, vitality and overall fermentation efficiency, hence beer quality and stability (Jin et al. 1996). In addition, wort FAN is used by the yeast to accomplish its metabolic activities such as the de novo synthesis of amino acids and ultimately structural and enzymatic proteins (Pierce 1987).

FAN is believed to be a good index for potential yeast growth and efficiency (O'Connor-Cox and Ingledew 1989), with most yeast strains consuming peptides no larger than tripeptides—details later. It is believed that only 40% of the total oligopeptides available are used for nitrogen metabolic activity and that the rest may contribute to the development of large polypeptide-polyphenol complexes (Leiper et al. 2003a) or to foam stability (Leiper et al. 2003b). Routine FAN measurement in wort has been used within the brewing and distilling industries for historical reasons because of the ease and availability of the analytical method (Lie 1973). However, FAN is only a general measurement; it is a "blunt instrument" for establishing wort and malt specifications. The objectives of the studies discussed here were to enable the establishment of more meaningful specifications for malt and wort through, in part, the development of more extensive analyses.

There are differences between lager and ale yeast strains with respect to wort-assimilable nitrogen uptake characteristics (Jones and Pierce 1965). Nevertheless, with all brewing and distilling strains, the amount of wort FAN content required by the yeast under normal brewing fermentation conditions is directly proportional to yeast growth. Also, the amount affects certain aspects of beer maturation, for example, diacetyl (VDK) management (Gorinstein et al. 1999; Krogerus and Gibson 2013). There has been considerable polemic regarding the minimal FAN required to achieve satisfactory yeast growth and fermentation performance in normal and in high-gravity (10–20°Plato) worts. It is now generally agreed to be approximately 130 mg FAN/L. For rapid attenuation of higher-gravity wort (>16°Plato), increased levels of FAN are required—details later (O'Connor-Cox and Ingledew 1989). However, optimum FAN levels differ between wort composition and fermentation conditions and from yeast strain to yeast strain. In particular, the optimum FAN values change with different wort sugar levels and type (Stewart et al. 2013).

During the 1960s, Margaret Jones and John Pierce (Fig. 7.18) (working in the Guinness Park Royal R&D laboratories in London, England) conducted notable brewing studies on nitrogen metabolism during malting, mashing and fermentation. They reported that the absorption and utilization of exogenous nitrogenous compounds are controlled by three principal factors:

Fig. 7.18 John Pierce and Margaret Jones

- The total wort concentration of assimilable nitrogen
- The concentration of individual nitrogenous compounds and their ratio one to another
- The competitive inhibition of the uptake of these components (mainly amino acids) via various permease systems (Jones and Pierce 1964; Jones et al. 1969).

Jones and Pierce (1964) established a unique classification of amino acids according to their uptake rates during *S. cerevisiae* (ale) wort fermentations (Table 7.3). When this classification was developed, the methodology employed (novel liquid chromatography for measuring individual amino acids within a spectrum of 20 amino acids) was iconic! Similar measurements currently employ gradient elution automated computerized high performance liquid chromatography (HPLC) (Wiedmeier et al. 1982), and it is difficult to envisage the challenges that Margaret Jones, John Pierce and their colleagues overcame 50 years ago compared to current analytical methodology (Stewart and Murray 2011)! The Jones and Pierce (1965) amino acid classification is the basis of our current understanding of the relative importance of individual wort amino acids during the fermentation and manipulation of wort nitrogen levels, by the addition of yeast extract on specific amino acids, particularly during high-gravity brewing. However, this FAN assimilable pattern is often specific to the conditions employed. An individual yeast strain's nutritional preferences are unique. Because of the differences in malting barley varieties and the other fermentable raw materials, brewing conditions and the yeast strains employed in the brewing- and malting-based distilling industries worldwide, a more detailed view is required (Stewart 2010b). There are four uptake groups of wort amino acids. Three groups of amino acids are taken up at different stages of the fermentation cycle, and the fourth group, which consists of only one amino acid, proline (the largest concentration of amino acid in wort), is not taken up during a typical fermentation (Fig. 7.19). This is because of the anaerobic conditions that prevail late in a brewing fermentation. Recent studies have confirmed the Jones and Pierce's amino acid classification, but it has been proposed that methionine uptake be moved to Group A from Group B (Lekkas et al. 2007) (Table 7.3).

A research project has recently been conducted in order to investigate the wort components that might play an essential role in brewer's yeast strain fermentation performance. However, it was concluded by perhaps posing more questions than answers (Lekkas et al. 2009)! A quantitative and qualitative identification and determination of malt nitrogen compounds that affect yeast metabolic activity, in terms of oligopeptides, ammonium salts and both total and individual amino acids, were conducted. Fermentation results indicated that the wort FAN spectrum correlated well with at least three fermentation performance indicators (Lekkas et al. 2005):

- High initial FAN content permitted a more efficient and rapid reduction of wort gravity.
- The gravity decrease during fermentation was proportional to the amount of FAN utilized.

Table 7.3 Order of wort amino acids and ammonia uptake during fermentation

Group A	Group B	Group C	Group D
Fast absorption	Intermediate absorption	Slow absorption	Little or no absorption
Glutamic acid	Valine	Glycine	Proline
Aspartic acid	Leucine	Phenylalanine	
Asparagine	Isoleucine	Tyrosine	
Glutamine	Histidine	Tryptophan	
Serine		Alanine	
Methionine		Ammonia	
Threonine			
Lysine			
Arginine			

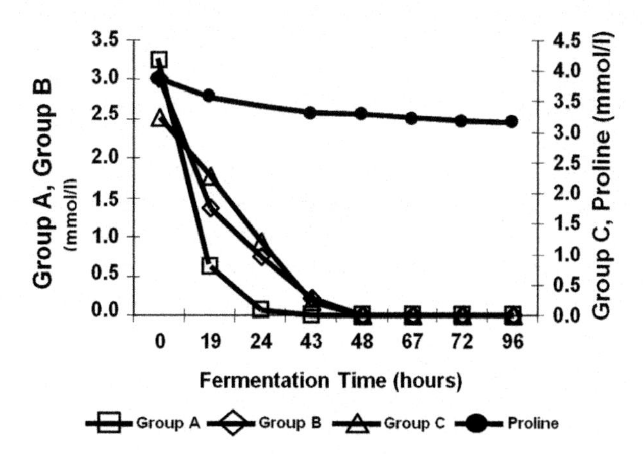

Fig. 7.19 Amino acid adsorption pattern during wort fermentation

- The FAN content has been suggested to be a useful index towards the formation of total VDKs (especially diacetyl), esters and higher alcohols during the later stages of fermentation—details later.

Wort FAN and its influence on yeast activity have a direct influence on beer quality, through its components and the metabolites that survive into the beer or spirit (Pickerell 1986). These determine key aspects of beer flavour (Chap. 15) and also impact on yeast performance. It has been suggested that the most useful index is the flavour compounds that display the most sensitive reactions to changes in one or more of wort FAN compounds. Currently, most brewers rely on wort FAN as an index of fermentation and yeast quality, assuming that the relative balance of nitrogenous materials remains constant. In terms of beer flavour, it is not only a matter of initial FAN wort content but equally the amino acid and ammonium ion equilibrium in the medium and a number of undefined fermentation parameters. Our knowledge of the roles of nitrogenous components of malt and wort in order to meet yeast requirements has substantially increased over the years. Nevertheless, the optimization of wort nitrogen content is a very complex issue owing to the large number of malt nitrogen compounds (Lekkas et al. 2009, 2014).

7.4 The General Amino Acid Permease (GAP) System

The regulation of amino acid uptake by brewer's and distiller's yeast strains is complex and involves carriers specific to certain amino acids (Grenson and Acheroy 1982; Jauniaux and Grenson 1990). Over 20 specific transport systems mediate the active transport of amino acids across the plasma membrane in both *S. cerevisiae* and *S. pastorianus* strains. Each is specific for just one, or a few, related L-amino acids. In addition to the specific systems, these yeasts possess a general amino acid transport system known as GAP (Surdin et al. 1965). The GAP system catalyses active transport of apparently all biological amino acids across the plasma membrane, and this is characterized as an enzyme amino acid permease (Palmquist and Ayrapaa 1969), and it has the ability to concentrate exogenous amino acids up to 1000-fold. The GAP transports wort amino acids by an active mechanism that involves proton cotransport. The internal pool of a particular amino acid inhibits its own transport system, a phenomenon known as transinhibition (Walker 1998). The amino acids compete with one another for transport, a process that follows Michaelis-Menten kinetics. Different yeast strains have different Michaelis-Menten constants. *GAP1* is believed to be the structural gene of the transporter. The product of the *GAP1* gene is a highly hydrophobic polypeptide, as expected for an integral membrane protein, consisting of 601 codons. Forty-one percent of these amino acids are hydrophobic, and the N-terminal of the polypeptide is hydrophilic. Acidic and basic amino acids are infrequent in the sequence. Mutations at the *GAP1* locus affect the general amino acid permease system by either modifying or repressing its affinity for specific substrates (Jauniaux and Grenson 1990).

The general permease system is very complex, and it has been found that its activity depends upon the nature and composition of the nitrogen sources in the medium. In addition, the level of assimilable nitrogen present in wort profoundly affects nutrient uptake and subsequent fermentation performance (Walker 1998). Also, the GAP transport system requires energy for operation, and consequently, the energy availability of the cell may limit the rate at which absorption can occur. The presence of an exogenous supply of certain nitrogenous nutrients inhibits the utilization of others by repressing the enzymes responsible for their assimilation (Grenson 1983).

Nitrogen limitation in yeast is known to result in a number of adaptations, which allows the cells to use alternate nitrogen sources (Woodward and Cirillo 1977). Maximum activity of the GAP system is only expressed when nitrogen is limiting. In other words, GAP becomes effective at transporting both D and L isomers of basic and neutral amino acids by yeast only under nitrogen-limiting conditions and during growth in poor nitrogen sources such as proline. During cell growth on proline, this amino acid is taken up by two permeases located in the plasma membrane, *Gap1p*, a low-affinity general amino acid permease, and *Put4p*, a high-affinity proline-specific permease (ter Schure et al. 2000). Once these two proline transport proteins are properly expressed, they are activated by phosphorylation. *Npr1p*, a protein kinase homologue, is also involved in the activation of both proline permeases. This enzyme is responsible for phosphorylation of the protein transporter and hence their regulation and activation. Finally, when proline enters the cell, it is then transferred from the cytoplasm into the mitochondria, for its further metabolism (ter Schure et al. 2000).

The inability of *Saccharomyces* sp. to assimilate proline under brewing fermentation conditions is the result of several phenomena (Stewart 2014). When the other amino acids or ammonium ions are still present in wort, the activity of proline permease, the enzyme that catalyses the transport of proline across the plasma membrane and into the cell, is repressed. Once proline is inside the cell, the first catabolic repression of proline involves proline oxidase, which requires the participation of cytochrome c and molecular oxygen. By this time, the other wort amino acids have been assimilated, thus removing the repression of the proline permease system; conditions are strongly anaerobic. As a result, the activity of proline oxidase is inhibited, and proline uptake does not occur. This amino acid (which is the amino acid present in the largest concentration in wort) therefore goes directly into the beer.

The reactivation of the general amino acid uptake system during cell growth in poor nitrogen sources, such as the presence of proline or under nitrogen starvation in the presence of glucose and/or fructose, results in an increase in the utilization rates of both hydrophobic and basic amino acids, with the simultaneous release of deaminated derivatives of hydrophobic amino acids, when the starved cells are inoculated into a nitrogen-rich medium. The deamination products released from the hydrophobic amino acids are presumed to be α-keto acids and fusel oil derivatives, respectively, of the specific amino acids. The manner in which nitrogen-starved yeast uses hydrophobic amino acids in the presence of glucose and fructose illustrates how the general amino acid permease, together with specific transaminases, accomplishes a selective retention of the amino acid moiety of hydrophobic amino acids. High transaminase activity is observed with tyrosine, phenylalanine, leucine and isoleucine as substrates, whereas aspartate, valine and methionine are only slowly transaminated. Glycine is not transaminated at all. Transinhibition may account for the variations in amino acid transport activity during cell growth in repressed and derepressed cultures. In addition, during nitrogen starvation conditions in the fermentation broth, an alternative control mechanism is also activated and known as the peptide transport system, which is described in greater detail shortly.

However, when yeast cells are growing in wort containing an adequate utilizable nitrogen source, the GAP system synthesis is inhibited, and the yeast appears to exhibit very high selectivity in terms of the amino acids that it utilizes as its major nitrogen sources. For instance, in the presence of excess concentrations of ammonia, glutamine and/or asparagine, the development of permease activity is prevented or decreased (Grenson and Acheroy 1982; Grenson 1983).

Two distinct control mechanisms are responsible for permease activity inhibition, and these are the repression of permease synthesis and the reversible permease inactivation (Jauniaux and Grenson 1990). Repression of GAP peptide synthesis involves the presence of ammonia in the medium, which regulates *GAP1* messenger formation and stability (Jauniaux and Grenson 1990). Hence, control of amino acid permease synthesis appears to result from transcription level repression. The *gdhCR* gene product is a repressor molecule, and it controls the transcription of permease and enzyme genes, which are subjected to nitrogen catabolite repression. Glutamine is an effector of this regulation (Wiame et al. 1985).

When the nitrogen nutrients are good sources to support efficient yeast proliferation and viability, then the GAP transport system is dephosphorylated and subsequently deactivated, and the *Npr1p* protein is responsible for triggering ubiquitin degradation. Whereas, the transporter membrane proteins via endocytosis are finally degraded inside the vacuole. In addition, the synthesis of raw permeases is blocked at the level of gene expression (ter Schure et al. 2000). Hence, by supplying the fermentation medium with ample amounts of assimilable nitrogen, this leads to the gradual deactivation of the general amino acid transport system with continuous activation of individual acid transporters.

7.5 Amino Acid Biosynthesis

Yeast cells are capable of synthesizing all the necessary amino acids needed for growth from suitable carbon and inorganic nitrogen sources. The biosynthesis of sulphur containing amino acids is a complex process involving carbon skeletons, nitrogen and sulphur (Fig. 7.20). The origins of these components vary according to the cultural conditions employed, each following an independent pathway (Jones et al. 1969; Stewart et al. 2013).

Regulation of amino acid biosynthesis takes place on two levels:

- The regulation of enzyme formation by control of gene expression.
- The regulation of enzymatic activity that controls metabolic flux. This regulation into selected pathways is affected by the enzymatic activity of coenzyme A (CoA) and glucose repression of certain enzymes of amino acid biosynthesis, each of which provides linkage to carbon metabolism (Jones and Fink 1982).

An important aspect of metabolic flow control is the exchange and interaction of metabolites between intracellular compartments (Fig. 2.1). Substantial portions of intracellular amino acids are compartmentalized within the cell, largely in the vacuole (Wiemken 1980). Compartmentalization of amino acids within the mitochondria (details of mitochondrial structures are in Chap. 14) has also been found for some anabolic enzymes concerning mainly arginine and some branched-chained amino acid pathways. With the exception of acidic amino acids, more than half of the concentrations of each amino acid is found in the vacuole. Moreover, the amino acid distribution within the cell varies with the wort amino acid composition and structure (Messenguy et al. 1980). If wort is supplemented with lysine, amino acids are transferred from the vacuole into the cytosol. The opposite effect is seen when the ammonia concentration of the medium decreases (Jones and Fink 1982).

7.6 Flavour Components

Although the major proportion of the FAN is utilized for the synthesis of new cellular and enzymatic proteins for the yeast, it has been shown that there is also a correlation between FAN and the synthesis of higher alcohols, esters, aldehydes and vicinal diketones, which affect and significantly contribute to final beer

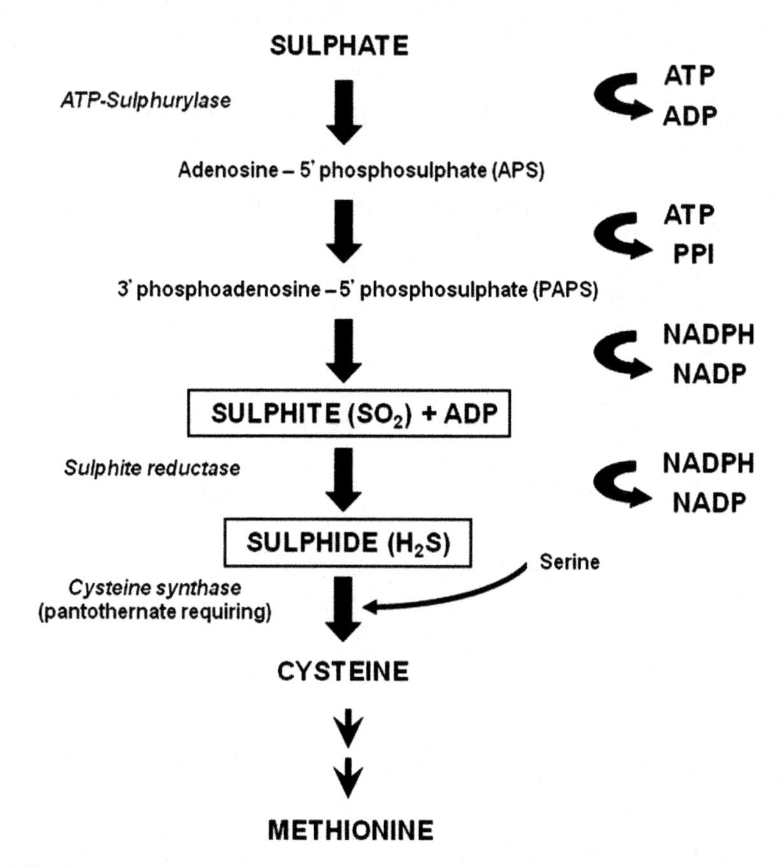

Fig. 7.20 Sulphur-containing amino acids from carbon, sulphur and nitrogen sources

characteristics, especially flavour (Stewart and Russell 2009; Stewart et al. 2013)—details in Chap. 15.

7.7 Ammonia

Brewer's and distiller's yeast strains are also capable of utilizing inorganic sources of nitrogen, such as ammonia. Ammonium ions are actively transported and readily assimilated by yeast. Ammonia uptake probably involves at least three permeases, *Meplp*, *MepZp* and *Mep3p* (ter Schure et al. 2000). *MEP2* exhibits the higher affinity for ammonia ions, followed closely by *MEP1* and finally *MEP2*. Growth on ammonium ions at concentrations higher than 20 mM does not require any of the ammonia permeases.

In yeast cultures growing in the presence of ammonia, almost all of the nitrogen is first assimilated into glutamate and glutamine (Holmes et al. 1989), as these amino acids donate the α-amino nitrogen that the cell requires for growth. Other amino acids are formed by transamination reactions. These amino acids themselves are precursors for the biosynthesis of other amino acids. Consequently, glutamate and glutamine are primary products of ammonium assimilation and are key compounds in both nitrogen and carbon metabolism.

The primary route of ammonium assimilation in yeasts is into glutamate through the action of an NADP-dependent glutamate dehydrogenase (GDH). In this pathway, glutamate is generated via NADPH-GDH by coupling ammonia to α-ketoglutarate at the expense of one NADPH molecule (Lacerda et al. 1990):

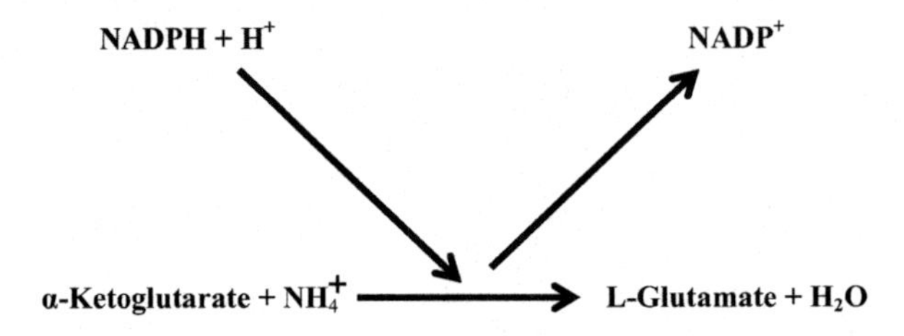

In the second pathway, glutamine is produced by glutamine synthetase (GS), which converts glutamate and ammonia into glutamine by consuming one ATP molecule, and glutamate is synthesized by glutamate synthase (GOGAT). This reaction converts one molecule of glutamine and one molecule of α-ketoglutarate into two molecules at the expense of one NADH molecule (ter Schure et al. 2000).

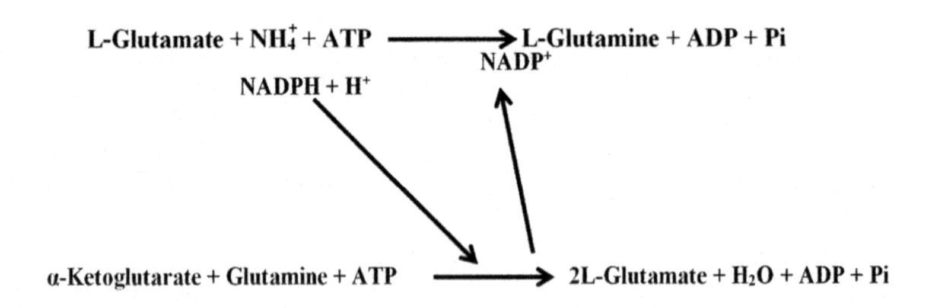

When glutamine synthetase is coupled with glutamate synthase, this glutamine pathway not only provides an alternative route to glutamate, compared with the glutamate dehydrogenase reaction, but also represents a highly efficient process for yeasts to assimilate ammonia into α-amino nitrogen. In addition, the glutamine pathway may also play an important role in yeast cell physiology such as the maintenance of Citric Acid Cycle intermediates (details in Chap. 6), cell growth and morphology (Magasanik 1992).

The GDH and GS-GOGAT pathways are highly regulated, and the particular route of ammonium assimilation adopted by yeasts will depend on various factors, not the least of which is the concentration of available ammonium ions and intracellular amino acid pools. Glutamate and glutamine can also be used as sole nitrogen sources. During growth on glutamate, glutamine synthetase produces glutamine using ammonia generated by NAD-GDH. When glutamine is the only nitrogen source in the growth medium, glutamate is synthesized by glutamate synthase or via NADPH-GDH; in the former case, ammonia is produced by glutaminases, which degrade glutamine to glutamate and ammonia.

Ammonia also inhibits the synthesis of numerous proteins involved in the assimilation of poorer nitrogen sources at the level of transcription (ter Schure et al. 2000). Growth on ammonia results in higher growth rates compared to growth on proline, and this is caused by a higher metabolic nitrogen flux towards the synthesis of glutamate and glutamine. In other words, ammonia yields high intracellular concentrations that will trigger transcription repression. During growth on ammonia, these effects would originate from either the nitrogen flow or the nitrogen concentration. Higher metabolic fluxes would result in inactivation of transcription. To be more precise, when continuous cultures were used, it was shown that the expression of the *GAP1* gene was not changed by the ammonia flux but rather by the ammonia concentration. However, the increase in the ammonium ion levels resulted in an increase of the intracellular glutamine concentration, and hence ammonia repression could be generated indirectly via glutamine. Another fact that supports this finding is that *GAP1* expression in continuous cultures correlated with the ammonia concentration, whereas the intracellular glutamine concentration remained constant (ter Schure et al. 2000).

7.8 Small Wort Peptides

Four hundred dipeptides and up to 8000 possible tripeptides may theoretically be found in wort, according to the number of amino acids in wort and their binding combination to form oligopeptides with two or three residues (MacWilliam and Clapperton 1969). Small peptides can be used by yeast as nutritional sources of amino acids, as carbon or nitrogen sources and as precursors of cell wall peptides during yeast growth, although growth is much slower when they are the sole nitrogen source (Ingledew and Patterson 1999). Polypeptides are also used as a substitute because yeasts can generate proteolytic enzymes extracellularly in order to provide additional assimilable nitrogen to the cells (Maddox and Hough 1970; Cooper et al. 2000). Approximately 40% of the wort oligopeptide fraction is removed by yeast during fermentation, and the peptides in beer differ from those found in wort (Calderbank et al. 1986). However, very little is known about the range of peptides found, and generated, in wort and about the order in which they are removed during fermentation.

Most brewing yeast strains transport peptides with no more than three amino acid residues, but this limit is strain dependent (Marder et al. 1977). Yokota et al. (1993) and Clapperton (1971) reported that the concentration of low molecular weight peptides was found to decrease during brewing fermentations. By studying fermentations that were conducted with a malt extract medium (a wort-type product prepared from malted cereals in a concentrated form), a very low attenuation rate was observed compared to that of a normal-gravity wort, even when the initial FAN levels were similar (200 mg/L). This suggested that, regardless of the high initial FAN content of the malt extract, peptides larger than tripeptides were not utilized by the yeast strains examined (Moneten et al. 1986). The inability of a larger peptide to enter the yeast cell suggests a size limit for peptide transport. It has also been demonstrated (Ingledew and Patterson 1999) that the L-stereoisomers, but not the D-forms, of amino acids in di- and tripeptides are preferred substrates and also that the basic amino acid-containing peptides are transported more rapidly than peptides containing acidic amino acid residues.

Studies by Ingledew and Patterson (1999) suggested that the phase of growth and the concentration of the non-peptide nitrogen might affect peptide utilization. Dipeptide transport in brewing yeast is affected by the presence of micromolar concentrations of amino acids in the growth medium (Island et al. 1987). The presence of amino acids in the growth medium increases the sensitivity of the yeast to small peptides, but not all amino acids produce the same response to the uptake of small peptides. Leucine and tryptophan appear to be the most effective peptide uptake regulators regardless of their concentration. On the other hand, asparagine appears to be a potential inhibitor of peptide utilization. For this reason, three categories of amino acids have been identified based on the effect that they induce on the uptake of small peptides. These are:

- Amino acids that promote only slight sensitivity.
- Amino acids that are good nitrogen sources (e.g. arginine) and repress the yeast's sensitivity to small peptides.
- Other amino acids that are considered to be inducers or accelerators of yeast peptide uptake sensitivity.

An examination of the intracellular pools of cells grown in a medium containing amino acid inducers and small peptides showed increased levels of peptide amino acid residues. This observation led to speculation that these accumulated residues might result in transinhibition of amino acid uptake (Inoue and Kashihara 1995).

In brewing yeast cultures, the absence of competition between individual amino acids and the uptake of simple peptides suggests that peptide transport is distinct from the system for amino acids. It has been shown that di- and tripeptides share the same transport system. Ingledew and Patterson (1999), using a synthetic medium containing amino acids and small peptides, found that extracellular proteinase activity was produced, which suggests that small peptides are taken up by the intact yeast, via a specific peptide transport system, which is mediated by a specific metabolic sensor that triggers rapid synthesis of an additional permease or catalytic activity that is capable of modifying the existing peptide transporter system. In accordance with its role as a system providing nutrients in the form of peptides, it has been verified that the activity of this mechanism is modulated by both the quality of the nitrogen sources and the presence of amino acids included in the growth medium. Initial peptide uptake rates were observed to be higher in cells that were grown on poor nitrogen sources, such as proline, than in cells grown on preferred nitrogenous materials, such as glutamine. Based on this criterion, it has been proposed that the peptide transport system of yeasts falls under the regulatory control of the nitrogen catabolite repression mechanism (Perry et al. 1994). Probably similar inactivation takes place in the GAP system may also occur for the peptide transport system, when yeast cells have been propagated in a medium with ammonium ions as the sole nitrogen source. Nitrogen catabolite repression most likely exerts a mild effect on the peptide transport rate, whereas amino acids result in changes in initial oligopeptide B uptake rates.

Single peptides are not necessarily as good a source of nitrogen for growth as the amino acids that constitute them. Therefore, growth on a particular amino acid could not be used to predict the growth characteristics on the homologous di- or tripeptides. For example, the enhanced growth of yeast cells with arginine supplementation was not requested with medium enrichment by an Arg-Arg dipeptide. Also, poor cell division was observed. The negative growth response with this dipeptide could have been caused by non-transfer into the site of the peptidase activity inside of the cell. Most of these peptidases are either located in intracellular organelles, such as the vacuole, or in the cytoplasm.

Numerous nitrogenous materials are released from yeast cells into wort during wort fermentation (Clapperton 1971). A significant number of these compounds are oligopeptides, which are formed during fermentation. Some of these compounds may be assimilated by the yeast; others will remain in the final beer and contribute to its initial flavour and long-term stability and/or instability. In addition, if the capacity of accumulated peptide residues is exceeded intracellularly, then some of these might be released as deaminated derivatives (Woodward and Cirillo 1977). These peptides remain in the fermented wort because they are probably too large to be assimilated by the yeast due to the yeast culture's preference for peptides with three or fewer amino acids (Lekkas et al. 2007).

A method has been developed for the detection, isolation and measurement of small peptides during a wort fermentation (Lekkas et al. 2007). Lager and ale strain fermentations have been conducted. All of the fermentable free amino acids in the wort were consumed by the yeast within the first 48 h of fermentation, as would have been expected based on previous experiments reported by Jones and Pierce (1964) and Lekkas et al. (2007). In all of the fermentations conducted, utilization of ammonia and proline was incomplete, as was reported in previous experiments (Jones and Pierce 1964; Lekkas et al. 2007).

Oligopeptide levels in wort can fluctuate throughout a fermentation. Three hypotheses for this fluctuation have been considered:

- The increase in small peptide wort concentrations could be induced by the low cell viability and vitality as a consequence of cell lysis and release of nitrogenous materials into the fermentation environment. This assumption was unproven because cell viability, even at the end of fermentation, was high (>90%). In addition, extracellular ATP levels (an index of yeast viability and cell autolysis) were found not to correlate with the pattern of oligopeptide adsorption.
- In order to clarify such a phenomenon, the yeast cells could possibly excrete the small peptides that have already been taken up, back into the fermenting wort, and then they are absorbed again! However, such a theory did not seem to be applicable because when the total wort FAN content was measured, no increase in FAN concentration was detected throughout the duration of the experiments.
- The last hypothesis that could provide a coherent explanation was that excretion/ secretion of yeast protease into the fermentation environment could hydrolyse larger peptides and proteins into smaller molecules in order to supply proliferating yeast cells with more available nitrogenous materials (Cooper et al. 2000).

It has been shown (Ingledew and Patterson 1999; Stewart et al. 2013) that a number of the lager and ale yeast strains studied, regardless of the experimental conditions employed (static or shaken fermentations), could simultaneously use amino acids and small peptides as sources of assimilable nitrogen. In addition, it is believed that extracellular proteolytic enzymes, produced by the yeast culture, are responsible for the degradation of larger wort peptides into smaller peptides in order to provide yeast cells with more available assimilable nitrogen sources.

The metabolism of wort sugars, dextrins, amino acids, peptides and ammonium ions is a series of dynamic processes. A yeast culture depends on consistent reactions in order to maintain high cell viability and vitality (Chap. 8) of the harvested/cropped cultures (Chap. 13) that will be used in subsequent wort fermentations (Chaps. 6 and 7) to produce metabolites with appropriate flavouring properties (details in Chap. 15).

References

Adams J (2004) Microbial evolution in laboratory environments. Res Microbiol 155:311–318

Alves SL Jr, Herberts RA, Hollatz C, Trichez D, Miletti LC, de Araujo PS, Stambuk BU (2008) Molecular analysis of maltotriose active transport and fermentation by *Saccharomyces cerevisiae* reveals a determinant role for the AGT1 permease. Appl Environ Microbiol 74:1494–1501

Bathgate GN (2016) A review of malting and malt processing for whisky distillation. J Inst Brew 122:197–211

Batistote M, Da Cruz SH, Ernandes JR (2006) Altered patterns of maltose and glucose fermentation by brewing and wine yeasts influenced by the complexity of nitrogen source. J Inst Brew 112:84–91

Baxter ED (1995) The application of genetics in brewing. Fermentation 8:307–311

Bisson LF, Fraenkel DG (1984) Expression of kinase-dependent glucose uptake in *Saccharomyces cerevisiae*. J Bacteriol 159:1013–1017

Bisson LF, Coons DM, Kruckeberg AL, Lewis DA (1993) Yeast sugar transporters. Crit Rev Biochem Mol Biol 28:259–308

Boulton CA, Quain DE (2001) Brewing yeast and fermentation. Blackwell Science, Oxford

Calderbank J, Rose AH, Tubb RS (1986) Peptide removal from all malt and adjunct worts by *Saccharomyces cerevisiae*. J Inst Brew 91:321–324

Clapperton JF (1971) Simple peptides of wort and beer. J Inst Brew 77:177–180

Cooper DJ, Stewart GG, Bryce JH (2000) Yeast proteolytic activity during high and low gravity wort fermentations and its effect on head retention. J Inst Brew 106:197–201

Cousseau FEM, Alves SL Jr, Trichezl D, Stambuk BU (2012) Characterization of maltotriose transporters from the *Saccharomyces eubayanus* subgenome of the hybrid *Saccharomyces pastorianus* lager brewing yeast strain Weihenstephan 34/70. Lett Appl Microbiol 56:21–29

Crumplen RM, Slaughter JC, Stewart GG (1996) Characteristics of maltose transporter activity in an ale and lager strain of the yeast *Saccharomyces cerevisiae*. Lett Appl Microbiol 23:448–452

Cruz SH, Batistote M, Ernandes JR (2003) Effect of sugar catabolite repression in correlation with the structural complexity of the nitrogen source on yeast growth and fermentation. J Inst Brew 109:349–355

D'Amore T, Russell I, Stewart GG (1989) Sugar utilization by yeast during fermentation. J Ind Microbiol 4:315–324

Day RE, Higgins VJ, Rogers PJ, Dawes IW (2002) Characterization of the putative maltose transporters encoded by YDL247w and YJR160c. Yeast 19:1015–1027

Ernandes JR, D'Amore T, Russell I, Stewart GG (1992) First online: Regulation of glucose and maltose transport in strains of *Saccharomyces*. J Ind Microbiol 9:127–130

Ernandes JR, Williams JW, Russell I, Stewart GG (1993) Effect of yeast adaptation to maltose utilization on sugar uptake during the fermentation of brewer's wort. J Inst Brew 99:67–71

Erratt JA, Stewart GG (1978) Genetic and biochemical studies on yeast strains able to use dextrins. J Am Soc Brew Chem 36:151–161

Erratt JA, Stewart GG (1980) Genetic and biochemical studies on glucoamylase from *Saccharomyces diastaticus*. In: Stewart GG, Russell I (eds) Advances in biotechnology, current developments in yeast research. Pergamon Press, Toronto, pp 177–183

Fukuyo S, Myojo Y (2014) Chap. 3: Japanese whisky. In: Russell I, Stewart GG (eds) Whisky: technology, production and marketing, 2nd edn. Elsevier, Boston, MA

Gibson BR, Londesborough J, Rautio J, Mattinen L, Vidgren V (2013) Transcription of α-glucoside transport and metabolism genes in the hybrid brewing yeast *Saccharomyces pastorianus* with respect to gene provenance and fermentation temperature. J Inst Brew 119:23–31

Gorinstein S, Zemser M, Vargas-Albores F, Ochoa JL, Paredes-Lopez O, Scheler C, Salnikow J, Martin-Belloso O, Trakhtenberg S (1999) Proteins and amino acids in beers, their contents and relationships with other analytical data. Food Chem 67:71–78

Gray A (2013) The Scotch Whisky Industry review, 36th edn. Sutherlands, Edinburgh

Grenson M (1983) Inactivation-reactivation process and repression of permease formation regulate several ammonia-sensitive permeases in the yeast *Saccharomyces cerevisiae*. Eur J Biochem 133:141–144

Grenson M, Acheroy B (1982) Mutations affecting the activity and the regulation of the general amino-acid permease of *Saccharomyces cerevisiae*. Localisation of the cis-acting dominant pgr regulatory mutation in the structural gene of this permease. Mol Gen Genet 188:261–265

Guimarães PMR, Virtanen H, Londesborough J (2006) Direct evidence that maltose transport activity is affected by the lipid composition of brewer's yeast. J Inst Brew 112:203–209

Hammond JRM (1995) Genetically-modified brewing yeasts for the 21st century: progress to date. Yeast 11:1613–1627

Han EK, Cotty F, Sottas C, Jiang H, Michels CA (1995) Characterization of AGT1 encoding a general a-glucoside transporter from *Saccharomyces*. Mol Microbiol 17:1093–1107

Hertrich J (2013) Topics in brewing: brewing adjuncts. MBAA Tech Quart 50:72–81

Holmes AR, Collings A, Farnden KJF, Shepherd MG (1989) Ammonium assimilation in *Candida albicans* and other yeasts: evidence for activity of glutamate synthase. J Gen Microbiol 135:1423–1430

Ingledew WM, Patterson CA (1999) Effect of nitrogen source and concentration on the uptake of peptides by a lager yeast in continuous culture. J Am Soc Brew Chem 57:9–17

Inoue T, Kashihara T (1995) The importance of indices related to nitrogen metabolism in fermentation control. MBAA Tech Quart 32:109–113

Inoue SB, Takewaki N, Takasuka T, Mio T, Adachi M, Fujii Y, Miyamoto C, Arisawa M, Furuichi Y, Watanabe T (1995) Characterization and gene cloning of 1,3-beta-D-glucan synthase from *Saccharomyces cerevisiae*. Eur J Biochem 231:845–854

Island MD, Naider F, Becker JM (1987) Regulation of dipeptide transport in *Saccharomyces cerevisiae* by micromolar amino acid concentrations. J Bacteriol 169:2132–2136

James TC, Campbell S, Donnelly D, Bond U (2003) Transcription profile of brewery yeast under fermentation conditions. J Appl Microbiol 94:432–448

Jauniaux JC, Grenson M (1990) GAP1, the general amino acid permease gene of *Saccharomyces cerevisiae*. Nucleotide sequence, protein similarity with the other bakers yeast amino acid permeases, and nitrogen catabolite repression. Eur J Biochem 31(190):39–44

Jespersen L, Cesar LB, Meaden PG, Jakobsen M (1999) Multiple α-glucoside transporter genes in brewer's yeast. Appl Environ Microbiol 65:450–456

Jin H, Ferguson K, Bond M, Kavanagh T, Hawthorne D (1996) Malt nitrogen parameters and yeast fermentation behavior. In: Proceedings of the 24th convention. The Institute of Brewing, Asia Pacific Section, Singapore, pp 44–50

Jones EW, Fink GR (1982) Regulation of amino acid and nucleotide biosynthesis in yeast. In: Strathern JN, Jones EW, Broach JR (eds) The molecular biology of the yeast Saccharomyces: metabolism and gene expression. Cold Spring Harbor Laboratory Press, Cold Spring Harbor, NY, pp 181–299

Jones M, Pierce J (1964) Absorption of amino acids from wort by yeasts. J Inst Brew 70:307–315

Jones M, Pierce J (1965) Nitrogen requirements in wort – practical applications. In: Proceedings of European brewery convention congress, Stockholm. Elsevier Scientific, Amsterdam, pp 182–194

Jones M, Pragnell MJ, Pierce JS (1969) Absorption of amino acids by yeasts from a semi-defined medium simulating wort. J Inst Brew 75:520–536

Jones RM, Russell I, Stewart GG (1986) The use of catabolite derepression as a means of improving the fermentation rate of brewing yeast strains. J Am Soc Brew Chem 44:161–166

Jones RM, Russell I, Stewart GG (1987) Classical genetic and protoplast fusion techniques in yeast. In: Berry DR, Russell I, Stewart GG (eds) Yeast biotechnology, pp 55–79

Kapral D (2008) Stratified fermentation – causes and corrective actions. MBAA Tech Quart 45:115–120

Kilonzo PM, Margaritis A, Bergougnou MA (2008) Effect of medium composition on glucoamylase production during batch fermentation of recombinant *Saccharomyces cerevisiae*. J Inst Brew 114:83–96

Kirsop BH (1974) Oxygen in brewery fermentations. J Inst Brew 80:252–259

Kodama Y, Fukui N, Ashikari T, Shibano Y, Morioka-Fujimoto K, Hiraki Y, Nakatani K (1995) Improvement of maltose fermentation efficiency: constitutive expression of *MAL* genes in brewing yeasts. J Am Soc Brew Chem 53:24–29

Krogerus K, Gibson BR (2013) Diacetyl and its control during brewing fermentation: 125th anniversary review. J Inst Brew 119:86–97

Lacerda V, Marsden A, Buzato J.B, Ledingham WM (1990) Studies on ammonium assimilation in continuous culture of *Saccharomyces cerevisiae* under carbon and nitrogen limitation, In: Christiansen C, Munck L, Villadsen J (eds) Proceedings of 5th European congress on biotechnology, 305. Munksgaard International, Copenhagen, pp 1075-1078

Latorre-García L, Adam AC, Polaina J (2008) Overexpression of the glucoamylase-encoding STA1 gene of *Saccharomyces cerevisiae* var. *diastaticus* in laboratory and industrial strains of *Saccharomyces*. World J Microbiol Biotechnol 24:2957–2963

Leiper KA, Stewart GG, McKeown IP (2003a) Beer polypeptides and silica gel. Part I. Polypeptides involved in haze formation. J Inst Brew 109:57–72

Leiper KA, Stewart GG, McKeown IP (2003b) Beer polypeptides and silica gel. Part II. Polypeptides involved in foam formation. J Inst Brew 109:77–79

Lekkas C, Stewart GG, Hill A, Taidi B, Hodgson J (2005) The importance of free amino nitrogen in wort and beer. MBAA Tech Quart 42:113–116

Lekkas C, Stewart GG, Hill AE, Taidi B, Hodgson J (2007) Elucidation of the role of nitrogenous wort components in yeast fermentation. J Inst Brew 113:3–11

Lekkas C, Hill AE, Taidi B, Hodgson J, Stewart GG (2009) The role of small peptides in brewing fermentation. J Inst Brew 115:134–139

Lekkas C, Hill AE, Stewart GG (2014) Extraction of FAN from malting barley during malting and mashing. J Am Soc Brew Chem 72:6–11

Lie S (1973) The EBC-ninhydrin method for determination of free alpha amino nitrogen. J Inst Brew 79:37–41

Lyness CA, Jones CR, Meaden PG (1993) The *STA2* and *MEL1* genes of *Saccharomyces cerevisiae* are idiomorphic. Curr Genet 23:92–94

Lyons TP (2014) Chap. 5, North American whiskies: a story of evolution, experience, and an ongoing entrepreneurial spirit. In: Russell I, Stewart G (eds) Whisky: technology, production and marketing, 2nd edn. Elsevier, Boston, MA, pp 39–48

Macwilliam JC, Clapperton JF (1969) Dynamic aspects of nitrogen metabolism in yeast. In: Proceedings of European brewery convention congress, Interlaken. Fachverlag Hans Carl, Nürnberg, pp 271–279

Maddox SJ, Hough JS (1970) Proteolytic enzyme of *Saccharomyces carlsbergensis*. Biochem J 117:843–852

Magasanik B (1992) Regulation of nitrogen utilization. In: Jones E, Pringle JR, Broach JR (eds) Molecular and cellular biology of the yeast, *Saccharomyces cerevisiae*, vol II, Gene expression. Cold Spring Harbor Laboratory Press, Plainview, NY, pp 283–317

Marder R, Becker JM, Naider F (1977) Peptide transport in yeast: utilization of leucine and lysine containing peptides by *Saccharomyces cerevisiae*. J Bacteriol 131:906–916

Messenguy F, Colin D, Ten Have J-P (1980) Regulation of compartmentation of amino acid pools in *Saccharomyces cerevisiae* and its effects on metabolic control. Eur J Biochem 108:439–447

Moneton P, Sarthou P, Le Goffic F (1986) Role of the nitrogen source in peptide transport in *Saccharomyces cerevisiae*. Fed Eur Microbiol Microbiol Lett 36:95–98

Murray D (2014) Chap. 10, Grain whisky distillation. In: Russell I, Stewart G (eds) Whisky: technology, production and marketing, 2nd edn. Elsevier, Boston, MA

Murray CR, Barich T, Taylor D (1984) The effect of yeast storage conditions on subsequent fermentations. MBAA Tech Quart 21:189–195

Narziss L (1984) The German beer laws. J Inst Brew 90:351–358

Naseeb S, James SA, Alsammar H, Michaels CJ, Gini B, Nueno-Palop C, Bond CJ, McGhie H, Roberts IN, Delneri D (2017) *Saccharomyces jurei sp.* isolation and genetic identification of a novel yeast species from *Quercus robur*. Int J Syst Evol Microbiol 67:2046–2052

Novak ST, D'Amore T, Russell I, Stewart GG (1990a) Characterization of sugar transport in 2-Deoxy-D-glucose resistant mutants of yeast. J Ind Microbiol 6:149–155

Novak ST, D'Amore T, Stewart GG (1990b) 2-Deoxy-D-glucose resistant yeast with altered sugar transport activity. FEBS Lett 269:202–204

Novak ST, D'Amore T, Russell I, Stewart GG (1991) Sugar uptake in a 2-deoxy-D-glucose resistant mutant of *Saccharomyces cerevisiae*. J Ind Microbiol 7:35–40

O'Connor-Cox ESC, Ingledew WM (1989) Wort nitrogenous sources – their use by brewing yeasts: a review. J Am Soc Brew Chem 47:102–108

Ogata T, Iwashita Y, Kawada T (2017) Construction of a brewing yeast expressing the glucoamylase gene *STA1* by mating. J Inst Brew 123:66–69

Owades JL, Koch CJ (1989) Preparation of low calorie beer. US Patent 4837034

Palmquist U, Ayrapaa T (1969) Uptake of amino acids in bottom fermentations. J Inst Brew 75:181–190

Panchal CJ, Russell I, Sills AM, Stewart GG (1984) Genetic manipulation of brewing and related yeast strains. Food Technol 38:99–106

Perry JR, Basrai MA, Steiner HY, Naider F, Becker JM (1994) Isolation and characterization of a *Saccharomyces cerevisiae* peptide transport gene. Mol Cell Biol 14:104–115

Pickerell ATW (1986) The influence of free alpha-amino nitrogen in sorghum beer fermentations. J Inst Brew 92:568–571

Pierce J (1987) The IBD Horace Brown Medal Lecture: The role of nitrogen in brewing. J Inst Brew 95:378–381

Piña B, Casso D, Orive M, Fité B, Torrent J, Vidal JF (2007) A new source to obtain lager yeast strains adapted to very high gravity brewing: 2-deoxy-D-glucose resistant colonies with fermentative capacity up to 25°Plato. In: Proceedings of European brewery convention congress, Venice. Verlag Hans Carl Getränke-Fachverlag, Nürnberg, CD ROM, Contribution 45

Pugh TA, Maurer JM, Pringle AT (1997) The impact of wort nitrogen limitation on yeast fermentation performance and diacetyl. MBAA Tech Quart 34:185–189

Quinn D (2014) Chap. 2: Irish whiskey. In: Russell I, Stewart G (eds) Whisky: technology, production and marketing, 2nd edn. Elsevier, Boston, MA, pp 7–16. isbn:978-0-12-401735-1

Rautio J, Londesborough J (2003) Maltose transport by brewer's yeasts in brewer's wort. J Inst Brew 109:251–261

Roberts T, Wilson R (2006) Chap. 7: Hops. In: Priest FG, Stewart GG (eds) Handbook of brewing, 2nd edn, pp 177–279

Russell I, Stewart GG (2014) Distilling yeast and fermentation. In: Russell I, Stewart GG (eds) Whisky: technology, production and marketing, 2nd edn. Elsevier, Boston, MA, pp 123–143

Russell I, Hancock IF, Stewart GG (1983) Construction of dextrin fermentative yeast strains that do not produce phenolic off-flavours in beer. J Am Soc Brew Chem 41:45–51

Russell I, Crumplen GM, Jones RM, Stewart GG (1986) Efficiency of genetically engineered yeast in the production of ethanol from dextinized cassava starch. Biotechnol Lett 8:169–174

Salema-Oom M, Pinto VV, Gonçalves P, Spencer-Martins I (2005) Maltotriose utilization by industrial *Saccharomyces* strains: characterization of a new member of the α-glucoside transporter family. Appl Environ Microbiol 71:5044–5049

Salema-Oom M, de Sousa HR, Assumcao M, Gonçalves P, Spencer-Martins I (2011) Derepression of a baker's yeast strain for maltose utilization is associated with severe deregulation of HXT gene expression. J Appl Microbiol 110:364–374

Sherman F (1991) Getting started with yeast. In: Guthrie C, Fink GR (eds) Methods in enzymology, vol 194 Guide to yeast genetics and molecular biology. Academic Press, New York, pp 3–21

Sills AM, Stewart GG (1985) Studies on cellobiose metabolism by yeasts. Dev Ind Microbiol 26:527–534

Stewart GG (1981) The genetic manipulation of industrial yeast strains. Can J Microbiol 27:973–990

Stewart GG (1987) The biotechnological relevance of starch-degrading enzymes. CRC Crit Rev Biotechnol 5:89–93

Stewart GG (2006) Studies on the uptake and metabolism of wort sugars during brewing fermentations. MBAA Tech Quart 43:264–269

Stewart GG (2009) The IBD Horace Brown Medal Lecture – forty years of brewing research. J Inst Brew 115:3–29

Stewart GG (2010a) Wort glucose, maltose or maltotriose – do brewer's yeast strains care which one? In: Proceedings of 31st convention of the Institute of Brewing (Asia Pacific Section), Gold Coast, Paper No. 4

Stewart GG (2010b) A love affair with yeast. MBAA Tech Quart 47:4–11

Stewart GG (2012) Flaked barley. The Oxford companion to beer. Oxford University Press, New York, pp 357–358

Stewart GG (2014) High-gravity brewing. In: Brewing intensification. American Society of Brewing Chemsits, St. Paul, MN, pp 7–39

Stewart GG (2016) Adjuncts. In: Bamforth CW (ed) Brewing materials and processes. Academic Press, Elsevier, Boston, MA, pp 27–46

Stewart GG, Murray J (2011) Using brewing science to make good beer. MBAA Tech Quart 48:13–19

Stewart GG, Priest FG (2006) Chap. 6: Adjuncts (pp 161–176) and Chap. 19: Beer stability (pp 715–728). In: Priest FG, Stewart GG (eds) Handbook of brewing, 2nd edn. Taylor and Francis, London

Stewart GG, Russell I (2009) An introduction to science and technology. Series III: Brewer's yeast, 2nd edn. The Institute of Brewing and Distilling, London

Stewart GG, Russell I, Panchal CJ (1988) Genetically stable allopolyploid somatic fusion product useful in the production of fuel alcohols. US Patent 4,772,556

Stewart GG, Zheng X, Russell I (1995) Wort sugar uptake and metabolism – the influence of genetic and environmental factors. In: Proceedings of 25th European brewery convention congress, Brussels, pp 403–410

Stewart GG, Hill A, Lekkas C (2013) Wort FAN – its characteristics and importance during fermentation. J Am Soc Brew Chem 71:179–185

Surdin Y, Sly W, Sire J, Bordes AM, Robichon-Szulmajster H (1965) Propriétés et contrôle génétique du système d'accumulation des acides aminés chez *Saccharomyces cerevisiae*. Biochim Biophys Acta 107:546–566

Tamaki H (1978) Genetic studies of ability to ferment starch in *Saccharomyces*: gene polymorphism. Mol Gen Genet 164:205–209

ter Schure EG, van Reil NAW, Verrips CT (2000) The role of ammonia metabolism in nitrogen catabolite repression in *Saccharomyces cerevisiae*. Fed Eur Microbiol Soc Microbiol Rev 24:67–83

Thorne RSW (1949) Nitrogen metabolism of yeast. A consideration of the mode of assimilation of amino acids. J Inst Brew 50:201–222

Vidgren V, Ruohonen L, Londesborough J (2007) Lager yeasts lack AGT1 transporters, but transport maltose at low temperatures faster than ale yeasts. In: Proceedings of 31st European brewery convention congress, Venice, vol 46, 6–10 May 2007. Fachverlag Hans Carl, Nurnberg, pp 438–444

Vidgren V, Huuskonen A, Virtanen H, Ruohonen L, Londesborough J (2009) Improved fermentation performance of a lager yeast after repair of its AGT1 maltose and maltotriose transporter genes. Appl Environ Microbiol 75:2333–2345

Vidgren V, Rautio J, Mattinen L, Gibson B, Londesborough J (2013) A new maltose/maltotriose transporter (*Saccharomyces eubayanus* –type Agt1) in lager yeast and its relevance to fermentation performance. In: 34th EBC congress, Luxembourg, pp 26–30

Walker GM (1998) Yeast physiology and biotechnology. Wiley, Chichester

Wang JJ, Xiu PH (2010) Construction of amylolytic industrial brewing yeast strain with high glutathione content for manufacturing beer with improved anti-staling capability and flavor. J Microbiol Biotechnol 20:1539–1545

Whitney GK, Murray CR, Russell I, Stewart GG (1985) Potential cost savings for fuel ethanol production by employing a novel hybrid yeast strain. Biotechnol Lett 7:349–354

Wiame JM, Grenson M, Arst HN Jr (1985) Nitrogen catabolite repression in yeasts and filamentous fungi. Adv Microb Physiol 26:1–88

Wickerham LJ (1971) Genus 7. Hanseluna H. et P. Sydow. In: Lodder J (ed) The yeasts, 2nd edn. North Holland, Amsterdam, London, pp 247–253

Wiedmeier VT, Porterfield SP, Hendrich CE (1982) Quantitation of DNS-amino acids from body tissues and fluids using high-performance liquid chromatography. J Chromatogr 231:410–417

Wiemken A (1980) Compartmentation and control of amino acid utilization in yeast. In: Nover L, Lynen F, Mothes K (eds) Cell compartmentation and metabolic channeling. VEB Gustav Fischer Verlag, Jena (GDR) and Elsevier/North Holland Biomedical Press, Amsterdam, pp 225–237

Winkelmann L (2016) 500 years of beer purity law. Brauwelt Int 34:188

Woodward JR, Cirillo VP (1977) Amino acid transport and metabolism in nitrogen-starved cells of *Saccharomyces cerevisiae*. J Bacteriol 130:714–723

Yokota H, Sahar H, Koshino S (1993) Fractionation and quantitation of oligopeptides in beer and wort. J Am Soc Brew Chem 51:54–57

Zheng X, D'Amore T, Russell I, Stewart GG (1994a) Transport kinetics of maltotriose in strains of Saccharomyces. J Ind Microbiol 13:159–166

Zheng X, D'Amore T, Russell I, Stewart GG (1994b) Factors influencing maltotriose utilization during brewery wort fermentations. J Am Soc Brew Chem 52:41–47

Chapter 8
Yeast Viability and Vitality

8.1 Basis of Viability and Vitality Measurements

It has already been, or will be, described (Chaps. 1, 2, 3, 14, 15, 16) that *Saccharomyces cerevisiae* is a model organism for studies of cellular responses to various types of stresses (Botstein and Fink 2011). The determination of cell viability is one of the most commonly used methods of cyto- or genotoxicity under different kinds of chemical, physical or environmental factors. An analysis of yeast viability is also important for industrial processes such as brewing and distilling (Nikolova et al. 2002). Generally, cell viability is reported as the percentage of live cells in a whole population (Minois et al. 2005). This parameter has been determined by a number of methods described in a plethora of research papers (details later). Kwolek-Mirek and Zadrag-Tecza (2014) have introduced a classification of available methods to determine viability according to their performance and the type of results obtained (Fig. 8.1) (details of vitality later). It is worthy of note that one of the first publications to discuss in detail techniques to determine yeast viability was authored by the notable brewing microbiologist from Guinness in Dublin, Brian Gilliland (Gilliland 1959) (Fig. 8.2). This paper succinctly analysed the various methods available at the time to determine yeast viability.

Despite the frequent use of the terms yeast viability and vitality in the brewing and distilling literature, the application of these terms is often confused. The Oxford English Dictionary defines viable as "capable of surviving or living successfully—especially under specific environmental conditions", whereas vitality is defined as "that state of being strong and active"; energy, in the context of living organisms such as yeast, is "the power giving continuance to life". Although the results of many brewing analyses do in fact show that a yeast population is essentially alive, they may or may not show that the yeast population, as a whole, is also active. Thus, data generated describing both yeast viability and vitality are often intertwined (Lloyd and Hayes 1995; Chan et al. 2016). When more than one method is employed, the resulting data should be used to describe "overall yeast fitness during brewing", which summarizes the complex

G.G. Stewart, *Brewing and Distilling Yeasts*, The Yeast Handbook,
https://doi.org/10.1007/978-3-319-69126-8_8

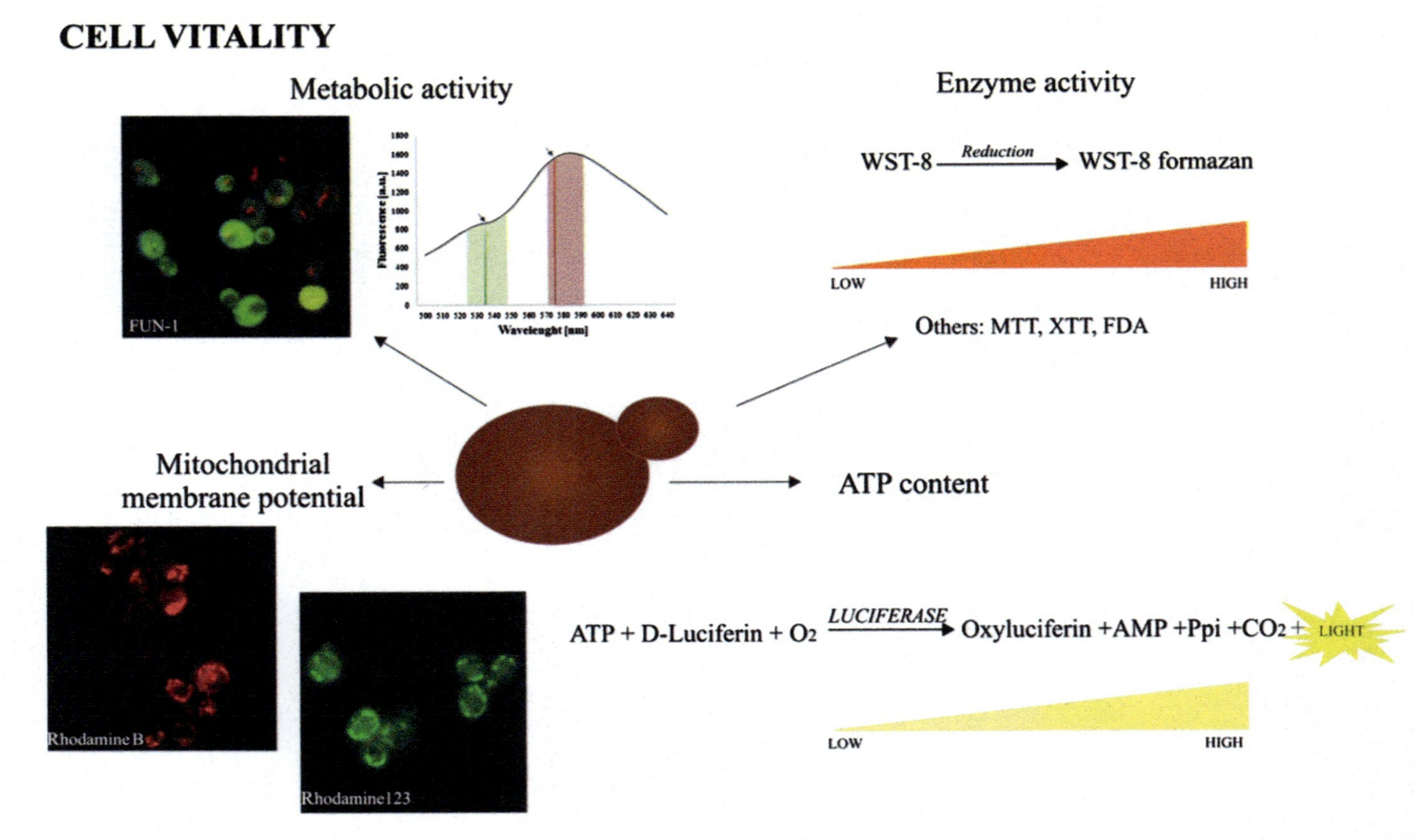

Fig. 8.1 Methods used to determine the viability of yeast cells. Kwolek-Mirek M and Zadrag-Tecza R (2014) Comparison of methods used for assessing the viability and vitality of yeast cells. FEMS Yeast Res. 14:1068–1079

Fig. 8.2 Brian Gilliland

and dynamic relationships between a cell's metabolism, the environment and the methods employed for its cultivation (Davey 2011):

- The first category includes methods based on the ability of yeast cells to grow on solid or in liquid media. One of the most commonly used methods in this category is an analysis of the number of colonies (CFUs, colony-forming units, Longo et al. 2012). There are also other methods in this category such as spot tests, measurement of a growth inhibition zone and cultures in liquid media. The advantages of these methods are the ease of use at a reasonable cost. The major disadvantage is the long waiting time for results. Importantly, even though these tests enable a measure of the degree of growth inhibition, they fail to provide an accurate estimation of viable cells or of cells' ability to reproduce (bud).
- The other category of viability measurements is strain-based systems (Levitz and Diamond 1985). In this case, both colorimetric and fluorescent dyes are employed (Nikolova et al. 2002). The action mechanism of these dyes depends on the properties of the plasma membrane that separates the inside of the cell plasma from the external environment (Chap. 5). Consequently, cell membrane damage usually leads to cell death. Yeast strains that are used to estimate cell viability may be divided into two categories:
  - Dyes that are dependent on the changes in the integrity and functionality of the cell membrane and that at physiological pH are anionic and do not penetrate living cells due to the negative charge of the cell membrane. They are blocked by the intact membrane of viable cells and penetrate into

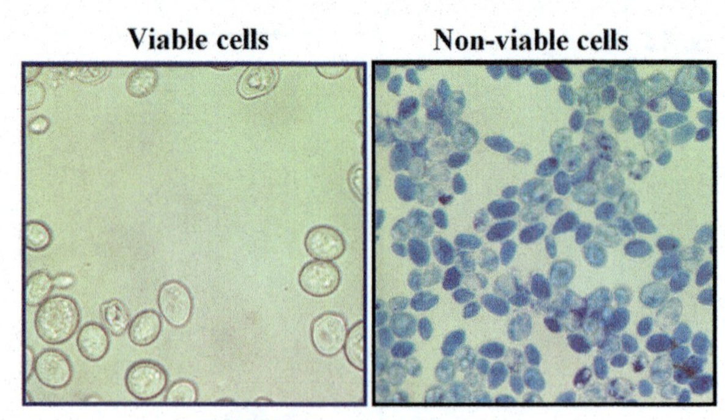

Fig. 8.3 Viability assay using methylene blue stain

dead or damaged cells only staining the nucleus or cytoplasm. The presence of dyes within cells indicates cell membrane damage and cell death. Commonly used dyes in this grouping include DNA-binding fluorescent dyes such as propidium iodide (Fannjiang et al. 2004), ethidium bromide (Aeschbacher et al. 1986), colorimetric dyes such as trypan blue (McGahon et al. 1995) and erythrosin B (Bochner et al. 1989).

– Dyes that penetrate into live and dead cells. Living cells are able to pump out the dye (e.g. phloxine B) (Minois et al. 2005) or reduce the dye (e.g. methylene blue; Painting and Kirsop 1990; Bapat et al. 2006 and methylene violet; Smart et al. 1999) and remain colourless (Fig. 8.3). There are variations on the methylene blue method such as the Gutstein stain (Maneval 1929), where dead cells are unable to affect this reduction and are stained red (phloxine B), blue (methylene blue) or violet (methylene violet). These methods provide rapid and objective results. Also, they enable observation of single cells, providing a distinction between alive and dead cells, and this permits measurement of these two cell categories within a yeast population (Fig. 8.3).

These methods for assessing cell viability (Fig. 8.4) only provide information on live and dead cells within the whole population (Mochaba et al. 1998). However, in many cases, toxic effects of chemicals (e.g. ethanol, acetaldehyde) and/or physical (e.g. temperature and osmotic mechanical stress) factors do not lead directly to cell death. Although such factors may carry a number of morphological, intracellular or metabolic alterations that will result in the inability of the cell to divide, the cell itself may still be metabolically alive. This aspect represents cell vitality, defined briefly as the physiological and metabolic capabilities of a cell (Kara et al. 1988; Stewart and Russell 2009; Kwolek-Mirek and Zadrag-Tecza 2014). The fermentation performance of yeast can be accurately predicted using the acidification power test (Kara et al. 1988). This is a test that can be performed with basic laboratory equipment and can be employed to select pitching yeast for prescribed fermentation purposes.

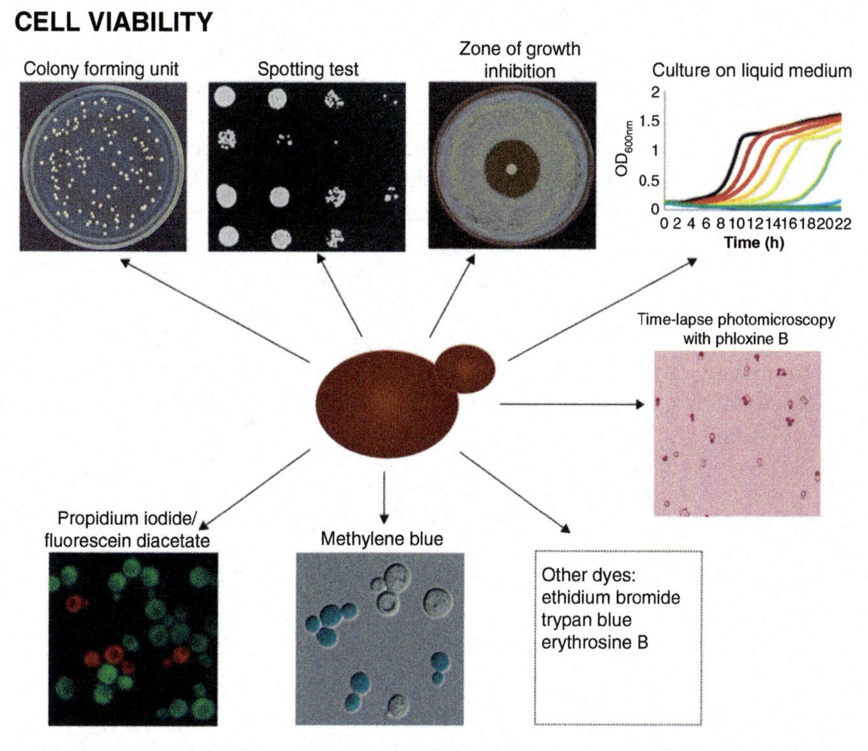

Fig. 8.4 Methods used to determine the viability of yeast cells. Kwolek-Mirek M and Zadrag-Tecza R (2014) Comparison of methods used for assessing the viability and vitality of yeast cells. FEMS Yeast Res. 14:1068–1079

Methods for the determination of yeast cell vitality are based on studies of various physiological and metabolic aspects of cells. These methods fall into the following categories:

- Determination of the intracellular ATP content based on the luciferin reaction (Ansehn and Nilsson 1984).
- Determination of the mitochondrial membrane potential based on staining with rhodamine 123 (Ludovico et al. 2001) or rhodamine B (Marchi and Cavalieri 2008).
- An acidification power test that measures the decrease in the extracellular pH of a suspension of yeast cells following the addition of glucose. Because the amount of maltose in most worts is the majority sugar (Chap. 7), this sugar should be used for the acidification power test instead of glucose, but an evaluation of this possibility is pending (Stewart 2015). This method is useful for detecting large differences in yeast metabolic activity but requires extensive yeast acid washing and the use of multiple sample points (Kara et al. 1988; Dinsdale et al. 1995; Cunningham and Stewart 2000).

- The magnesium release test is based on the observation that low molecular weight species such as magnesium, potassium and phosphate ions are released by yeast immediately following its inoculation into a glucose-containing medium (Mochaba et al. 1997; Walker et al. 1995). Trials performed on a brewer's yeast ale strain showed that cells that released greater quantities of magnesium immediately after inoculation (pitching) into high-gravity (>16°Plato) wort had higher vitality and fermentation performance than yeast that released lower amounts of magnesium. Subsequent fermentations were performed using a more vital yeast, which then exhibited a shorter lag phase, a higher cell count, a greater final ethanol concentration and a lower diacetyl level (D'Amore et al. 1991). The magnesium release test takes less than 15 min to perform and can use a commercially available magnesium test kit (Sigma-Aldrich Cat. No. MAK026). However, I believe that this method is questionable to determine yeast vitality.
- Determination of enzyme activity:
  - Esterases—staining with fluorescein diacetate (FDA) (Chrzanowski et al. 1984; Breeuwer et al. 1995)
  - Oxidoreductases—staining with tetrazolium salts (MTT) (Levitz and Diamond 1985)
  - Redox enzymes—staining with resazurin (O'Brien et al. 2000; Czekanska 2011)

These methods (and others) require the use of advanced equipment; however, they give objective and measurable results! They permit detailed characterization of the metabolic state of the cells, providing a superior basis for developing conclusions regarding the action mechanism(s) of the factors under consideration.

The recent publications by Kwolek-Mirek and Zadrag-Tecza (2014) and by Li-Ying et al. (2016) aimed to develop a clear classification of the methods for analysing both the viability and vitality of yeast cells. A comparative analysis of the results obtained by different methods has been performed based on the effect of selected oxidants that cause oxidative stress. Hydrogen peroxide, allyl alcohol and menadione cause oxidative stress in yeast cells mainly through decreasing the levels of reduced glutathione and increasing the levels of reactive oxygen species (ROS). These compounds can cause a number of negative changes in yeast cells, resulting in cell viability reduction (Castro et al. 2008; Marchi and Cavalieri 2008; Kwolek-Mirek et al. 2009). A number of methods to determine viability have been employed. These included the number of CFUs, the spotting test and the use of a micromanipulator in tandem with a microscope (Fig. 2.11) to observe cells able to reproduce. In this particular study, the spotting test was not found to be a reliable quantitative method, unlike the CFU and micromanipulation methods.

Compared to cell viability measurements based on cell growth, methods based on colorimetric or fluorescent dyes are much more rapid and produce increased measurable results. The commonly employed methods involve labelling cells with dyes such as methylene blue (sometimes together with safranin, as the counterstain) or propidium iodide (P1) and FDA. The use of these dyes allows for the analysis of individual yeast cells. Methylene blue penetrates into every cell (Fig. 8.3). Living cells enzymatically reduce the dye to a colourless product and become unstained, whereas dead cells are stained blue (Bapat et al. 2006). The number of cells considered dead in the case of the CFU method is significantly higher than the number obtained by labelling cells with dyes. However, the results in the study by Kwolek-Mirek and Zadrag-Tecza (2014) showed that after exposure to oxidants, there were many cells unable to reproduce but were still alive. It seems that methods based on staining with dyes provide more objective results on the cells' death rate (Fig. 8.4).

Although the above methods may be used to assess yeast cell viability, which in practice means the determination of the percentage of dead cells in a population, these methods do not characterize the physiological state of cells after exposure to different types of stress. This parameter is very important, especially to focus on cells that are unable to reproduce. In this situation, the determination of cell vitality in the sense of physiological capabilities of cells may assist in the explanation of different types of stress. The level of metabolic activity is closely related to the cell's ATP content. As already discussed, one of the methods for measuring ATP content is based on the reaction of luciferin with ATP in the presence of luciferase, Mg^{++} ions and oxygen resulting in light emission. The luminescent signal is proportional to the cellular ATP content, which is dependent on the number of living cells in suspension (Millard et al. 1997).

This chapter has discussed that there are a wide range of methods for the assessment of both viability and vitality of yeast cells. Yeast strains for use in brewing and distilling are selected, in part, with respect to their viability and thus their capacity to produce ethanol and carbon dioxide, together with other end products. An in-depth characterization of cells, not only in terms of their viability but also their vitality, could be helpful during characterization regarding their potential use on a production scale. It has been suggested by Kwolek-Mirek et al. (2011) that only one method for determining yeast viability and vitality should be employed. A number of selected methods from both categories are preferable. In this way, a full and comprehensive view will be obtained that will help explain the action mechanism(s) of stress agents.

Yeast vitality can be attributed to yeast activity and fermentation performance (Heggart et al. 1999; Lentini 1993). It has already been discussed that yeast vitality is crucial to a rapid and healthy fermentation. However (unlike yeast viability) vitality has to be precisely defined (Heggart et al. 1999), and many different approaches exist that can be used to measure it. Layfield and Sheppard (2015) have published a mini-review entitled "What brewers should know about viability, vitality, and overall production fitness". It discusses the numerous factors that affect overall production fitness in brewing strains.

There is currently no absolutely preferred single method for determining the viability or vitality of a yeast population (Figs. 8.1 and 8.4). However, some methods are recommended by both the American Society of Brewing Chemists (Methods of Analysis, 9th ed 2004) and the European Brewery Convention (Methods of Analysis 2014). Each method for assessing viability, vitality and/or fitness for wort fermentation is usually based on only one parameter. For example, cell composition and/or metabolic activity may both contribute to fermentation performance. As a result, one technique is limited to its usefulness in determining overall brewing fitness. Li-Ying et al. (2016) have recently described a fluorescence method to determine both viability and vitality.

Viability and vitality are characteristics of an individual cell (Davey 2011), although most approaches mean values generated from brewing-related methods obtained from a population of cells, and it describes the average cell(s) in a population. Methods employed included cell count (Knudsen 1999), pH (Mochaba et al. 1998; Kaneda et al. 1997), capacitance (Carvell et al. 2000; Carvell and Turner 2003) and agar plating (Grant 1999; Jett et al. 1997).

8.2 Factors That Influence Viability and Vitality

A number of factors are available with overlapping parameters:

- There is a direct relationship between yeast metabolism and the environment to which a yeast population is being subjected. Wort composition will vary from batch to batch, brewery to brewery and operator to operator (Kennedy et al. 1997). Also, new materials can influence wort composition that will impact on environmental variability. The influence of physical and chemical conditions that exist in the environment can result in variations in cell physiology, composition, metabolism, replication ability and ultimately the cell's overall fermentation fitness (Sheppard and Dawson 1999).
- The importance of carbohydrates (particularly fermentable sugars) during wort fermentation has been discussed in detail in Chaps. 6 and 7, and other aspects are discussed further in Chap. 11. In the context of yeast viability and vitality, monitoring carbohydrate composition can be a useful tool to obtain information on the fermentation fitness of a particular yeast population.

  In brewing, wort fermentable carbohydrates (sugars) are the major nutrient energy source and, consequently, have the capacity to greatly affect cell division, and overall yeast growth, lifespan potential together with fermentation fitness and capacity (Pratt-Marshall et al. 2003) (further details are in Chap. 7). Wort produced with high glucose adjuncts (unmalted cereals) may experience fermentation difficulties as a result of excessive yeast growth, attenuation and flavour irregularities due to enhanced glucose repression and high ester formation (together with other congeners) (Ernandes et al. 1993). Brewer's yeast strains (ale and lager) vary in their sensitivity to glucose (D'Amore et al. 1989). It has already been discussed (Chaps. 6 and 7) that the disaccharide maltose is the primary fermentable sugar in brewer's and distiller's wort (Stewart 2006). However, in yeast, the uptake of maltose and maltotriose will be repressed in favour of utilizing glucose (glucose repression) (Novak et al. 1990; Ernandes et al. 1992). The rate of maltose and maltotriose uptake will remain low until the glucose level falls below approximately 0.4% (w/v), depending on the yeast strain (and other conditions) being employed (Zheng et al. 1994a, b).

Monitoring carbohydrate composition and its utilization can be a further brewing tool that helps in determining the overall brewing fitness when looking for more information regarding a particular yeast culture. In order to obtain this information, high-performance liquid chromatography (HPLC) analysis is an invaluable analytical tool for monitoring the carbohydrate composition of the wort. This yields information on the initial carbohydrate ratio and strength together with carbohydrate utilization patterns during fermentation. HPLC is an invaluable analytical tool for brewing and distilling laboratories that employ trained personnel (for instrument maintenance, data analysis and results interpretation) with an appropriate budget (Stewart and Murray 2010).

- Nitrogen composition is the other major class of wort nutrients that affects brewing yeast performance. Wort nitrogenous constituents influence healthy yeast growth and development as well as having effects on beer haze formation, head retention and biological stability. It has been discussed in Chap. 7, and will be in Chap. 16, that nitrogen utilization can have a profound effect on the overall fitness of brewing yeast cultures. The formation of flavour-active compounds in beer during fermentation by yeast is fundamentally affected by the yeast's ability to grow and utilize wort nitrogenous compounds. Esters, higher alcohols, vicinal diketones and H_2S (together with other sulphur compounds) formation are all influenced by overall nitrogenous compound levels and amino acid metabolism (Chap. 15) (Piskur et al. 2006; Lekkas et al. 2007; Lodolo et al. 2008). The uptake of wort amino acids has already been discussed in detail (Chap. 7) (Jones and Pierce 1964; Stewart et al. 2013). Similar to carbohydrates (Fig. 7.2), amino acids are used sequentially in groups, and the traditional model includes four groups (Table 7.3). This classification of amino acid uptake depends on the criteria used. However, generally amino acid utilization is similar irrespective of the brewing conditions (e.g. the fermentation temperature, wort gravity and vessel type and geometry) but will differ between ale and lager species and from strain to strain (Piskur et al. 2006).

As a facultative aerobe, brewer's yeast cultures (both ale and lager) can shift between aerobic and anaerobic growth phases (the Pasteur and Crabtree effects—details in Chap. 6). It has been discussed in Chaps. 6 and 7 that oxygen has a multifaceted role for yeast in that it can be both helpful and harmful for both yeast viability and vitality. During aerobic growth (yeast propagation), energy is produced through oxidative phosphorylation (Chap. 6). This allows ample energy to be available to generate sufficient yeast cells of acceptable viability and vitality through multiplication (Zitomer and Lowry 1992; Nielsen 2005, 2010). It has already been discussed (Chap. 7) that oxygen is also required by brewing yeast cells to synthesize sterols and unsaturated fatty acids (UFAs), which are essential for the plasma membrane's integrity and functionality (Kirsop 1974; Murray et al. 1984). The presence or absence of these compounds can have wide-ranging effects on the transport of molecules in and out of the cell, regulation of membrane-bound enzymes, ethanol tolerance and the levels of active flavour compounds in beer. Although sterols and UFAs are abundant in malt, they can be lost during mashing and are not normally transferred into wort. Consequently, they must be produced by the yeast under semi-aerobic conditions. The situation with whisky wort is somewhat different. This wort does contain more sterols and UFA because mashing is less controlled than in the brewing process, as the wort is not boiled and is unfiltered (Russell and Stewart 2014).

Although oxygen and aerobic respiration are desired for increasing cell numbers and healthy cell development, oxygen is also a highly reactive molecule that can be reduced to numerous reactive oxygen species (ROS). Yeast generates ROS endogenously as a consequence of aerobic respiration and is consequently subjected to slow, continuous damage to their cellular components owing to free radical stress (Bamforth and Lentini 2009). The free radical theory of cell ageing is based on a cell's inherent antioxidant defences (both enzymatic and non-enzymatic), which would normally quench ROS or repair damaged molecules, gradually being depleted over the course of a cell's replicative lifespan (Jamieson 1998; Powell et al. 2003). To efficiently operate the two alternate physiological states (aerobic and anaerobic), there are a number of genes that are expressed in response to oxygen. This permits the cell to more efficiently utilize oxygen (Zitomer and Lowry 1992). Decreased yeast viability (determined by methylene blue staining) was observed during acid washing but was accompanied by only small decreases in the yeast culture's intracellular pH (Cunningham 2000). These observations by Simpson and Hammond (1989), who proposed that it was only when yeast was in poor physiological conditions, suggested that it reacted poorly to acid washing (details in Chap. 11).

Cell age can be determined from the number of times individual cells, within a population, have divided (variable number of bud scars and only one birth scar per cell) (Fig. 2.6). It should be remembered that a yeast culture is composed of individual cells with varying cell ages. Cell wall analysis (more details in Chap. 13) has shown that, in general, stationary cultures consist of 50% new (virgin) cells, 25% first-generation cells and 12.5% second-generation cells (Powell et al. 2003). The replicative capacity of a particular yeast strain is called the Hayflick limit (Hayflick 1965; Hayflick and Moorhead 1961). This limit is influenced by both genetic (Austriaco 1996) and environmental factors (Barker and Smart 1996; Pratt-Marshall et al. 2003)—also described as nature-nurture effects (Galton 1869; Stewart et al. 1975; Stewart 2014a, b). In brewing, culture age is calculated from the number of times a culture has been repitched in the plant; this is also known as "back sloped" by some brewers, which is a term really relating to the production of sourdough bread (Gänzle 2014).

S. cerevisiae has been shown to be a good model organism to study mechanisms that regulate eukaryotic cells' lifespan, and it has provided a great deal of relevant knowledge on this topic (Longo et al. 2012; Jiménez et al. 2015). The yeast model has shown how the cell cycle impacts lifespan and ageing based on the knowledge generated during these studies. Nutrient-limiting conditions have been involved in lifespan extension, especially in the case of calorie restriction, which also has a

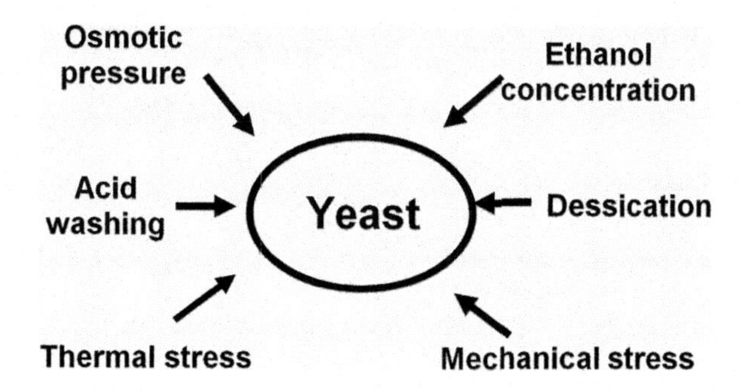

Fig. 8.5 Stress factors which promote proteinase A release

direct impact on cell cycle progression. Other environmental stresses (e.g. osmotic pressure, ethanol, temperature, dessication, oxidative influences) (Fig. 8.5) that interfere with normal cell progression also influence cell lifespan. These facts indicate a relationship between this lifespan and cell cycle control.

Serial repitching subjects a yeast population to repeated stresses that may cause reversible and/or irreversible damage, depending on the stress tolerance of a particular strain (Jenkins et al. 2003; Smart and Whisker 1996). Chronological age is only a factor when yeast is stored for a prolonged period in stationary growth phase leading to compromised cell integrity and eventual cell death.

As yeast cells age, several cellular phenotypic and metabolic changes occur (Barker and Smart 1996; Pratt-Marshall et al. 2003; Powell et al. 2000, 2003; Smart et al. 1999). Phenotypic changes associated with ageing include increases in cell size and the number of bud scars per cell, cell chitin composition, vacuole size, cell surface topography (wrinkling, etc.), decreases in cell turgor and overall modifications in cell shape (Powell et al. 2000).

Of particular importance in this regard is the influence of temperature on the cell's intracellular glycogen level during storage of culture yeast between wort fermentations. It has already been discussed (Chap. 6) that glycogen is the major reserve carbohydrate stored within the cell and that glycogen is similar in structure to plant amylopectin because it contains a number of α-1,4 glucose linkages and α-1,6 branch points (Fig. 6.18). Glycogen serves as a store of potential biochemical energy during the lag phase of fermentation when the energy demand is intense for the synthesis of sterols and unsaturated fatty acids (Fig. 6.19) (Quain and Tubb 1982). It is important that appropriate levels of glycogen and trehalose (Table 8.1) are maintained during storage so that during the initial stages of fermentation, the yeast cell is able to synthesize sterols and unsaturated fatty acids together with trehalose. Trehalose is a nonreducing disaccharide (Fig. 6.20) that serves a protective role against various stresses such as osmoregulation, by protecting the cells during conditions of nutrient depletion and starvation, and by improving cell resistance to high and low temperatures and elevated ethanol levels (Gadd et al. 1997; D'Amore et al. 1991; Odumeru et al. 1992; Bolat 2008). Yeast stress is discussed in greater detail in Chap. 11.

Table 8.1 Concentration of trehalose and glycogen in a lager yeast culture following one, four and eight cycles after fermentation in 15°Plato wort

	Fermentations (cycles)		
	One	Four	Eight
Trehalose[a]	8.8	9.2	11.6
Glycogen[b]	14.6	12.6	9.2

[a]µg/g dry weight of yeast
[b]mg/g dry weight of yeast

There is generally a standard number of times that a yeast crop (generations or cycles) is used for wort fermentation of a particular brewery specification. This is usually standard practice in a particular brewery in order to maintain a culture's appropriate viability and vitality. Typically, a lager yeast culture is now used 6–15 times (cycles) before reverting to a fresh culture of the same strain from a pure yeast culture plant depending on the original gravity of the wort. If a particular yeast culture is used beyond its cropping specification, the culture's overall viability and vitality may well decrease, and fermentation difficulties (wort fermentation rate and extent together with flocculation characteristics are typical examples) are often encountered (Stewart and Russell 2009). An example of this effect is seen when a brewery increases its wort gravity (details in Chap. 11) and adopts high-gravity brewing practices. A particular brewing operation in Canada, over a 15-year period, increased its wort gravity incrementally. In order to avoid fermentation difficulties, it reduced the specification number of yeast cycles with the same strain from a single propagation to:

12° Plato wort OG > 20 yeast cycles (generations)
14° Plato – 16 yeast cycles
16° Plato – 12 yeast cycles
18° Plato – 8 yeast cycles

A limited number of breweries have adopted a revised specification with as few as four or six yeast cycles! Also, with higher wort gravities, the dissolved oxygen concentration at pitching was increased (details in Chap. 11) (Stewart 2014a).

The reason why multiple yeast cycles have decreased viability and vitality leading to a negative effect on a culture's fermentation performance is unclear. However, multiple cycles will result in reduced levels of intracellular glycogen and an increase in trehalose, indicating additional stress conditions during the cycles' progress (Table 8.1) (Boulton and Quain 2001; Stewart 2015).

Storage conditions between brewing fermentations can affect yeast viability and vitality and thus its fermentation efficiency (rate of sugar uptake and extent), together with yeast and beer quality. Good yeast handling practices should encompass collection and storage procedures such as avoiding inclusion of oxygen in the slurry, cooling the slurry to 0–4 °C soon after collection and, perhaps most importantly, ensuring that intracellular glycogen levels are maintained because, as already discussed (Fig. 6.19), this is a critical property at the start of the subsequent wort fermentation:

- Stress is a condition to which yeast is exposed in various forms throughout and between fermentations. These stresses include osmotic stress primarily at the beginning of fermentation, reduced pH and ethanol stress later in a fermentation and at its completion (Stewart 1995). These stresses are primarily due to high concentrations of wort sugars particularly during the use of high-gravity worts (Pratt-Marshall 2002; Stewart 2010, 2014a; Stewart and Murray 2010). Other forms of yeast stress are desiccation, mechanical, acid washing and thermal stress (both hot and cold) (Fig. 8.5) (further details in Chap. 11). A yeast culture is expected to maintain its metabolic activity during stressful conditions, not only by surviving these stresses but by rapidly responding to ensure continued acceptable yeast cell viability and vitality (Casey et al. 1985).

  Stress has a profound and varied effect on yeast cells, including:

- A negative effect on overall fermentation performance, resulting in decreased attenuation rates, sluggish fermentations and a marked reduction in cell volume with a concomitant loss of cell viability and vitality.
- Cell autolysis can occur also with the loss in viability and vitality with the cell contents being excreted into the fermenting wort. This has consequences on beer flavour and stability—especially foam stability and drinkability.
- Stress can also result in the excretion of intracellular enzymes, particularly proteinase A (PrA), which can also negatively affect beer foam stability. This is especially the case when high-gravity worts (>16°Plato) are employed where elevated PrA excretion occurs (Cooper et al. 1998). This excreted PrA hydrolyses hydrophobic polypeptides and reduces the foam-inducing properties of these polypeptides (Cooper et al. 2000). Details in Chap. 11.

8.3 Intracellular pH

Research on yeast physiology has indicated that the cell's ability to maintain and regulate its intracellular pH (pHi) is crucial in order to ensure proper functioning of cellular processes. The initial research on intracellular pH was conducted in the early 1980s by Slavik (1982), who reported that the pHi in a number of *Saccharomyces* spp. strains increased in nutrient-rich anaerobic conditions and decreased as the nutrients were depleted. This implied that pHi was related to the cell's metabolic activity. Maintaining pHi is crucial for cellular equilibrium and to regulate biological pathways. The initial techniques employed to measure pHi were inaccurate and unreliable (Kotyk 1963). However, the advent of flow cytometry and specific probes has permitted the development of fluorescence methods that are more accurate and reproducible to determine pHi in yeast cells (Chlup et al. 2007).

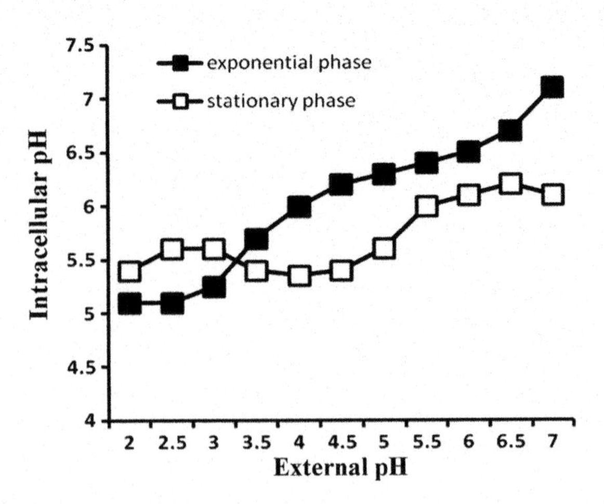

Fig. 8.6 Effect of the external pH on the pHi of *S. cerevisiae* cells analysed by flow cytometry. Cells were harvested in exponential (filled square) and stationary phase (open square), loaded with the fluorochrome and incubated at different pH values

Intracellular pH in yeast cells is an important parameter regarding a culture's viability and vitality. It is regulated by the plasma membrane ATPase, which acts as a proton [H^+] pump (Imai and Ohno 1994; Imai et al. 1994). This proton pump plays a significant role in cell proliferation activity and fermentation regulation (Heggart et al. 2000). The transmembrane [H^+] gradient that is generated is the driving force for the transport of important nutrients (e.g. maltose, maltotriose and amino acids) across the cell membrane. Intracellular pH homeostasis is very important, as key enzymes in glycolysis and gluconeogenesis (details in Chap. 6) are regulated by cascade reactions, which are pH dependent (c-AMP-dependent protein kinases: phosphorylase, glycogen synthase, trehalose, fructose-1,6 bisphosphatase, 6-phosphofructo-2-kinase). Imai and Ohno (1995) found that, under acidic conditions, cells possessing high [H^+] extrusion activity were more vital than cells having low extrusion activity. The results are depicted in Fig. 8.6. Using the flow cytometric method for measuring intercellular pH developed by Valli et al. (2005), it has been found possible to determine reproducibly the pHi distribution within a cell population.

More specifically, the intracellular pH (pHi) of *S. cerevisiae* was determined by a fluorescence microscopic image processing technique. The method enables not only to measure the intracellular pH of dilute suspensions but also to obtain two-dimensional information. In resting cells, the intracellular pH is dependent upon the extracellular pH, and this value is constant when the extracellular pH is constant. However, in the case of actively growing cells, the intracellular pH changes, even when the extracellular pH was constant. The values observed were an intercellular pH of 5.7 during lag phase, intracellular pH of 6.8 during exponential phase and intracellular pH of 5.5 during stationary phase. These results for intercellular pH indicate that the yeast proton pump was activated during growth (Imai and Ohno 1995; Valli et al. 2005).

Ethanol and temperature treatments influence the pHi of *Saccharomyces* strains. A clear relationship has been identified between elevated ethanol concentrations and temperature levels resulting in decreasing cell pHi. The yeast strain's initial pHi was 7.0. This strain was subjected to 20% (v/v) ethanol for 20, 40 and 60 min, and the culture's pHi decreased to 6.2 (11%), 5.8 (18%) and 5.6 (20%), respectively. Also, a 10% decrease to 6.3 pHi was exhibited with cells heated at 60 °C for 30 min, whereas cells heated at 40 °C for 30 min experienced a pH nominal decrease.

There are many factors that influence overall yeast activity (a number of which are discussed in other chapters of this text, e.g. Chaps. 6, 7, 11, 12, 14 and 15). However, yeast viability and vitality are critical parameters. It is imperative that a culture has a high vitality (>95%) and, at the same time, possesses the appropriate intracellular pH (pHi). Yeast pHi has been investigated for nearly 50 years, but metabolic factors controlling pHi are still unclear.

8.4 Conclusions

Viability and vitality of industrial yeast strains are critical parameters. This is particularly important when considering stress effects (high-gravity conditions, high and low temperatures, mechanical stress, etc.). Methods of assessing both viability and vitality (employing flow cytology methodology) have improved, and meaningful results are now obtained. Other chapters in this text (e.g. Chaps. 7, 9, 11, 12 and 13) employ viability and vitality measurements in order to assess overall yeast culture quality.

References

Aeschbacher M, Reinhardt CA, Zbinden G (1986) A rapid cell membrane permeability test using fluorescent dyes and flow cytometry. Cell Biol Toxicol 2:247–255

Ansehn S, Nilsson L (1984) Direct membrane-damaging effect of ketoconazole and tioconazole on Candida albicans demonstrated by bioluminescent assay of ATP. Antimicrob Agents Chemother 26:22–25

Austriaco NR (1996) Review: To bud until death: the genetics of ageing in the yeast *Saccharomyces*. Yeast 12:623 630

Bamforth C, Lentini A (2009) The flavor instability of beer. In: Bamforth CW (ed) Beer: a quality perspective. Elsevier, Boston, MA, pp 85–109

Bapat P, Nandy SK, Wangikar P, Venkatesh KV (2006) Quantification of metabolically active biomass using Methylene Blue dye Reduction Test (MBRT): measurement of CFU in about 200 s. J Microbiol Methods 65:107–116

Barker MG, Smart KA (1996) Morphological changes associated with the cellular ageing of a brewing yeast strain. J Am Soc Brew Chem 54:121–126

Bochner BS, McKelvey AA, Schleimer RP, Hildreth JEK, DW MG Jr (1989) Flow cytometric methods for the analysis of human basophil surface antigens and viability. J Immunol Methods 125:265–271

Bolat I (2008) The importance of trehalose in brewing yeast survival. Innov Roman Food Biotech 2:1–10

Botstein D, Fink GR (2011) Yeast: an experimental organism for 21st century biology. Genetics 189:685–704

Boulton C, Quain D (2001) Brewing yeast and fermentation. Blackwell Science, Oxford

Breeuwer P, Jean-Louis D, Bunschoten N (1995) Characterization of uptake and hydrolysis of fluorescein diacetate and carboxyfluorescein diacetate by intracellular esterases in *Saccharomyces cerevisiae*, which result in accumulation of fluorescent product. Appl Environ Microbiol 61:1614–1619

Carvell JP, Turner K (2003) New applications and methods utilizing radio-frequency impedance measurements for improving yeast management. MBAA Tech Quart 40:30–38

Carvell JP, Austin G, Matthee A, Van de Spiegle K, Cunningham S, Harding C (2000) Developments in using off-line radio frequency impedance methods for measuring viable cell concentration in the brewery. J Am Soc Brew Chem 58:57–62

Casey G, Chen E, Ingledew W (1985) High gravity brewing: production of high levels of ethanol without excessive concentrations of esters and fusel alcohols. J Am Soc Brew Chem 43:179–182

Castro FAV, Mariani D, Panek AD, Eleutherio ECA, Pereira MD (2008) Cytotoxicity mechanism of two naphthoquinones (menadione and plumbagin) in *Saccharomyces cerevisiae*. PLoS One 3(12):e3999

Chan L, Driscoll D, Kuksin D, Saldi S (2016) Measuring lager and ale yeast viability and vitality using fluorescence-based image cytometry. MBAA Tech Quart 53:49–54

Chrzanowski TH, Crotty RD, Hubbard JG, Welch RP (1984) Applicability of the fluorescein diacetate method of detecting active bacteria in freshwater. Microb Ecol 10:179–185

Cooper DJ, Stewart GG, Bryce JH (1998) Some reasons why high gravity brewing has a negative effect on head retention. J Inst Brew 104:221–228

Cooper DJ, Stewart GG, Bryce JH (2000) Yeast proteolytic activity during high and low gravity wort fermentations and its effect on head retention. J Inst Brew 106:197–202

Cunningham (2000) The reaction of brewer's yeast from different fermentation conditions to acid washing. PhD thesis, Heriot Watt University, Edinburgh

Cunningham S, Stewart GG (2000) Acid washing and serial repitching a brewing ale strain of *Saccharomyces cerevisiae* in high gravity wort and the role of wort oxygenation conditions. J Inst Brew 106:389–402

Czekanska EM (2011) Assessment of cell proliferation with resazurin-based fluorescent dye. Methods Mol Biol 740:27–32

D'Amore T, Russell I, Stewart GG (1989) Sugar utilization by yeast during fermentation. J Ind Microbiol 4:315–324

D'Amore T, Crumplen R, Stewart GG (1991) The involvement of trehalose in yeast stress tolerance. J Ind Microbiol 7:191–196

Davey HM (2011) Life, death, and in-between: meanings and methods in microbiology. Appl Environ Microbiol 77:5571–5576

Dinsdale MG, Loyd D, Jarvis B (1995) Yeast vitality during cider fermentation: two approaches to the measurement of membrane potential. J Inst Brew 101:453–458

Ernandes JR, D'Amore T, Russell I, Stewart GG (1992) Regulation of glucose and maltose transport in strains of *Saccharomyces*. J Ind Microbiol 9:127–130

Ernandes JR, Williams JW, Russell I, Stewart GG (1993) Respiratory deficiency in brewing yeast strains – effects on fermentation, flocculation and beer flavour components. J Am Soc Brew Chem 51:16–20

Fannjiang Y, Cheng WC, Lee SJ, Qi B, Pevsner J, McCaffery JM, Hill RB, Basañez G, Hardwick JM (2004) Mitochondrial fission proteins regulate programmed cell death in yeast. Genes Dev 18:2785–2797

Gadd GM, Chalmers K, Reed RH (1997) The role of trehalose in dehydration resistance in *Saccharomyces cerevisiae*. FEMS Microbiol Lett 48:249–254

Galton F (1869) Heredity genius: an enquiry into its laws and consequences. Macmillan, London

Gänzle MG (2014) Enzymatic and bacterial conversions during sourdough fermentation. Food Microbiol 37:2–10

Gilliland RB (1959) Determination of yeast viability. J Inst Brew 65:424–429

Grant HL (1999) Hops. In: McCabe JT (ed) The practical brewer. Master Brewers Association of the Americas, Wauwatosa, WI, pp 201–219

Hayflick L (1965) The limited *in vitro* lifespan of human diploid cell strains. Exp Cell Res 37:614–636

Hayflick L, Moorhead PS (1961) The serial cultivation of human diploid cell strains. Exp Cell Res 25:585–621

Heggart H, Margaritis A, Pilkington H, Stewart RJ, Dowhanick TM, Russell I (1999) Factors affecting yeast viability and vitality characteristics: a review. MBAA Tech Quart 36:383–406

Heggart H, Margaritis A, Pilkington H, Stewart RJ, Sobczak H, Russell I (2000) Measurement of brewing yeast viability and vitality: a review of methods. MBAA Tech Quart 37:409–430

Imai T, Ohno T (1994) Measurement of yeast intracellular pH by image processing and the change it undergoes during growth phase. J Biotechnol 38:165–172

Imai T, Ohno T (1995) The relationship between viability and intracellular pH in the yeast *Saccharomyces cerevisiae*. Appl Environ Microbiol 61:3604–3608

Imai T, Nakajima I, Ohno T (1994) Development of a new method of evaluation of yeast vitality by measuring intracellular pH. J Am Soc Brew Chem 52:5–8

Jamieson DJ (1998) Oxidative stress responses of the yeast *Saccharomyces cerevisiae*. Yeast 14:1511–1527

Jenkins CL, Kennedy AI, Hodgson JA, Thurston P, Smart KA (2003) Impact of serial repitching on lager brewing yeast quality. J Am Soc Brew Chem 61:1–9

Jett BD, Hatter KL, Huycke MM, Gilmore MS (1997) Simplified agar plate method for quantifying viable bacteria. BioTechniques 23:648–650

Jiménez J, Bru S, Ribeiro M, Clotet J (2015) Live fast, die soon: cell cycle progression and lifespan in yeast cells. Microb Cell 2:62–67

Jones M, Pierce JS (1964) Absorption of amino acids from wort by yeasts. J Inst Brew 70:307–315

Kaneda H, Tokashio M, Tomaki T, Osawa T (1997) Influence of pH on flavour staling during beer storage. J Inst Brew 103:21–23

Kara BV, Simpson WJ, Hammond JRM (1988) Prediction of the fermentation performance of brewing yeast with the acidification power test. J Inst Brew 94:153–158

Kennedy AI, Taidi B, Dola JL, Hodsgon JA (1997) Optimisation of a fully defined medium for yeast fermentation studies. Food Technol Biotechnol 35:261–265

Kirsop BH (1974) Oxygen in brewery fermentation. J Inst Brew 80:252–259

Knudsen FB (1999) Fermentation, principles and practices. In: McCabe JT (ed) The practical brewer. Master Brewers Association of the Americas, Wauwatosa, WI, pp 235–261

Kotyk A (1963) Folia. Microbiol (Prague) 8:27–31

Kwolek-Mirek M, Zadrag-Tecza R (2014) Comparison of methods used for assessing the viability and vitality of yeast cells. FEMS Yeast Res 14:1068–1079

Kwolek-Mirek M, Bednarska S, Bartosz G, Bilinski T (2009) Acrolein toxicity involves oxidative stress caused by glutathione depletion in the yeast *Saccharomyces cerevisiae*. Cell Biol Toxicol 25:363–378

Kwolek-Mirek M, Bednarska S, Zadrąg-Tęcza R, Bartosz G (2011) The hydrolytic activity of esterases in the yeast *Saccharomyces cerevisiae* is strain dependent. Cell Biol Int 35:1111–1119

Layfield JB, Sheppard JD (2015) What brewers should know about viability, vitality, and overall brewing fitness: a mini-review. MBAA Tech Quart 52:132–140

Lekkas C, Stewart GG, Hill AE, Taidi B, Hodgson J (2007) Elucidation of the role of nitrogenous wort components in yeast fermentation. J Inst Brew 113:3–8

Lentini A (1993) A review of the various methods available for monitoring the physiological status of yeast: yeast viability and vitality. Fermentation 6:321–327

Levitz SM, Diamond RD (1985) A rapid colorimetric assay of fungal viability with the tetrazolium salt MTT. J Infect Dis 152:938–945

Lloyd D, Hayes AJ (1995) Vigour, vitality and viability of microorganisms. FEMS Microbiol Lett 133:1–7

Lodolo EJ, Kock JLF, Axcell BC, Brooks M (2008) The yeast *Saccharomyces cerevisiae* – the main character in beer brewing. FEMS Yeast Res 8:1018–1036. Special Issue: Thematic issue: Alcoholic fermentation: beverages to biofuel

Longo VD, Shadel GS, Kaeberlein M, Kennedy B (2012) Replicative and chronological aging in *Saccharomyces cerevisiae*. Cell Metab 16:18–31

Ludovico P, Sansonetty F, Corte-Real M (2001) Assessment of mitochondrial membrane potential in yeast cell populations by flow cytometry. Microbiology 147:3335–3343

Maneval WE (1929) Some staining methods for bacteria and yeasts. Stain Technol 4:21–25

Marchi E, Cavalieri D (2008) Yeast as a model to investigate the mitochondrial role in adaptation to dietary fat and calorie surplus. Genes Nutr 3:159–166

McGahon AJ, Martin SJ, Bissonnette RP, Mahboubi A, Shi Y, Mogil RJ, Nishioka WK, Green DR (1995) The end of the (cell) line: methods for the study of apoptosis *in vitro*. In: Schwartz LM, Osborne BA (eds) Methods in cell biology: vol 46. Cell death. Academic Press, New York, pp 153–187

Millard PJ, Roth BL, Thi HPT, Yue ST, Haugland RP (1997) Development of the FUN-1 family of fluorescent probes for vacuole labeling and viability testing of yeasts. Appl Environ Microbiol 63:2897–2905

Minois N, Frajnt M, Wilson C, Vaupel JW (2005) Advances in measuring lifespan in the yeast *Saccharomyces cerevisiae*. Proc Natl Acad Sci USA 102:402–406

Mochaba FM, O'Connor-Cox ESC, Axcell BC (1997) A novel and practical yeast vitality method based on magnesium ion release. J Inst Brew 103:99–102

Mochaba FM, O'Connor-Cox ESC, Axcell BC (1998) Practical procedures to measure yeast viability and vitality prior to pitching. J Am Soc Brew Chem 56:1–6

Murray CR, Barich T, Taylor D (1984) The effect of yeast storage conditions on subsequent fermentations. MBAA Tech Quart 21:189–194

Nielsen O (2005) Control of the yeast propagation process – how to optimize oxygen supply and minimize stress. MBAA Tech Quart 42:128–132

Nielsen O (2010) Status of the yeast propagation process and some aspects of propagation for re-fermentation. Cerevisia 35:71–74

Nikolova M, Savova I, Marinov M (2002) An optimised method for investigation of the yeast viability by means of fluorescent microscopy. J Cult Collect 3:66–71

Novak S, D'Amore T, Stewart GG (1990) 2-Deoxy-D-glucose resistant yeast with altered sugar transport activity. FEBS Lett 269:202–204

O'Brien J, Wilson I, Orton T, Pognan F (2000) Investigation of the alamar blue (resazurin) fluorescent dye for the assessment of mammalian cell cytotoxicity. Eur J Biochem 267:5421–5426

Odumeru JA, D'Amore T, Russell I, Stewart GG (1992) Change in protein composition of *Saccharomyces* brewing in response to heat shock and ethanol stress. J Ind Microbiol 9:229–234

Painting K, Kirsop B (1990) A quick method for estimating the percentage of viable cells in a yeast population, using methylene blue staining. World J Microbiol Biotechnol 6:346–347

Piskur J, Rozpedowska E, Polakova S, Merico A, Compagno C (2006) How did *Saccharomyces* evolve to become a good brewer? Trends Genet 22:183–186

Powell CD, Van Zandycke SM, Quain DE, Smart KA (2000) Replicative ageing and senescence in *Saccharomyces cerevisiae* and the impact on brewing fermentations. Microbiology 146:1023–1034

Powell CD, Quain DE, Smart KA (2003) The impact of brewing yeast cell age on fermentation performance, attenuation and flocculation. FEMS Yeast Res 3:149–157

Pratt-Marshall PL (2002) High gravity brewing—an inducer of yeast stress. Its effect on cellular morphology and physiology. Ph.D. thesis, Heriot-Watt University, Edinburgh, Scotland

Pratt-Marshall PL, Bryce JH, Stewart GG (2003) The effects of osmotic pressure and ethanol on yeast viability and morphology. J Inst Brew 109:218–228

Quain DE, Tubb RS (1982) The importance of glycogen in brewing yeasts. MBAA Tech Quart 19:29–33

Russell I, Stewart GG (eds) (2014) Whisky: Technology, Production and Marketing, 2nd edn. Academic Press (Elsevier), Boston, MA

Sheppard JD, Dawson PSS (1999) Cell synchrony and periodic behavior in yeast populations. Can J Chem Eng 77:893–902

Simpson WJ, Hammond JRM (1989) Cold ATP extractants compatible with constant light signal firefly luciferase reagents. In: Stanley PE, McCarthy BJ, Smither R (eds) ATP luminescence: rapid methods in microbiology. Society for Applied Bacteriology technical series, vol 26. Blackwell Scientific Publications, Oxford, pp 45–52

Slavik J (1982) Intracellular pH of yeast cells measured with fluorescent probes. FEBS Lett 140:22–26

Smart A, Whisker S (1996) Effect of serial repitching on the fermentation properties and condition of brewing yeast. J Am Soc Brew Chem 54:41–44

Smart KA, Chambers KM, Lambert I, Jenkins C (1999) Use of methylene violet staining procedures to determine yeast viability and vitality. J Am Soc Brew Chem 57:18–23

Stewart GG (2006) Studies on the uptake and metabolism of wort sugars during brewing fermentations. MBAA Tech Quart 43:265–269

Stewart GG (2010) High gravity brewing and distilling – past experiences and future prospects. J Am Soc Brew Chem 68:1–9

Stewart GG (2014a) Brewing intensification. American Society of Brewing Chemists, St Paul, MN

Stewart GG (2014b) The concept of nature-nurture applied to brewer's yeast and wort fermentation. MBAA Tech Quart 51:69–80

Stewart GG (2015) Seduced by yeast. J Am Soc Brew Chem 73:1–21

Stewart GG, Murray J (2010) A selective history of high gravity and high alcohol beers. MBAA Tech Quart 47: TQ-47-2-0416-01

Stewart GG, Russell I, Goring TE (1975) Nature-nurture anomalies – further studies in yeast flocculation. Am Soc Brew Chem Proc 33:137–147

Valli M, Sauer M, Branduardi V, Borth N, Porro D, Mattanovich D (2005) Intracellular pH distribution in *Saccharomyces cerevisiae* cell populations, analyzed by flow cytometry. Appl Environ Microbiol 71:1515–1521

Walker GM, Chandrasena G, Birch RM, Maynard A (1995) Proceedings of the 4th Aviemore malting, brewing and distilling conference, 185–192

Zheng X, D'Amore T, Russell I, Stewart GG (1994a) Transport kinetics of maltose and maltotriose in strains of *Saccharomyces*. J Ind Microbiol Biotechnol 13:159–166

Zheng X, D'Amore T, Russell I, Stewart GG (1994b) Factors influencing maltotriose utilization during brewery wort fermentations. J Am Soc Brew Chem 52:41–47

Zitomer RS, Lowry CV (1992) Regulation of gene expression by oxygen in *Saccharomyces cerevisiae*. Microbiol Rev 56:1–11

Chapter 9
Bioethanol

9.1 Introduction

As well as its use in alcoholic beverages, ethanol is employed as a liquid fuel, most often in combination (blended) with gasoline. For the most part, it is used in a 9:1 ratio of gasoline to ethanol to reduce (in part) the negative environmental effects and enhance the fuel's octane value.

In the United States, there is increasing interest in the use of an 85% fuel ethanol blended with 15% gasoline. This fuel blend, called E85, has a higher fuel octane than most premium gasolines. Ethanol used for gasoline and industrial purposes may be considered as a fossil fuel because it is often synthesized from ethylene, which is usually less expensive than bioethanol produced from cereals or cane sugar. However, this chapter will only consider the production of ethanol by fermentation. Although the proliferation of hydraulic fracturing wells in the United States (over two million oil and gas wells in 2013) (Montgomery and Smith 2010) has influenced gasoline prices, bioethanol production as a liquid fuel is still in demand, but it does vary with prevailing economic conditions.

The global population is estimated to increase by approximately 3 billion persons by the mid-twenty-first century. With this population increase, the demand and cost of fossil fuels will increase considerably (Johnson 2007). New technologies are needed for fuel extraction using feedstocks that do not threaten food security, cause minimal or no loss of animal's natural habitats and reduce soil erosion. Also, waste management has to be improved, and environmental pollution should be minimized or eliminated. Many of the current oil reserves are located in politically unstable regions (e.g. the Middle East and parts of Africa), which could result in considerable fluctuations in the global oil supply and the price of a barrel of crude oil (Rommer 2010).

Any fuel produced from biological materials (e.g. agricultural residues and municipal waste) is generally referred to as a biofuel. More specifically, the term generally refers to liquid transportation fuels. Over time, gasoline costs have

G.G. Stewart, *Brewing and Distilling Yeasts*, The Yeast Handbook,
https://doi.org/10.1007/978-3-319-69126-8_9

increased significantly, although lately there has been a large decrease in world oil prices. How long this will last is impossible to predict! There has been an interest in biofuels since the early 1970s particularly, but not exclusively, in the United States and Brazil (Walker 2011a, b). This has resulted in a large number of publications on this topic (e.g. Bellissimi and Ingledew 2005; Searchinger et al. 2008; Jacobucci 2008; Rommer 2010).

Liquid biofuels offer a promising alternative to fossil fuels (Kline and Dale 2008). The primary characteristics of a suitable renewable biofuel are:

- It has the potential to replace fossil fuels but should not affect global food supplies.
- The production process must have a net positive energy balance.
- It should have minimal negative environmental impact (Johnson 2007).

The possible consequences of an increasing mean global temperature include rising sea levels as polar ice caps melt, redistribution of rainfall, more intense storms and an enhanced rate of species extinction. While there has been debate about the likelihood of these and other possible consequences of increasing atmospheric CO_2 levels, the uncertainty associated with large-scale anthropogenic (human characteristics) changes of this nature represents an unacceptable risk to our planet and the environment as a whole, especially if alternatives are found. It is important that carbon-neutral biofuels be widely adopted to eventually replace fossil fuels.

It has already been stated that any fuel produced from biological materials—for example, cereals, sugar (beet and cane), agricultural residues and municipal residues—is generally referred to as a biofuel. More specifically, the term generally (but not always) refers to liquid transportation fuels. Over time, gasoline costs have increased substantially although recently there has been a substantial decrease in world oil prices (currently [2017] US$48/barrel down from the high of US$110/barrel in 2011). How long this reduced price structure will continue is impossible to predict (Ingledew 2005).

9.2 Biofuel in Brazil

In the early 1970s, as crude oil became more expensive and sugar less expensive, the Brazilian government established a national programme (*Pró-Álcool*), which was aimed at the substitution of gasoline with fuel ethanol (Goldemberg 2008) produced by the fermentation of sucrose from sugar cane with yeast. The primary aim of this initiative was to reduce Brazil's requirement for US dollars and other major currencies. This initiative ensured that Brazil was about 30 years ahead of other countries in terms of expertise for the development of first-generation (1G) bioethanol production. This gave Brazil a privileged position in first-generation bioethanol (1G) production. Indeed, Brazilian sugar cane fuel ethanol

Table 9.1 Annual fuel ethanol production by country (2007–2011) (millions of US liquid gallons per annum)

	2011	2010	2009	2008	2007
United States	13.9	13.2	10.9	9.2	6.5
Brazil	5.6	6.9	6.6	6.5	5.1
EU	1.2	1.2	1.1	0.7	0.6
China	0.6	0.5	0.5	0.5	0.5
Canada	0.5	0.4	0.3	0.2	0.2
Global total	22.4	22.9	19.5	17.3	13.1

is highly competitive when compared with production processes from other crops (e.g. corn, sorghum and sugar beet—details later). The sugar cane ethanol industry (particularly in Brazil) shows the highest percentage of greenhouse gas emission reduction, the highest energy balance and yield per hectare, and lower production costs (Garoma et al. 2012).

During fermentation, the key step in Brazilian bioethanol production is the conversion of sucrose (glucose and fructose) by *S. cerevisiae* to ethanol, in fermentation vessels with millions of litres of capacity (Basso et al. 2011b). Assuming there are no real operational issues (such as rain, contamination or power failures), yields as high as 92% of the stoichiometric conversion (0.511 g ethanol/g hexose equivalent) can be achieved. However, the average industrial ethanol yield in Brazil has decreased slightly during the past decade (Table 9.1). Since more than half of ethanol's final production cost is due to the variable price of sugar cane, any increment on the high-yield value would represent economic gains. For example, with a Brazilian annual production of 30 billion litres of ethanol, a 1% increase in fermentation yield would allow for the production of an extra 300 million litres of ethanol from approximately the same amount of raw material or, in other words, from the same cultivation area (Zanin et al. 2000).

One of the ways to achieve increased production yields is through the use of appropriate yeast strains (details later) and modifications to the process, as has been traditionally conducted in the wine, potable spirit and brewing industries (Miranda et al. 2009; Stewart 2014). Strains that have adapted to the harsh environments encountered in industrial ethanolic fermentations capable of delivering higher alcohol yields (Russell et al. 1987). Increasing productivity has become an essential task for optimizing the process or having strains genetically modified—not only for ethanol fermentation but also for the production of other biofuels, chemical precursors and/or higher value compounds, from sugar cane-based substrates (details later) (Slapack et al. 1987).

In Brazil, sugar cane is the principal raw material of choice for bioethanol production (Herrero et al. 2003). In the United States, starch (and latterly lignocellulosics) is the primary raw material—details later. Sugar cane provides readily fermentable sugars (11–20% w/w with 90% being sucrose and 10% glucose and fructose) because it does not require prior treatment to be metabolized into ethanol by *S. cerevisiae* (Fig. 9.1). Moreover, nitrogen-fixing bacteria can supply up to 60% of the cane sugar's nitrogen demand in low fertility soils, reducing the need for nitrogen fertilizers, which are known to need large amounts of fossil energy for

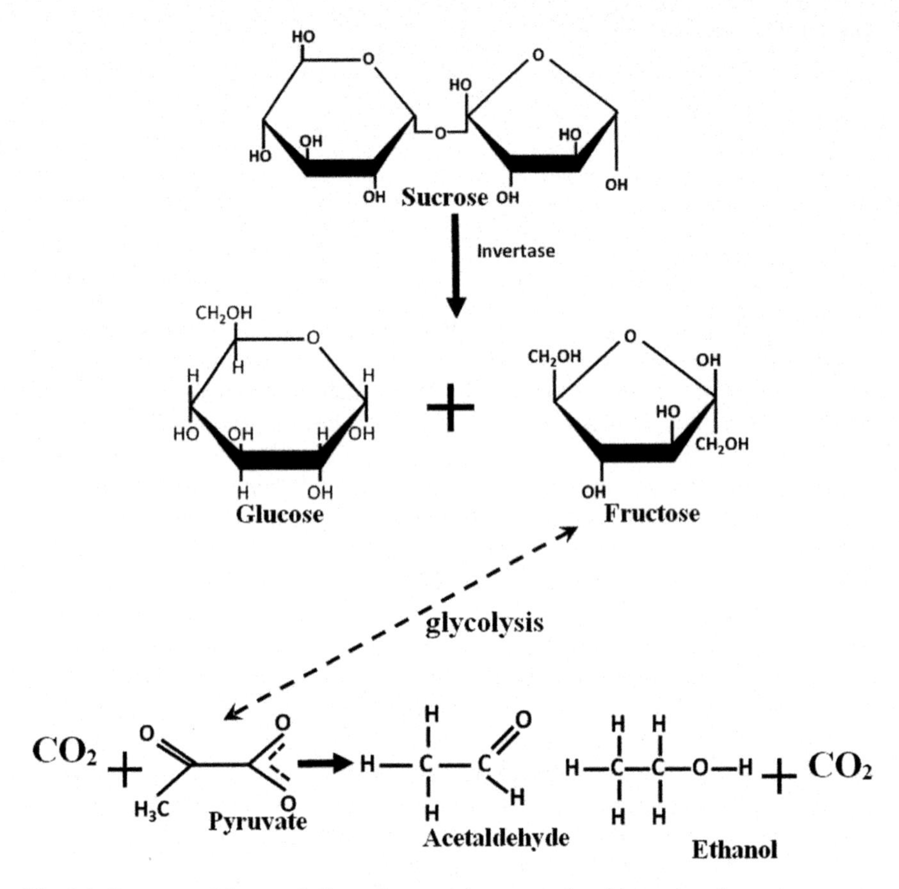

Fig. 9.1 Summary of the metabolism of sucrose (sugar cane/beet) into ethanol

their production (Della-Bianca et al. 2013). These features contribute to a favourable energy balance (output/input ratio). The sugar cane overall ethanol energy balance in Brazil for the 2005/2006 crop season was estimated to be 9.3:1 and is predicted to be at least 11:1 by 2020 (Macedo et al. 2008), while for corn starch ethanol, it varies from less than 1:1 (Shapouri et al. 2008).

The Melle-Boinot process (cell recycle batch fermentation—CRBF) is the method of choice in Brazil for ethanol fermentation (Zanin et al. 2000). It comprises three main features:

- A fed-batch mode is used due to higher production stability, simpler flow adjustment and low equipment costs compared to the continuous mode (Amorim et al. 2009). The usual low-cost adaptation of plants from batch to continuous modes may hide the benefits from the latter option.
- Up to 90–95% of the yeast cells are recycled by centrifugation, and this allows high cell densities during the fermentation of 10–18% sugar, so that no intensive yeast propagation is needed before (or during) each fermentation cycle. This,

combined with the low nitrogen content needed for sugar cane-based substrates, allows for an increase of only 5–10% yeast biomass during one fermentation cycle, which is sufficient to replace the loss of cells that can occur during centrifugation (Chlup and Stewart 2011).

- Lastly, recycled yeast is usually acid treated to reduce bacterial contamination, very similar to the process originally developed by Pasteur and Faulkner (1879) with tartaric acid for the treatment of wine and brewing yeast cultures (details in Chap. 6) (Cunningham and Stewart 2000). Many breweries have employed acid washing of yeast for over 100 years. Today some large breweries, though fewer than historically, still conduct this procedure (Stewart and Russell 2009), and many craft brewers seem to favour this procedure. Acid treatment lasts for 1–3 h and consists of dilution with cold water and the addition of either phosphoric acid or sulphuric acid (pH 2.0–2.5) with stirring, after which the cells must be immediately reused (Simpson and Hammond 1989). This acid washing step can take place at least twice a day during the bioethanol production season, which may last for 250 days (Basso et al. 2011a).

In summary, fermentation for the production of bioethanol starts by adding a substrate, containing 18–22% (w/w) of total reducing sugars (TRS) to a 30% (wet basis) yeast suspension, which represents approximately 25–30% of a fermenter's total volume. Substrate infusion time normally lasts for 4–6 h, and the fermentation is complete within 6–10 h, resulting in the production of ethanol concentrations of 8–12% (v/v). Ethanol concentrations could be higher, but this would involve increasing fermentation times to unacceptable levels and could compromise yeast viability and, as a result, negatively affect the yeast's fermentation fitness during the subsequent fermentation cycle (Ercan and Demirci 2015).

The fermentation process for fuel ethanol production in Brazil (also in the United States and other countries) presents far from the optimal physiological conditions for the yeast. Several stress factors alternate during the process, amongst which the following are the most relevant:

- High sugar and ethanol concentrations (Bauer and Pretorius 2000; Gibson 2011)
- Elevated temperatures (Slapack et al. 1987; Piper 1997; Piper et al. 1997)
- Media concentrations (Stewart 2014)
- pH variations
- Presence of toxic compounds (such as acetic acid, acetaldehyde, diacetyl, etc.) (Herrero et al. 2003)
- Contamination with non-*Saccharomyces* microorganisms (Abbott and Ingledew 2005).

In this atypical environment, yeast exhibits stress responses, and a good industrial strain must be sufficiently robust to respond efficiently to these environmental (nurture) variations, without altering its fermentation characteristics over the cycles to which it is exposed during the entire cropping season. Sugar cane media are not sterilized, and therefore this is an environment that does not preclude the proliferation of contaminants, mainly bacteria, because of its high nutrient concentration, high water activity and favourable pH for some species (not all) (Amorim et al. 2004). Besides bacteria, wild yeasts (including unwanted *S. cerevisiae* strains) can contaminate and dominate the medium. After isolation, followed by molecular and physiological characteristics (Basso et al. 2008) (nature-nurture—further details in Chaps. 6 and 8), some of these contaminating strains were found to:

- Be more adapted to the process than the starter cultures
- Be able to promote high ethanol yields
- Not usually result in technological issues such as yeast flocculation and/or foaming—there are exceptions (e.g. bacteria—yeast co-flocculation) (Zarattini et al. 1993)

Until the mid-1990s, baker's yeast strains, together with strains IZ-1904, and TA, were the most used strains for industrial fuel ethanol production in Brazil. However, in laboratory studies, their physiological parameters were determined after just a single fermentation cycle (without cell recycle procedures being practised). When these same strains were evaluated under cell recycle conditions, valuable physiological data was obtained. Indeed, when compared with typical baker's yeast strains, IZ-1904 produced a higher ethanol yield and lower biomass concentrations together with higher glycerol concentrations, lower cell viability and very low levels of intracellular trehalose (Fig. 6.21). These results indicated that the IZ-1904 cultures would be unable to endure cell recycling because the higher ethanol yield obtained was at the expense of biomass formation and reduced trehalose accumulation (Alves 1994). Further studies demonstrated that the only strain able to persist in the industrial-scale fermentations was JA-1—a strain previously isolated from an ethanol plant in Brazil (Basso and Amorim 1994).

The PE-2 and VR-1 yeast strains of *S. cerevisiae* were initially introduced into 24 Brazilian distilleries, and they exhibited remarkable fermentation efficiencies. These strains represented 80–100% of the total yeast biomass in the fermentation of 12 distilleries and were implanted into 63% of the plants, accounting for 42% of the yeast biomass at the end of the season (Basso et al. 2011a, b). Since then, the PE-2 strain has been extensively used as a reference industrial strain compared with baker's yeast strains. This revealed physiological traits that could be related to the superior fermentation performance of strain PE-2. This strain has also been found to be suitable for the microvinification of raspberry juice (Duarte et al. 2010) and sweet potato hydrolysate fermentations (Pavlak et al. 2011). The high ethanol tolerance of PE-2 rendered this strain appropriate for very high-gravity (VHG) fermentation, with ethanol concentrations >19% (v/v) being obtained (Pereira et al. 2010, 2011). Also, a recombinant product of PE-2 containing the *FLO1* gene, responsible for a flocculation protein (Stewart et al. 2013) (details in Chap. 13), performed successfully in VHG and flocculation-sedimentation recycle conditions (Choi et al. 2009; Gomes et al. 2012).

These industrial strains invariably accumulate high levels of the storage carbohydrates trehalose and glycogen (details of their overall metabolism in Chaps. 6 and 7), which can account for 20% (w/w) of yeast dry weight towards the end of fermentation (Paulillo et al. 2003). The degradation kinetics of trehalose and glycogen in the strain PE-2 has been analysed under high temperature and high biomass conditions (details in Chap. 11). This resulted in additional ethanol formation and a consequent increase in protein levels, which rose from 35 to 47% (w/w). This strategy has been accepted as an important way to enhance the economic value of the excess yeast slurry when used as an animal feed, in which the protein levels of the culture are expected to be at least 40% (Amorim and Basso 1991).

Regarding the patterns of sugar utilization by industrial fuel ethanol yeasts, their ability to ferment sucrose (Miranda et al. 2009), maltose and maltotriose (Amorim Neto et al. 2009; Duval et al. 2010) has been evaluated. The results indicated that some industrial ethanol-producing strains ferment maltose but not maltotriose and exhibited higher thermotolerance when compared to standard commercial whisky-distilling yeast strains (Amorim Neto et al. 2009). Whisky yeast strains can invariably ferment maltose and maltotriose (Russell and Stewart 2014). Most whisky, wine and brewing yeast strains can ferment glucose, fructose, sucrose, maltose and maltotriose in high concentrations, up to 330 g/l for strain PE-2 (Pereira et al. 2011)—further details on sugar uptake by *Saccharomyces* spp. can be found in Chap. 7.

The majority of the publications considering the physiology of industrial yeast strains employed in Brazil reported on investigations with these strains and the stress conditions associated with high ethanol conditions, elevated temperatures, high osmotic pressure environments due to sugar and salts, low pH, possible inhibitors and bacterial contaminations (Walker 2011a, b) (further details in yeast stress are in Chap. 11). These conditions have been investigated in more detail for other fuel ethanol-related strains, such as those involved the corn-based ethanol industry in the United States—details discussed later (Zhao and Bai 2009). With a number of ethanol-producing strains, the effects of temperature, pH, sugar concentration, nitrogen (Alves 1994; Gutierrez et al. 1989, 1991a), potassium, sulphite (Alves 1994), nitrite (Gutierrez and Orelli 1991), 2,4-dinitrophenol (Gutierrez et al. 1989, 1991b), benzoic acid (Gutierrez et al. 1991b), acetic acid (Gutierrez 1993) and octanoic acid (Gutierrez et al. 1991a) have been investigated. Up to 300 g/l fermentable sugar increased glycerol formation but did not usually affect the ethanol yield, yeast viability and intracellular trehalose levels, possibly because of reduced bacterial growth and, consequently, lactic and acetic acid concentrations in the culture medium (Alves 1994).

Excessive yeast growth, which occurs in some distilleries, usually compromises the ethanol yield. For this reason, increasing the ethanol yield by lowering the biomass and/or by-product formation (mainly glycerol) was achieved by the addition of 2,3-dinitrophenol (Gutierrez et al. 1989), acetic acid (Gutierrez et al. 1991b) or benzoic acid (Gutierrez et al. 1991a). The latter treatment was more efficient to reduce biomass formation and maintained the yeast culture's high viability.

Bacterial contamination is often regarded as a major drawback during industrial ethanol fermentation—this is also the case in the production of potable ethanol such as whisk(e)y (Wilson 2014). Besides diverting fermentable sugars from ethanol formation, there are also the detrimental effects of some bacterial metabolites (such as lactic and acetic acids) upon yeast fermentation performance, resulting in a reduced ethanol yield, early yeast flocculation with fewer yeast cells in suspension and lowered yeast viability. Most of the bacterial contaminants during the fermentation steps of ethanol production are lactic acid bacteria, probably because these bacteria are more able to tolerate low pH values and high ethanol concentrations compared to other microorganisms. Indeed, *Lactobacillus* was the most abundant bacterial genus isolated from the majority of Brazilian ethanol plants, both homo- and heterofermentative types (Basso et al. 2011a).

There are many aspects regarding Brazilian yeast ethanol strains that remain unknown. This is why they dominate fermenters so quickly in some distilleries, but in 40% of the plants, none of them can be replaced (Della-Bianca et al. 2013). These questions should be the object of further research, and there is a need to isolate and/or produce new strains because the number of commercially available strains for this industry is limited (unlike the situation in the fuel ethanol industry in the United States—details later in this chapter). One possibility is to identify strain(s) versatile enough to be introduced into a number of distilleries. This can be achieved by isolating customized yeast strains and subsequently selecting strains that are already employed in plants. These strains are probably more suitable to the distinct environment that they encounter naturally.

The possibility of using GM yeast strains with specific traits for industrial fuel ethanol production in Brazil is dependent on the introduction of more sanitary conditions in the different stages of the process because wild yeasts with a higher specific growth rate and viability will probably outcompete the GM strains (Lynd et al. 2008). Studies are ongoing in order to understand what makes these industrial strains dominate and persist (or otherwise) inside production-scale fermenters. Detailed genome sequencing and annotation, targeted physiological studies and also field trials are required. Finally, we need to know the source of these wild yeasts. Do they originate in the sugar cane fields, are they brought in by insects that occur in the production plant or, since they are more related to baker's strains than to wine or brewing strains (Stambuk et al. 2009), did they just evolve from commercial baker's yeast strains (Maybee 2007)? The answer is probably all of the above!

9.3 Biofuel in North America

In the United States (Searchinger et al. 2008) and Canada, bioethanol is produced mainly from corn (or maize, *Zea mays*) while in Europe wheat (Chin and Ingledew 1993), and sucrose from sugar beet (Leiper et al. 2006) is used as the main fermentation substrate. Cereal conversion to bioethanol with cereals basically comprises the following steps: milling, starch liquefaction and hydrolysis, fermentation and distillation.

In North America, corn-to-bioethanol processes are differentiated into two process types: dry and wet milling (Fig. 9.2) (Abbas 2007; O'Brien and Woolverton

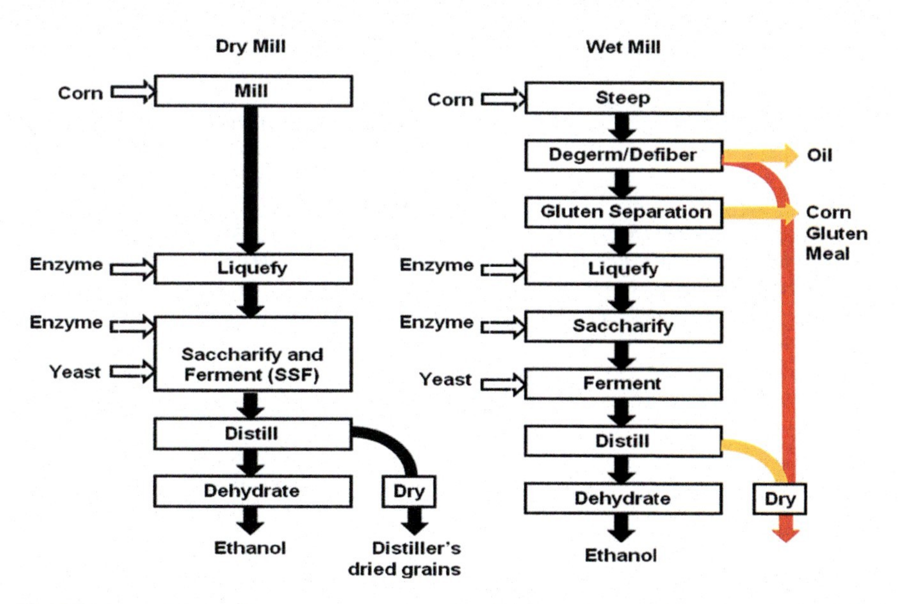

Fig. 9.2 Dry and wet milling corn processes for bioethanol production. Abbas CA (2010) Abbas C. Going against the grain: food versus fuel uses of cereals. In: Distilled Spirits. New Horizons: Energy, Environment and Enlightenment. Proc of the Worldwide Distilled Spirits Conference, Edinburgh (2008) Walker GM, Hughes PS, eds. Nottingham University Press: Nottingham, 2010; p. 9–18

2009). Dry milling processes involve fine grinding of corn kernels, which are further processed without fractionation. In this context, wet milling processers first soak the kernels in water (sometimes dilute acid), which separates the cereal into starch, gluten, protein, oil and fibre. In both dry and wet milling processes, the starch is liquefied and saccharified by amylolytic enzymes (Fig. 9.2) into fermentable sugars prior to fermentation (Shapouri et al. 2008).

Starch is not directly fermented by most strains of *S. cerevisiae*—there are some strains that used to be classified as *S. diastaticus* that can hydrolyse dextrins with their extracellular glucoamylase but not actual starch (Erratt and Stewart 1978, 1981; Ogata et al. 2017—details in Chap. 7). Consequently, corn starch requires cereal cooking, starch liquefaction and amylolysis. Complete starch hydrolysis is required using exogenous amylolytic enzymes including α- and β-amylases (for liquefaction), proteinase (de-branching enzyme) to debranch amylopectin, glucoamylase to produce glucose and glucanases (for viscosity reduction) (Fig. 7.1). Industrial enzymes used in starch-to-ethanol bioconversions are produced by specialist companies from microbial fermentations using bacteria such as *Bacillus* spp. and fungi such as *Aspergillus* spp. (Khew et al. 2008; Nair et al. 2008).

Wheat-to-bioethanol processes share similarities with corn processes. However, wheat has the potential to produce higher alcohol yields when compared to corn when residual biomass (spent grains) saccharification using selected commercial enzymes is taken into account (Green et al. 2015). Wheat and corn processing

Table 9.2 Microorganisms employed for bioethanol fermentations

Microbe	Characteristics
Saccharomyces cerevisiae	Capable of fermenting glucose, fructose, sucrose, maltose and usually maltotriose. Genetically modified xylose, arabinose, and cellobiose
Pichia stipitis, Candida shehatae, Kluyveromyces marxianus, Pachysolen tannophilus, Hansenula polymorpha, Dekkera bruxellensis, Candida krusei	Non-*Saccharomyces* yeasts capable of fermenting pentose sugars—second-generation lignocellulose feedstocks. Some are high-temperature fermenters
Non-GM bacteria	Ethanolic bacteria such as *Zymomonas mobilis*
GM bacteria	*Geobacillus stearothermophilus* is a thermophilic microorganism that ferments C5Ml6 sugars
Microalgae	Blue-green algae (cyanobacteriae) that can be engineered to produce ethanol

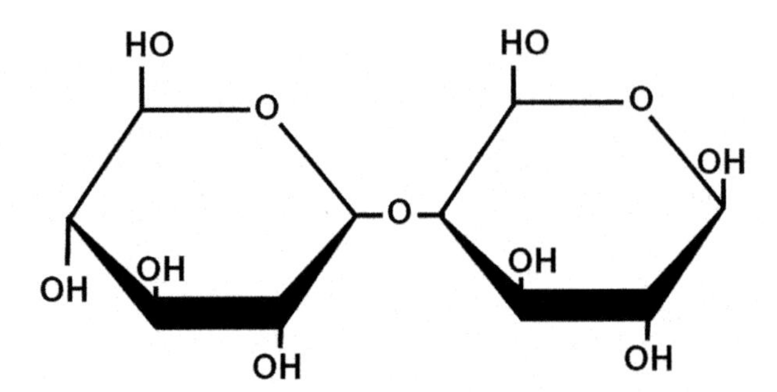

Fig. 9.3 Structure of cellobiose

temperatures and the use of processing aids are of potential economic benefit to bioethanol producers seeking to understand the factors influencing the processing properties of different cereals (Miedl et al. 2007).

As with sugar cane fermentation, *S. cerevisiae* is the predominant microorganism responsible for the fermentation of corn and wheat starch hydrolysates in the production of industrial bioethanol. Other yeasts [e.g. genetically manipulated or GM varieties of *S. cerevisiae* (Den Haan et al. 2007), pentose-fermenting yeast strains (Hahn-Hägerdal et al. 2007a, b), cellulose degradation (Ito et al. 2004), starch utilization (Khaw et al. 2006), *Pichia* sp. (Russell et al. 1987), *Candida tropicalis* (Jamai et al. 2007), whey fermentation by *Kluyveromyces marxianus* (Ariyanti and Hadiyanto 2013), certain bacteria *Zymomonas mobilis* (Rogers et al. 2007) and hemicellulose degradation by *Thermoanaerobacterium saccharolyticum* (Curry et al. 2015)] have potential to be used in this process (further details of many of these microenzymes are in Chap. 17). Table 9.2 summarizes the major microorganisms (yeast and bacteria) that are employed during bioethanol fermentations for example, with cellobiose (Fig. 9.3).

9.4 Second-Generation Bioethanol

Starch and sucrose are regarded as first-generation feedstocks for fermentation bioethanol production. They are regarded by some as unsustainable due to their food security and land-use importance. Second-generation bioethanol (or advanced fuels) refers to fuel alcohol produced from non-food biomass sources, such as lignocellulose (Solomon et al. 2007). Lignocellulosic biomass encompasses two main categories of bioethanol feedstocks:

- Biowaste materials—straw, corn residues, woody wastes/chippings, forestry residues, waste paper, cardboard, bagasse (cane sugar residues), spent grains, municipal solid waste (MSW), agricultural residues (oil seed pulp, sugar beet pulp)
- Energy crops—short rotation crops, energy grasses, giant reeds, switchgrass, alfalfa, etc.

Cellulose-based agricultural and industrial biomass (or biowastes) represents suitable and ethically acceptable materials and processes for bioethanol production (Fig. 9.4). These materials also offer greater cost reductions compared to the utilization of starch and sugar crops (Abbas 2007; Kasi and Ragauskas 2010; Sanchez and Cardona 2008; Shi et al. 2009; Yan et al. 2010).

A typical lignocellulosic material is of the following: cellulose $C_6H_{10}O_5$, hemicellulose $C_5H_8O_4$ and lignin $C_6H_{11}O_2$. The former two macromolecules can both be hydrolysed to fermentable sugars, but lignin cannot. Cellulose is

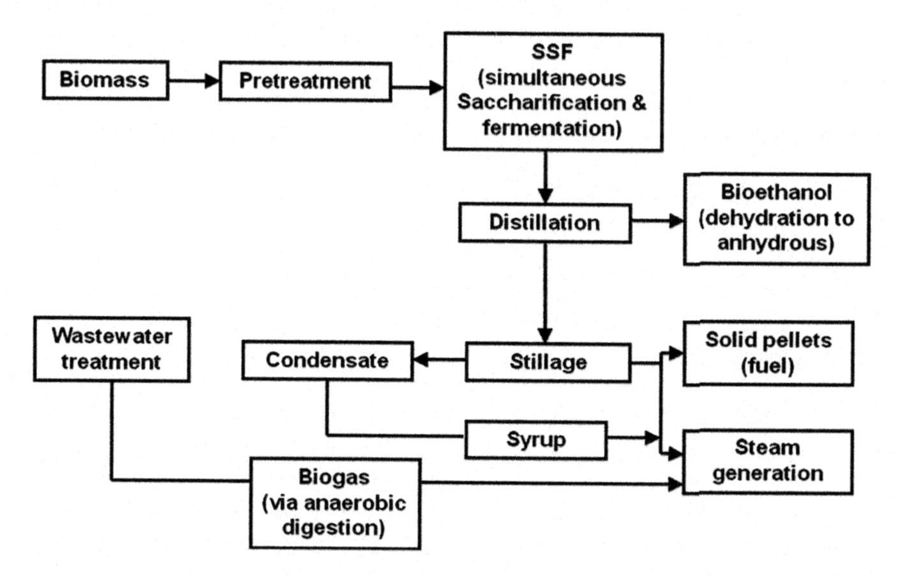

Fig. 9.4 Generalized lignocellulose to bioethanol process. Sassner P, Galbe M Zacchi G. (2008) Techno-economic evaluation of bioethanol production from three different lignocellulosic materials. Biomass Bioenergy 32:422–430

derived from D-glucose units, which condense through $\beta(1 \rightarrow 4)$-glycosidic bonds. This linkage motif contrasts with that of the $\alpha(1 \rightarrow 4)$-glycosidic bonds present in starch, glycogen and other carbohydrates (details in Chap. 6). Cellulose is a straight chain polymer, unlike amylose. Compared to starch, cellulose is much more crystalline. Hemicellulose (also known as polyose) is any of several heteropolymers (mature polysaccharides), such as arabinoxylans that are present, along with cellulose, in almost all plant cell walls (Scheller and Ulvskov 2010). Unlike cellulose, it has a random, amorphous highly branched structure with little strength. It is easily hydrolysed by dilute acid or alkali as well as a myriad of hemicellulases. It is comprised of both pentose (xylose and arabinose) and hexose (glucose, mannose and galactose) sugars (Lennartsson et al. 2014).

Lignin is a tough, recalcitrant plant cell wall material, which is comprised of di- and mono-methoxylated and non-methoxylated phenylpropanoid units in a three-dimensional network (Davin and Lewis 2005). Acid hydrolysis of lignocellulosic biomass will leave behind acid-insoluble lignin, but some acid-soluble lignin may be released into the hydrolysate liquor. For bioethanol production processes, acid-soluble lignin components, including phenolic degradation products, can inhibit cellulase activity and yeast fermentation. Other minor components of lignocellulosic biomass for bioethanol production include ash (inorganic materials), pectins (highly branched polysaccharides of galacturonic acid and its methyl esters), acids and extractives (extracellular non-cell wall material).

It has already been discussed that *S. cerevisiae* is the most desirable yeast species to produce bioethanol (Walker 1998). The important characteristics necessary include the ability of yeast strains to tolerate stresses due to physicochemical and biological factors during the rigours of industrial fermentation processes (Walker 2011a, b). There is a need to develop stress-resistant yeasts for fuel alcohol fermentations, especially strains able to withstand substrate and product toxicity (Bettiga et al. 2008). Some commercially available bioethanol yeast strains can produce ethanol at $>10\%$ (v/v) in high solids and $>20\%$ (w/v) in mashes. However, it is currently possible (with the aid of appropriate nutrition) to produce over 20% (v/v) ethanol during high-gravity wheat fermentations (Thomas and Ingledew 1992; Miedl et al. 2007).

In addition, several physiological approaches can be adopted to improve stress tolerance of yeasts for bioethanol production. As well as genetic manipulation techniques (details later and Chap. 17), strategies include sterol pre-enrichment (preoxygenation, mild aeration), mineral preconditioning of yeast (Mg^{++}, Zn^{++} enrichment), ethanol adaption, and pre-heat shock to confer thermotolerance and/or osmotolerance (Logothetis et al. 2010). Naturally robust indigenous yeast cultures (e.g. isolated from distilleries) have been isolated for industrial fermentations (Amorim et al. 2004; Basso et al. 2008; Chandrasena et al. 2006).

It is important to emphasize that *S. cerevisiae* is the predominant organism responsible for the production of bioethanol. Other yeasts (e.g. genetically manipulated or mutational variants of *S. cerevisiae*, *Pichia*, *Candida*, *Schizosaccharomyces* and *Kluyveromyces* spp.) and certain bacteria [e.g. *Zymomonas mobilis* (Rogers et al. 2007), *Thermoanaerobacterium* spp.] have potential in this regard (Table 9.3). Further discussion of non-*Saccharomyces* yeast species involved in the production of potable and industrial ethanol can be found in Chap. 17.

It has already been discussed in this chapter that the principal sugars derived from second-generation feedstocks are glucose, cellobiose (Fig. 9.3), xylose and arabinose (from lignocellulose hydrolysis). Although *S. cerevisiae* can ferment glucose without difficulty, it cannot ferment cellobiose or the pentose sugars xylose and arabinose (Bettiga et al. 2008; Sills and Stewart 1985), without genetic manipulation (Hahn-Hägerdal and Gorwa-Grauslund 2007). This has led to various microbiological and molecular genetic approaches to enable efficient fermentation of these sugars (Bauer and Pretorius 2001). Yeasts and bacteria with an ability to ferment a lignocellulosic hydrolysate have been subject to intense research activity (Mousdale 2008).

The failure of *S. cerevisiae* to metabolize pentoses presents the bioethanol industry with a significant challenge. If this yeast species could be engineered to ferment xylose and/or arabinose, this would present the bioethanol industry with considerable benefits in the production of fuel alcohol from lignocellulose hydrolysates. A number of *S. cerevisiae* strains have been manipulated to metabolize pentoses (Jin et al. 2003; Karhumaa et al. 2005), but their stability and fermentation efficiency with pentoses leave a lot to be desired (Chu and Lee 2007)! Some non-*Saccharomyces* yeasts (e.g. *Candida shehatae* (Tanimura et al. 2012), *K. marxianus* (Fonseca et al. 2008), *Pachysolen tannophilus* (Slininger et al. 1987) and *Pichia stipitis* (Silva et al. 2011)) are able to ferment xylose (the pathway is outlined in Fig. 9.5), but they do so inefficiently (Hahn-Hägerdal et al. 2007a). Also, these yeast species cannot metabolize xylose to ethanol under anaerobic conditions (Skoog and Hahn-Hägerdal 1990). In addition, these yeast species are not regarded as very alcohol tolerant for use in bioethanol production (Chandel et al. 2015).

Genetic manipulation strategies with microorganisms that produce bioethanol aim to:

- Expand appropriate metabolic pathways.
- Alleviate metabolic blocks.
- Circumvent sugar transport limitations (e.g. glucose repression, novel sugar transport permeases).
- Employ concentrated fermentation media (high-gravity fermentations) (Stewart 2014).

Various approaches have been considered to overcome the challenges presented by the dilemma of xylose fermentation. These include:

- Co-fermentations with C6- and C5-fermenting yeast species (e.g. *S. cerevisiae* and *P. stipitis*).
- Metabolic engineering of *S. cerevisiae* to enable cultures to ferment xylose.
- Use of genetically engineered bacteria such as *E. coli*, *Zymomonas*, *Klebsiella oxytoca*, *Thermoanaerobacterium* and *Geobacillus* (with xylose-utilizing genes).
- Immobilization of xylose isomerize with *S. cerevisiae*.

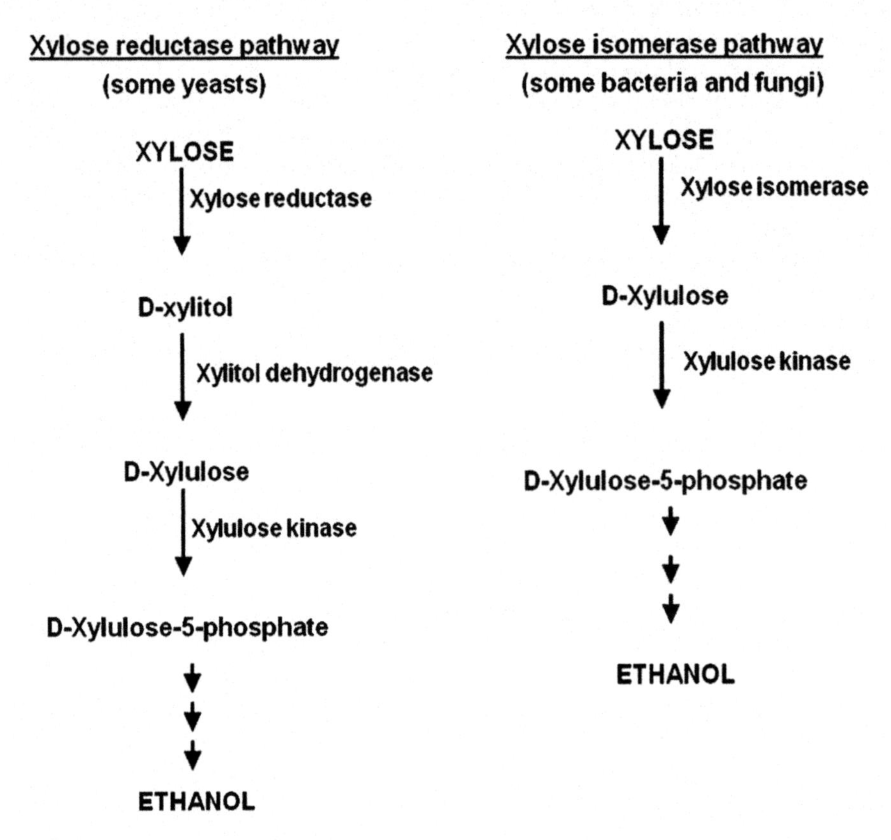

Fig. 9.5 Pathway of microbial xylose fermentation

Regarding recombinant DNA approaches to construct strains of *S. cerevisiae* able to ferment pentose sugars, successful clonings of xylose isomerase genes from the following organisms into this yeast have been accomplished (Ho et al. 1998; Gonçalves et al. 2013):

- Fungi (e.g. *Piromyces*).
- Other yeasts (e.g. *Pichia stipitis*).
- Bacteria (e.g. *Clostridium phytofermentans*).

The expression of xylose isomerase genes, rather than xylose reductase and xylitol dehydrogenase, avoids accumulation of xylitol and an imbalance of NADP and NAD (Hahn-Hägerdal et al. 2007b; Kuyper et al. 2003). Bacteria have also been engineered to ferment lignocellulose hydrolysates, some at higher temperatures. For both yeast and bacterial processes, significant technological challenges remain for commercial lignocellulose-to-bioethanol processes. For example, the presence of toxic chemicals in hydrolysates can seriously inhibit fermentation activity (Palmqvist and Hahn-Hägerdal 2000).

Regarding the improvement of yeast strains for lignocellulose hydrolysate fermentations, major advances in *S. cerevisiae* metabolic engineering have been made in recent years. For example, the following characteristics have been conferred on *S. cerevisiae* for bioethanol production: expression of cellulolytic activity, expression of xylose (and arabinose)-fermenting enzymes, reduction of glycerol, xylitol and arabitol biosynthesis, tolerance to chemical inhibitors and reduced glucose repression (Medina et al. 2010; Nevoigt 2008). Research is currently ongoing to develop robust GM yeast strains that can survive the rigours and stresses of large-scale lignocellulose fermentations. Indeed, future challenges in biotechnology technology also centre on the bioconversion of feedstocks other than first- and second-generation biomass sources (Wilhelm et al. 2007).

9.5 Third-Generation Technologies

There are also third-generation bioethanol technologies. Third-generation bioethanol refers to fuel alcohol produced from nonterrestrial feedstocks such as microalgae, particularly the giant brown seaweeds such as kelp. A number of examples of these technologies do not involve yeast. Nevertheless, they will be mentioned briefly here (Goh and Lee 2010). The growth rate of marine plants of this type far exceeds that of terrestrial plants, and macroalgal cultivation does not encroach on the land required for food crops. Also, microalgae only need seawater, sunlight and carbon dioxide for their growth. They have great ethanol potential compared with conventional bioethanol feedstocks (Adams et al. 2009; Han et al. 2015).

In conclusion to this chapter considering bioethanol production, primarily from yeast fermentation, recent and current research has illustrated the potential of *S. cerevisiae* to produce other types of biofuel. For example, n-butanol and isobutanol (Steen et al. 2008) can be produced by a GM *S. cerevisiae* strain that expresses appropriate *Clostridium* spp. genes. Butanol, a C4 alcohol, exhibits several advantages over ethanol as a fuel, including good combustibility, amenability to storage and transportation and miscibility with diesel (Qureshi et al. 2010). However, butanol is a very flammable liquid and should be treated with great care! *S. cerevisiae* can also be engineered to produce hydrocarbons (e.g. farnesene) with potential to be used as a biodiesel.

Much of the information contained in this chapter was obtained from the following review (with permission)—Walker GM (2011) Fuel alcohol: current production and future changes. J Inst Brew 117:3–22.

The following is added as an epilogue:

> The key to optimizing alcohol production from cereals is a full understanding of the physiology and processing characteristics of different cereals. This study examined the maximum alcohol yields that can be obtained from wheat and maize using different processing technologies. Lower processing temperatures (85 °C) resulted in high alcohol yields from wheat (a temperate crop), whereas higher processing temperatures (142 °C)

gave maximum alcohol yields from maize (a tropical crop). Similar trends were also observed when the spent grains from these cereals were processed using commercial enzymes. Mill settings were additional factors in influencing alcohol production. Wheat has the potential to produce higher alcohol yields when compared with maize, when residual biomass (i.e. spent grains) saccharification using selected commercial enzymes is taken into account. While this approach is not applicable for the Scotch whisky industry owing to strict legislation forbidding the use of exogenous enzymes, this is pertinent for bioethanol production to increase the alcohol yield obtained from both starch and lignocellulosic components of whole cereal grains. Wheat and maize processing temperatures and the use of processing aids are of potential economic benefit to bioethanol producers and to beverage alcohol producers seeking to understand the factors influencing the processing properties of different cereals (Copyright © 2015 The Institute of Brewing & Distilling).

References

Abbas CA (2007) Yeast as ethanologens for biofuel production: limitations and prospects for continued biocatalytic improvements. In: Proceedings of the 26th international specialised symposium on yeasts, Sorrento, Italy, p165

Abbott DA, Ingledew WM (2005) The importance of aeration strategy in fuel alcohol fermentations contaminated with *Dekkera/Brettanomyces* yeasts. Appl Microbiol Biotechnol 69:16–21

Adams JM, Gallagher JA, Donnison IS (2009) Fermentation study on Saccharina latissima for bioethanol production considering variable pre-treatments. J Appl Phycol 21:569–574

Alves DMG (1994) Fatores que afetam a formação de ácidos orgânicos bem como outros parâmetros da fermentação alcoólica. MS Thesis, ESALQ Universidade de São Paulo, Piracicaba, SP

Amorim HV, Basso LC (1991) PI9102738–1—processo para aumentar os teores alcoólicos do vinho e protéico da leve-dura após o térmico da fermentação. National Institute for Industrial Property of Brazil. MarcaPatente/jsp/pate ntes/patente Se arch Basico.jsp

Amorim Neto HB, Yohannan DK, Bringhurst TB, Brosnan JM, Pearson SY, Walker JW, Walker GM (2009) Evaluation of a Brazilian fuel ethanol yeast strain for Scotch Whisky fermentations. J Inst Brew 115:198–207

Amorim HV, Basso LC, Lopes ML (2004) Evolution of ethanol production in Brazil. In: Bryce JH, Stewart GG (eds) Distilled spirits—tradition and innovation. Nottingham University Press, Nottingham, pp 143–148

Amorim HV, Basso LC, Lopes ML (2009) Sugar cane juice and molasses, beet molasses and sweet sorghum: composition and usage. In: Ingledew WM, Kelsall DR, Austin GD, Kluhspies C (eds) The alcohol textbook: a reference for the beverage, fuel, and industrial alcohol industries, vol 1. Nottingham University Press, Nottingham, pp 39–46

Ariyanti D, Hadiyanto H (2013) Ethanol production from whey by *Kluyveromyces marxianus* in batch fermentation system: kinetics parameters estimation. Bull Chem React Eng Catal 7:179–184

Basso LC, Amorim HV (1994) Estudo comparativo entre diferentes leveduras. Relat Anu Pesqui Ferment Alcool 14:71–114

Basso LC, Amorim HV, Oliveira AJ, Lopes ML (2008) Yeast selection for fuel ethanol production in Brazil. FEMS Yeast Res 8:1155–1163

Basso LC, Basso TO, Rocha SN (2011a) Ethanol production in Brazil: the industrial process and its impact on yeast fermentation. In: Bernardes MAS (ed) Biofuel production—recent developments and prospects. Intech, Rijeka, pp 85–100

Basso TO, de Kok S, Dario M, Schlölg PS, Silva CP, Tonso A, Daran J-M, Gombert AK, van Maris AJA, Pronk JT, Stambuk BU (2011b) Engineering topology and kinetics of sucrose metabolism in *Saccharomyces cerevisiae* for improved ethanol yield. Metab Eng 13:694–703

Bauer FF, Pretorius IS (2000) Yeast stress response and fermentation efficiency: how to survive the making of wine. S Afr J Enol Vitic 21:27–51
Bauer FF, Pretorius IS (2001) Pseudohyphal and invasive growth in *Saccharomyces Cerevisiae*. Focus on biotechnology book series. In: Durieux A, Simon JP (eds) Applied microbiology, vol 2. Kluwer Academic, Dordrecht, pp 109–133
Bellissimi E, Ingledew WM (2005) Metabolic acclimatization: preparing active dry yeast for fuel ethanol production. Process Biochem 40:2205–2213
Bettiga M, Hahn-Hagerdal B, Gorwa-Grauslund MF (2008) Comparing the xylose reductase/ xylitol dehydrogenase and xylose isomerase pathways in arabinose and xylose fermenting *Saccharomyces cerevisiae* strains. Biotechnol Biofuels 1:16–22
Chandel AK, Gonçalves BC, Strap JL, da Silva SS (2015) Biodelignification of lignocellulose substrates: an intrinsic and sustainable pretreatment strategy for clean energy production. Crit Rev Biotechnol 35:281–293
Chandrasena G, Keerthipala AP, Walker GM (2006) Isolation and characterization of Sri Lankan yeast germplasm and its evaluation for alcohol production. J Inst Brew 112:302–307
Chin PM, Ingledew WM (1993) Effect of recycled laboratory backset on fermentation of wheat mashes. J Agric Food Chem 41:1158–1163
Chlup PH, Stewart GG (2011) Centrifuges in brewing. MBAA Tech Quart 48:46–50
Choi GW, Kang HW, Moon SK (2009) Repeated-batch fermentation using flocculent hybrid, *Saccharomyces cerevisiae* CHFY0321 for efficient production of bioethanol. Appl Microbiol Biotechnol 84:261–269
Chu BC, Lee H (2007) Genetic improvement of *Saccharomyces cerevisiae* for xylose fermentation. Biotechnol Adv 25:425–441
Contribuições dos espaços não-formais de educação para a formação da cultura científica. Jacobucci DFC (2008) Indexadores e Bases de Dados: Clase; Diadorim; EBSCO; Geodados; Latindex. Periódico incluído na Rede CARINIANA de Preservação Digital. Qualis: B4 (Enfermagem e Odontologia)
Cunningham S, Stewart GG (2000) Acid washing and serial repitching of brewing ale strains of *Saccharomyces cerevisiae* in high gravity wort and the role of wort oxygenation conditions. J Inst Brew 106:389–402
Curry DH, Raman B, Gowen CM, Tschaplinski TJ, Land ML, Brown SD, Covalla SF, Klingeman DM, Yang ZK, Engle NL, Johnson CM, Rodriguez M, Shaw AJ, Kenealy WR, Lynd LR, Fong SS, Mielenz JR, Davison BH, Hogsett DA, Herring CD (2015) Genome-scale resources for *Thermoanaerobacterium saccharolyticum*. B.M.C. Syst Biol 9:10–19
Davin L, Lewis N (2005) Lignin primary structures and dirigent sites. Curr Opin Biotechnol 16:407–415
Della-Bianca BE, Basso TO, Stambuk BU, Basso LC, Gombert AK (2013) What do we know about the yeast strains from the Brazilian fuel ethanol industry? Appl Microbiol Biotechnol 97:979–991
Den Haan R, Rose SH, Lynd LR, Van Zyl WH (2007) Hydrolysis and fermentation of amorphous cellulose by recombinant Saccharomyces cerevisiae. Metab Eng. 9:87–94
Duarte WF, Dragone G, Dias DR, Oliveira JM, Teixeira JA, Silva JB, Schwan RF (2010) Fermentative behavior of *Saccharomyces* strains during microvinification of raspberry juice (Rubus idaeus L.) Int J Food Microbiol 143:173–182
Duval EH, Alves SL Jr, Dunn B, Sherlock G, Stambuck BU (2010) Microarray karyotyping of maltose-fermenting *Saccharomyces cerevisiae* yeasts with differing maltotriose utilization profiles reveals copy number variation in genes involved in maltose and maltotriose utilization. J Appl Microbiol 109:248–259
Ercan D, Demirci A (2015) Current and future trends for biofilm reactors for fermentation processes. Crit Rev Biotechnol 35:1–14
Erratt JA, Stewart GG (1978) Genetic and biochemical studies on yeast strains able to utilize dextrins. J Am Soc Brew Chem 36:151–161

Erratt JA, Stewart GG (1981) Fermentation studies using *Saccharomyces diastaticus* yeast strains. Dev Ind Microbiol 22:577–586
Fonseca GG, Heinzle E, Wittmann C, Gombert AK (2008) The yeast *Kluyveromyces marxianus* and its biotechnological potential. Appl Microbiol Biotechnol 79:339–354
Garoma T, Ben-Khaled M, Beyene A (2012) Comparative resource analyses for ethanol produced from corn and sugarcane in different climatic zones. Int J Energy Res 36:1065–1076
Gibson BR (2011) Improvement on higher gravity brewery fermentation via wort enrichment and supplementation. J Inst Brew 117:268–284
Goh CS, Lee KT (2010) A visionary and conceptual macroalgae-based third-generation bioethanol (TGB) biorefinery in Sabah, Malaysia as an underlay for renewable and sustainable development. Renew Sust Energ Rev 14:842–848
Goldemberg J (2008) The Brazilian biofuels industry. Biotechnol Biofuels 1:6
Gomes DG, Guimarães PMR, Pereira FB, Teixeira JA, Domingues L (2012) Plasmid-mediate transfer of FLO1 into industrial *Saccharomyces cerevisiae* PE-2 strain creates a strain useful for repeat-batch fermentations involving flocculation sedimentation. Bioresour Technol 108:162–168
Gonçalves GAL, Prazeres DMF, Monteiro GA, Prather KLJ (2013) De novo creation of MG1655-derived E. coli strains specifically designed for plasmid DNA production. Appl Microbiol Biotechnol 97:611–620
Green DI, Agu RC, Bringhurst TA, Brosnan JM, Jack FR, Walker GM (2015) Maximizing alcohol yields from wheat and maize and their co-products for distilling or bioethanol production. J Inst Brew 121:332–337
Gutierrez LE (1993) Changes in trehalose content of baker's yeast as affected by octanoic acid. Sci Agric 50:460–463
Gutierrez LE, Orelli VFM (1991) Efeito do nitrito sobre a fermentação alcoólica realizada por *Saccharomyces cerevisiae*. Anais da Escola Superior de Agricultura "Luiz de Queiroz". Piracicaba 48:41–54
Gutierrez C, Ardourel M, Bremer E, Middendorf A, Boos W, Ehmann U (1989) Analysis and DNA sequence of the osmoregulated treA gene encoding the periplasmic trehalase of Escherichia coli K12. Mol Gen Genet 217:347–354
Gutierrez LE, Annicchino AVKO, Lucatti L, Leite da Silva SB (1991a) Aumento da produção de etanol a partir de melaço de cana-de-açúcar pela adição de benzoato. Anais da Escola Superior de Agricultura "Luiz de Queiroz". Piracicaba 48:1–21
Gutierrez LE, Annicchino AVKO, Lucatti L, Stipp JMS (1991b) Effects of acetic acid on alcoholic fermentation. Arq Biol Tecnol 34:235–242
Hahn-Hägerdal B, Gorwa-Grauslund MF (2007) Comparison of the xylose reductase-xylitol dehydrogenase and the xylose isomerase pathways for xylose fermentation by recombinant *Saccharomyces cerevisiae*. Microb Cell Factor 6:5
Hahn-Hägerdal B, Karhumaa K, Fonseca C, Spencer-Martins I, Gorwa-Grauslund MF (2007a) Towards industrial pentose-fermenting yeast strains. Appl Microbiol Biotechnol 74:937–953
Hahn-Hägerdal B, Karhumaa K, Jeppsson M, Gorwa-Grauslund MF (2007b) Metabolic engineering for pentose utilization in *Saccharomyces cerevisiae*. Adv Biochem Eng Biotechnol 108:147–177
Han S-F, Jin W-B, Tu R-J, Wu W-M (2015) Biofuel production from microalgae as feedstock: current status and potential. Crit Rev Biotechnol 35:255–268
Herrero EM, Lopez Gonzalvez A, Ruiz MA, Lucas-García JA, Barbas C (2003) Uptake and distribution of zinc, cadmium, lead and copper in Brassica napus var. oleifera and Helianthus annus grown in contaminated soils. Int J Phytoremediation 5:153–167
Ho NW, Chen Z, Brainard AP (1998) Genetically engineered *Saccharomyces* yeast capable of effective cofermentation of glucose and xylose. Appl Environ Microbiol 64:1852–1859
Ingledew WM (2005) Improvements in alcohol technology through advancements in fermentation technology. Getreidetechnologie 59:308–311

Ito S, Takeyama K, Yamamoto A, Sawatsubashi S, Shirode Y, Kouzmenko A, Tabata T, Kato S (2004) In vivo potentiation of human oestrogen receptor α by Cdk7-mediated phosphorylation. Genes Cells 9:983–992

Jamai L, Ettayebi K, Yamani ELJ, Ettayebi M (2007) Production of ethanol from starch by free and immobilized *Candida tropicalis* in the presence of α-amylase. Bioresour Technol 98:2765–2770

Jin YS, Ni HY, Laplaza JM, Jeffries TW (2003) Optimal growth and ethanol production from xylose by recombinant *Saccharomyces cerevisiae* require moderate D-xylulokinase activity. Appl Environ Microbiol 69:495–503

Johnson FX (2007) Bioenergy and the Sustainability Transition: from Local Resource to Global Commodity. World Energy Congress (WEC) Rome, 2007

Karhumaa K, Hahn-Hagerdal B, Gorwa-Grauslund MF (2005) Investigation of limiting metabolic steps in the utilization of xylose by recombinant *Saccharomyces cerevisiae* using metabolic engineering. Yeast 22:359–368

Kasi D, Ragauskas AJ (2010) Switchgrass as an energy crop for biofuel production: a review of its ligno-cellulosic chemical properties. Energy Environ Sci 3:1182–1190

Khaw TS, Katakura Y, Koh J, Kondo A, Ueda M, Shioya S (2006) Evaluation of performance of different surface-engineered yeast strains for direct ethanol production from raw starch. Appl Microbiol Biotechnol 70:573–579

Khew ST, Yang QJ, Tong YW (2008) Enzymatically crosslinked collagen mimetic dendrimers that promote integrin-targeted cell adhesion. Biomaterials 29:3034–3035

Kline KL, Dale VH (2008) Biofuels: effects on land and fire. Science 321:199–201

Kuyper M, Harhangi HR, Stave AK, Winkler AA, Jetten MS, de Laat WT, den Ridder JJ, Op den Camp HJ, van Dijken JP, Pronk JT (2003) High-level functional expression of a fungal xylose isomerase: the key to efficient ethanolic fermentation of xylose by *Saccharomyces cerevisiae*? FEMS Yeast Res 4:69–78

Leiper KA, Schlee C, Tebble I, Stewart GG (2006) The fermentation of beet sugar syrup to produce bioethanol. J Inst Brew 112:122–133

Lennartsson PR, Erlandsson P, Taherzadeh MJ (2014) Integration of the first and second generation bioethanol processes and the importance of by-products. Bioresour Technol 165:3–8

Logothetis NK, Augath M, Murayama Y, Rauch A, Sultan F, Goense J, Oeltermann A, Merkle H (2010) The effects of electrical microstimulation on cortical signal propagation. Nat Neurosci 13:1283–1291

Lynd LR, Laser MS, Bransby D, Dale BE, Davison B, Hamilton R, Himmel M, Keller M, McMillan JD, Sheehan J, Wyman CE (2008) How biotech can transform biofuels. Nat Biotechnol 26:169–172

Mabee WE (2007) Policy options to support biofuel production. Adv Biochem Eng Biotechnol 108:329–357

Macedo IC, Seabra JEA, Silva JEAR (2008) Green house gases emissions in the production and use of ethanol from sugarcane in Brazil: the 2005/2006 averages and a prediction for 2020. Biomass Bioenergy 32:582–595

Medina VG, Almering MJ, van Maris AJ, Pronk JT (2010) Elimination of glycerol production in anaerobic cultures of a *Saccharomyces cerevisiae* strain engineered to use acetic acid as an electron acceptor. Appl Environ Microbiol 76:190–195

Miedl M, Cornfine S, Leiper KA, Shepherd M, Stewart GG (2007) Low-temperature processing of wheat for bioethanol production. J Am Soc Brew Chem 65:183–191

Miranda M Jr, Batistote M, Cilli EM, Ernandes JR (2009) Sucrose fermentation by Brazilian ethanol production yeasts in media containing structurally complex nitrogen sources. J Inst Brew 115:191–197

Montgomery CT, Smith MB (2010) Hydraulic fracturing – history of an enduring technology. J Petrol Technol 62:26–32

Mousdale DM (2008) Biofuels. Biotechnology, chemistry and sustainable development. CRC Press, Boca Raton, FL

Nair SG, Sindhu R, Shankar S (2008) Purification and biochemical characterization of two xylanases from *Aspergillus sydowii SBS* 45. Appl Biochem Biotechnol 149:229–243
Nevoigt E (2008) Progress in metabolic engineering of *Saccharomyces cerevisiae*. Microbiol Mol Biol Rev 72:379–412
O'Brien D, Woolverton M. (2009) Recent trends in U.S. Wet and dry corn milling production. AgMRC Renewable Energy Newsletter, February 2009
Ogata T, Iwashita Y, Kawada T (2017) Construction of a brewery yeast expressing the glucoamylase gene *STA1* by mating. J Inst Brew 123:66–69
Palmqvist E, Hahn-Hägerdal B (2000) Fermentation of lignocellulosic hydrolysates. II: Inhibitors and mechanisms of inhibition. Bioresour Technol 74:25–33
Pasteur L (1879) Studies on fermentation: the diseases of beer, their causes and the means of preventing them. Translated by Frank Faulkner. MacMillan
Paulillo SCL, Yokoya F, Basso LC (2003) Mobilization of endogenous glycogen and trehalose of industrial yeasts. Braz J Microbiol 34:249–254
Pavlak MCM, de Abreu-Lima TL, Carreiro SC, de Lima Paulillo SC (2011) Study of fermentation of the hydrolyzate sweet potato using different strains of *Saccharomyces cerevisiae*. Química Nova 34:82–86
Pereira FB, Guimaraes PM, Teixeira JA, Domingues L (2010) Selection of *Saccharomyces cerevisiae* strains for efficient very high gravity bio-ethanol fermentation processes. Biotechnol Lett 32:1655–1661
Pereira FB, Gomes DG, Guimaraes PM, Teixeira JA, Domingues L (2011) Cell recycling during repeated very high gravity bio-ethanol fermentations using the industrial *Saccharomyces cerevisiae* stain PE-2. Biotechnol Lett 34:45–53
Piper PW (1997) The yeast heat shock response. In: Hohmann S, Mager WH (eds) Yeast stress responses. R. G. Landes, Austin, TX, pp 75–99
Piper PW, Ortiz-Calderon C, Holyoak C, Coote P, Cole M (1997) Hsp30, the integral plasma membrane heat shock protein of *Saccharomyces cerevisiae*, is a stress-inducible regulator of plasma membrane H(+)-ATPase. Cell Stress Chaperones 2:12–24
Qureshi N, Saha B, Dien B, Cotta MA (2010) Production of butanol (a biofuel) from agricultural residues: Part I – Use of barley straw hydrolysate. Biomass Bioenergy 34:559–565
Rogers PL, Jeon YJ, Lee KJ, Lawford HG (2007) *Zymomonas mobilis* for fuel ethanol and higher value products. Adv Biochem Eng Biotechnol 108:263–288
Rommer T (2010) World biofuels production potential energy policies, politics and prices series. Nova Science
Russell I, Stewart GG (eds) (2014) Whisky: technology, production and marketing, 2nd edn. Academic Press (Elsevier), Boston
Russell I, Jones RM, Stewart GG (1987) Yeast – the primary industrial microorganism. In: Stewart GG, Russell I, Klein RD, Hiebsch RR (eds) CRC biological research on industrial yeasts. CRC Press, Boca Raton, pp 1–20
Sanchez OJ, Cardona CA (2008) Trends in biotechnological production of fuel ethanol from different feedstocks. Bioresour Technol 99:5270–5295
Scheller HV, Ulvskov P (2010) Hemicelluloses. Annu Rev Plant Biol 61:263–289
Searchinger T, Heimlich R, Houghton RA, Dong F, Elobeid A, Fabiosa J, Tokgoz S, Hayes D, Yu T-H (2008) Use of U.S. croplands for biofuels increases greenhouse gases through emissions from land-use change. Science 319(5867):1238–1240
Shapouri H, Gallagher PW, Nefstead W, Schwartz RH, Noe S, Conway R (2008) 2008 Energy balance for the corn-ethanol industry. United States Department of Agriculture
Shi DJ, Wang CL, Wang KM (2009) Genome shuffling to improve thermotolerance, ethanol tolerance and ethanol productivity of *Saccharomyces cerevisiae*. J Ind Microbiol Biotechnol 36:139–147
Sills AM, Stewart GG (1985) Studies on cellobiose metabolism by yeasts. Dev Ind Microbiol 26:527–534

Silva JPA, Mussatto SI, Roberto IC, Teixeira JA (2011) Ethanol production from xylose by *Pichia stipitis* NRRL Y-7124 in a stirred tank bioreactor. Braz J Chem Eng 28:151–156

Simpson WJ, Hammond JRM (1989) The response of brewing yeasts to acid washing. J Inst Brew 95:347–354

Skoog K, Hahn-Hägerdal B (1990) Effect of oxygenation on xylose fermentation by *Pichia stipitis*. Appl Environ Microbiol 56:3389–3394

Slapack GE, Russell I, Stewart GG (1987) Thermophilic microbes in ethanol production. CRC Press, Boca Raton, FL

Slininger PJ, Bolen PL, Kurthman CP (1987) Pachysolen tannophilus: properties and process considerations for ethanol production from D-xylose. Enzyme Microb Technol 9:5–15

Solomon BD, Barnes JR, Halvorsen KE (2007) Grain and cellulosic ethanol: history, economics, and energy policy. Biomass Bioenergy 31:416–425

Stambuk BU, Dunn B, Alves SL, Duval EH, Sherlock G (2009) Industrial fuel ethanol yeasts contain adaptive copy number changes in genes involved in vitamin B1 and B6 biosynthesis. Genome Res 19:2271–2278

Steen EJ, Chan R, Prasad N, Myers M, Petzold RA, Ouellet M, Keasling JD (2008) Metabolic engineering of *Saccharomyces cerevisiae* for the production of n-butanol. Microb Cell Factor 7:36

Stewart GG (2014) Brewing intensification. American Society for Brewing Chemists, St. Paul, MN

Stewart GG, Russell I (2009) An introduction to brewing science and technology. Series lll, Brewer's yeast, 2nd edn. The Institute of Brewing and Distilling, London

Stewart GG, Hill AE, Russell I (2013) 125th Anniversary review – developments in brewing and distilling yeast strains. J Inst Brew 119:202–220

Tanimura A, Nakamura T, Watanabe I, Ogawa J, Shima J (2012) Isolation of a novel strain of Candida shehatae for ethanol production at elevated temperature. Springer Plus 27:1–7

Thomas KC, Ingledew WM (1992) Production of 21% (v/v) ethanol by fermentation of very high gravity (VHG) wheat mashes. J Ind Microbiol 10:61–68

Walker GM (1998) Yeast physiology and biotechnology. Wiley, England

Walker GM (2011a) 125th Anniversary review: fuel alcohol: current production and future challenges. J Inst Brew 117:3–22

Walker R (2011b) The impact of Brazilian biofuel production on Amazônia. Ann Assoc Am Geogr 101:929–938

Wilhelm WW, Johnson JMF, Karlen DL, Lightle DT (2007) Corn stover to sustain soil organic carbon further constrains biomass supply. Agron J 99:1665–1667

Wilson N (2014) Contamination: bacteria and wild yeast in a whisky fermentation. In: Russell I, Stewart GG (eds) Whisky: technology, production and marketing. Elsevier, London, pp 147–154

Yan Z, Delannoy M, Ling C, Daee D, Osman F, Muniandy PA, Shen X, Oostra AB, Du H, Steltenpool J, Lin T, Schuster B, Décaillet C, Stasiak A, Stasiak AZ, Stone S, Hoatlin ME, Schindler D, Woodcock CL, Joenje H, Sen R, de Winter JP, Li L, Seidman MM, Whitby MC, Myung K, Constantinou A, Wang W (2010) A histone-fold complex and FANCM form a conserved DNA-remodeling complex to maintain genome stability. Mol Cell 37:865–878

Zanin G, Santana C, Bon E, Giordano R, de Moraes F, Andrietta S, Neto C, Macedo I, Lahr FD, Ramos L, Fontana J (2000) Brazilian bioethanol program. Appl Biochem Biotechnol 84–86:1147–1161

Zarattini RA, Williams JW, Ernandes JR, Stewart GG (1993) Bacterial-induced flocculation in selected brewing strains of *Saccharomyces*. Cerevisiae Biotechnol 4:65–70

Zhao XQ, Bai FW (2009) Mechanisms of yeast stress tolerance and its manipulation for efficient fuel ethanol production. J Biotechnol 144:23–30

Chapter 10
Killer (Zymocidal) Yeasts

Killer yeast strains secrete toxins that are lethal to sensitive strains of the same or related yeast species. All the known killer toxins produced by killer yeasts are proteins that kill sensitive cells via a two-step mode of action (Liu et al. 2015), while antibiotics are bioactive substances that are produced by any organism and possess activity against selective fungi, bacteria, viruses and cancer cells. To date, it has been known that under competitive conditions, the killer phenomenon offers considerable advantage to these yeast strains against other sensitive microbial cells in their ecological niches (Wang et al. 2012). Toxin-producing strains are termed "killers" and susceptible strains are termed "sensitives". However, there are strains that do not kill and are not themselves killed, and these are called "resistant" (Fig. 10.1).

Although the original terminology for this factor was "killer" (Woods and Bevan 1968), it has been renamed "zymocide" (Young and Yagiu 1978) to indicate that it is only lethal towards yeasts, fungi, a few bacteria and not higher organisms. Zymocidal yeasts have been recognized to be a problem in both batch and continuous fermentation systems (Maule and Thomas 1973). An infection can completely eliminate all the yeast culture from a fermenter (Hammond and Eckersley 1984).

Killer yeasts and their toxins have several applications. For example, killer yeasts have been used to combat contaminating wild yeasts in food and to control pathogenic fungi in plants (Schmitt and Breinig 2002). In the medical field, these yeasts have been used for the development of novel antimycotics used in the treatment of human and animal fungal infections and in the biotyping of pathogenic yeasts and yeast-like fungi (Magliani et al. 2008). Moreover, killer yeasts have been used to control contaminating wild yeasts in the winemaking, brewing and other fermentation industries (Schmitt and Breinig 2002).

The killer yeast system was first described in 1963 (Bevan and Makower 1963). This study of killer toxins helped to better understand the reaction pathway(s) of yeast (further details in Chap. 16). The phenomenon was first observed by Louis Pasteur (Pasteur and Joubert 1877) (Fig. 2.13). The best characterized toxin system is from *Saccharomyces cerevisiae*, which was found to spoil beer during

G.G. Stewart, *Brewing and Distilling Yeasts*, The Yeast Handbook,
https://doi.org/10.1007/978-3-319-69126-8_10

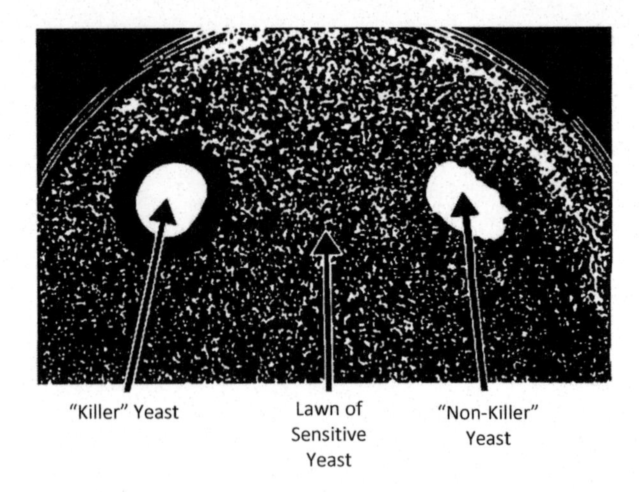

Fig. 10.1 *Saccharomyces* brewing yeasts with and without zymocidal ("killer") activity

fermentation and maturation (Hammond and Eckersley 1984). In *S. cerevisiae*, killer toxins are encoded by a double-stranded RNA virus, translated to a precursor protein, cleaved and secreted outside the cells, where they may affect a susceptible yeast (Schmitt and Breinig 2006). There are other killer systems in *S. cerevisiae*, such as *KHR* (Goto et al. 1990a, b) and *KHS* (Goto et al. 1991) genes, encoded on 1 or other of the 16 nuclear chromosomes.

A brewer can protect the process from this occurrence in one of two ways:

- Maintain vigorous standards of hygiene to ensure that contamination with a wild yeast possessing zymocidal activity is prevented.
- Genetically modify the brewery yeast so that it is not susceptible to the zymocidal toxin.

The first method is the one that brewers have been invoking for many years to protect their process. However, genetic manipulation can also be employed to produce a brewing yeast strain that is less vulnerable to destruction by a zymocidal yeast infection.

It has already been discussed that the killer character of *Saccharomyces* spp. is determined by the presence of two cytoplasmically located double-stranded (ds) RNA (Woods and Bevan 1968). M-dsRNA (killer plasmid), which is killer strain-specific, codes for a killer toxin and also for a protein or proteins that renders the host immune to the toxin (Hammond and Eckersley 1984). The L-dsRNA, which is also present in many non-killer yeast strains, specifies a capsid protein that encapsulates batch forms of dsRNA, thereby yielding virus-like particles. Although the killer plasmid is contained within these virus-like particles, the killer genome is not naturally transmitted from cell to cell by any infection process. The killer plasmid behaves as a true cytoplasmic element (further details in Chap. 16) and requires at least 29 different chromosomal genes (*mak* for its maintenance in the cell). In addition, three other chromosomal genes (*kex1*, *kex2* and *rex*) are required for production and resistance to the toxin (Woods and Bevan 1968).

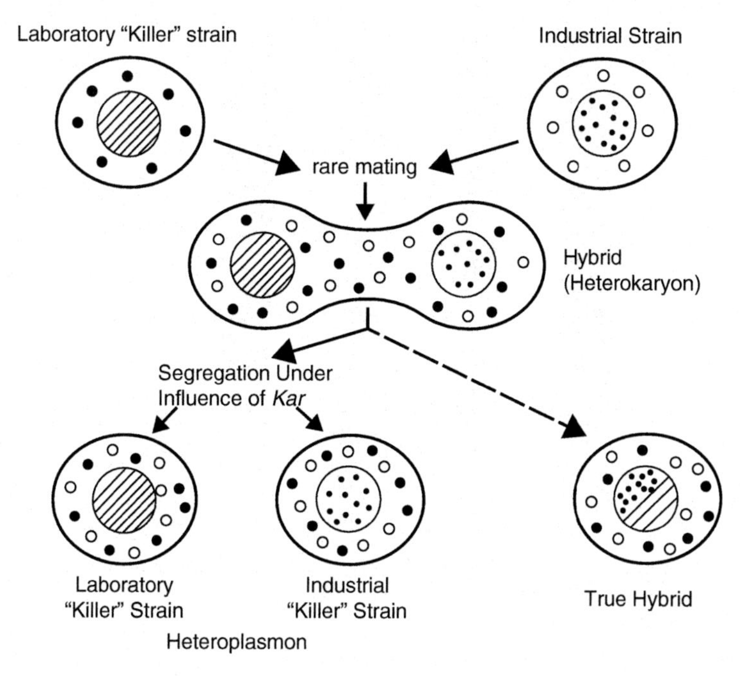

Fig. 10.2 Rare mating protocol to produce industrial strains with zymocidal activity

The technique of rare mating has been employed to produce hybrids with enhanced fermentation ability (e.g. Stewart and Russell 1986). A modification of this technique, employing the *kar* mutation, has been successfully employed by a number of research laboratories (Conde and Fink 1976). The *kar* mutation prevents nuclear fusion, and hybrids from such a rare mutation can be selected that contain only the brewing strain's nucleus (Fink and Styles 1972). Such hybrids contain the cytoplasm of both parental cells, thereby permitting the introduction of cytoplasmically transmitted characteristics such as killer toxin production into brewing strains, without modifying the nucleus of brewing strains with a brewing polyploidy lager strain and several rare mating products isolated (Fig. 10.2).

In addition to biochemical tests to characterize rare mating products, agarose gel electrophoresis demonstrated that some rare mating products contained the 2 μm plasmid (from the parental brewing lager strain) and the L- and M-dsRNA plasmids (from the haploid partner), which encode for killer toxin production (Fig. 10.3) (Russell and Stewart 1985).

The effect of zymocidal lager strains on a typical brewery fermentation was studied by mixing it at a concentration of 10% with an ale brewing strain (Fig. 10.4). The yeast culture was sampled throughout the fermentation and viable cells determined by plating onto nutrient agar plates. The plates were incubated at 37 °C (a temperature that inhibits the growth of lager yeast but permits the growth of ale strains—details in Chap. 3). Within 10 h after inoculation, the killer lager stain had almost totally eliminated the ale strain. When the concentration of killer yeast was reduced to 1% within 24 h, the ale yeast was again eliminated (Fig. 10.4).

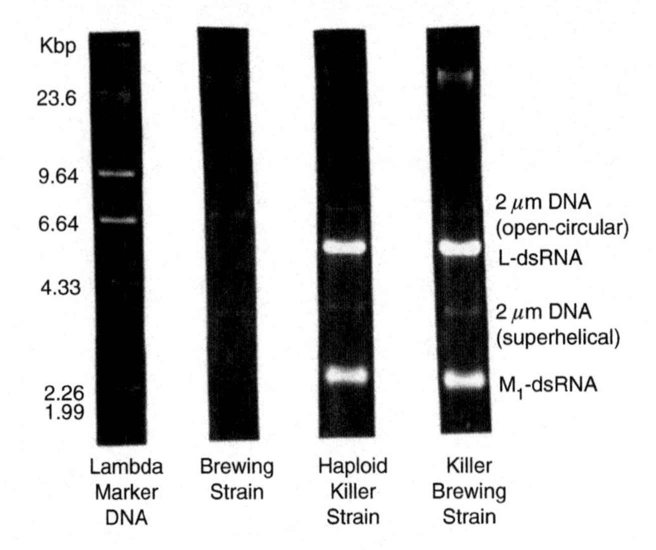

Fig. 10.3 Agarose gel electrophoresis of nucleic acid extracts from rare mating partners and the rare mating produce (the "killer" brewing strain)

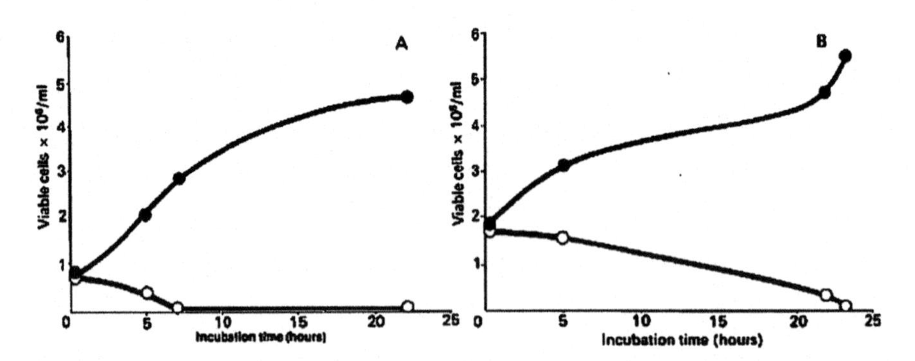

Fig. 10.4 Effect of a "killer" lager yeast on growth of an ale yeast during a 12°P wort fermentation. The yeast was sampled throughout the fermentation, and viable cells were determined by plating on nutrient agar plates incubated at 37 °C to inhibit growth of lager yeast but allow growth of ale yeast. (**a**) Control—growth of ale yeast from 90% ale yeast and 10% non-killer lager yeast (Line with filled circle). Test—growth of ale yeast from 90% ale yeast and 10% killer lager yeast (Line with open circle). (**b**) Control—growth of ale yeast from 99% ale yeast and 1% non-killer lager yeast (Line with filled circle). Test—growth of ale yeast from 99% ale yeast and 1% killer lager yeast (Line with open circle)

The speed at which death of the brewing yeast occurred made the brewing community apprehensive about employing killer yeasts during fermentation, particularly in a brewery where several yeast strains are employed for the production of different beers (e.g. an ale, a lager and two licenced beers). An error by an operator in maintaining lines and yeast tanks separate could result in serious consequences!

In a brewery where only one yeast strain is employed (increasingly the situation currently), this would not be a cause for concern.

An alternative to employing a killer strain would be to produce a yeast strain that does not kill but is killer resistant. It has to receive the genetic complement that rendered a brewing strain immune to zymocidal activity. The construction of such a strain would be an interesting compromise because it does not itself kill. This allays the fear of the brewer that this yeast might kill all other production strains in the plant and, at the same time, it is not itself killed by a contaminating yeast with killer activity. The initial protein product from translation of the M-dsRNA is called the preprotoxin which is targeted to the yeast secretory pathway. The preprotoxin is processed and cleaved to produce an α/β dimer, which is the active form of the toxin and is released into the environment (Bussey 1991).

Since they were first described in *S. cerevisiae*, toxin-producing yeasts have been identified in other yeast genera, such as *Candida*, *Cryptococcus*, *Debaryomyces*, *Hanseniaspora*, *Hansenula*, *Kluyveromyces*, *Metschnikowia*, *Pichia*, *Ustilago*, *Torulopsis*, *Williopsis*, *Zygosaccharomyces*, *Aureobasidium*, *Zygowilliopsis* and *Mrakia*, indicating that the killer phenomenon is widespread amongst yeasts (Liu et al. 2015; Buzdar et al. 2011; Naumov et al. 2011; Wang et al. 2008; Comitini et al. 2004a).

S. cerevisiae killer strains produce and secrete protein toxins that are lethal to sensitive strains of the same or related yeast species. These toxins have been grouped into three types: killer toxin 1(K1), killer toxin 2(K2) and killer toxin 28 (K28), based on their killing profiles and lack of cross-immunity (Marquina et al. 2001). In addition, it has been found that some wine *S. cerevisiae* strains produce a new killer toxin (Klus) that kill all the previously known *S. cerevisiae* killer strains, in addition to other yeast species including *Kluyveromyces lactis* and *Candida albicans* (Rodriguez-Cousino et al. 2011). The killer phenotype is conferred by a medium-sized double-stranded RNA (dsRNA) virus in *S. cerevisiae* (Rodriguez-Cousino et al. 2011). After the killer toxin-sensitive *S. cerevisiae* HAU-1 was fused with the killer toxin-producing *S. cerevisiae* MTCC 475, the fusant obtained could stably produce both ethanol and killer toxin (Bajaj and Sharma 2010).

The zymocin produced by *K. lactis* inhibits the growth of a wide range of susceptible yeasts in the yeast genera *Candida*, *Kluyveromyces*, *Saccharomyces*, *Torulopsis* and *Zygosaccharomyces*, as well as non-killer strains of *K. lactis* (Marquina et al. 2002; Comitini et al. 2004b). It has also been reported that *Brettanomyces*/*Dekkera* yeasts that cause haze, turbidity and strong off-flavours in some wines and beers can be effectively killed by the killer toxin (Kwkt) produced by *Kluyveromyces wickerhamii* (Comitini and Ciani 2011).

More species of the genus *Pichia* have been identified that produce killer toxins than those of any other yeast genus. The killer toxin pmKT1 produced by *Pichia membranifaciens* CYC1086 acts by disrupting plasma membrane electrochemical gradients, leading to the death of sensitive cells (Santos et al. 2009), and the killer toxin produced by *P. membranifaciens* CYC 1106 has activity against *Botrytis cinerea* (Santos and Marquina 2004). Within the genus *Pichia*, *P. acacia*, *Wickerhamomyces anomalus*, *P. farinosa*, *P. inositovora*, *P. kluyveri* and

P. membranifaciens produce different toxins (Santos et al. 2009). For example, *P. membranifaciens* CYC 1086 also secretes a killer toxin (PMKT2) that is inhibitory to a variety of spoilage yeasts and fungi and is of interest to the beverage (alcoholic and non-alcoholic) and food industries. PMKT2 is able to inhibit a number of wild yeasts (e.g. *Botrytis bruxellensis*), whereas *S. cerevisiae* is fully resistant to this toxin, indicating that PMKT2 could be used in alcoholic fermentations to avoid the development of spoilage yeast strains without deleterious effects on the strains conducting the fermentation.

Killer activity has also been observed in *P. holstii* and *P. jadinii* against a spectrum of clinical and industrial yeast strains (Antunes and Aguiar 2012). *P. kudriavzevii* RY55 toxin exhibited excellent antibacterial activity against a number of pathogens of human health significance such as *Escherichia coli*, *Enterococcus faecalis*, *Klebsiella* sp., *Staphylococcus aureus*, *Pseudomonas aeruginosa* and *Pseudomonas alcaligenes* (Bajaj et al. 2013).

The amylolytic yeast *Schwanniomyces occidentalis* produces a killer toxin lethal to sensitive strains of *S. cerevisiae*, and killer activity is destroyed after pepsin and papain (proteinase) treatment, suggesting that the toxin is a protein (Chen et al. 2000). *Tetrapisispora phaffii* (previously known as *K. phaffii*) secretes a glycoprotein called Kpkt that is lethal to spoilage yeasts under winemaking conditions (Ciani and Fatichenti 2001; Comitini et al. 2009).

In the osmotolerant yeast *Zygosaccharomyces bailii*, killer phenotype expression is caused by the secretion of a non-glycosylated protein toxin that rapidly kills a broad spectrum of yeasts and filamentous fungi, including human pathogenic strains of *C. albicans*, *C. glabrata* and *Sporothrix schenckii* as well as phytopathogenic strains of the filamentous fungi *Fusarium oxysporum* and *Colletotrichum graminicola* (Weiler and Schmitt 2003). This confirms that some zymocidal toxins, produced by selected yeasts, exhibit killer activity against filamentous fungal cells.

Most killer toxins are stable and act only at acidic pH values and low temperatures. For example, at pH 4, the optimal killer activity of the killer toxin produced by *P. membranifaciens* CYC 1106 was observed at temperatures up to 20 °C (Santos and Marquina 2004). Meanwhile, the killer activity was higher in acidic media. Above pH 4.5, its activity decreased sharply and was barely detectable at pH 6.

The most stable killer toxins are those produced by *H. mrakii* which is stable between pH 2 and 11 and is unaffected by heating at 60 °C for 1 h. Indeed, the killer toxins produced by marine yeasts also function best in acidic media. Nevertheless, it is difficult to apply marine killer toxins to the biocontrol of pathogenic yeasts in marine environments.

There are two kinds of receptor for killer toxins—primary and secondary receptors. The former is usually located in the cell wall and the latter on the plasma membrane [details of yeast surface architecture (wall and membrane) are in Chaps. 5 and 13]. It has been shown that cell walls are necessary for most killer toxin activity, and different components of the cell wall can be receptors for killer toxins (Peng et al. 2010).

The toxins K1 and K2, produced by *S. cerevisiae*, kill their sensitive yeast cells in receptors mediated by a two-step process. The first step of killer toxin activity involves fast and energy-independent binding to a toxin receptor within the cell wall of their sensitive target cells. In the case of K1 and K2 toxins, this primary receptor has been identified as β-1,6-D-glucan. It has been reported that the glucans (β-1,3 and β-1,6 branched glucans) also represent the first receptor sites of the toxin from *K. phaffii* located on the envelope of the sensitive target (Comitini et al. 2004a), and the killer toxin (WmKT) from *W. saturnus* var. *mrakki* MUcl 41,968 has the receptor of β-glucan (Guyard et al. 2002). Indeed, the purified killer toxins from marine yeasts could not kill the protoplasts (discussed in Chap. 16) of sensitive yeast cells. This implies that the binding receptor of the killer toxin from marine yeasts also exists in the cell wall of sensitive yeasts (Buzdar et al. 2011; Hua et al. 2010; Peng et al. 2010).

The cell wall receptor for the viral toxin K28 produced by *S. cerevisiae* has a high molecular mass of β-1,3-mannoprotein. The mannoproteins are also the receptors for *Z. bailii* killer toxin (Santos et al. 2000). The *RHK1* gene, encoding for a hydrophobic protein, is composed of 458 amino acids in *S. cerevisiae* and does not participate directly in the synthesis of β-1,3-glucan. However, it is involved in the synthesis of the receptor for the HM-1 killer toxin produced by *H. mrakii* because after disruption of the *RHK1* gene in *S. cerevisiae* cells, the disruptants obtained have complete killer resistance to the *HM-1* killer toxin (Kimura et al. 1997).

It has been well documented that chitin, which is located in the birth and bud scars, (structural details in Chap. 5) in sensitive cell walls is a receptor for the *K. lactis* killer toxin (zymocin) (Santos et al. 2000). It is regarded that the α subunit of the zymocin has a chitin-binding domain (CBD) and a chitinase motif, which contains a key catalytic residue (E466). Therefore, the zymocin binds chitin *in vivo* and displays exo-chitinase activity. In addition to the G1 block, the zymocin may cause cell wall damage. This may explain why the exo-zymocin induces the death of sensitive cells, whereas inhibition by intracellular toxin can be reversed (Schaffrath and Meinhardt 2005).

Only cell wall mannan prevented sensitive yeast cells from being killed by the killer protein from *S. occidentalis* suggesting that mannan may interact with the killer protein (Chen et al. 2000). The receptors for the killer toxins from *S. cerevisiae* (K28 type) and *Zygosaccharomyces bailii* are also mannans. In fact, almost all fungal cell wall structural components can function as the primary killer toxin receptors (Magliani et al. 2008).

The K28 killer toxin produced by *S. cerevisiae* enters a sensitive target by endocytosis (Marquina et al. 2002). After receptor-mediated entry into the cell, the toxin enters the secretion pathway in reverse via the Golgi and endoplasmic reticulum. Subsequently, it enters the cytosol and transduces its toxic signal into the yeast cell's nucleus where the lethal events occur. As a result, DNA synthesis is rapidly inhibited, cell viability is reduced more slowly and cells in early S phase of the cell cycle (further details of the cell cycle are in Chap. 5) with a nucleus in the mother cell. The *K. lactis* killer toxin is similar to the K28 toxin discussed above. It causes sensitive yeasts to arrest proliferation as unbudded cells, suggesting that it blocks the cell cycle in the G1 phase. This zymocin is a heterotrimeric protein toxin consisting of three subunits (Klassen et al. 2004).

10.1 Summary

Killer yeast toxins have many potential applications in fermentation, taxonomy, medicine, agriculture and the mariculture industry. These possible applications have been reviewed by Schmitt and Breinig (2002), Marquina et al. (2002) and Polonelli et al. (2011). Killer toxins kill sensitive cells by inhibition of DNA replication, induction of membrane permeability changes and the arrest of the cell cycle. In some cases, a toxin can interfere with cell wall synthesis by inhibiting β-1,3-glucan synthetase or by hydrolyzing the major cell wall components, β-1,3 glucans and β-1,6 glucans. However, it is still unknown what are the receptors of many other killer toxins on the sensitive cells and how the killer toxins act on the sensitive cells. In addition, little is known about the relationship between the structure of killer toxins, their killer activity and binding to targets on sensitive cells (Liu et al. 2015).

References

Antunes J, Aguiar C (2012) Search for killer phenotypes with potential for biological control. Short Communications. Ann Microbiol 62:427–433

Bajaj BK, Sharma S (2010) Construction of killer industrial yeast *Saccharomyces cerevisiae* hau-1 and its fermentation performance. Braz J Microbiol 4:477–485

Bajaj B, Raina S, Singh S (2013) Killer toxin from a novel killer yeast *Pichia kudriavzevii* RY55 with idiosyncratic antibacterial activity. J Basic Microbiol 53:645–656

Bevan EA, Makower M (1963) The physiological basis of the killer character in yeast. In: Geerts SJ (ed) Genetics today. Proceedings of eleventh international congress on genetics (Haig 1963). Pergamon Press, Oxford, pp 202–203

Bussey H (1991) K1 killer toxin, a pore-forming protein from yeast. Mol Microbiol 5:2339–2343

Buzdar MA, Chi Z, Wang Q, Hua MX, Chi ZM (2011) Production, purification, and characterization of a novel killer toxin from *Kluyveromyces siamensis* against a pathogenic yeast in crab. Appl Microbiol Biotechnol 91:1571–1579

Chen WB, Han YF, Jong SC, Chang SC (2000) Isolation, purification, and characterization of a killer protein from *Schwanniomyces occidentalis*. Appl Environ Microbiol 66:5348–5352

Ciani M, Fatichenti F (2001) Killer toxin of *Kluyveromyces phaffii* DBVPG 6076 as a biopreservative agent to control apiculate wine yeasts. Appl Environ Microbiol 67:3058–3063

Comitini F, Ciani M (2011) *Kluyveromyces wickerhamii* killer toxin: purification and activity towards *Brettanomyces/Dekkera* yeasts in grape must. FEMS Microbiol Lett 316:77–82

Comitini F, De Ingeniis J, Pepe L, Mannazzu I, Ciani M (2004a) *Pichia anomala* and *Kluyveromyces wickerhamii* killer toxins as new tools against Dekkera/Brettanomyces spoilage yeasts. FEMS Microbiol Lett 238:235–240

Comitini F, Di Pietro N, Zacchi L, Mannazzu I, Ciani M (2004b) *Kluyveromyces phaffii* killer toxin active against wine spoilage yeasts: purification and characterization. Microbiology 150:2535–2541

Comitini F, Mannazzu I, Ciani M (2009) *Tetrapisispora phaffii* killer toxin is a highly specific β-glucanase that disrupts the integrity of the yeast cell wall. Microbial Cell Fact 8:55

Conde J, Fink GR (1976) A mutant of *Saccharomyces cerevisiae* defective for nuclear fusion. Proc Natl Acad Sci U S A 73:3651–3655

Fink GR, Styles C (1972) Curing of a killer factor in *Saccharomyces cerevisiae*. Proc Natl Acad Sci U S A 69:2846–2849

Goto K, Iwatuki Y, Kitano K, Obata T, Hara S (1990a) Cloning and nucleotide sequence of the KHR killer gene of *Saccharomyces cerevisiae*. Agric Biol Chem 54:979–984

Goto K, Iwase T, Kichise K, Kitano K, Totuka A, Obata T, Hara S (1990b) Isolation and properties of a chromosome-dependent KHR killer toxin in *Saccharomyces cerevisiae*. Agric Biol Chem 54:505–509

Goto K, Fukuda H, Kichise K, Kitano K, Hara S (1991) Cloning and nucleotide sequence of the KHS killer gene of *Saccharomyces cerevisiae*. Agric Biol Chem 55:1953–1958

Guyard C, Séguy N, Cailliez JC, Drobecq H, Polonelli L, Dei-Cas E, Mercenier A, Menozzi FD (2002) Characterization of a *Williopsis saturnus var. mrakii* high molecular weight secreted killer toxin with broad-spectrum antimicrobial activity. J Antimicrob Chemother 49:961–971

Hammond JRM, Eckersley KW (1984) Fermentation properties of brewing yeast with killer character. J Inst Brew 90:167–177

Hua MX, Chi Z, Liu GL, Buzdar MA, Chi ZM (2010) Production of a novel and cold-active killer toxin by *Mrakia frigida* 2E00797 isolated from sea sediment in Antarctica. Extremophiles 14:515–521

Kimura T, Kitamoto N, Kito Y, Iimura Y, Shirai T, Komiyama T, Furuichi Y, Sakka K, Ohmiya K (1997) A novel yeast gene, RHK1, is involved in the synthesis of the cell wall receptor for the HM-1 killer toxin that inhibits beta-1,3-glucan synthesis. Mol Gen Genet 254:139–147

Klassen R, Teichert S, Meinhardt F (2004) Novel yeast killer toxins provoke S-phase arrest and DNA damage checkpoint activation. Mol Microbiol 53:263–273

Liu G-L, Chi Z, Wang G-Y, Wang Z-P, Li Y, Chi Z-M (2015) Yeast killer toxins, molecular mechanisms of their action and their applications. Crit Rev Biotechnol 35:222–234

Magliani W, Conti S, Travssos LR, Polonelli L (2008) From killer yeast toxins to antibodies and beyond. FEMS Microbiol 288:1–8

Marquina D, Barroso V, Santos A, Peinado JM (2001) Production and characteristics of *Debaryomyces hansenii* killer toxin. Microbiol Res 156:387–391

Marquina D, Santos A, Peinado JM (2002) Biology of killer yeasts. Int Microbiol 5:65–71

Maule AP, Thomas PD (1973) Strains of yeast lethal to brewery yeasts. J Inst Brew 79:137–141

Naumov GI, Kondratieva VI, Naumova ES, Chen G-Y, Li CF (2011) Polymorphism and species specificity of killer activity formation in the yeast *Zygowilliopsis californica*. Biotekhnologiya 3:29–33

Pasteur L, Joubert JF (1877) Charbon et septicémie. C R Hebd Seances Acad Sci 85:101–115

Peng Y, Chi ZM, Wang XH, Li J (2010) β-1,3-Glucanase inhibits activity of the killer toxin produced by the marine-derived yeast *Williopsis saturnus* WC91-2. Mar Biotechnol 12:479–485

Polonelli L, Magliani W, Ciociola T, Giovati L, Conti S (2011) From *Pichia anomala* killer toxin through killer antibodies to killer peptides for a comprehensive anti-infective strategy. Antonie Van Leeuwenhoek 99:35–41

Rodriguez-Cousino N, Maqueda M, Ambrona J, Zamora E, Esteban R, Ramírez M (2011) A new wine *Saccharomyces cerevisiae* killer toxin (Klus), encoded by a double-stranded RNA virus, with broad antifungal activity is evolutionarily related to a chromosomal host gene. Appl Environ Microbiol 77:1822–1832

Russell I, Stewart GG (1985) Valuable techniques in the genetic manipulation of industrial yeast strains. J Am Soc Brew Chem 43:84–90

Santos A, Marquina D (2004) Ion channel activity by *Pichia membranifaciens* killer toxin. Yeast 21:151–162

Santos A, Marquina D, Leal JA, Peinado JM (2000) (1→6)-β-d-glucan as cell wall receptor for *Pichia membranifaciens* killer toxin. Appl Environ Microbiol 66:1809–1813

Santos A, San Mauro M, Bravo E, Marquina D (2009) PMKT2, a new killer toxin from, and its promising biotechnological properties for control of the spoilage yeast *Brettanomyces bruxellensis*. Microbiol 155:624–634

Schaffrath R, Meinhardt F (2005) *Kluyveromyces lactis zymocin* and other plasmid-encoded yeast killer toxins. Topics in Current Genetics, 11:133–155, Microbial Protein Toxins (Schmitt MJ & Schaffrath R, eds). Springer, Berlin

Schmitt MJ, Breinig F (2002) The viral killer system in yeast: from molecular biology to application. FEMS Microbiol Rev 26:257–276

Schmitt MJ, Breinig F (2006) Yeast viral killer toxins: lethality and self-protection. Nat Rev Microbiol 4:212–221

Stewart GG, Russell I (1986) One hundred years of yeast research and development in the brewing industry. J Inst Brew 92:537–558

Wang L, Yue L, Chi Z, Wang X (2008) Marine killer yeasts active against a yeast strain pathogenic to crab *Portunus trituberculatus*. Dis Aquat Org 80:211–218

Wang Y, Zhang X, Zhang H, Lu Y, Huang H, Dong X, Chen J, Dong J, Yang X, Hang H, Jiang T (2012) Coiled-coil networking shapes cell molecular machinery. Mol Biol Cell 23:3911–3922

Weiler F, Schmitt MJ (2003) Zygocin, a secreted antifungal toxin of the yeast *Zygosaccharomyces bailii*, and its effect on sensitive fungal cells. FEMS Yeast Res 3:69–76

Woods DR, Bevan EA (1968) Studies on the nature of the killer factor produced by *Saccharomyces cerevisiae*. J Gen Microbiol 51:115–126

Young TW, Yagiu M (1978) A comparison of the killer character in different yeasts and its classification. Antonie Van Leeuwenhoek 44:59–77

Chapter 11
Stress Effects on Yeast During Brewing and Distilling Fermentations: High-Gravity Effects

11.1 Introduction

All organisms have evolved to cope with changes in their environmental conditions, ensuring the optimal combinations of metabolism, cell proliferation and survival (Mitchell et al. 2009). In order to survive in a changing external environment, microbes have to rapidly sense, respond to and adapt their physiology to new conditions. In yeast, exposure to a mild stress leads to increased tolerance to other stresses (Hallsworth 1998; Bauer and Pretorius 2000). It is suggested that yeast uses information from the environment to prepare for future threats (Zakrzewska et al. 2011).

Brewing and distilling fermentations exert a number of stresses on yeast cultures, and this chapter attempts to identify and elaborate on them. As a consequence, extensive research on this subject has been conducted resulting in a large number of publications (e.g. Gibson et al. 2007; Stewart et al. 1988a, b; Bose et al. 2005; Piper 1995; Powell et al. 2003; You et al. 2003; Blieck et al. 2007; Stewart 2014a, b). It has already been discussed in this text (Chaps. 1, 8 and 9) that this book does not describe in detail the actual brewing and distilling process. However, during the past 50 years or so, the question of process efficiency has become a major focus of both the brewing and distilling industries, and over time, this focus has intensified. This has included various aspects of wort fermentation, including reduced processing times and an increase in beer volume produced per unit time, primarily to shorten overall vessel residence time (including fermenter use). This improvement process has been called "brewing intensification" (Soanes and Hawker 2000). This development (it is actually a number of related developments) has imposed a variety of stresses on the yeast culture.

Brewing intensification is a form of lean manufacturing (Holweg 2007). Lean is centred on "preserving (or enhancing) value with less overall work". Lean manufacturing ensures that appropriate improvements are chosen for the relevant reasons without sacrificing product quality and integrity. However, lean

G.G. Stewart, *Brewing and Distilling Yeasts*, The Yeast Handbook,
https://doi.org/10.1007/978-3-319-69126-8_11

manufacturing, as applied to wort fermentation, must involve a number of stress effects on yeast. This is why lean manufacturing is being discussed here. Dudbridge (2011) has discussed lean manufacturing techniques as they apply to the food industry in general. He proposes that such techniques can be employed to achieve extra output, lower overhead costs (e.g. utilities, labour and raw materials), improved yields, enhanced systems and a superior working environment. Also, in many instances (including brewing), a product with enhanced product stability is the result (not enhanced beer foam stability—details later). At the same time, the quality of the product (beer or spirit in this case) should remain constant or is even enhanced!

The primary brewing technique (not the only one) that is susceptible to intensification procedures is high-gravity processing, particularly, in this context, high-gravity wort fermentation processes (Stewart 2014a, b) together with procedures to produce high-gravity (HG) worts (Murray and Stewart 1991). The question has often been asked: "Does the use of high-gravity wort inflict unusual and cruel punishment on yeast?" (Stewart 2005a, b, 2009, 2010a). As well as HG conditions, there are a number of other parameters that persist during both brewing and distilling fermentations that exert stresses upon yeast. These stress parameters include ethanol, osmotic pressure, temperature, cell surface shear, continuous fermentation compared to batch fermentation, wort ionic balance and some other minor wort components. Nevertheless, the conditions that prevail in high-gravity worts are the primary factors that exert stresses on yeast during both the brewing and distilling processes. Consequently, HG will be initially discussed and its effects on yeast followed by other parameters that exert stress on yeast during both brewing and distilling Thomas and Ingledew (1992).

What is high-gravity brewing? It is a procedure that employs wort at higher than normal concentrations and consequently requires dilution with water (usually deoxygenated in order to enhance beer stability) at a later stage in processing (Stewart 2004, 2010a; Pfisterer and Stewart 1976). By this means, increased production demands can be met without radically expanding brewing, fermenting and storage facilities.

The benefits of high-gravity processing have been extensively documented (Stewart 2005a, b, 2009, 2014a; Saerens et al. 2008; Gibson 2011). Also, there are a number of brewing disadvantages to this procedure (Pfisterer and Stewart 1976; Cooper et al. 1998; Brey et al. 2002; Bryce et al. 1997; Stewart 2012a, b, c). It has cynically been stated that "high-gravity brewing is a technique that waters down beer!" This is not the case! The high-gravity brewing procedure changes the point in the process where water is added. The point of water addition is modified from the beginning of the brewing process (usually during mash-in) to later in the process. Essentially, the same volume of water is employed for a standard beer. However, a detailed discussion of the brewing factors that do not involve yeast is beyond the scope of this publication (Murray and Stewart 1991). However, stress, product flavour and other metabolic effects are discussed here.

The negative effects of HG wort fermentation on beer foam stability are, in part, due to stress effects on yeast. Also, there can be difficulties in achieving product

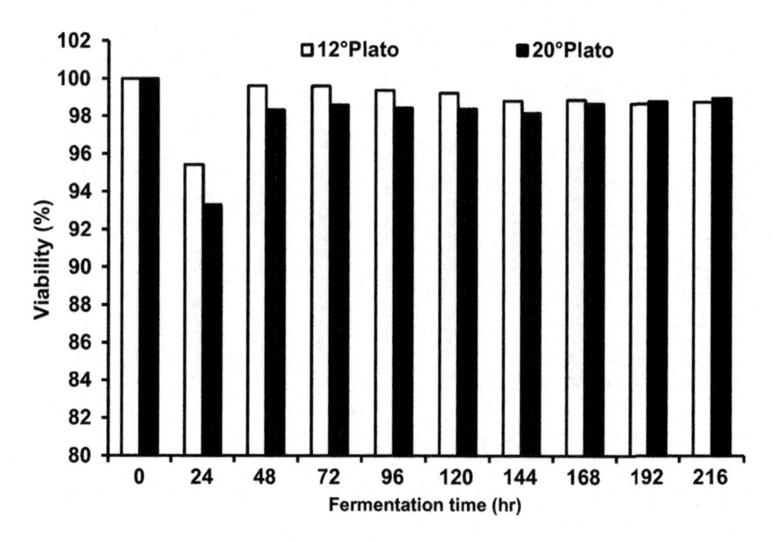

Fig. 11.1 Effect of wort gravity on the viability of a lager yeast strain. Viability determined by the methylene blue staining method

flavour match when compared to lower-gravity-produced beers and undistilled fermented worts (Russell and Stewart 2014). This is usually (but not always) because of increased ester formation—details later (Anderson and Kirsop 1975; Stewart et al. 2008; Stewart 2013). With high-gravity wort, there is a requirement for increased wort dissolved oxygen (DO) at pitching and higher yeast pitching levels at the start of fermentation. High-gravity worts can also influence overall yeast performance (Pratt-Marshall 2002). More specifically, the phenotypic effects of the stresses imposed by HG worts on yeast are as follows:

- A disproportionate decrease in yeast growth (Pratt et al. 1999).
- Decrease in yeast viability necessitating a reduced number of yeast generations (cycles) (Figs. 11.1 and 11.2).
- The production of disproportionately higher levels of esters (Table 11.1).
- Greater "leakage" (secretion/excretion) of specific intracellular enzymes (e.g. proteinase A) resulting in a decrease in foam stability and other effects—details later.
- Alterations in yeast vacuolar size (Fig. 2.1), overall cell volume and cell surface morphology (Pratt et al. 2003).

There are a number of parameters that can be employed to improve the fermentation performance of high-gravity worts (>16°Plato) through wort supplementation. These supplements include metal ions, lipids and "yeast foods". The impact of such additions on wort fermentation performance is considered in a review by Gibson (2011) and in other sections of this text (Chap. 7).

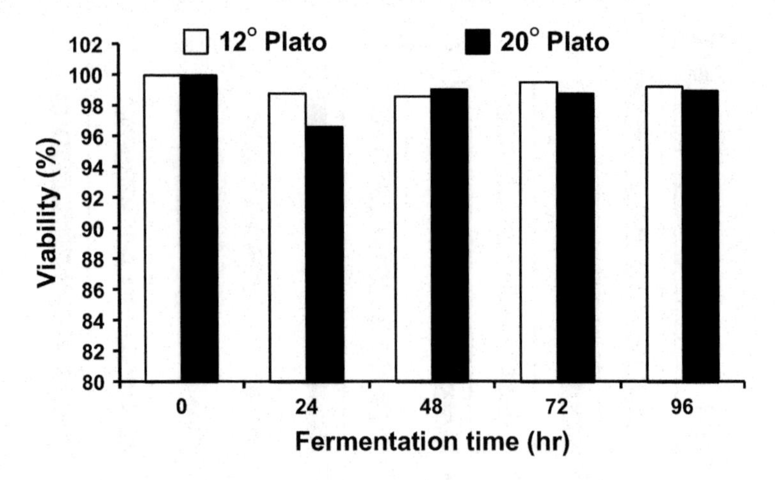

Fig. 11.2 Effect of wort gravity on the viability of an ale yeast strain. Viability determined by the methylene blue staining method

Table 11.1 Influence of wort gravity on beer ester levels

	12°Plato	20°Plato
Ethanol (v/v)	5.1	5.0
Ethyl acetate (mg/L)	14.2	21.2
Isoamyl acetate (mg/L)	0.5	0.7

11.2 Influence of High-Gravity Worts on Yeast Viability

When yeast is first inoculated (pitched) into a high-gravity wort (>16°Plato worts), passive diffusion of water out of the cell occurs, and this produces a decrease in cell viability. Experiments comparing 12 and 20°Plato worts fermented with either lager or ale yeast strains are illustrated in Figs. 11.1 and 11.2. Cell viability decreased in both yeast strains within the first 24 h of fermentation. However, this decrease in viability was exacerbated with the 20°Plato wort compared to the 12°Plato wort. With both types of yeast, the viability is usually recovered later in the fermentation. For reasons that are unclear, the ale strains maintained higher viability than lager strains. Nevertheless, aspects of these differences have already been discussed in this text (Chap. 8).

11.3 Stress Effects on Yeast Intracellular Storage Carbohydrates

Brewer's and distiller's yeast strains contain four major carbohydrates: glucan and mannan plus two intracellular storage carbohydrates—trehalose and glycogen (details in Chap. 6). Trehalose is a disaccharide containing glucose moieties

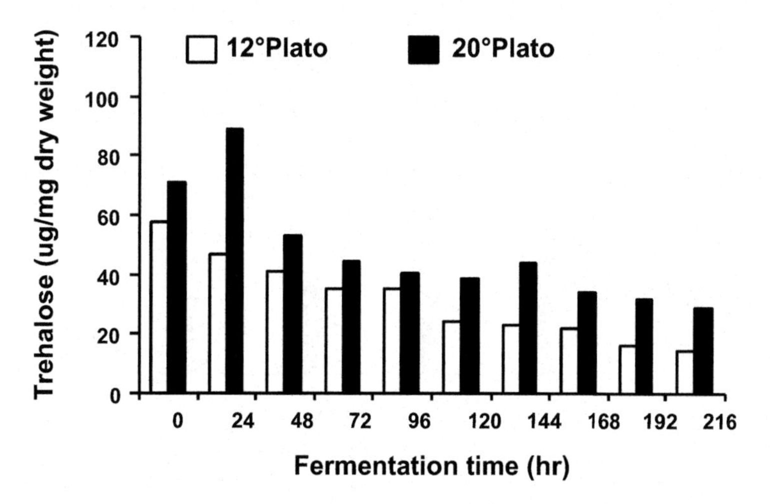

Fig. 11.3 Effect of wort gravity on trehalose metabolism in a lager yeast strain

(Fig. 6.20). It protects the cell against stresses [e.g. metal ions, osmotic pressure, ethanol, high and low temperatures, desiccation and mechanical stress (Gadd 2010; Odumeru et al. 1993; Mansure et al. 1994; van Dijck et al. 1995; Hounsa et al. 1998; Sano et al. 1999; Bolat 2008; Zhang et al. 2013)]. Trehalose has been correlated with cell survival under adverse conditions and is also an important stress indicator in brewing yeast cultures during high-gravity fermentation (Figs. 11.3 and 11.4). There was more rapid synthesis of intracellular trehalose in 20°Plato wort during the initial 24 h of fermentation than was the case with 12°Plato wort. As the cultures acclimatized to the stress conditions imposed by the 20°Plato wort, the intracellular trehalose levels decreased. It is interesting to note that lager strains generally maintained higher trehalose levels than ale strains (D'Amore et al. 1991)—details discussed later and also in Chap. 6.

Glycogen has already been discussed in this text (Chap. 11). It is an intracellular polysaccharide with a structure similar to amylopectin (the branched form of starch) (Fig. 6.18). It consists of α-1,4 linkages and 1,6 branch points. Glycogen has been reported to accumulate under conditions of nutrient limitation indicating its role as a provider of carbon and energy for the maintenance of multiple cellular activities (Thurston et al. 1982). During the first 6–8 h of wort fermentation, there is rapid utilization of intracellular glycogen (Fig. 6.19). This utilization is directly proportional to the synthesis of lipids [mainly unsaturated lipids (UFA) and sterols (ergosterol)]. These lipids are employed by the cells to produce *de novo* membrane material during cell division. Once cell division begins and decreased accumulation of glycogen occurs, it is important that maximum levels of intracellular glycogen are present in the yeast culture when it is harvested for storage prior to being repitched into a subsequent fermentation. It is critical that glycogen levels in yeast are conserved during storage between fermentations because depleted glycogen levels will lead to incomplete fermentation probably as a result of poor

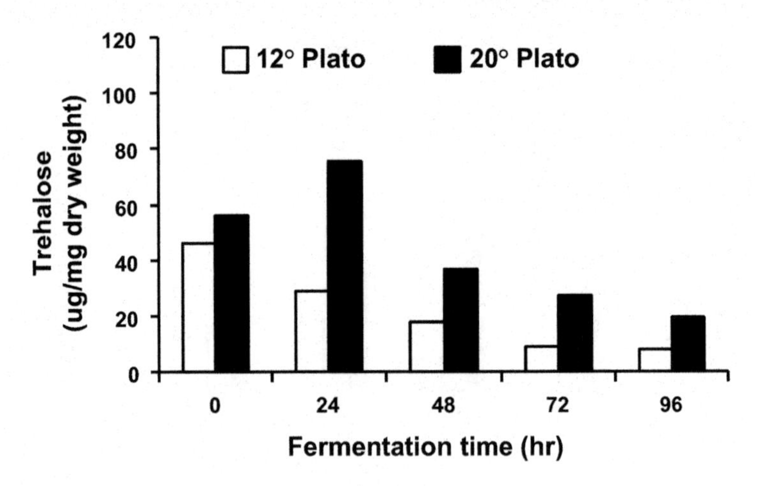

Fig. 11.4 Effect of wort gravity on trehalose metabolism in an ale yeast strain

fermentation efficiency and low concentrations of yeast in suspension (Fig. 6.21) (Thurston et al. 1981).

11.4 Yeast Morphological Changes Induced by High-Gravity Worts and Other Stress Factors

There is a relationship between yeast stress tolerance and growth because the stress response enables yeast cells to continue growing and ferment under adverse conditions (Pratt et al. 2003). For this growth to occur, it not only requires various physiological adaptive changes (Pratt et al. 2007) but also produces distinct cellular morphological changes. The vacuole has been reported to function during periods of both osmotic pressure and ethanol stress in order to ensure continued metabolic activity and yeast cell vitality (Pratt-Marshall 2002). Because both of these stresses are characteristic of high-gravity wort fermentations, morphological studies into the role of the vacuole during high-gravity processing have been conducted (Pratt et al. 2007).

The vacuole designates an optically empty space within the cytoplasm (Figs. 2.1 and 11.5) (Wickner 2002). The space (which can be as much as one third of the cell's volume) is encircled by the vacuolar membrane called the tonoplast. The membrane contains specific membrane-bound proteins and several permeases (Meaden et al. 1999). The tonoplast plays an important role in the metabolic processes associated with the vacuole. The tonoplast remains intact when yeast protoplasts lyse under hypo-osmotic conditions. However, tonoplasts will disintegrate under prolonged nutrient starvation and other conditions that result in the autolysis of the cell by release of vacuolar enzymes. Yeast vacuoles are primarily storage compartments for basic amino acids, polyphosphate, a number of metal ions

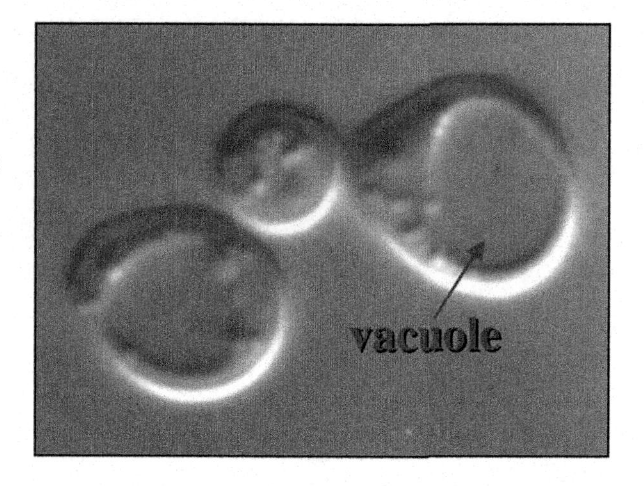

Fig. 11.5 The yeast vacuole—the most prominent organelle in a yeast cell. Permission to reproduce this photograph has gratefully been received from Daniel Lionsky of the University of Michigan

(e.g. Ca^{++}, Zn^{++}, Mg^{++} and Mn^{++}) and specific enzymes including a number of proteinases (details later).

The vacuole is an inherited yeast organelle. Vacuolar segregation begins with elongation of the vacuole towards or into the emerging bud which is a tribute or a line of vesicles (Wickner 2002). Its volume changes with growth phase and environmental conditions. Morphological changes of yeast vacuoles from the lager yeast culture have also been examined at specific times during fermentation in both 12 and 20°Plato all-malt worts employing fluorescence microscopy and a fluorescent dye (FM 4–64) specific for vacuoles. These changes are depicted in Fig. 11.6). The 20°Plato wort produced cells containing enlarged vacuoles compared with the 12°Plato wort (Pratt-Marshall 2002). The diameter of yeast vacuoles of these lager strains has been measured at specified times during static fermentation in both 12 and 20°Plato all-malt worts. Figure 11.6 shows the effect of wort gravity on vacuolar morphology of one of the three lager yeast strains studied. At 0.5 h after pitching, there was a passive flow of water from yeast cells fermenting the 12°Plato wort. Vacuolar morphology remained relatively constant during the lag growth phase, during which the yeast cells adapted to the new environment. When budding commenced, approximately 6 h after pitching, small fragmented vesicles were distributed between mother and daughter cells, fusing to form large vacuoles. Consequently, vacuolar volumes increased until approximately 48 h into fermentation with a 12°Plato wort. There were no further changes in vacuolar dynamics after this time (Zalewski and Buchholz 1996) (Fig. 11.7).

A similar trend in vacuolar dynamics was observed in yeast cells fermenting the 20°Plato wort (Fig. 11.7). Budding was not initiated until 10 h into fermentation, indicating that the ability of these cells to adapt to high solute concentrations, to ensure continued metabolic activity and cell growth, was hindered by the high-gravity wort. In the presence of elevated ethanol concentrations (from the 20°Plato wort), there were continuous increases in vacuolar volumes between 48 and 96 h in all six yeast (ale and lager) strains studied (Fig. 11.8 shows data for a single strain).

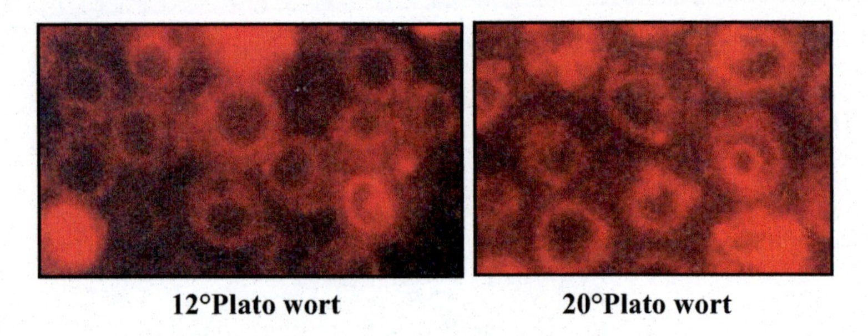

Fig. 11.6 Effect of wort gravity on vacuole size in a brewer's yeast strain

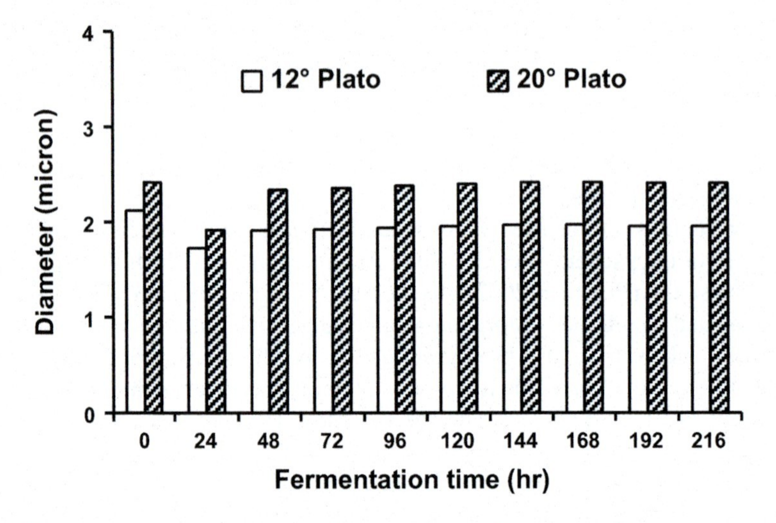

Fig. 11.7 Effect of wort gravity on vacuolar size of lager industrial strain B

These findings confirmed that the yeast vacuole plays an important role in the ability of yeast cells to successfully ferment HG worts. The reduction in vacuolar volume at 24 h during normal (12°Plato) and high (20°Plato)-gravity wort lager fermentations was consistent with accepted theories of stress effects on vacuolar volumes, and a relationship exists between vacuole integrity and yeast viability (Pratt et al. 2003). The findings confirmed that wort gravity has a significant negative effect on yeast cell vacuolar volume of both lager and ale strains during fermentation (Pratt et al. 2007).

In addition to studies on vacuolar volume, the effect of wort gravity on cell surface morphology with ale and lager yeast strains has also been studied. To this end, the surface morphology of yeast strains was determined with a SEM during the static fermentation of 12 and 20°Plato worts. During late stationary growth phase, the cell surface was observed in HG fermentations, resulting in a wrinkly prune-like, crenellated surface with numerous invaginations compared with cells fermenting normal gravity worts (Fig. 11.9).

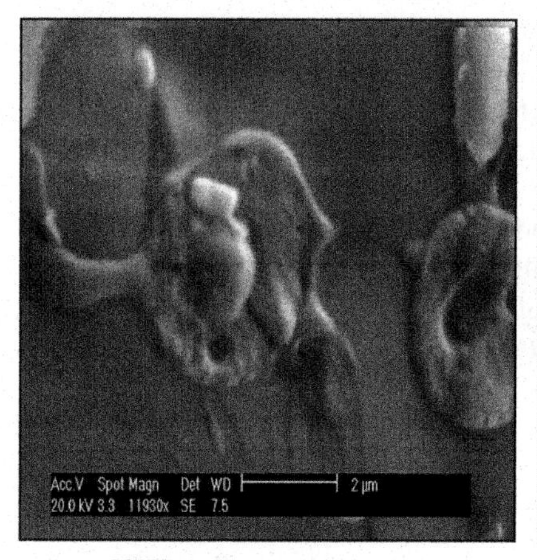

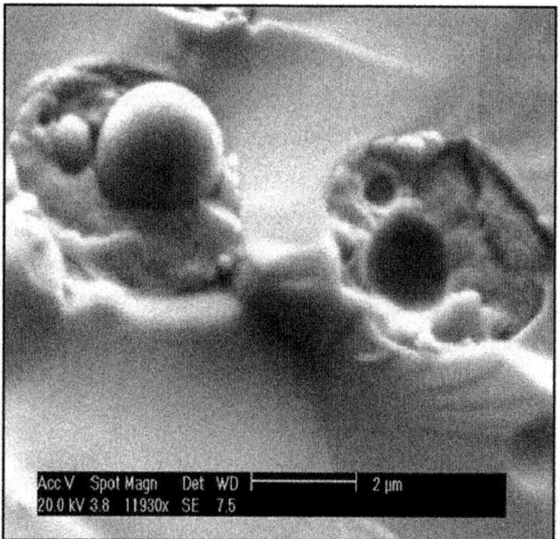

Fig. 11.8 Changes in the vacuolar morphology during high-gravity wort fermentations with an ale yeast strain

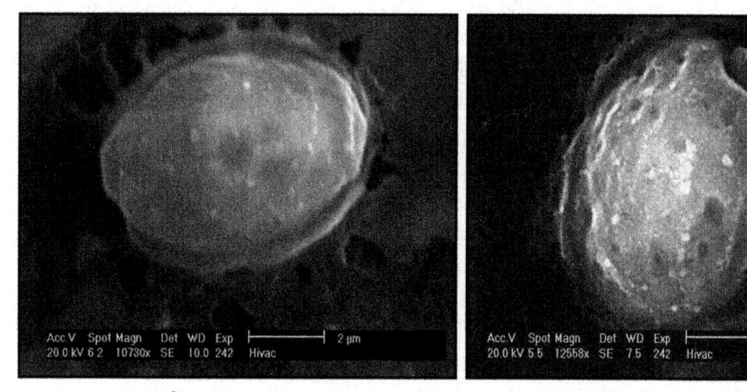

Fig. 11.9 Effect of wort gravity on the cell surface morphology of an ale yeast strain

11.5 Influence of High-Gravity Worts on Yeast Culture Viability

It has already been discussed that when yeast is first pitched (inoculated) into HG wort, passive diffusion of water out of the cell occurs, and this diffusion results in decrease in cell viability (determined by methylene blue or methylene violet staining). Figures 11.1 and 11.2 illustrate experiments with 12 and 20°Plato worts fermented with either a lager or an ale yeast strain. Cell viability decreased in both strains during the first 24 h of fermentation (Pratt et al. 2003). However, the decrease in viability was exacerbated in the 20°Plato wort compared with the 12°Plato wort. With both yeast types [which were representative of a number of lager and ale strains studied—data not shown (Pratt et al. 2003)], the cell viability recovered later in the fermentation. In addition, ale strains maintained higher viabilities than lager strains (Pratt-Marshall 2002). Although the reasons for this difference are unclear, the whole question of ale/lager yeast differences has already been discussed in this text Stewart (1975) (Chap. 3).

11.6 Effects of Stress on Yeast Intracellular Storage Carbohydrates

It has already been discussed that *Saccharomyces* and related species, including brewer's and distiller's yeast strains, contain the two major intracellular storage carbohydrates: trehalose (Figs. 11.3 and 11.4) and glycogen (Figs. 6.19 and 6.21). It has also already been discussed (Chaps. 8 and 12) that trehalose protects the cell against stress [e.g. osmotic pressure, ethanol (and other alcohols), high and low temperature, desiccation and mineral and organic acids (Odumeru et al. 1993)]. It has been correlated with cell survival under adverse conditions and is an important indicator of stress in brewing yeast cultures, for example, during HG wort fermentation. There was rapid synthesis of trehalose in 20°Plato wort during the first 24 h of fermentation. As the cultures acclimatized to the stress conditions imposed by this concentrated wort, the intracellular trehalose levels decreased. It is interesting to note that lager strains maintained higher trehalose levels than ale strains (Pratt et al. 2003; Bolat 2008) (Figs. 11.3 and 11.4).

It has also already been discussed (Chap. 6) that glycogen is an intracellular glucose polysaccharide with a structure similar to starch consisting of α-1,4 linkages with 1,6 branch points (Fig. 6.18). Glycogen is the major reserve energy storage material in yeast cells and many other organisms and tissues (including the muscles of humans). Glycogen accumulates in yeast under nutrient-limiting conditions. Its major role is to provide carbon and energy for the maintenance of cellular activities (Thurston et al. 1982). During the first 6–8 h of wort fermentation, there is rapid utilization of intracellular glycogen (Fig. 6.19). This utilization is directly proportional to the synthesis of lipids [mainly unsaturated lipids (UFA) and sterols (ergosterol)]. These lipids are employed by the cells to produce *de novo* membrane material during cell division. As soon as cell division begins to decrease and the culture enters stationary phase, glycogen accumulates. It is important that maximum levels of intracellular glycogen are present in the yeast culture when it is harvested for storage prior to being stored and repitched into a subsequent fermentation. It is critical that glycogen levels in yeast are conserved during storage between fermentations because cells with depleted glycogen levels will lead to incomplete fermentation probably as a result of depleted yeast growth and low numbers of yeast cells in suspension (Fig. 6.21) (Thurston et al. 1981).

Another form of stress imposed upon brewer's yeast is the process of acid washing (details in Sect. 11.12). This treatment has been shown, since the studies of Pasteur (1876), to be an effective procedure to remove bacterial contaminants from yeast slurries but not wild yeast cells (Bah and McKee 1965). The physiological state of the yeast, together with various environmental circumstances, has been shown to exacerbate the resistance of brewer's yeast strains (and contaminating yeasts—called wild yeasts) to acid washing conditions (Simpson and Hammond 1989). One such environmental condition is HG wort (>16°Plato) (Cunningham and Stewart 1998). Acid washing adversely affected yeast viability from a 20°Plato wort fermentation, whereas yeast from a 12°Plato wort fermentation was not affected to the same extent (Cunningham and Stewart 2000). Strain variations were observed between lager yeast strains in their resistance to HG worts and acid washing (Cunningham and Stewart 1998). The resistance to acid washing was also influenced by the yeast storage conditions, with the yeast that was improperly stored having the lowest viability. Yeast management procedures must be optimized when repitching (reinoculating) yeast from HG fermentation to ensure that the yeast is in good physiological condition and can maintain its resistance to acid washing (Simpson and Hammond 1989). It is also important to emphasize that yeast from a high-alcohol environment during HG fermentation (>6.5% alcohol by volume ABV) should not be acid-washed until it is diluted (<5% ABV) (Stewart 2009).

11.7 Effect of High-Gravity Wort on the Secretion of Yeast Proteinase Activity

There are a number of stress factors that promote release of intracellular yeast proteinase A activity (Brey et al. 2002) (Fig. 11.10). A decrease in beer foam stability due to proteinase A activity has been reported by several research groups (Cooper et al. 2000; Dreyer et al. 1983; Muldbjerg et al. 1993). Proteinase A is a vacuolar aspartic proteinase, which is encoded by the *PEP4* gene. Maddox and Hough (1970) first demonstrated leakage of intracellular proteolytic enzymes through the cell wall and membrane of living yeast cells into the fermenting wort. This proteinase leakage out of living yeast cells, particularly under stress conditions, is now an accepted phenomenon. Proteinase A from brewer's yeast strains may digest the wort proteins/polypeptides responsible for beer foam stability leaving peptides with reduced or no foam-enhancing properties. This results in beer with poor foam stability (Cooper et al. 1998).

Yeast "secretes" proteolytic enzymes into the fermenting wort, and these enzymes appear to have a negative effect on beer foam stability because of polypeptide degradation (hydrolysis) that occurs during wort fermentation and the

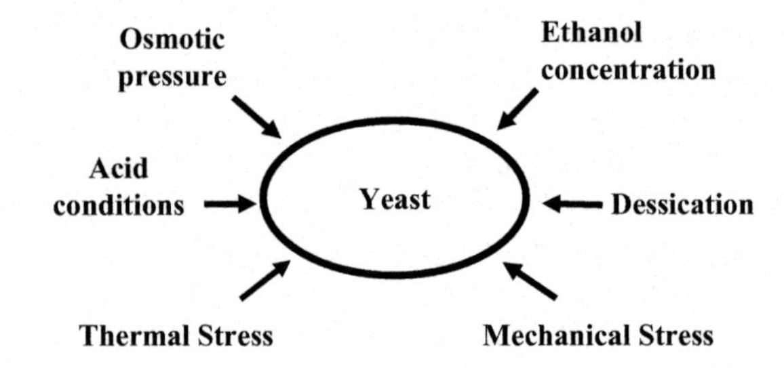

Fig. 11.10 Stress factors which promote proteinase A release

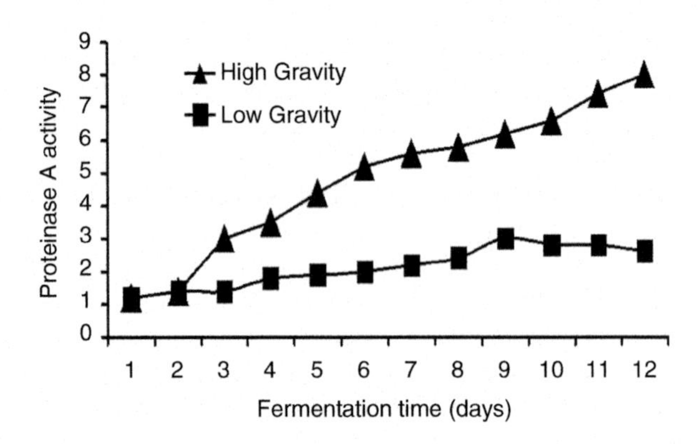

Fig. 11.11 The effect of wort gravity on proteinase A release during the fermentation of low (12°Plato)- and high (12°Plato)-gravity worts

storage of immature beer. Higher amounts of proteinase A were released during a 20°Plato wort fermentation compared with a 12°Plato wort fermentation (Cooper et al. 2000) (Fig. 11.11). During high-gravity wort fermentations, the increased stress on the yeast, in the form of elevated osmotic pressure and ethanol (Beaven et al. 1982), appears to stimulate the secretion of proteinase A into the fermenting wort (Brey et al. 2003; Muldbjerg et al. 1993). As well as osmotic pressure and ethanol, other stresses such as high and low temperature, desiccation, acid washing and mechanical stress will also stimulate yeast proteinase A secretion (Brey et al. 2003) (Fig. 11.10)—further details later.

Because proteinase A is detrimental to beer foam stability, a proteinase A-deficient transformant of a brewing yeast has been developed. The coding region of the *PEP4* gene (proteinase A coding) has been deleted by PCR-mediated gene disruption (a procedure for generating a gene description construct—Kuwayama et al. 2002). Proteinase A activity could not be detected in the mutant cell gene cells or following fermentation in a 12°Plato wort. Also, the beer foam stability was considerably improved compared to beer produced from the unmutated yeast strain (Wang et al. 2007).

There have been many publications concerning the involvement of barley (malt) proteins in beer foam development and stability (Evans and Bamforth 2009). Beer foam formation and stability depends on several factors, such as the presence of proteins, hop iso-α-acids, metallic ions, polysaccharides and melanoidins

(Leisegang and Stahl 2005). The proteins originate mainly from barley (Blasco et al. 2011; Sørensen et al. 1993).

The yeast cell wall mannoproteins (Chaps. 5 and 13) are released into wort during fermentation, and they confer foam stability as a result of their hydrophobic nature (Núñez et al. 2005). When gas bubbles are formed, the mannoproteins orient their hydrophobic protein moiety towards the inner side of the bubble, whereas their hydrophilic glycosylated moiety faces towards the surrounding liquid. The hydrophobic molecule increases the surface tension of the bubbles, whereas the glycosylated areas increase the liquid viscosity, thus enhancing foam stability and retarding liquid drainage (Núñez et al. 2006).

The involvement of yeast in foam formation was initially focused on enology not brewing (Marchal et al. 1996). However, recently (Blasco et al. 2012), the isolation and characterization of a novel fermentation gene *CFG1* (*C*arlsbergensis *f*oaming *g*ene) from the lager yeast species (*S. pastorianus*) have produced interesting results. *CFG1* encodes for the cell wall protein Cfg1p, a 105 kDa protein. This protein is highly homologous to *S. cerevisiae* cell wall mannoproteins, particularly those involved in foam formation, such as Awa1p and Fpl1p. Studies with Cfg1p revealed that this protein is responsible for beer foam stabilization. It is also interesting to note that during wort fermentation studies with a yeast strain that lacked Cfg1p compared to a strain containing Cfg1p, no differences were observed in the foam level during fermentation. Further characterization of Cfg1p has demonstrated that this protein is responsible for beer foam stabilization by *S. pastorianus* Weihenstephan 34/70. This particular lager yeast strain is employed for a large percentage of the lager beer produced in Germany and other countries in Central Europe.

11.8 Yeast Ethanol Toxicity in Distilling

Whisky fermentations commonly employ worts with gravities 1.048–1.070 (12–17.5°Plato) (Cheung et al. 2012). Table 11.9 summarizes the expected ethanol yield of worts from different original gravities. HG fermentations with distilling worts, prior to distillation, are usually expected to yield at least 8.5% (v/v) ethanol. Recent trends in the industry suggest that very-high-gravity (VHG) fermentations, employing worts with a gravity of 22°Plato (or higher), have begun to be employed by the sector because of their capacity to yield at least 12% (v/v) ethanol (Bai et al. 2004; Stewart 2010b) during grain distilling procedures. The advantages of HG and VHG wort fermentations in distilling share many of the advantages of brewing. These advantages centre on the capacity to produce more ethanol during fermentation and to maximize plant capacity. In addition, there are reductions in the volume of waste produced per unit of ethanol and the labour, energy and space required in the plant, which, in turn, reduce overall production costs (Inoue et al. 1998; Russell and Stewart 2014).

It has already been discussed in this chapter (Sects. 11.3 and 11.4) that one of the main drawbacks of HG and VHG fermentations is the elevated stresses experienced by the yeast cells throughout wort fermentation. HG wort exhibits increased osmolarity at the beginning of fermentation due to the higher concentration of wort sugars and other solutes, and this imposes an osmotic challenge to the yeast culture (Gibson et al. 2007). Similarly, the higher concentration of ethanol that accumulates towards the end of fermentation can be toxic to the yeast culture as the expected ethanol yield of VHG fermentations can be twofold higher than that from 12°Plato wort fermentations (Table 11.9). Ethanol stress imposed on *S. cerevisiae* has been extensively reviewed (Gibson et al. 2007; Ding et al. 2009; Stanley et al. 2010a, b). Unlike brewing, the Scotch whisky industry does not practise serial repitching, where yeast cells are collected at the end of fermentation, sometimes washed in acid, stored and repitched into a subsequent fermentation (details in Chaps. 8 and 12). However, yeast exposed to high levels of ethanol is challenged with respect to its viability and vitality maintenance (Chap. 8). This affects the uptake of wort sugars and, critically for the distilling industry, the overall fermentation rate (Fernandes et al. 1997; Pascual et al. 1988). Yeast strains suitable for either potable distilling or industrial (bio) ethanol production for HG and VHG fermentation conditions (details in Chap. 9) must be able to tolerate both osmotic and ethanol stresses. In order to identify strains that might be capable of surviving stress, knowledge regarding strain-dependent tolerance is critical, so that target sites leading to cell damage and death can be effectively identified and monitored. The primary target site for ethanol has been proposed to be the plasma membrane, where ethanol interacts with the phospholipid head group and causes increased membrane fluidity, leading to leakage of the cellular content into the fermentation medium (Marza et al. 2002; Weber and de Bont 1996).

Palmitoleic acid (16:1) and oleic acid (18:1) are the two major monounsaturated fatty acids present in the yeast plasma membrane with the former fatty acid being predominant. Alexandre et al. (2001) observed that, with elevated ethanol stress, the unsaturation index of yeasts (the ratio between saturated fatty acids and unsaturated fatty acids in the plasma membrane) increased. It is also widely agreed that the proportion and capacity to modify the concentration of unsaturated fatty acids (unsaturation index) in the plasma membrane is one of the determining elements in ethanol tolerance (Alexandre et al. 2001; Odumeru et al. 1993). In addition, the presence of unsaturated fatty acids in wort may affect the ethanol tolerance of *S. cerevisiae* strains. You et al. (2003) reported that supplementation of different unsaturated acids conferred improved levels of ethanol tolerance of a desaturase-deficient yeast mutant in the presence of 5% (v/v) ethanol. An increase in unsaturated fatty acids counteracts the effects of ethanol on the phospholipid bilayer of the plasma membrane but also increases the fluidity of the plasma membrane (Weber and de Bont 1996). Elevated fluidity is generally considered to be undesirable. It has been suggested that the rigidity of the plasma membrane is critical for ethanol tolerance (Ding et al. 2009; Gibson et al. 2007). To compensate, unsaturated fatty acids and rigid lipids such as ergosterol are incorporated in order to stabilize the plasma membrane (D'Amore et al. 1988; Inoue et al. 2000).

An unstable plasma membrane causes leakage of the cellular content and disrupts cytoplasmic pH (Stanley et al. 2010a, b). This then impairs the activity of the plasma membrane H^+-ATPase, which is responsible for stabilizing the cytoplasmic pH by pumping protons out of the cell (Jeffries and Jin 2000). Other factors, such as the synthesis of general stress protectants that protect key proteins [heat-shock proteins (hsp)], space hydrophilic heads of the phospholipids that comprise the plasma and cellular membranes [trehalose (Fig. 6.20)] (Odumeru et al. 1993), compatible solutes (glycerol) and intracellular storage carbohydrates [glycogen (Fig. 6.18)] (D'Amore et al. 1990; François and Parrou 2001; Hohmann 2002), have all been implicated in ethanol tolerance.

Ethanol toxicity has been reported to be an inducer of respiratory-deficient (petite) mutants (Ephrussi and Hottinguer 1951; Gibson et al. 2008b; Stewart 2009, 2014b) (details in Chap. 14). However, it must be emphasized that the propensity of distilling and brewing yeasts to form petite mutants is strain dependent (Jimenez et al. 1988; Ibeas and Jimenez 1997). The mechanism by which ethanol induces a petite mutation is not clear, and several hypotheses have been proposed. The majority of petite-inducing agents have the ability to bind, cleave and affect mitochondrial DNA synthesis. Ethanol may also damage mitochondrial DNA via binding and cleavage (Ferguson and Vonborstel 1992). In support of this hypothesis, mitochondrial DNA has been proposed as a target for mutation (Sia et al. 2003). Cheung et al. (2012) speculate that this is because mtDNA is less well protected by structural proteins than chromosomal DNA. In this context, it is important to emphasize that petites possess damaged mtDNA (Castrejon et al. 2002). Ethanol can damage mitochondrial membranes including the cristae (Stewart 2014b), and this may adversely affect replication and/or repair leading to the formation of petite mutants (Gibson et al. 2009; Castrejon et al. 2002). Acetaldehyde (a precursor of ethanol metabolism—Fig. 6.16) has been shown to be a more potent inducer of petites than ethanol (Gibson et al. 2009). These findings could explain why endogenously generated ethanol has a more toxic effect than when exogenously delivered ethanol is used for *in vitro* stress tests (Obe et al. 1977). Petite mutants typically constitute a small proportion of distilling yeast populations (usually less than 1% of the total population) (Stewart 2014b).

However, it has been demonstrated that yeast handling can exacerbate petite levels in brewing production slurries (further details in Chap. 14) (Powell and Diacetis 2007; Jenkins et al. 2003; Stewart 2010b). An increase in petite mutants in a yeast culture has been associated with the following: poor flocculation (D'Amore et al. 1990), changes in flavour profiles (Ernandes et al. 1993; Good et al. 1993) and slower rates of wort sugar uptake (particularly maltose and maltotriose) (Stewart 2014a, b) leading to slower growth and fermentation rates (Spencer et al. 1983). In addition, petite mutants have been demonstrated to be more sensitive to stress than their equivalent wild-type industrial yeast counterparts (Aguilera and Benitez 1985; Hutter and Oliver 1998; Stewart 2014a, b).

Stressful conditions have been simulated in typical Scotch whisky fermentations (Cheung et al. 2012). The conditions maximized product formation and minimized energy and water inputs. This approach increased ethanol concentrations at the end of fermentation, creating stressful conditions for the yeast culture. The relative tolerance of four *S. cerevisiae* distilling strains (supplied in dried, creamed, cake and slurry format) to ethanol under CO_2 induced anaerobic conditions have been studied no beer assessment. The cultures were assessed for their capacity to recover and grow on inhibition spot plates in order to maintain cell viability in ethanol-dosed suspensions. Variations in ethanol tolerance were observed between the cultures and with the same strain supplied in different formats. The creamed yeast typically exhibited a higher ethanol tolerance (Cheung et al. 2012). One possible explanation for the observation is that cells surviving dehydration and subsequent rehydration might incur sublethal genome damage. Thus, the genetic integrity of the most ethanol-tolerant strain was assessed as a function of the supply format (two dried and one creamed). The mitochondrial DNA was examined using mitochondrial restriction fragment length polymorphism (RFLP) (Nguyen et al. 2000) and the chromosomal DNA using pulsed-field gel electrophoresis (PFGE) and polymerase chain reaction (PCR) with a number of specific primers. In one dried yeast sample, the genetic integrity was compromised, highlighting the requirement for yeast intake quality to be closely monitored.

11.9 Yeast Stress: Basic Concepts

It has already been discussed (Sects. 11.1 and 11.2) that, in order to survive in a changing environment, microorganisms (including yeast) have to rapidly sense, respond to and adapt their physiology to new conditions. The conditional changes often occur simultaneously, or in a recurring order (Mitchell et al. 2009). First there are the direct physicochemical effects of the change, which either do or do not kill the cell. If the initial shock is survived, there is usually a response phase and phenotypic adaptation and/or acclimatization and finally a growth phase (Hohmann and Mager 2003; Smits and Brul 2005). As will be discussed later, the responses are, to some extent, specific, so that cells can withstand ethanol (and other chemicals, e.g. acetaldehyde and acetic acid), osmolarity, heat, cold, nutrient depletion, acidic and oxidative environments and/or attacks by other microbes. These specific responses (some of which will be discussed later) are described in the transcriptional and protein synthesis profiles as a result of exposure to various stresses. Many nonlethal stresses were shown to induce common transcriptional responses of approximately 900 genes—the environmental stress response (ESR) (Causton et al. 2001). The ESR entails the down-regulation of many genes involved in ribosome biogenesis, translation and transcription and the upregulation of genes controlled by the general stress transcription factors Msn2 and Msn4 (Gasch et al. 2000). On the basis of these transcriptional data, a strong overlap exists between growth-regulated and stress-regulated genes (Castillo et al. 2008; Brauer et al. 2008).

It has already been discussed that different stresses provoke a similar series of responses. The terms "*g*eneral *s*tress *r*esponse" (GSR) and "*e*nvironmental *s*tress *r*esponse" (ESR) have been adopted for these events. As a physiological result, exposure to a single stress condition at a sublethal dose (e.g. exposure to ethanol or osmotic pressure) confers production protection not only against higher doses of the same stress but also different areas (e.g. temperature tolerance). This effect is commonly known as "cross protection" (CP) (Heinisch and Rodicio 2009).

Control of gene expression in the GSR pathway is mediated by the *s*tress *r*esponse *e*lement (*STRE*) sequences with the respective promoters. Induction also occurs in response to oxidative stress, nutrient limitation, heavy metals and DNA damage (Ruis and Schüller 1995; Treger et al. 1998). *STRE* sequences are recognized by a redundant pair of transcription factors, Msn2 and Msn4, whose subcellular location depends on CAMP/PKA signalling. The latter also plays a role in sugar sensing. Activated protein kinase A (PKA) phosphorylates transcription factors, and this results in their export into the cytosol. During stress (or sugar limitation), the PKA pathway is down-regulated, and the dephosphorylated transcription factors enter the nucleus in order to activate STRE-dependent gene expression (Estruch 2000).

Msn2 and Msn4 control expression of genes to GSR and also of genes encoding heat-shock proteins, oxidative stress detoxification enzymes and trehalose metabolism (Odumeru et al. 1993). The latter is of particular interest. It has already been discussed that trehalose (Fig. 6.20) is a major stress protectant (Gancedo and Flores 2004). It has been implicated in protection against dehydration (Sano et al. 1999), freezing, heating (Attfield 1987) and toxic compounds such as ethanol, acetaldehyde, acids (e.g. acetic acid), oxygen radicals and heavy metals (D'Amore et al. 1991; Van Dijck et al. 1995; Estruch 2000). Also, it has already been discussed (Chap. 8) that research on the carbohydrate glycogen (Fig. 6.18) illustrates that it confers increased yeast cell viability and vitality during a plethora of alcoholic fermentations (D'Amore et al. 1990; Pérez-Torrado et al. 2002).

The metabolism of glycogen and trehalose is summarized in Fig. 11.12. Carbohydrate moieties for their synthesis (anabolism) are provided by UDP-glucose. For glycogen, the first glucose molecules are covalently linked to glycogen, and the glycogen synthase isoenzymes (Gsy1, Gsy2) subsequently elongate the chains. They are allosterically activated by glucose-6-phosphate and inactivated by PKA-dependent phosphorylation. Branching activity is provided by Glc3. For glycogen mobilization, glycogen phosphorylase (Gph1) and a debranching enzyme (Gdb1) liberate glucose-1-phosphate and glucose, respectively. Inversely to glycogen synthase, Gph1 is inhibited by glucose-6-phosphate and activated by a PKA-dependent phosphorylation.

Trehalose is synthesized by a multienzyme complex, composed of two catalytic subunits, Tps1 (trehalose-6-phosphate synthase) and Tps2 (trehalose-6-phosphate phosphatase), and two regulatory subunits (Tps3 and Ts11). This disaccharide can also be imported from the medium by Agt1. Hydrolysis is catalysed by either of two trehalases [an acid isoform encoded by ATH1 turnover in a laboratory strain (Lillie and Pringle 1980)]. Nth1 activity is high in log-phase cells, decreases during the diauxic shift and remains low in stationary phase. Similar kinetics for trehalose are

Trehalose

- Trehalose is a disaccharide consisting of glucose units (Fig **6.20**). It protects yeast cells against stress (Fig. **11.10**).
- It has been correlated with cell survival. Under adverse conditions and is an indicator of stress during HG wort fermentations (Fig. **11.3**).
- Lager yeast cultures maintain higher trehalose levels than ale cultures (Figs. **11.3** and **11.4**).

Glycogen

- Glycogen is an intracellular glycogen polymer with a similar structure to amylopectin. It consists of α-1.4 linkages with 1,6 branch points (Fig. **6.18**).
- It provides carbon with energy for the maintenance of cell activities.
- During the first 6-8 hours of wort fermentation there is rapid metabolism of glycogen and its utilization is directly proportional to the synthesis of unsaturated fatty acids and sterols (Fig. **6.19**).

Fig. 11.12 Summary of trehalose and glycogen metabolism

observed under vinification conditions, i.e. accumulation after ammonium depletion and degradation in the lag phase of initial growth (Novo et al. 2003).

It has already been discussed that during stressful conditions (e.g. heat shock), trehalose synthesis occurs rapidly and is degraded soon after stress relief (Singer and Lindquist 1998). Genes for the four biosynthetic subunits are induced by heat stress and repressed by the CAMP/PRA pathway and mediated by STRE sequences in their promoters. Post-translationally, Nth1 is also activated by a PKA-dependent phosphorylation in response to external glucose (Alexandre et al. 2001). A model for trehalose's role in yeast stress protection may lend physiological significance to these findings. In the early stages of fermentation, trehalose stabilizes protein structure and prevents aggregation of denatured proteins. As a consequence, heat-shock proteins assume this function, and trehalose needs to be degraded to avoid interference. This also explains why trehalose responds to a plethora of stresses that affect protein folding (e.g. heat, cold, ethanol and osmotic pressure). Importantly, this disaccharide has been shown to stabilize the plasma membrane during such stresses (Heinisch and Rodicio 2009).

11.10 Influence of Wort Sugar Spectrum and Gravity on Ester Formation

One disadvantage of the use of high-gravity procedures in a brewing environment, which was recognized in the early stages of R&D on this project, is that the use of concentrated wort induces synthesis of disproportionately high levels of esters that occur in the beer (Table 11.1) (Anderson and Kirsop 1975). It has been known for many years (Stewart 2009; 2014a) that varying the wort sugar profile will modify the level of many metabolites including the spectrum of esters, although reasons for these differences are still unclear. Entry of the hexose sugars (glucose and fructose) into the cell is facilitated by the same protein transport system, although the utilization of glucose occurs more rapidly than fructose when the two sugars are fermented separately. This is possibly because of differing affinities of the two sugars and for the transporters (Younis and Stewart 1998; Guillaume et al. 2007). Also, differences in the rates of phosphorylation between glucose and fructose occur. The wort disaccharide maltose (Fig. 7.5) is internalized by the yeast only when 40–50% of the glucose has been removed from the wort (Stewart 2006; Stewart and Russell 2009) (details in Chap. 7) and occurs via an active transport system, whereas the uptake of glucose and fructose is by passive transport (Bisson et al. 1993).

In order to conduct an initial investigation of glucose/maltose effects, 4% glucose and 4% maltose in a synthetic medium (yeast extract—peptone) were fermented separately following pitching at 21 °C in order to eliminate inhibition of sugar uptake, and the production of the esters ethyl acetate and isoamyl acetate is monitored (Younis and Stewart 1998). The fermentation performance of three ale and three lager brewing yeast strains employed in this study was similar. Tables 11.2 and 11.3 show the viabilities (determined by methylene blue staining) and vitalities (determined by the acidification power test, also called the proton efflux rate) (Siddique and Smart 2000) of the cells, respectively, following 4 days of fermentation. All six strains, when the cells were cultured in maltose, consistently had higher viabilities and enhanced vitalities compared with their glucose-cultured counterparts. Reasons for these differences are not immediately apparent. It may be the result of slower initial uptake rates of maltose compared with glucose and the consequent reduced growth rates (Younis and Stewart 1998). In addition, maltose uptake occurs by active transport (conversion of ATP to ADP and ADP to AMP) (Fig. 7.6), and glucose uptake by passive transport is no doubt relevant (Rautio and Londesborough 2003).

Despite the apparent sturdiness of the maltose-grown cells, the production of ethyl acetate and isoamyl acetate was lower than in the glucose-grown cells (Table 11.4). The lower levels of ester production, with maltose (compared to glucose) as the substrate, could be due to a number of reasons. It is possible that fermentation with maltose inhibits the transport of esters out of the cell, perhaps by modifying the plasma membrane, thus giving the impression that fewer esters are produced. However, in the light of the enhanced viability and vitality of the

Table 11.2 Percentage viability of brewing yeast strains after 96 h fermentation of synthetic media[a]

	Glucose	Maltose
Ale 1	96	98
Ale 2	92	98
Ale 3	94	98
Lager 1	97	99
Lager 2	96	98
Lager 3	95	99

[a]Peptone—yeast extract—4% sugar medium
Methylene violet stains employed

Table 11.3 Vitality of brewing yeast strains after 96 h fermentation of synthetic media[a]

	Glucose	Maltose
Ale 1	0.8	1.3
Ale 2	0.9	1.3
Ale 3	1.1	1.4
Lager 1	0.7	0.9
Lager 2	0.8	1.2
Lager 3	0.9	1.0

[a]Peptone—yeast extract—4% sugar medium
Acidification power test

maltose-grown cells, this possibility is unlikely. Another possibility is that maltose, compared to glucose, metabolism produces lower levels of acetyl-coenzyme A, which has been suggested as resulting in fewer esters because of a lack of intermediate metabolites. It has been proposed that ester production is linked to lipid metabolism (Thurston et al. 1982). If this is the case or if for some reason maltose metabolism produces fewer toxic fatty acids, it would seem reasonable to assume that reduced toxic fatty acids would be produced in wort containing elevated levels of maltose (Thurston et al. 1982; Piddocke et al. 2009).

It has been generally agreed for a long time (Anderson and Kirsop 1975; Pfisterer and Stewart 1976) that a reduction in ester levels, particularly ethyl acetate and isoamyl acetate, from high-gravity brewed beers would be welcome by most brewers. In order to study the influence of maltose and glucose levels in high-gravity worts, two 20°Plato worts were prepared, one containing 30% maltose syrup (MS) and the other containing 30% very high MS (VHMS). The sugar composition of these two brewing syrups is shown in Table 11.5. In addition, a 12°Plato wort containing 70% (w/v) MS was prepared and used as the control. The sugar spectra of the three worts are shown in Fig. 11.13. The maltose plus maltotriose concentration in the 20°Plato VHMS wort increased compared with the 20°Plato MS wort, with a corresponding decrease in the concentration of glucose plus fructose.

Table 11.4 Ethyl acetate and isoamyl acetate produced by brewing yeast strains during fermentation of synthetic media[a]

	Ethyl acetate (mg/L)		Isoamyl acetate (mg/L)	
	Glucose	Maltose	Glucose	Maltose
Ale 1	4.13	2.79	0.14	0.14
Ale 2	2.97	2.59	0.06	0.04
Ale 3	3.13	2.71	0.05	0.03
Lager 1	6.00	5.22	0.22	0.21
Lager 2	3.75	3.28	0.26	0.22
Lager 3	4.13	3.51	0.23	0.17

[a]Peptone—yeast extract—4% sugar medium

The three worts were fermented in a 2hL pilot brewery (Fig. 11.14) by a lager yeast strain at 13 °C and the concentrations of ethyl acetate and isoamyl acetate determined throughout the fermentations (Figs. 11.15 and 11.16, respectively). The profiles were similar for both esters. The concentrations of both esters in the 20°Plato MS fermented wort were twice those observed in the 12°Plato fermented wort. However, the ester concentration in the fermented 20°Plato VHMS wort was approximately 25% reduced compared with the 20°Plato MS wort (Younis and Stewart 1999). This observation with wort confirms the findings employing synthetic media with single sugars, where maltose fermentations produced less ethyl acetate and isoamyl acetate than glucose fermentations. In addition, similar to synthetic media fermentations, the wort with elevated maltose concentrations produced higher yeast viabilities and vitalities than the wort containing lower levels of maltose (Tables 11.6 and 11.7).

In a separate recent study (Piddocke et al. 2009), the lager strain Weihenstephan 34/70 has been characterized at three wort gravities by adding either glucose or maltose syrups to a basic all-malt wort at 14°Plato. Fermentations with the higher-gravity worts resulted in a lower specific growth rate, a longer lag phase prior to initiation of ethanol production, an incomplete sugar utilization and an increase in the concentrations of ethyl acetate and isoamyl acetate in the final beer. However, increasing the gravity by adding maltose syrup instead of glucose syrup resulted in greater balanced fermentation performance in terms of higher cell numbers and enhanced wort fermentability, and the resulting beer exhibited a more favourable flavour profile. This study underlines the results discussed above—namely, the effect of various stress factors (Pratt-Marshall 2002; Stewart 2008) on brewer's yeast metabolism particularly during high-gravity conditions and the influence of sugar type on fermentation performance and the beer's flavour profile.

11.11 High-Gravity Yeast Varieties

A successful strategy has been adopted (Blieck et al. 2007) for yeast variants with significantly improved fermentation capacity under high-gravity conditions. Improved performing variants of a lager strain have been isolated by subjecting a

Table 11.5 Sugar composition of brewing syrups

	Maltose syrup (MS)	Very high maltose syrup (MS)
Glucose	15[a]	5
Maltose	55	70
Maltotriose	10	10
Dextrins	20	15

[a]% (w/v) composition

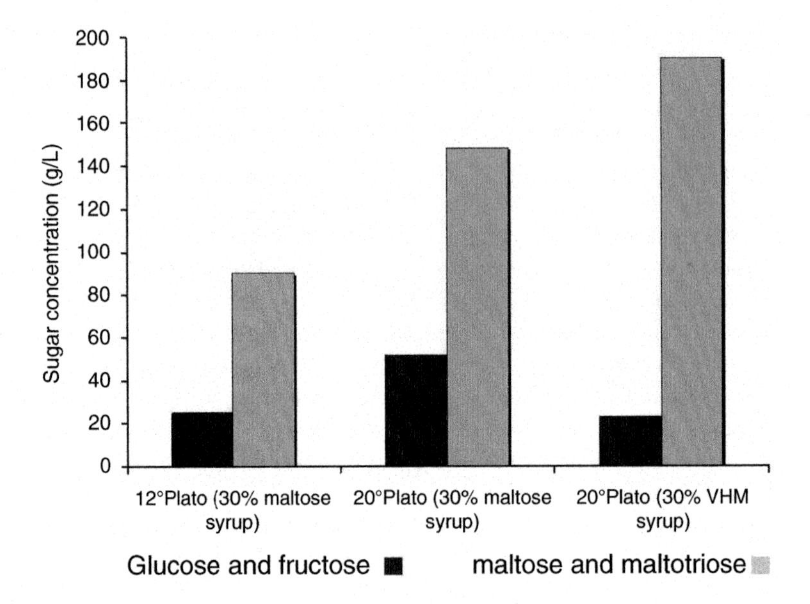

Fig. 11.13 Wort sugar profiles

pool of ultraviolet-induced variants in very high-gravity wort (>20°Plato). Two variants showing faster fermentation rates with more complete attenuation as well as improved viability under high ethanol conditions have been identified. These variants displayed the same advantages in pilot-scale stirred fermenters under high-gravity conditions at 11 °C. Microarray analysis (a method to measure the expression levels of a large number of genes simultaneously—Moran et al. 2004) identified several genes whose modified expression may be responsible for the superior fermentation performance of the variants. The role of some of these candidate genes has been confirmed by genetic transformation. Proper selection conditions allow the isolation of production brewer's yeast variants with superior fermentation characteristics.

In addition to yeast genotypic effects on the fermentation of high-gravity wort, phenotypic effects are equally important. It has already been discussed in this chapter that, for example, osmotic stress influences high-gravity wort fermentation (Thomas et al. 1993). In addition, it has been observed that Mg^{++} (Hu et al. 2003), Zn^{++} (Rees and Stewart 1998), yeast extract (Bafrncova et al. 1999), glycine

Fig. 11.14 The Heriot-Watt University, Edinburgh, Scotland, brewing pilot plant

(Thomas et al. 1993), biotin (Alfenore et al. 2002) and peptone (Stewart et al. 1988b) exhibit protective effects on yeast growth and viability and also improve the final ethanol concentration.

11.12 Stress Parameters on the Production of Grain Whisky in Scotland

The production of whisky in Scotland is regulated by the Scotch Whisky Regulations 2009. There are two distinct types of whisky, malt and grain, each of which has different characteristics:

- Malt whisky has a pronounced bouquet and taste and is made exclusively from malted barley and yeast by the pot still method (Fig. 11.17). This is a batch process that does not enable continuous production. Consequently, this type of whisky is made in separate batches (fermentation and distillation), each of which is similar although not identical. The average annual capacity of a typical malt distillery is approximately 2.5 million litres of pure alcohol, although it may be considerably higher. In 2015, there were 98 functioning malt distilleries in Scotland. The average concentration of unfermented wort is 12–14°Plato (10.48–10.56 OG), and the average yield in 2012 was approximately 412 litres of alcohol per metric ton of malted barley employed (Gray 2013).

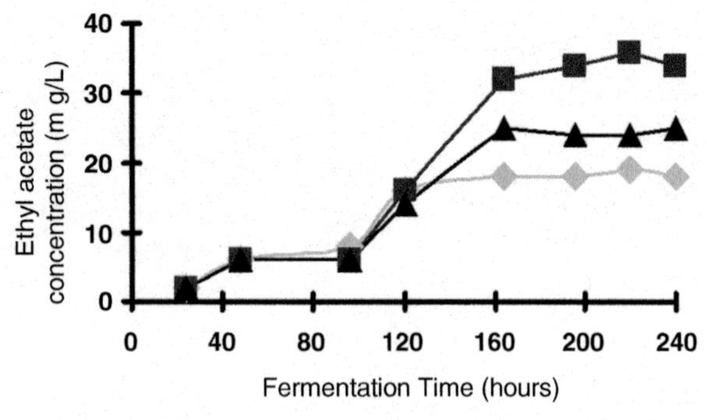

Fig. 11.15 Ethyl acetate concentration in worts of differing gravities and sugar composition

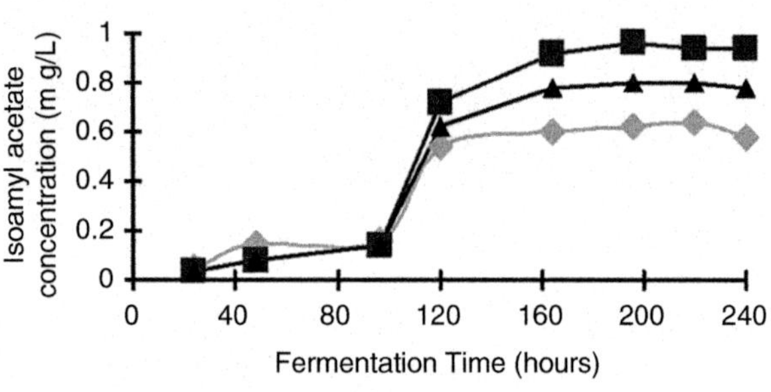

Fig. 11.16 Isoamyl acetate concentration (mg/L) in worts of differing gravities and sugar composition

Table 11.6 Percentage viability of an ale and a lager brewing yeast strain after fermentation in 12°Plato and 20°Plato[a]

	12°Plato		20°Plato	
	MS[b]	VHMS[c]	MS	VHMS
Ale	95	98	93	96
Lager	94	97	95	98

[a]Methylene blue and methylene violet stains employed
[b]Maltose (55) syrup
[c]Very high maltose (70) syrup

Table 11.7 Vitality of an ale and a lager brewing yeast strain after fermentation in 12°Plato and 20°Plato worts[a]

	12°Plato		20°Plato	
	MS [b]	VHMS [c]	MS	VHMS
Ale	0.9	1.1	0.6	0.8
Lager	0.7	0.9	0.6	0.9

[a]Acidification power test
[b]Maltose (55) syrup
[c]Very high maltose (70) syrup

Fig. 11.17 A typical pot still employed for the batch distillation of malt whisky

- Grain whisky is produced from a mixture of malted barley, maize (corn), wheat, yeast and water in the average proportion of approximately 16% (w/v) barley malt and 84% (w/v) maize, wheat or a combination of both, although this proportion varies from one distillery to another. Unlike malt whisky, the grain product is produced by a continuous distilling process (modelled on the Coffey still) that lends itself to large-scale production (Fig. 11.18). In addition, some plants employ a continuous fermentation process, whereas others are batch fermented (details later in this chapter). Grain whisky has less well-defined characteristics than malt whisky, which makes it suitable for blending purposes. Unlike malt whisky, grain whisky varies little in taste from one distillery to another. Consequently, the industry regards it as a commodity product, and it is traded from one distilling company to another at a set price. In 2015, there were seven operating grain distilleries in Scotland. The average concentration of unfermented wort is 18–20°Plato, and the average alcohol yield of a typical grain whisky distillery in 2008 was 383 litres of alcohol per metric ton of cereal employed (Gray 2013).

There are essentially three types of Scotch whisky (Table 11.8). Only strains of *S. cerevisiae* (ale type) are employed for the fermentation of whisky worts (further details are in Chap. 3). Fermentation is conducted at higher temperatures (28–32 °C) than brewing in different geometry fermenters (also called washbacks). Unlike in brewing, the yeast is only used once; it is not reused. At the end of

Fig. 11.18 Coffey still employed for the continuous distillation of grain whisky (Courtesy of F.O. Robson)

fermentation, the fermented wort plus yeast goes directly into the batch (malt whisky) (Fig. 11.17) or continuous (grain whisky) stills (Fig. 11.18).

As already discussed, grain whisky is produced by both batch and continuous fermentation processes. A schematic of a continuous fermentation process is shown

Table 11.8 Scotch whisky types

Malt whisky	100% malted barley
96 distilleries in Scotland	Batch process
	Copper pot stills
Grain whisky—	Malted barley (10–15%)
7 distilleries in Scotland	Corn or wheat (85–90%)
	"Continuous stills"
Blended whisky	Malt whisky (15–50%)[a]
	Grain whisky (50–85%)

[a]Blended after at least 3-year maturation in oak casks

Table 11.9 Fermentation characteristics of 19°Plato and 21°Plato grain worts during 2008

	Alcohol (v/v)	Residual mutual (g/L)	Residual maltotriose (g/L)
21°Plato[a] wort	9.6	5.8	19.6
19°Plato wort	10.2	4.3	6.5

[a]Problems with distiller's dried grain (DDG) consistency in 2007 and 2008

in Fig. 11.19. Yeast is purchased (in cream, cake or dried form) from a yeast supplier. The yeast is usually grown on a molasses-based medium, containing ammonium ions, where the predominant carbohydrate source is sucrose. To acclimatize (liven) the yeast to a cereal fermentation environment (where the predominant fermentable sugars are glucose, maltose and maltotriose, plus dextrins, with a broad spectrum of amino acids), the yeast is incubated in grain (wheat or corn/maize) wort in a tub vessel for 24 h. The acclimatized yeast is incubated in grain wort (also termed wash) contained in a continuous fermenter (also termed a washback) in flow-through mode at 30–32 °C for approximately 36 h. When a steady state has been established, the rate of wort addition is in balance with the rate at which the fermented wort leaves the fermenter to be held in a holding tank prior to distillation in a continuous (Coffey-type) still (Fig. 11.19).

In 2005, a grain distillery in Scotland employing continuous fermentation was successfully fermenting 21°Plato grain wort yielding 11% (v/v) alcohol in the fully fermented wort. This situation prevailed until early 2006, but in late 2006, problems began to be encountered with a decreased yield of alcohol to 9.6% (v/v) because of incomplete utilization of wort maltose and particularly maltotriose (Fig. 11.20). This decreased sugar utilization equated to a reduced alcohol yield from 385 to 370 L of alcohol per metric ton of grain. In addition, because of the residual wort maltose and maltotriose, the resulting distiller's dried grain had a sticky consistency and was not acceptable to use as an animal feed. In an attempt to overcome this problem, the original gravity of the wort was reduced to 19°Plato (Table 11.9) (Stewart 2010a). This reduction resulted in complete fermentation of the wort with no residual maltose and maltotriose and improved the consistency of the distilled dried grain. However, as the gravity of the wort had to be reduced, the distillery's overall alcohol yield was reduced below budgeted productivity levels. The reasons

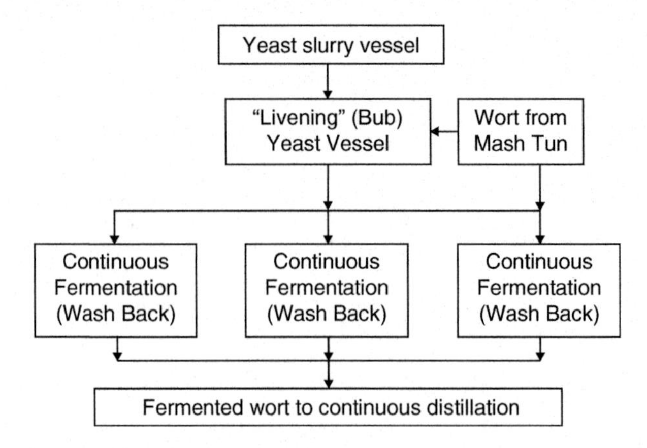

Fig. 11.19 Schematic of a continuous fermentation process for grain whisky production

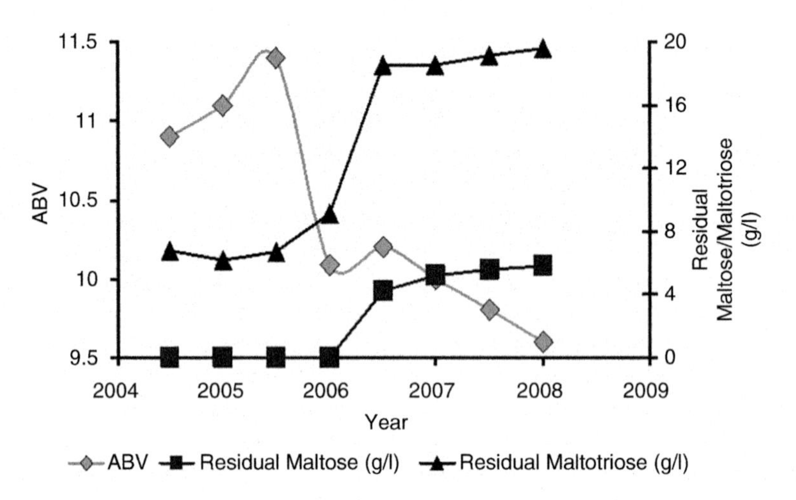

Fig. 11.20 Fermentation trends with a 21°Plato grain wort (July 2005 to September 2008)

for the deterioration in yeast efficiency regarding maltose and maltotriose uptake are still unclear.

It appears that the 21°Plato wort exerted stress effects on this particular pitching yeast (this is strain to strain varieties in this regard—Stewart 2006), with inhibitory effects on maltose and maltotriose uptake. Stress effects on sugar uptake, particularly maltose and maltotriose, have been discussed previously (Chap. 7) (Stewart 2010b).

It appears that the 21°Plato wort exerts stress effects on the pitched yeast, with inhibitory effects on maltose and maltotriose having been described previously (Chap. 7) (Bisson et al. 1993). It is worthy of note that in 2013, this distillery began using maize (corn) instead of wheat. The problems with incomplete fermentation have been largely eliminated; reasons for this improvement in fermentation efficiency are largely unclear (*personal communication*).

11.13 Stress Effects on Yeast as a Result of Acid Washing

The most common source of batch contamination in a brewery is probably from infected pitching yeast (Simpson 1987a), which can transfer bacterial contaminants from fermentation to fermentation. Contaminating bacteria can cause beer spoilage which may result in the production of undesirable flavour characteristics and beer containing haze which also lacks foam stability (Hill 2015).

The process of acid washing brewer's yeast to remove contaminating bacteria from yeast slurries prior to pitching was first practised by Louis Pasteur (Fig. 2.14) (Pasteur 1876). He recommended the use of tartaric acid solution, whereas today phosphoric acid at pH 2–2.2 is the predominant acid employed. Also, hydrochloric, sulphuric and nitric acids are employed. In addition, acidified ammonium persulphate has been identified as an effective washing medium for removing bacteria (Bah and McKee 1965; Simpson 1987a). This improved efficiency was probably due to the decomposition of ammonium persulphate into peroxide and ozone—both have enhanced bactericidal properties (Simpson and Hammond 1989).

By the early part of the last century, routine acid washing pitched yeast had been introduced to many British (and other) countries' breweries (Brown 1916). He reported that acid washing (pH 2.0–2.2) yeast decreased many of the problems associated with brewing beer during the summer months. It produced a superior beer with increased microbiological stability and also rendered changes in pitching yeast cultures unnecessary or, at least, changes in the yeast cultures less often. Therefore, this process had financial and quality benefits for the brewer and consequently for the consumer. As a result of improved refrigeration in breweries, seasonal variation is not a current problem in many countries.

Simpson (1987b) reported that acid washing was a relatively inexpensive and effective method of eliminating a wide range of brewery bacterial contaminants. The acid washing regime will vary amongst breweries, with some using acid washing every fermentation cycle, while with other brewers, their pitching yeast is only acid-washed when there is a significant contamination problem. However, many brewers, for reasons that will be discussed later, never acid wash their yeast because of the stress effects on the culture! If their yeast culture is found to be contaminated with bacteria, a fresh culture of the same yeast strain is propagated and introduced into the brewing process.

Bacteria that contaminate a brewery (or a distillery—details later), for example, *Lactobacilli*, *Pediococcus*, *Acetobacter*, *Pectinatus*, *Zymomonas*, etc., are more sensitive to acid treatment than yeasts. It should be emphasized that as well as brewing yeast strains, wild yeasts (e.g. *Brettanomyces*, *Zygosaccharomyces*, *Pichia*, *Hansenula*, etc.) can contaminate a pitching yeast culture and are resistant to acid washing. When brewing yeast cultures are exposed to other environmental stress conditions during acid washing [e.g. ethanol, temperature (hot and cold), osmotic pressure, desiccation and mechanical stress], the physiological condition of the yeast culture may deteriorate further (Simpson and Hammond 1989). The adverse effects that acid washing and other stress conditions have upon pitching yeast primarily affect the cell surface and the physiological systems associated with the cell wall and membrane (Fernandez et al. 1993). The physiological condition of the yeast prior to acid washing will determine this effect. Acid washing has been shown to affect the flocculation characteristics of yeast (Simpson and Hammond 1989) although the effect of deflocculating or acid addition would more efficiently expose contaminating bacteria to the acid conditions (Simpson 1987b). The effect that acid washing has on pitching yeast tends to vary with some brewer's yeast strains that were grown in 10°Plato worts exhibiting low viability after acid washing (Ogden 1987). Other studies (Simpson and Hammond 1989; Cunningham and Stewart 1998) have found that acid washing has no negative effect on cell viability when acid-washed and repitched into 12°Plato wort. When yeast from high-gravity wort fermentations (20°Plato) was acid-washed, decrease in viability was observed during the first 24 h of the fermentation, which suggested that stress (either increased osmotic pressure or increased ethanol concentrations) was exerted on the yeast. Casey and Ingledew (1983), Casey et al. (1983) showed that decreased viability over the first 12 h of a fermentation increased with elevated wort original gravity but increasing the pitching rates in high-gravity worts significantly reduced the extent of cell death in the first few hours of the fermentation. It has been reported that increased osmotic pressure decreases yeast viability (Owades 1981; Panchal and Stewart 1980) but that it can be alleviated by modifying the wort composition. Casey and colleagues (1983, 1984) showed that by adding a supplementary nitrogen source, ergosterol and oleic, a high viability could be maintained throughout the fermentation. The combined effects on cell viability of acid washing and serial repitching or acid washing and pitching into 12°Plato wort are more interesting! The combination of increased osmotic pressure and ethanol concentrations in conjunction with the low pH during acid washing exerted excess stress which some yeast strains are unable to tolerate. It is also interesting to note that supplementing wort with inorganic ions or with an added nitrogen source, ergosterol and a source of unsaturated fatty acids prevented the decrease in viability that was noted upon pitching before and after acid washing (Cunningham and Stewart 2000).

Various effects of acid washing on subsequent fermentation performance have been observed. Some breweries have noted that the fermentation performance of the yeast culture was stimulated in the fermentation immediately after the acid wash (Brown 1916; Russell and Stewart 1995). These results have been supported by Simpson and Hammond (1989), where it was found that there was no significant difference in the fermentation profile between washed and unwashed cells, even when the cells were serially repitched in 20°Plato wort and when the guidelines on acid washing (the do's and do not's) (Simpson and Hammond, 1989) were carefully followed (Table 11.10).

Table 11.10 Effects of yeast acid washing[a]

Acid washing can influence yeast performance, including: • Reduced yeast viability • Reduced yeast vitality • Reduced rate and/or degree of fermentation • Changes in yeast quality parameters such as flocculation, fining, size of yeast crop and excretion of cell components
Acid washing of yeast can be summarized into the do's and do not's. The do's of acid washing are: • Use food grade acid • Chill the acid and the yeast slurry before use to less than 5 °C • Wash the yeast as a beer slurry or as a slurry in water • Ensure constant stirring, while the acid is added to the yeast and preferably throughout the wash • Ensure that the temperature of the yeast slurry does not exceed 5 °C during washing • Verify the pH of the yeast slurry • Pitch the yeast immediately after washing
The do not's of acid washing are: • Do not wash for more than 2 h • Do not store washed yeast • Do not wash unhealthy yeast • Avoid washing yeast from high-gravity fermentations prior to dilution
There are a number of options to acid washing brewer's yeast: • Never acid wash yeast • Low yeast generation (cycle) specification • Discard yeast when there is evidence of contamination (bacteria or wild yeast) • Acid wash every cycle; this procedure can have adverse effects on yeast • Acid wash when bacterial infection levels warrant the procedure

[a]Simpson WJ and Hammond TRM (1989) The response of brewing yeasts to acid washing. J Inst Brew 95:347–354

The vitality of the yeast cultures has been determined with the acidification power test (Kara et al. 1988), and the results obtained are less clear. On some occasions, acid washing stimulated the acidification response during glucose addition. Although the trend in this study was that yeast vitality was similar for acid-washed and non-acid-washed cells at the end of each fermentation, the vitality after washing did not mirror the performance of the yeast in the subsequent fermentation. Surprisingly, the vitality of the yeast culture in 20°Plato fermentations was greater than that of the culture in 12°Plato fermentations, although this may be explained by the fact that increased ethanol concentrations stimulated the plasma membrane ATPase (Rosa and Sá-Correia 1994) which may account for the observed increased vitality. The stimulating effect of acid washing yeast may also be due to the removal of extraneous materials associated with the cell wall, which Simpson (1987b) described as predominantly proteinaceous in nature.

As the acidification power test measures the proton efflux from the cell, any extracellular conditions that could cause perturbations in the plasma membrane, H^+-ATPase activity may account for the varied results obtained. This enzyme has been widely studied and has been reported to be affected by glucose (Serrano 1983), temperature (Coote et al. 1995), ethanol (Rosa and Sá-Correia 1994) and an inadequate nitrogen source (Benito et al. 1992), and it may be considered to be a cellular protective mechanism to environmental stress. The effect of H^+-ATPase on maltose compared to glucose activity has not been reported. Its major role is to control the intracellular pH of yeast cells and to allow the intracellular environment to remain constant and consistent, permitting normal metabolic function. During acid washing, the yeast is exposed to pH 2.0–2.2. The plasma membrane ATPase is therefore important to maintain the intracellular pH, allowing the yeast to survive acid exposure. When the plasma membrane ATPase is inhibited with diethylstilbestrol (Van der Rest et al. 1995) during acid washing, the yeast viability [measuring with methylene (blue staining)] decreased from 98.4% at the start of acid washing to 30% after 11 h incubation, whereas the viability of the untreated cells was 95.0%. Diethylstilbestrol is an inhibitor of plasma membrane H^+-ATPase activity. Yeast death is probably due to intracellular acidification as the cell was unable to maintain its intracellular pH concentration. This suggests that the plasma membrane ATPase is an important mechanism to allow the yeast to survive exposure to acid conditions. Eraso and Gancedo (1987) showed that the plasma membrane ATPase was stimulated by low pH. However, a study by Carmelo et al. (1996) has reported that when a strong acid (HCl) was used to acidify the medium, a decrease in ATPase activity was observed. There are no reports regarding the effect of phosphoric acid on the plasma membrane ATPase, but the combination of these different environmental conditions encountered during and after each fermentation would have contributed to the vitality measurements making analysis difficult. It has already been discussed (Chap. 8) that the advent of flow cytometry has rendered intracellular pH measurements, with this instrument, possible and reliable (Valli et al. 2005; Chlup et al. 2008).

It has already been discussed that when yeast is exposed to elevated osmotic pressure, heat and increased ethanol concentrations at the end of fermentation, stress responses are the result (Odumeru et al. 1992a, b, 1993). Casey et al. (1984) have shown that pitching yeast into high-gravity wort exerts stress upon the pitching yeast. This results in decreased viability over the first few hours of a fermentation. When yeast from a 20°Plato wort fermentation was acid-washed and subsequently exposed to a third environmental stress (e.g. pH), these multiple stresses appeared to have a cumulative negative effect on the yeast where it accentuated the viability decrease and inability to utilize glucose over the first 24 h of the fermentation at a similar rate to that of the first cycle culture. It has also been reported that increased ethanol concentrations reduced the acid resistance of a strain of *S. cerevisiae* (Kalathenos et al. 1995), suggesting that the increased ethanol concentrations at the end of the 20°Plato wort fermentation reduced the resistance of the yeast to acid washing. Ethanol denatures proteins and increases the permeability of the cell membrane (Jones et al. 1988), but Beaven et al. (1982) reported that ethanol tolerance was increased when the proportion of unsaturated fatty acids within the membrane was increased, and this is thought to be an adaptation in order to maintain membrane integrity. If the proportion of unsaturated fatty acids in the plasma membrane was low before acid washing, yeast tolerance to the low pH may be affected, and this could explain the reduced viability and lower the glucose utilization rates observed upon repitching.

It is worth reiterating that acid washing and serial repitching of yeast in 12°Plato wort did not exhibit any negative effects on the yeast's fermentation ability. This is probably due to the yeast not being exposed to environmental stresses sufficient to cause cell damage. Yeast from 20°Plato fermentations was exposed to increased osmotic pressure upon pitching and increased ethanol concentrations at the end of fermentation. Thus, the ability to tolerate the conditions encountered following acid washing was diminished. The importance of yeast pitching rate with high-gravity worts has already been discussed (Casey and Ingledew 1983) together with the importance of wort dissolved oxygen concentrations and the relationship to wort gravity considered (Cunningham and Stewart 2000) (Table 11.1).

11.14 Yeast Responses to Acetic Acid Stress

Acetic acid represents one of the most significant by-products of yeast metabolism, particularly during the hydrolysis of lignocelluloses. This contributes to reduced ethanol yield and productivity when predominantly these waste materials are used as substrates for biofuel production (Maiorella et al. 1983; Palmqvist and Hahn-Hägerdal 2000)—further details in Chap. 9. The undissociated form of acetic acid is freely membrane permeable and therefore enters the cell by simple diffusion. Once in the cytosol, the pHi is near neutral, determined by flow cytometry (Valli et al. 2005; Chlup et al. 2008), and it dissociates, leading to the release of protons (H^+) and acetate (CH_3COO^-) (Guldfeldt and Arneborg 1998). Proton accumulation determines the intracellular acidification, which inhibits many metabolic activities, while acetate may cause turgor pressure and free radical production, including severe oxidative stress (Piper et al. 2001; Semchyshyn et al. 2011a, b; Ullah et al. 2012). Acetic acid has also been shown to induce programmed cell death (defined as AA-DCD) in *S. cerevisiae* cells characterized by chromatin condensation, a TUNEL-positive phenotype and ROS accumulation, resulting in lipid peroxidative, protein oxidation together with carbonylation and genetic damage (Morano et al. 2012; Rego et al. 2012). Acetic acid will impair cellular metabolism and growth reducing the productivity of the process (Martani et al. 2013). Also, an important question has been asked about the relevance of acetic acid effects concerning the relevance in the yeast chronological ageing model to ageing in higher eukaryotes (Burhans and Weinberger 2009). It has been reported that accumulation of acetic acid in stationary-phase cultures stimulates highly conserved growth signalling pathways and increases oxidative and replication stress, all of which have been implicated in ageing and/or age-related diseases in more complex organisms. Low pH also stimulates growth signalling pathways in mammals. Although the reduced production of acetic acid, identified by Erasmus et al. (2004) as a factor in the CLS-extending effects of calorie restruition in yeast, may be specific for this organism. The underlying mechanism of this protects against chronological ageing and is likely to be the same for calorie restruction in higher eukaryotes. It is reduced signalling that inhibits both oxidative and replication stresses. The remarkable parallels between the regulation of chronological ageing in yeast and of ageing in more complex higher organisms suggests that conserved growth signalling pathways impact upon ageing in all eukaryotes via dual effects on oxidative and replication stress. The yeast chronological ageing model will likely continue to provide insights that will provide a better understanding of these, and other, aspects of ageing and age-related diseases in mammals. This is yet another example of the importance of *S. cerevisiae* (and other yeasts) as an experimental eukaryote, the application of results from it to mammals including humans (another example is discussed in Chap. 14).

11.15 Summary

Brewing and distilling fermentations exert a number of stresses on yeast cultures. The question of process efficiency and intensification has recently become a major focus. The primary brewing technique (not the only one) that is susceptible to intensification is high-gravity (HG) processing, particularly HG wort fermentation processes. As well as HG conditions, there are a number of other parameters that persist during brewing and distilling fermentations, which exerts stresses upon yeast. These parameters include ethanol, osmotic pressure, temperature, cell surface shear, pH, continuous fermentation compared to batch fermentation, wort ionic balance and some minor wort components. Nevertheless, the conditions that prevail in high-gravity worts are the principal factors that exert stresses on yeast.

References

Aguilera A, Benitez T (1985) Role of mitochondria in ethanol tolerance of *Saccharomyces cerevisiae*. Arch Microbiol 142:389–392

Alexandre H, Ansanay-Galeote V, Dequin S, Blondin B (2001) Global gene expression during short-term ethanol stress in *Saccharomyces cerevisiae*. FEBS Lett 498:98–103

Alfenore S, Molina-Jouve C, Guillouet SE, Uribelarrea JL, Goma G, Benbadis L (2002) Improving ethanol production and viability of *S. cerevisiae* by vitamin feeding strategy during fed-batch process. Appl Microbiol Biotechnol 60:67–72

Anderson RG, Kirsop BH (1975) Quantitative aspects of the control of oxygenation on acetate ester concentration in beer obtained from high gravity. J Inst Brew 81:286–301

Attfield PV (1987) Trehalose accumulates in *Saccharomyces cerevisiae* during exposure to agents that induce heat shock response. FEBS Lett 225:259–263

Bafrncova P, Smogrovicova D, Salvikova I, Patkova J, Domeny Z (1999) Improvement of very high gravity ethanol fermentation by media supplementation using *S. cerevisiae*. Biotechnol Lett 21:337–341

Bah S, McKee WE (1965) Beer-spoilage bacteria and their control with a phosphoric acid – ammonium persulfate wash. Canadian J Microbiol 11:309–318

Bai FW, Chen LJ, Zhang Z, Anderson WA, Moo-Young M (2004) Continuous ethanol production and evaluation loss under very high gravity medium conditions. J Biotechnol 110:287–293

Bauer FF, Pretorius IS (2000) Yeast stress response and fermentation efficiency: how to survive the making of wine—a review. S Afr J Enol 21:27–51

Beaven MJ, Charpentier C, Rose AH (1982) Production and tolerance of ethanol in relation to phospholipid fatty-acyl composition in *Saccharomyces cerevisiae* NCYC 431. J Gen Microbiol 128:1447–1455

Benito B, Portillo F, Lagunas R (1992) *In vivo* activation of the yeast plasma membrane ATPase during nitrogen starvation. FEBS Lett 300:271–274

Bisson LF, Coons DM, Frankel AL, Lewis DA (1993) Yeast sugar transporters. CRC Crit Rev Biochem Mol Biol 284:269–308

Blasco L, Vinas M, Villa T (2011) Proteins influencing foam formation in wine and beer: the role of yeast. Int Microbiol 14:61–71

Blasco L, Veiga-Crespo P, Sánchez-Pérez A, Villa TG (2012) Cloning and characterization of the beer foaming gene CFG1 from *Saccharomyces pastorianus*. J Agric Food Chem 60:10796–10807

Blieck L, Toye G, Dumortier F, Vertrepen KJ, Delvaux FD, JM T, Vandijck P (2007) Isolation and characterization of brewer's yeast variants with improved fermentation performance under high-gravity conditions. Appl Environ Microbiol 73:815–824

Bolat I (2008) The importance of trehalose in brewing yeast survival. Innov Roman Food Biotechnol 2:1–10

Bose S, Dutko JA, Zitomer RS (2005) Genetic factors that regulate the attenuation of the general stress response of yeast. Genetics 169:1215–1226

Brauer MJ, Huttenhower C, Airoldi EM, Rosenstein R, Matese JC, Gresham D, Boer VM, Troyanskaya OG, Botstein D (2008) Coordination of growth rate, cell cycle, stress response, and metabolic activity in yeast. Mol Biol Cell 19:352–367

Brey SE, Bryce JH, Stewart GG (2002) The loss of hydrophobic polypeptides during fermentation and conditioning of high gravity and low gravity brewed beer. J Inst Brew 108:424–433

Brey SE, de Costa S, Rogers PJ, Bryce JH, Morris PC, Mitchell WJ, Stewart GG (2003) The effect of proteinase A on foam-active polypeptides during high and low gravity fermentation. J Inst Brew 109:194–202

Brown HT (1916) Reminiscences of fifty years' experience of the application of scientific method in brewing practice. J Inst Brew 13:265–354

Bryce JH, Cooper D, Stewart GG (1997) High gravity brewing and its negative effect on head retention. In: Proceedings of 26th congress of European Brewery Convention, Maastricht, pp 357–365

Burhans WC, Weinberger M (2009) Acetic acid effects on aging in budding yeast: are they relevant to aging in higher eukaryotes? Cell Cycle 8:2300–2302

Carmelo V, Bogaerts P, Sá-Correia I (1996) Activity of plasma membrane H+-ATPase and expression of PMA1 and PMA2 genes in *Saccharomyces cerevisiae* cells grown at optimal and low pH. Arch Microbiol 166:315–320

Casey GP, Ingledew WM (1983) High gravity brewing: Influence of pitching rate and wort gravity on early yeast viability. J Am Soc Brew Chem 41:148–153

Casey GP, Magnus CA, Ingledew WM (1983) High gravity brewing: nutrient enhanced production of high concentrations of ethanol by brewing yeast. Biotechnol Lett 5:429–434

Casey GP, Magnus CA, Ingledew WM (1984) High-gravity brewing: effects of nutrition on yeast composition, fermentative ability, and alcohol production. Appl Environ Microbiol 48:639–646

Castillo L, Martínez AI, Gelis S, Ruiz-Herrera J, Valentín E, Sentandreu R (2008) Genomic response programs of *Saccharomyces cerevisiae* following protoplasting and regeneration. Fungal Genet Biol 45:253–265

Castrejon F, Codon AC, Cubero B, Benitez T (2002) Acetaldehyde and ethanol are responsible for mitochondrial DNA (mtDNA) restriction fragment length polymorphism (RFLP) in flor yeasts. Syst Appl Microbiol 25:462–467

Causton HC, Ren B, Koh SS, Harbison CT, Kanin E, Jennings EG, Lee TI, True HL, Lander ES, Young RA (2001) Remodeling of yeast genome expression in response to environmental changes. Mol Biol Cell 12:323–337

Cheung AWY, Brosnan JM, Phister T, Smart KA (2012) Impact of dried, creamed and cake supply formats on the genetic variation and ethanol tolerance of three *Saccharomyces cerevisiae* distilling strains. J Inst Brew 118:152–162

Chlup PH, Bernard D, Stewart GG (2008) Disc stack centrifuge operating parameters and their impact on yeast physiology. J Inst Brew 114:45–61

Cooper DJ, Stewart GG, Bryce JH (1998) Some reasons why high gravity brewing has a negative effect on head retention. J Inst Brew 104:221–228

Cooper DJ, Stewart GG, Bryce JH (2000) Yeast proteolytic activity during high and low gravity wort fermentations and its effect on head retention. J Inst Brew 106:197–201

Coote PJ, Billon CM-P, Pennell S, McClure PJ, Ferdinando DP, Cole MB (1995) The use of confocal scanning laser microscopy (CSLM) to study the germination of individual spores of *Bacillus cereus*. J Microbiol Methods 21:193–208

Cunningham S, Stewart GG (1998) Effects of high-gravity brewing and acid washing on brewers' yeast. J Am Soc Brew Chem 56:12–18

Cunningham S, Stewart GG (2000) Acid washing and serial repitching a brewing ale strain of *Saccharomyces cerevisiae* in high gravity wort and the role of wort oxygenation conditions. J Inst Brew 106:389–402

D'Amore T, Panchal CJ, Stewart GG (1988) Intracellular ethanol accumulation in *Saccharomyces cerevisiae* during fermentation. J Appl Environ Microbiol 54:1471–1510

D'Amore T, Panchal CJ, Russell I, Stewart GG (1990) A study of ethanol tolerance in yeast. Crit Rev Biotechnol 9:287–304

D'Amore T, Crumplen R, Stewart GG (1991) The involvement of trehalose in yeast stress tolerance. J Ind Microbiol 7:191–195

Ding J, Huang X, Zhang L, Zhao N, Yang D, Zhang K (2009) Tolerance and stress response to ethanol in the yeast *Saccharomyces cerevisiae*. Appl Microbiol Biotechnol 85:253–263

Dreyer T, Biedermann K, Ottesen M (1983) Yeast proteinase in beer. Carlsb Res Commun 48:249–255

Dudbridge M (2011) The food industry, handbook of lean manufacturing in the food industry. Wiley-Blackwell, Chichester

Ephrussi B, Hottinguer H (1951) On an unstable cell state in yeast. Cold Spring Harb Symp Quant Biol 16:75–85

Erasmus DJ, Cliff M, van Vuuren HJJ (2004) Impact of yeast strain in the production of acetic acid, glycerol, and the sensory attributes of ice wine. Am J Enol Vitic 55:371–378

Eraso P, Gancedo F (1987) Activation of yeast plasma membrane ATPase by acid pH during growth. FEBS Lett 224:187–192

Ernandes JR, Williams JW, Russell I, Stewart GG (1993) Respiratory deficiency in brewing yeast strains – effects on fermentation, flocculation, and beer flavor components. J Am Soc Brew Chem 151:16–20

Estruch F (2000) Stress-controlled transcription factors, stress-induced genes and stress tolerance in budding yeast. FEMS Microbiol Rev 24:469–486

Evans E, Bamforth CW (2009) Beer foam, achieving a suitable head. In: Bamforth CW (ed) Beer: a quality perspective. Academic Press, Burlington, MA, pp 7–66

Ferguson LR, Vonborstel RC (1992) Induction of the cytoplasmic petite mutation by chemical and physical agents in *Saccharomyces cerevisiae*. Mutat Res 265:103–148

Fernandes L, Corte-Real M, Loureiro V, Loureiro-Dias MC, Leao C (1997) Glucose respiration and fermentation in *Zygosaccharomyces bailii* and *Saccharomyces cerevisiae* express different sensitivity patterns to ethanol and acetic acid. Lett Appl Microbiol 25:249–253

Fernandez E, Fernandez M, Moreno F, Rodicio R (1993) Transcriptional regulation of the isocitrate lyase encoding gene in *Saccharomyces cerevisiae*. FEBS Lett 333:238–242

François J, Parrou JL (2001) Reserve carbohydrates metabolism in the yeast *Saccharomyces cerevisiae*. FEMS Microbiol Rev 25:125–145

Gadd GM (2010) Metals, minerals and microbes: geomicrobiology and bioremediation. Microbiology 156:609–643

Gancedo C, Flores CL (2004) The importance of a functional trehalose biosynthetic pathway for the life of yeasts and fungi. FEMS Yeast Res 4:351–359

Gasch AP, Spellman PT, Kao CM, Carmel-Harel O, Michael B, Storz G, Botstein D, Brown Patrick O (2000) Genomic expression programs in the response of yeast cells to environmental changes. Mol Biol Cell 11:4241–4257

Gibson BR (2011) 125th Anniversary review: improvement of higher gravity brewery fermentation via wort enrichment and supplementation. J Inst Brew 117:268–284

Gibson BR, Lawrence SJ, Leclaire JP, Powell CD, Smart KA (2007) Yeast responses to stresses associated with industrial brewery handling. FEMS Microbiol Revs 31:535–569

Gibson BR, Boulton CA, Box WG, Graham NS, Lawrence SJ, Linforth RST, Smart KA (2008a) Carbohydrate utilization and the lager yeast transcriptome during brewery fermentation. Yeast 25:549–562

Gibson BR, Prescott KA, Smart KA (2008b) Petite mutation in aged and oxidatively stressed ale and lager brewing yeast. Lett Appl Microbiol 46:636–642

Gibson DG, Young L, Chuang R-Y, Venter JC, Hutchison CA III, Smith HO (2009) Enzymatic assembly of DNA molecules up to several hundred kilobases. Nat Methods 6:343–345

Good L, Dowhanick TM, Ernandes JE, Russell I, Stewart GG (1993) Rho-mitochondrial genomes and their influence on adaptation to nutrient stress in lager yeast strains. J Am Soc Brew Chem 5:35–39

Gray AS (2013) The scotch whisky industry review, 36th edn. Sutherlands, Edinburgh, Scotland

Guillaume C, Delobel P, Sablayrolles JM, Blondin B (2007) Molecular basis of fructose utilization by the wine yeast *Saccharomyces cerevisiae*: a mutated HXT3 allele enhances fructose fermentation. Appl Environ Microbiol 73:2432–2439

Guldfeldt LU, Arneborg N (1998) Measurement of the effects of acetic acid and extracellular pH on intracellular pH of nonfermenting individual *Saccharomyces cerevisiae* cells by fluorescence microscopy. Appl Environ Microbiol 64:530–534

Hallsworth J (1998) Ethanol-induced water stress in yeast. J Ferment Bioeng 85:125–137

Heinisch JJ, Rodicio R (2009) Physical and stress factors in yeast. In: Konig H, Unden G, Frolich J (eds) Biology of microorganisms on grapes, in must and in wine. Springer, New York, NY, pp 275–292

Hill AE (2015) Traditional methods of detection and identification of brewery spoilage organisms. In: Hill AE (ed) Brewing microbiology – managing microbes, ensuring quality and valorising waste. Woodhead Publishing, Cambridge, pp 271–318

Hohmann S (2002) Osmotic stress signaling and osmoadaptation in yeasts. Microbiol Mol Biol Rev 66:300–372

Hohmann S, Mager WH (2003) Yeast stress responses. Springer, Berlin

Holweg CL (2007) Living markers for actin block myosin-dependent motility of plant organelles and auxin. Cell Motil Cytoskeleton 64:69–81

Hounsa CG, Brandt EV, Thevelein J, Hohmann S, Prior BA (1998) Role of trehalose in survival of *Saccharomyces cerevisiae* under osmotic stress. Microbiology 144:671–680

Hu CK, Bai FW, An LJ (2003) Enhancing ethanol tolerance of a self-flocculating fusant of *Schizosaccharomyces pombe* and *Saccharomyces cerevisiae* by Mg^{2+} via reduction in plasma membrane permeability. Biotechnol Lett 25:1191–1194

Hutter A, Oliver SG (1998) Ethanol production using nuclear petite yeast mutants. Appl Microbiol Biotechnol 49:511–516

Ibeas JI, Jimenez J (1997) Mitochondrial DNA loss caused by ethanol in *Saccharomyces* flor yeasts. Appl Environ Microbiol 63:7–12

Inoue Y, Tsujimoto Y, Kimura A (1998) Expression of the glyoxalase I gene of *Saccharomyces cerevisiae* is regulated by high osmolarity glycerol mitogen activated protein kinase pathway in osmotic stress response. J Biol Chem 273:2977–2983

Inoue T, Iefuji H, Fujii T, Soga H (2000) Cloning and characterization of a gene complementing the mutation of an ethanol-sensitive mutant of sake yeast. Biosci Biotechnol Biochem 64:229–236

Jeffries T, Jin Y-U (2000) Ethanol and thermotolerance in the bioconversion of xylose by yeasts. Adv Appl Microbiol 47:221–268

Jenkins CL, Kennedy AI, Thurston P, Hodgson JA, Smart KA (2003) Serial repitching fermentation performance and functional biomarkers. In: Smart KA (ed) Brewing yeast fermentation performance. Blackwell Science, Oxford, pp 257–271

Jimenez J, Longo E, Benitez T (1988) Induction of petite yeast mutants by membrane-active agents. Appl Environ Microbiol 54:3126–3132

Jones JS, Weber S, Prakash L (1988) The *Saccharomyces cerevisiae* RAD18 gene encodes a protein that contains potential zinc finger domains for nucleic acid binding and a putative nucleotide binding sequence. Nucleic Acids Res 16:7119–7131

Kalathenos SP, Baranyi J, Sutherland JP, Roberts TA (1995) A response surface study on the role of some environmental factors affecting the growth of *Saccharomyces cerevisiae*. Int J Food Microbiol 25:63–74

Kara BV, Simpson WJ, Hammond JRM (1988) Prediction of the fermentation performance of brewing yeast with the acidification power test. J Inst Brew 94:153–158

Kuwayama H, Obara S, Morio T, Katoh M, Urushihara H, Tanaka Y (2002) PCR-mediated generation of a gene disruption construct without the use of DNA ligase and plasmid vectors. Nucleic Acids Res 30(2):e2

Leisegang R, Stahl U (2005) Degradation of a foam-promoting barley protein by a proteinase from brewing yeast. J Inst Brew 111:112–117

Lillie SH, Pringle JR (1980) Reserve carbohydrate metabolism. In: *Saccharomyces cerevisiae*: responses to nutrient limitation. J Bacteriol 143:1384–1394

Maddox IS, Hough JS (1970) Effect of zinc and cobalt on yeast growth and fermentation. J Inst Brew 76:262–264

Maiorella B, Blanch HW, Wilke CR (1983) By-product inhibition effects on ethanolic fermentation by *Saccharomyces cerevisiae*. Biotechnol Bioeng 25:103–121

Mansure JJC, Panek AD, Crowe LM, Crowe JH (1994) Trehalose inhibits ethanol effects on intact yeast cells and liposomes. Biochim Biophys Acta 1191:309–316

Marchal R, Bouquelet S, Maujean A (1996) Purification and partial bio-chemical characterization of glycoproteins in champenois Chardonnay wine. J Agric Food Chem 44:1716–1722

Martani F, Fossati T, Posteri R, Signori L, Porro D, Branduardi P (2013) Different response to acetic acid stress in *Saccharomyces cerevisiae* wild-type and l-ascorbic acid-producing strains. Yeast 30:365–378

Marza E, Camougrand N, Manon S (2002) Bax expression protects yeast plasma membrane against ethanol induced permeabilization. FEBS Lett 521:47–52

Meaden PG, Arneborg N, Guldfeldt LU, Siegumfeldt H, Jakobsen M (1999) Endocytosis and vacuolar morphology in *Saccharomyces cerevisiae* are altered in response to ethanol stress or heat shock. Yeast 15:1211–1222

Mitchell A, Romano GH, Groisman B, Yona A, Dekel E, Kupiec M, Dahan O, Yitzhak P (2009) Adaptive prediction of environmental changes by microorganisms. Nature 460:220–224

Moran MA, Buchan A, González JM, Heidelberg JF, Whitman WB, Kiene RP, Henriksen JR, King GM, Belas R, Fuqua C, Brinkac L, Lewis M, Johri S, Weaver B, Pai G, Eisen JA, Rahe E, Sheldon WM, Ye W, Miller TR, Carlton J, Rasko DA, Paulsen IT, Ren Q, Daugherty SC, Deboy RT, Dodson RJ, Durkin AS, Madupu R, Nelson WC, Sullivan SA, Rosovitz MJ, Haft DH, Selengut J, Ward N (2004) Genome sequence of *Silicibacter pomeroyi* reveals adaptations to the marine environment. Nature 432:910–913

Morano KA, Grant CM, Moye-Rowley WS (2012) The response to heat shock and oxidative stress in *Saccharomyces cerevisiae*. Genetics 190:1157–1195

Muldbjerg M, Meldal M, Breddam K, Sigsgaard P (1993) Protease activity in beer and correlation of foam. Proc Congr Eur Brew Conv 25:357–364
Murray CR, Stewart GG (1991) Experiments with high gravity brewing. Birra et Malto 44:52–64
Nguyen HV, Pulvirenti A, Gaillardin C (2000) Rapid differentiation of the closely related *Kluyveromyces lactis* var. *lactis* and *K. marxianus* strains isolated from dairy products using selective media and PCR/RFLP of the rDNA non transcribed spacer 2. Can J Microbiol 46:1115–1122
Novo MT, Beltran TMJ, Poblet M, Rozès N, Guillamón JM, Mas J (2003) Changes in wine yeast storage carbohydrate levels during preadaptation, rehydration and low temperature fermentations. Int J Food Microbiol 86:153–161
Núñez YP, Carrascosa AV, González R, Polo MC, Martínez-Rodríguez AJ (2005) Effect of accelerated autolysis of yeast on the composition and foaming properties of sparkling wines elaborated by champenoise method. J Agric Food Chem 53:7232–7237
Núñez YP, Carrascosa AV, González R, Polo MC, Martínez-Rodríguez AJ (2006) Isolation and characterization of a thermally extracted yeast cell wall fraction potentially useful for improving the foaming properties of sparkling wines. J Agric Food Chem 54:7898–7903
Obe G, Ristow H, Herha J (1977) Chromosomal damage by alcohol *in vitro* and *in vivo*. Adv Exp Med Biol 85A:47–70
Odumeru JA, D'Amore T, Russell I, Stewart GG (1992a) Changes in protein composition of *Saccharomyces* brewing strains in response to heat shock and ethanol stress. J Ind Microbiol 9:229–234
Odumeru JA, D'Amore T, Russell I, Stewart GG (1992b) Effects of heat shock and ethanol stress on the viability of a *Saccharomyces uvarum (carlsbergensis)* brewing yeast strain during fermentation of high gravity wort. J Ind Microbiol 10:111–116
Odumeru JA, D'Amore T, Russell I, Stewart GG (1993) Alterations in fatty acid composition and trehalose concentration of *Saccharomyces* brewing strains in response to heat and ethanol shock. J Ind Microbiol Biotechnol 11:113–119
Ogden K (1987) Cleansing contaminated pitching yeast with nisin. J Inst Brew 93:302–307
Owades JL (1981) The role of osmotic pressure in high and low gravity fermentations. MBAA Tech Quart 18:163–165
Palmqvist E, Hahn-Hägerdal B (2000) Fermentation of lignocellulosic hydrolysates. I: Inhibition and detoxification. Bioresour Technol 74:17–24
Panchal CJ, Stewart GG (1980) The effect of osmotic pressure on the production and excretion of ethanol and glycerol by a brewing yeast strain. J Inst Brew 86:207–210
Pascual C, Alonso A, Garcia I, Romay C, Kotyk A (1988) Effect of ethanol on glucose transport, key glycolytic enzymes, and proton extrusion in *Saccharomyces cerevisiae*. Biotechnol Bioeng 32:374–378
Pasteur L (1876) Studies on fermentation. Macmillan, London
Pérez-Torrado R, Carrasco P, Gimeno-Alcañiz AAJ, Perez-Ortin JE, Matallana E (2002) Study of the first hours of microvinification by the use of osmotic stress-response genes as probes. Syst Appl Microbiol 25:153–161
Pfisterer E, Stewart GG (1976) High gravity brewing. Brew Dig 51:34–42
Piddocke M, Kreisz S, Heldt-Hansen HP, Nielsen KF, Olsson L (2009) Physiological characterization of brewer's yeast in high gravity beer fermentations with glucose or maltose syrups as adjuncts. Appl Microbiol Biotechnol 84:453–464
Piper PW (1995) The heat shock and ethanol stress responses of yeast exhibit extensive similarity and functional overlap. FEMS Microbiol Lett 134:121–127
Piper P, Calderon CO, Hatzixanthis K, Mollapour M (2001) Weak acid adaptation: the stress response that confers yeasts with resistance to organic acid food preservatives. Microbiology 147:2635–2642
Powell CD, Diacetis AN (2007) Long term serial repitching and the genetic and phenotypic stability of brewer's yeast. J Inst Brew 113:67–74

Powell CD, Quain DE, Smart KA (2003) The impact of brewing yeast cell age on fermentation performance, attenuation and flocculation. FEMS Yeast Res 3:149–157

Pratt PL, Bryce JH, Stewart GG (1999) High gravity brewing – an inducer of yeast stress. Brewer's Guardian 131:28–31

Pratt PL, Bryce JH, Stewart GG (2003) The effects of osmotic pressure and ethanol on yeast viability and morphology. J Inst Brew 109:218–228

Pratt PL, Bryce JH, Stewart GG (2007) The yeast vacuole—its role during high gravity wort fermentations. J Inst Brew 113:55–60

Pratt-Marshall PL (2002) High gravity brewing—an inducer of yeast stress. Its effect on cellular morphology and physiology. Ph.D. thesis, Heriot-Watt University, Edinburgh, Scotland

Rautio J, Londesborough J (2003) Maltose transport by brewer's yeasts in brewer's wort. J Inst Brew 109:251–261

Rees EMR, Stewart GG (1998) Strain specific response of brewer's yeast strains to zinc concentrations in conventional and high gravity wort. J Inst Brew 104:221–228

Rego A, Costa M, Chaves SR, Matmati N, Pereira H, Sousa MJ, Moradas-Ferreira P, Hannun YA, Costa V, Côrte-Real M (2012) Modulation of mitochondrial outer membrane permeabilization and apoptosis by ceramide metabolism. PLoS One 7(11):e48571

Rosa MF, Sá-Correia I (1994) Limitations to the use of extracellular acidification for the assessment of plasma membrane H^+-ATPase activity and ethanol tolerance in yeasts. Enzyme Microb Technol 16:808–812

Ruis H, Schuller C (1995) Stress signaling in yeast. BioEssays 17:959–965

Russell I, Stewart GG (1995) Brewing. Biotechnology 9:419–462 (ed. by H-J Rehm and G. Reed, VCH Publishers, Weinheim, Germany)

Russell I, Stewart GG (eds) (2014) Whisky: technology, production and marketing, 2nd edn. Academic Press, Oxford

Saerens SM, Verbelen PJ, Vanbeneden N, Thevelein JM, Delvaux FR (2008) Monitoring the influence of high-gravity brewing and fermentation temperature on flavour formation by analysis of gene expression levels in brewing yeast. Appl Microbiol Biotechnol 80:1039–1051

Sano F, Asakawa N, Inoue Y, Sakurai M (1999) A dual role for intracellular trehalose in the resistance of yeast cells to water stress. Cryobiology 39:80–87

Semchyshyn HM, Abrat OB, Miedzobrodzki J, Inoue Y, Lushchak VI (2011a) Acetate but not propionate induces oxidative stress in bakers' yeast *Saccharomyces cerevisiae*. Redox Rep 16:15–23

Semchyshyn HM, Lozinska LM, Miedzobrodzki J, Lushchak VI (2011b) Fructose and glucose differentially affect aging and carbonyl/oxidative stress parameters in *Saccharomyces cerevisiae* cells. Carbohydr Res 346:933–938

Serrano R (1983) *In vivo* activation of yeast plasma membrane ATPase. FEBS Lett 156:11–14

Sia RAL, Urbonas BL, Sia EA (2003) Effects of ploidy, growth conditions and the mitochondrial nucleoid-associated protein Ilv5p on the rate of mutation of mitochondrial DNA in *Saccharomyces cerevisiae*. Curr Genet 44:26–37

Siddique R, Smart KA (2000) An improved acidification power test. In: Smart KA (ed) Brewing yeast fermentation performance. Blackwell Science, Oxford

Simpson WJ (1987a) Synergism between hop resins and phosphoric acid and its relevance to the acid washing of yeast. J Inst Brew 93:405–406

Simpson WJ (1987b) Kinetic studies of the decontamination of yeast slurries with phosphoric acid and added ammonium persulphate and a method for the detection of surviving bacteria involving solid medium repair in the presence of catalase. J Inst Brew 93:313–318

Simpson WJ, Hammond JRM (1989) The response of brewing yeasts to acid washing. J Inst Brew 95:347–354

Singer MA, Lindquist S (1998) Thermotolerance in *Saccharomyces cerevisiae:* the Yin and Yang of trehalose. Trends Biotechnol 16:460–468

Smits GJ, Brul S (2005) Stress tolerance in fungi—to kill a spoilage yeast. Curr Opin Biotechnol 16:225–230

Soanes C, Hawker S (2000) Compact Oxford English dictionary of current English. Oxford University Press, Oxford

Sorensen SB, Bech LM, Muldbjerg M, Beenfeldt T, Breddam K (1993) Barley lipid transfer protein 1 is involved in beer foam formation. MBAA Tech Quart 30:136–145

Spencer JFT, Spencer DM, Miller R (1983) Inability of petite mutants of industrial yeasts to utilize various sugars and a comparison with the ability of the parent strains to ferment the same sugars microaerophically. Z Naturforsch C 38:405–407

Stanley D, Bandara A, Fraser S, Chambers PJ, Stanley GA (2010a) The ethanol stress response and ethanol tolerance of *Saccharomyces cerevisiae*. J Appl Microbiol 109:13–24

Stanley D, Fraser S, Chambers PJ, Rogers P, Stanley GA (2010b) Generation and characterization of stable ethanol-tolerant mutants of *Saccharomyces cerevisiae*. J Ind Microbiol Biotechnol 37:139–149

Stewart GG (1975) Yeast flocculation – practical implications and experimental findings. Brew Dig 50:42–62

Stewart GG (2004) The chemistry of beer instability. J Chem Edu 81:963–968

Stewart GG (2005a) Does the use of high gravity brewing inflict unusual and cruel punishment on yeast? Brauwelt Int 23:422–428

Stewart GG (2005b) High gravity brewing as stress for the yeast. Brauwelt 145:920–921

Stewart GG (2006) Studies on the uptake and metabolism of wort sugars during brewing fermentations. MBAA Tech Quart 43:265–269

Stewart GG (2008) Esters – the most important group of flavour-active compounds in alcoholic beverages. In: Bryce JH, Piggott JR, Stewart GG (eds) Distilled spirits. Production, technology and innovation. Nottingham University Press, Nottingham, pp 243–250

Stewart GG (2009) The IBD Horace Brown Medal Lecture – forty years of brewing research. J Inst Brew 115:3–29

Stewart GG (2010a) The ASBC Award of Distinction Lecture – high gravity brewing and distilling – past experiences and future prospects. J Am Soc Brew Chem 68:1–9

Stewart GG (2010b) Glucose, maltose and maltotriose. Do brewer's yeast strains care which one? In: Proceedings of 31st convention of the Institute of Brewing and Distilling, Asia Pacific Section, Paper 03

Stewart GG (2012a) Biochemistry of brewing. In: NAM E, Shahidi N (eds) Biochemistry of foods. Elsevier, New York, pp 291–318

Stewart GG (2012b) Brewing intensification – successes and failures. World Brewing Congress, Portland, Ore. Paper 39

Stewart GG (2012c) Fermentation – The black box of the brewing process. In: 32rd Convention of the IBD Asia Pacific Convention, Melbourne, Australia, Paper No 14

Stewart GG (2013) Yeast management – culture handling between fermentations. In: Proceedings of 14th convention of the Institute of Brewing and Distilling, Africa Section, Paper No 13–22

Stewart GG (2014a) Brewing intensification. American Society for Brewing Chemists, St. Paul, MN

Stewart GG (2014b) Yeast mitochondria – their influence on brewer's yeast fermentation and medical research. MBAA Tech Quart 51:3–11

Stewart GG, Russell I (2009) An introduction to brewing science and technology. Series lll, Brewer's yeast, 2nd ed. The Institute of Brewing and Distilling, London

Stewart GG, D'Amore T, Panchal CJ, Russell I (1988a) Factors that influence the ethanol tolerance of brewer's yeast strains during high gravity wort fermentations. MBAA Tech Quart 25:47–53

Stewart GG, D'Amore T, Panchal CJ, Russell I (1988b) Regulation of sugar uptake in yeasts. In: Proceedings of 20th convention of the Institute of Brewing (Australia and New Zealand Section), Brisbane, pp 169–180

Stewart GG, Leiper KA, Miedl M (2008) Bioethanol – the current situation. In: Proceedings of 30th convention of the Institute of Brewing and Distilling, Asia Pacific Section, Paper 10

Thomas KC, Ingledew WM (1992) Production of 21% (v/v) ethanol by fermentation of very high gravity (VHG) wheat mashes. J Ind Microbiol 10:61–68

Thomas KC, Hynes SH, Jones AM, Ingledew WM (1993) Production of fuel alcohol from wheat by VHG technology. Appl Biochem Biotechnol 43:211–226

Thurston PA, Quain DE, Tubb RS (1981) The structural and storage carbohydrates of *Saccharomyces cerevisiae*; changes during the fermentation of wort and a role for glycogen catabolism in lipid biosynthesis. J Inst Brew 87:108–111

Thurston PA, Quain DE, Tubb RS (1982) Lipid metabolism and the regulation of volatile ester synthesis in *Saccharomyces cerevisiae*. J Inst Brew 88:90–94

Treger JM, Schmitt AP, Simon JR, McEntee K (1998) Transcriptional factor mutations reveal regulatory complexities of heat shock and newly identified stress genes in *Saccharomyces cerevisiae*. J Biol Chem 273:26875–26879

Ullah A, Orij R, Brul S, Smits GJ (2012) Quantitative analysis of the modes of growth inhibition by weak organic acids in *Saccharomyces cerevisiae*. Appl Environ Microbiol 78:8377–8387

Valli M, Sauer M, Branduardi P, Borth N, Porro D, Mattanovich D (2005) Intracellular pH distribution in *Saccharomyces cerevisiae* cell populations, analyzed by flow cytometry. Appl Environ Microbiol 71:1515–1521

van der Rest ME, Kamminga AH, Nakano A, Anraku Y, Poolman B, Konings WM (1995) The plasma membrane of *Saccharomyces cerevisiae:* structure, function, and biogenesis. Microbiol Rev 59:304–322

Van Dijck P, Colavizza D, Smet P, Thevelein M (1995) Differential importance of trehalose in stress resistance in fermenting and nonfermenting *Saccharomyces cerevisiae* cells. Appl Environ Microbiol 61:109–115

Wang ZY, He XP, Zhang BR (2007) Over-expression of GSH1 gene and disruption of PEP4 gene in self-cloning industrial brewer's yeast. Int J Food Microbiol 119:192–199

Weber FJ, de Bont JA (1996) Adaptation mechanisms of microorganisms to the toxic effects of organic solvents on membranes. Biochim Biophys Acta 1286:225–245

Wickner W (2002) Yeast vacuoles and membrane fusion pathways. EMBO J 21:1241–1247

You KM, Rosenfield C-L, Knipple DC (2003) Ethanol tolerance in the yeast *Saccharomyces cerevisiae* is dependent on cellular oleic acid content. Appl Environ Microbiol 69:1499–1503

Younis OS, Stewart GG (1998) Sugar uptake and subsequent ester and alcohol production in *Saccharomyces cerevisiae*. J Inst Brew 104:255–264

Younis OS, Stewart GG (1999) Effect of malt wort, very-high-gravity malt wort and very-high-gravity adjunct wort on volatile production in *Saccharomyces cerevisiae*. J Am Soc Brew Chem 52:38–45

Zakrzewska A, van Eikenhorst G, Burggraaff JE, Vis DJ, Hoefsloot H, Delneri D, Oliver SG, Brul S, Smits GJ (2011) Genome-wide analysis of yeast stress survival and tolerance acquisition to analyze the central trade-off between growth rate and cellular robustness. Mol Biol Cell 22:4435–4446

Zalewski K, Buchholz R (1996) Morphological analysis of yeast cells using an automated image processing system. J Biotechnol 48:243–249

Zhang CY, Liu Y-L, Qi Y-N, Zhang J-W, Dai L-H, Lin X, Xiao D-G (2013) Increased esters and decreased alcohols production by brewer's yeast strains. Eur Food Res Technol 236:1009–1014

Chapter 12
Yeast Management

12.1 Introduction

One of the major differences (not the only one) between brewer's and distiller's yeast is that brewer's yeast cultures are usually recycled from one fermentation to another, whereas distiller's yeast is rarely recycled. Following fermentation, the yeast goes directly into the still (batch, pot or continuous). In brewing, procedures between wort fermentations are collectively described as yeast management (Stewart 2015). This process includes strain storage [in a culture collection (Chap. 3)], cell propagation (biomass formation), cell cropping (Chap. 13), culture storage and acid washing (if required) followed by wort fermentation itself (Chaps. 6 and 7). The use of dried yeast is becoming popular in a limited number of brewing operations and will be discussed later in this chapter (Gosselin and Fels 1998; Finn and Stewart 2002; Jenkins et al. 2010; Jenkins 2011).

It is normal procedure in most (not all) brewing companies to propagate fresh yeast themselves (particularly lager yeast) every 8–10 generations (fermentation cycles) or less. Prolonged yeast cycles can result in sluggish fermentations (Hill and Stewart 2009), usually due to lower rates of wort maltose and maltotriose uptake (Zheng et al. 1994), reduced extent of wort amino acid metabolism (Miller et al. 2013), higher levels of sulphur dioxide and hydrogen sulphide (Samp and Sedlin 2017), prolonged diacetyl reduction times (Krogerus and Gibson 2013) and increased flocculation and sedimentation rates (Stewart 2009).

The long-term preservation of a brewing (and distilling) yeast culture requires that not only is optimal survival important but it is imperative that no changes in the characteristics of the yeast strain occur. Many (not all) strains are difficult to maintain in a stable state, and the long-term presentation by lyophilization (freeze drying), which has proved useful for mycelial fungi and bacteria, has been found to give poor results with most brewing and distilling yeast strains (Wellman and Stewart 1973). However, some pathogenic yeasts, such as *Candida albicans*, have been reported to be successfully stored freeze-dried (Douglas 2003). Storage

G.G. Stewart, *Brewing and Distilling Yeasts*, The Yeast Handbook,
https://doi.org/10.1007/978-3-319-69126-8_12

studies have been conducted with a number of ale and lager brewing strains together with two distilling strains (*S. cerevisiae*) (Russell and Stewart 1981).

It is important to emphasize that although considerable information is available about brewer's yeast fermentation per se (e.g., Boulton and Quain 2001; Sofie et al. 2010; Stewart et al. 2013), by comparison, basic detailed information on yeast management processes between wort fermentations has been lacking. Indeed, although the overall fermentation procedures and control have become very sophisticated, yeast management was, until recently, the "poor relation" segment of the brewing fermentation process!

Brewing yeast management can be divided into a number of overlapping procedures:

- Prior to propagation (the production of yeast biomass) but after fermentation and yeast cropping, most (but not all) yeast strains are stored under standard conditions in brewing R&D/QC laboratories or in an accredited culture collection—sometimes local hospital laboratories have also been employed in a few circumstances, for security.
- Yeast propagation (biomass formation) in wort under semi-aerobic conditions (Stewart 2017).
- Following propagation, the yeast is pitched into wort. This is the first generation (cycle) of a multi-generation sequence. The yeast is usually underpitched at this point due to a shortage of biomass.
- At the end of fermentation (attenuation) (Chaps. 6 and 7), yeast cropping occurs, followed by storage before repitching. Cropping occurs using the flocculation properties of the strain or with a centrifuge (Chap. 13).
- To eliminate contaminating bacteria, the yeast slurry can be acid washed (Chap. 11). Also, sometimes (but less frequently these days), the yeast slurry is sieved to remove trub (i.e. coagulated wort protein-phenol solid material) (Boulton and Quain 2001; Kunze 2014).

12.2 Storage of Yeast Stock Cultures Between Propagations

The advent of pure yeast strain fermentation has been discussed elsewhere in this volume (Chap. 4), and it dates from the virtuoso studies of Emil Christian Hansen (Fig. 3.1), working in the Carlsberg Laboratory in Copenhagen at the end of the nineteenth century (Holter and Moller 1976). Hansen, together with a coppersmith W.E. Jansen, developed a specific apparatus for large-scale yeast propagation (Fig. 2.2). The practice of using a pure yeast strain for lager production was soon adopted by breweries worldwide, particularly in the United States and Canada. Ale-producing brewers initially met this innovation with opposition and scepticism. It was only regarded as a means of reducing wild yeast and bacterial infection! It was not until the middle of the last century that pure ale strain methods were adopted. Even today, a number of ale breweries still confidently use a mixed strain culture (Stewart 2009, 2017).

12.3 Preservation of Brewer's Yeast Strains

The long-term preservation of a brewing yeast strain requires that not only is optimal survival important but it is imperative that no change in the fermentation characteristics of the yeast strain occurs. Hansen's studies resulted in storage of his strains in liquid nutrient media prior to propagation (Hansen 1883). This evolved where many breweries and independent culture collections maintained their yeast strains on nutrient media, solidified initially with gelatin and subsequently with agar. Some yeast strains are difficult to maintain in a stable state, and, as already discussed (Chap. 4), long-term preservation by lyophilization (freeze drying), which has proved useful for mycelial fungi and bacteria (Kirsop and Doyle 1990), has been found to produce poor survival results with brewing yeast strains. However, the use of dried yeast (not lyophilized) cultures for pitching into wort is increasing in popularity (Debourg and Van Nedervelde 1999; Jenkins et al. 2010; Jenkins 2011).

Storage studies have been conducted with a number of ale and lager brewing yeast strains (Russell and Stewart 1981). The following storage conditions were investigated:

- Low temperature as a result of storage in liquid nitrogen (−196 °C). With the advent of −70 °C refrigeration in the 1980s, liquid nitrogen has been largely replaced for this purpose with similar results.
- Lyophilization (freeze drying).
- Storage in distilled water.
- Storage under oil.
- Repeated direct transfer on solid culture media, subcultured once a week for 2 years.
- Long-term storage at 21 °C on solid nutrient medium, subcultured every 6 months for 2 years.
- Long-term storage at 4 °C on solid medium; subcultured every 6 months.

After a 2-year storage period, wort fermentation tests that included fermentation rate and wort sugar uptake efficiency, flocculation characteristics (details in Chap. 13), sporulation ability (details in Chap. 16), formation of respiratory-deficient mutants (details in Chap. 14) and the rate of cell survival (details in Chap. 8) were conducted. The results were compared to the characteristics of the stored control culture. Low temperature storage in liquid nitrogen appears to be the storage method of choice. However, there are capital and ongoing cost considerations associated with this method. Storage at 4 °C on nutrient agar slopes, subcultured every 6 months, was the next method of preference to low temperature storage as this method is simple to perform and relatively inexpensive. Lyophilisation and other storage methods revealed yeast instability which varied from strain to strain. Many breweries today store their strains (or contract store them) at −70 °C. The advent of −70 °C refrigeration replacing liquid nitrogen storage has reduced ongoing costs. Routine subculturing of yeast cultures on solid media every 6 months or so, albeit a less desirable method, is still an acceptable method. Freeze drying should be avoided as a storage method (Finn and Stewart 2002), but, as already discussed, the use of dried yeast (not freeze-dried) as a pitching culture is becoming increasingly popular (Fels et al. 1998).

12.4 Yeast Propagation

Yeast propagation is a traditional and well-established process in most large breweries (Nielsen 2010). Also, some multi-brewery operations propagate their yeast centrally and distribute the cultures to individual breweries. Nevertheless, development is constantly ongoing and a number of questions remain to be answered about this process (Stewart et al. 2013). The important requirement for a freshly propagated yeast culture is that it is not stressed unduly. The culture should be in a highly vital and viable condition and be free of contaminating organisms (Chan et al. 2016). The route to these objectives involves use of a carefully designed sanitary propagation plant with an aeration (oxygenation) system that is able to supply sufficient oxygen to all cells in the propagation, without causing mechanical (hydrodynamic) stress to the cells (Stoupis et al. 2003), which are grown in wort of the appropriate nutrient composition (details of oxygenation procedures are in Chap. 4) (Stewart 2017).

No matter how much the conditions are optimized, it is only possible to obtain relatively low cell numbers (approximately 100–200 million cells/mL, equivalent to 2.5 $\times$ 5.0 g dry matter per litre) because of the Crabtree effect (details of this effect are in Chap. 6) (Crabtree 1928). In order to avoid losing time during the wait for the yeast to consume all of the wort sugars, a complementary process is usually employed. This process has been adapted from the baker's yeast propagation procedure (Young and Cauvain 2007). This is conducted in a fed batch reactor, whereby the sugar concentration is maintained at a consistently low level, but not too low! This will avoid the yeast growing aerobically (the Crabtree effect) and thereby potentially losing some of its fermentation characteristics during propagation. Consequently, a hybrid process between traditional brewery propagation and the aerobic yeast propagation process employed for baker's yeast propagation is the preferred compromise (Boulton and Quain 1999). It should be noted that distiller's yeast cultures are rarely grown in a distillery—they are grown off-site, usually in a yeast factory that also produces baker's yeast (Young and Cauvain 2007). Also, instead of wort, the growth medium usually consists of molasses (sometimes hydrolysed whey) and ammonium ions. Also, a fed batch reactor with a continuous supply of dilute substrate and intense aeration (oxygenation) is used to produce both distiller's and baker's yeast. When propagation in a brewery is carried out in a batch reactor, the use of wort limits aerobic yeast growth in a concentrated sugar solution, making it very difficult to produce theoretical quantities of biomass (Nielsen 2005). However, the brewing industry has decided to tolerate this problem (unlike the distilling industry) because optimizing yeast growth in a molasses/nitrogen medium would jeopardize wort fermentation properties and lead to poorer beer quality and stability [particularly during the early yeast generations (cycles)].

Also, brewing must focus on strict sanitary conditions in order to avoid infection (the production of both distiller's and baker's yeast is not completely aseptic) and to minimize yeast stress during propagation to avoid negative effects on fermentation. It is worth reiterating that the propagation of brewer's yeast strains is based on aerobic conditions with widespread use of sterile air or oxygen throughout the process. This process differs extensively from brewing fermentations in which oxygen is only required at the beginning of the process in order for lag phase cells to begin synthesizing unsaturated fatty acids and sterols (Fig. 11.19) which are important membrane constituents. This synthesis occurs largely from glycogen as the carbon substrate (details of glycogen metabolism are in Chaps. 6 and 11) (Quain et al. 1981) (Fig. 6.19).

Parenthetically, it is worth noting that oxygen is only required at specific stages in the malting and brewing process:

- During barley germination as part of malting
- For biomass formation during yeast propagation
- At the beginning of fermentation when the yeast is pitched to wort

At any other point in the brewing process, oxygen can have a negative effect on beer quality, particularly when there is dissolved oxygen in the packaged product leading to the development of stale characteristics in the beer (Stewart 2004; Bamforth and Lentini 2009).

12.5 Yeast Collection

Yeast collection (also termed cropping) is discussed in detail in Chap. 13. Techniques vary depending on whether one is dealing with a traditional ale top cropping fermentation system, a traditional lager bottom fermentation system, a cylindroconical fermentation process (Stewart and Russell 2009) or a non-flocculent culture where the yeast is cropped with a centrifuge—details later. With traditional ale top fermentation procedures, although there are many variations to this process, for example, a simple, dual or multi-strain yeast system, can be employed (Anderson 2012). The timing of the skimming process can be critical in order to maintain the flocculation characteristics of the ale strains in subsequent fermentations. Traditionally, the first skim or "dirt skim", with the trub present, is discarded, as is the final skim, with the middle skim usually being maintained for repitching. With the traditional lager bottom fermentation system, the yeast is deposited on the bottom of the vessel at the end of fermentation. This type of yeast collection is essentially non-selective, and the flocculated yeast will normally contain entrained trub. Cylindroconical fermenters (Fig. 13.2) have now been widely adopted for both ale and lager fermentation, and the cone angle at the bottom of the fermenter allows for effective removal of the yeast crop plus some trub (details are in Chap. 13).

The use of centrifuges for the removal of yeast and the collection of pitching yeast is now commonplace (O'Rourke et al. 1996). There are a number of advantages to centrifuge use, these include shorter process time, cost reduction (after significant initial capital costs), increased productivity and reduced wort shrinkage (Chlup and Stewart 2011). Care must be taken to ensure that elevated temperatures (above 20 °C) are not generated during centrifugation and that the design ensures low dissolved oxygen pickup and a high throughput (Chlup et al. 2007a). In addition, centrifugation can (under certain circumstances) cause physical damage to yeast cells and, consequently, this negatively affects beer physical stability (haze) (Fig. 12.1). This is dependent on centrifuge operating parameters (Fig. 12.2). Hydrodynamic forces and yeast cell interaction within the gap of the centrifuge's disc stack (Fig. 12.3) creates collisions amongst yeast cells producing kinetic energy which can cause cellular damage (Stoupis et al. 2002) (Fig. 12.4). Release of cell wall mannan during mechanical agitation of yeast slurries in conjunction with an increase in beer haze has been well documented (Harrison et al. 1997; Chlup et al. 2007b, 2008). The advantages and disadvantages of the use of centrifuges during the brewing process are summarized in Table 12.1.

12.6 Yeast Storage

At the end of fermentation, the yeast is cropped for further use by employing the flocculating characteristics of the yeast strain or with a centrifuge. However, in this discussion, yeast cropping out of the wort is considered to be part of fermentation, not yeast management between fermentations (details of flocculation are in Chap. 13). It has already been described that another method of yeast cropping, that is increasing in popularity, is the use of centrifuges, although their use has not been without its problems (Table 12.1) (Chlup and Stewart 2011).

If a cropped yeast culture after fermentation is not stored properly, cell consistency and quality will suffer, which will subsequently adversely affect wort fermentation and beer quality. Following cropping, the yeast is stored in a specialized room at 0–2 °C that is appropriately sanitized and contains a plentiful supply of sterile water and a separate filtered air supply with positive pressure to prevent the entry of contaminants. Alternatively, insulated tanks in a dehumidified room can be used. In addition, "off the shelf" yeast storage facilities with various working capacities are available from appropriate suppliers (Nielsen 2010).

Yeast is usually stored under 6 inches of beer (sterile water has been used in the past but is unpopular these days). When high-gravity brewing procedures are practised, it is important to ensure that the ethanol level of the storage beer is decreased to 4–6% (v/v) ethanol to maintain the viability and vitality of the stored yeast. As more sophisticated systems become available, storage tanks with external cooling (0–2 °C), equipped with low-shear stirring devices, have become popular.

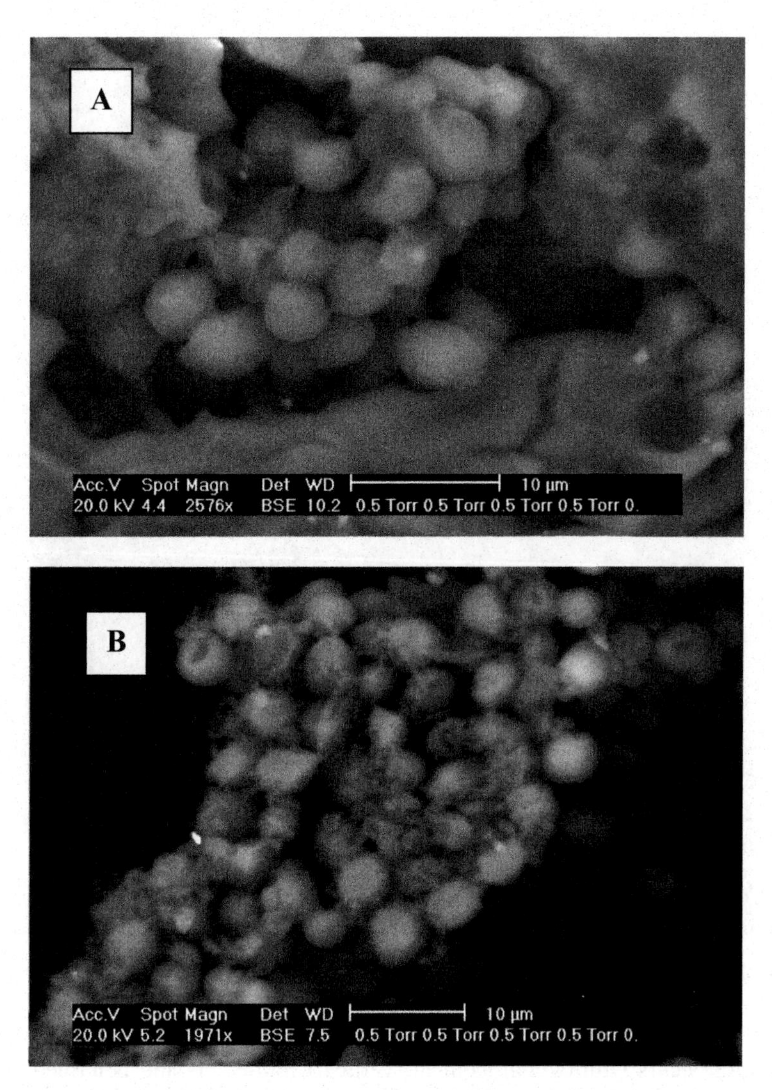

Fig. 12.1 Environment scanning electron microscope (ESEM) of ale yeast strain damage. (**a**) Cells before passage through a disc centrifuge and (**b**) cells following passage through a disc centrifuge operating at high G-force

The need for low-shear stirring systems has been shown to be important and has already been discussed in this chapter (Stoupis et al. 2003). With high-velocity agitation in a yeast storage tank, the yeast cell surface can become disrupted, and intracellular proteinases (particularly proteinase A [PrA]) are excreted—this can result in unfilterable beer mannan hazes (Stoupis et al. 2002) and poor head retention due to proteinase hydrolytic activity on foam stability-enhancing peptides—further details in Chap. 11 (Cooper et al. 2000). There are brewing procedures whereby the yeast is not stored between fermentations. In this case, the

Fig. 12.2 A disc stack centrifuge

yeast is pitched directly from one fermenter to another. This yeast handling procedure occurs with cylindroconical (vertical) fermenters and is termed "cone to cone pitching" (Shardlow and Thompson 1971; Shardlow 1972). This procedure was employed by some breweries in the 1970s, 1980s and 1990s, but currently it has limited application due to lack of flexibility and time to conduct quality and contamination studies on the pitching yeast culture between fermentations. However, with the advent of craft brewing, cone to cone repitching has been resurrected (Strevey 2014)!

One of the factors that will affect wort fermentation rate is the condition under which the yeast culture is stored between fermentations. Of particular importance in this regard is the influence of temperature during storage on the cell's intracellular glycogen level. It has already been discussed that glycogen is the major reserve carbohydrate stored within the yeast cell and is similar in structure to plant

Fig. 12.3 The disc stack

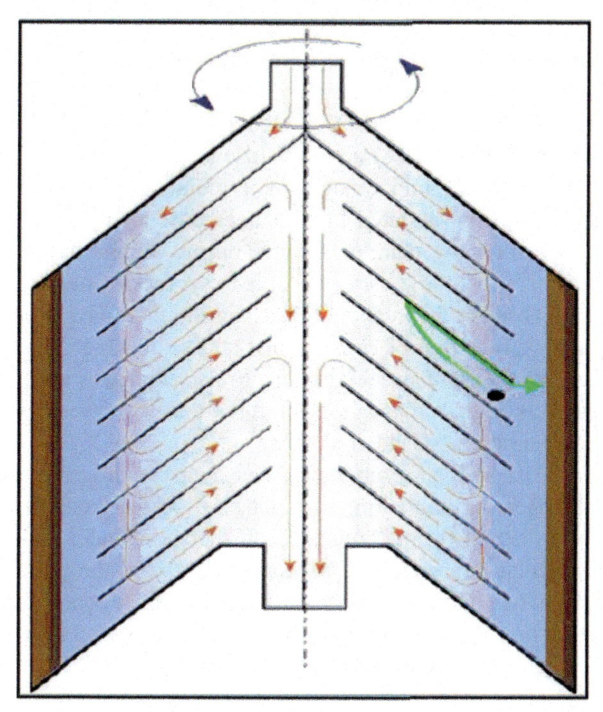

Fig. 12.4 Flow pattern in the stack centrifuge

Table 12.1 The use of centrifuges in breweries—advantages and disadvantages

Advantages
• Rapid and efficient clarification
• Most consistent clarity of beer
• Equipment can be sterilized
• Filter aids are not required or reduced
• Space requirements are small
• Most are self-cleaning
• Operate continuously
• Lower beer losses with minimal oxygen pick-up
Disadvantages
• High maintenance costs
• Beer temperature may increase
• Mechanical break-up of large particles—incurred in finer particles
• Increased mechanical stress or increased temperature may influence yeast quality—details later
• There may be oxygen pick-up and high noise levels
• Removal of cold break (with some yeast) at the end of maturation
• Removal of solid material from effluents with both brewing and packaging operations prior to treatment in house or by municipal treatment systems

amylopectin (Chap. 6) (Fig. 6.18). Glycogen serves as a store of biochemical energy (Quain and Tubb 1982) during the lag phase of fermentation when the energy demand is intense for the synthesis of compounds such as sterols and unsaturated fatty acids (UFA). Also, it is important that appropriate levels of glycogen (Fig. 11.18) and trehalose (Fig. 11.20) are maintained during storage so that during the initial stages of fermentation, the yeast cell is able to synthesize sterols and unsaturated fatty acids. Consequently, it is important that appropriate levels of glycogen are maintained during storage so that during the initial stages of fermentation, the yeast cell is able to synthesize sterols and unsaturated fatty acids. Trehalose is a nonreducing disaccharide that plays a protective role in osmoregulation, protection of cells against stresses during conditions of nutrient depletion and starvation and dehydration. It also improves cell resistance to other stresses such as high and low temperatures and osmotic pressure and elevated ethanol concentrations (Gadd et al. 1987).

Yeast storage temperature has a direct influence on the rate and extent of glycogen dissimulation, as might be expected, considering the effect that temperature has an overall influence on metabolic rate. Although strain dependent, of particular interest is the fact that within 48 h, the yeast stored semi-aerobically at 15 °C has only 15% of the original glycogen concentration remaining. Glycogen reduction to this extent will have a profound effect on the rate and extent of wort fermentation (Fig. 11.13). This is not always the case. In some instances, little change has been noted when a lager yeast culture was serially repitched 135 times and also stored at ambient temperatures between fermentations (Powell and Diacetis 2007; Speers and Stokes 2009).

The number of times that a yeast crop (generations or cycles) is used for wort fermentations is usually standard practice in a particular brewery or brewing operation. Typically, currently, a lager yeast culture (*Saccharomyces pastorianus*) is currently used 6–10 times prior to reverting to a fresh culture of the same strain obtained from a pure yeast culture plant. If a particular yeast culture is used beyond the agreed cropping specification (this is strain dependent), fermentation difficulties (fermentation rate and extent are typical examples) are sometimes (Stewart and Russell 2009) encountered. An example of fermentation problems is when a brewery increases its wort gravity as part of the adoption of high-gravity brewing procedures. A particular brewing operation in Canada, over a 15-year period, increased its wort gravity incrementally. In order to avoid fermentation difficulties, it reduced the number of yeast cycles from a single propagation as follows:

12°Plato wort > 20 yeast cycles
14°Plato wort – 16 yeast cycles
16°Plato wort – 12 yeast cycles
18°Plato wort – 8 yeast cycles

Some breweries have currently adopted a lager yeast reuse specification as few as four to six cycles!

The reasons why multiple yeast cycles can have a negative effect on a culture's fermentation performance is unclear. However, multiple generations will result in reduced levels of intracellular glycogen and an increase in trehalose, indicating reduced endogenous substrate and additional stress conditions during the culture's cycle progress (Table 12.2) (Boulton and Quain 2001).

Yeast storage conditions between wort fermentations can affect fermentation efficiency and beer quality. Good handling practices should encompass collection and storage procedures, avoid the inclusion of oxygen in the slurry, cool the slurry to 0–4 °C soon after collection and, perhaps most importantly, ensure that intracellular glycogen levels are maintained because of its critical importance at the start of a subsequent wort fermentation as the substrate for UFA and sterols (details in Chap. 11). Acid washing (see below and Chap. 11) can be part of the management protocol in order to eliminate bacterial contamination of a yeast slurry, but many brewing microbiologists and brewers frown upon this procedure (Simpson and Hammond 1989)!

Table 12.2 Concentration of trehalose and glycogen in lager yeast following one, four and eight cycles after fermentation in 15°Plato wort

	Generations (cycles)		
	One	Four	Eight
Trehalose[a]	8.8	9.2	11.6
Glycogen[b]	14.6	12.6	9.2

[a]μg/dry weight of yeast
[b]mg/dry weight of yeast

12.7 Yeast Washing

Acid-washing pitching yeast at pH 2.0–2.2 (with either phosphoric, tartaric, hydrochloric, sulphuric and nitric acid solutions), usually during the later stages of storage, just prior to being pitched into wort for fermentation, has been employed by many breweries (not all) for the past 100 years. This procedure has been discussed in Chap. 11, but because of its importance for brewing (and some distilling) yeast, it is also discussed briefly here. Pasteur (biographical details in Chap. 2) introduced this practice into the wine and brewing industries using tartaric acid solution (Pasteur 1876) as an effective method of eliminating contaminating bacteria (not wild yeasts) without adversely affecting the physiological quality of the yeast culture. The acid-washing regime differs between breweries with some routinely acid washing their yeast after each fermentation cycle, whereas other brewers only acid wash their pitching yeast when there is significant culture bacterial contamination and some completely refrain from acid washing. Also, some breweries use the acid (usually phosphoric acid) in combination with ammonium persulphate (Bah and McKee 1965). Brewer's yeast strains are normally resistant to acidic conditions when the washing is conducted properly. However, if other environmental and operating conditions are modified, then the acid resistance of the culture will vary. Simpson and Hammond (1989) demonstrated that if the temperature of acid washing was greater than 5 °C and/or the ethanol concentration greater than 8% (v/v) acid, washing had a detrimental effect on the yeast culture causing a decrease in viability and fermentation performance. The physiological condition of the yeast prior to acid washing is an important factor in acid tolerance with yeast in poor physiological condition, prior to washing, being more adversely affected by acid washing than a healthy yeast culture. It should be emphasized that acid washing eliminates most bacteria but not yeast cultures (brewing or wild).

Acid washing primarily affects the yeast cell envelope with the physiological systems associated with both the cell wall and the plasma membrane, subsequently decreasing yeast vitality as measured by the acidification power test (Kara et al. 1988). Studies in the Heriot Watt University brewing research laboratory (Cunningham and Stewart 2000) have reported that acid-washing pitching yeast from high-gravity (20°Plato) wort fermentations did not affect the fermentation performance of cropped yeast if it was maintained in good physiological condition. Oxygenation of the yeast at the start of fermentation stimulated yeast growth leading to a more efficient wort fermentation and equally important in the context of yeast management between fermentations produced yeast that was in good physiological condition permitting it to tolerate exposure to acid-washing conditions (phosphoric acid solution at pH 2.2). These data support the findings of Simpson and Hammond (1989) that are still appropriate. Acid washing of yeast can be summarized into the do's and do not's, and they are discussed in Chap. 11 (Table 11.10).

12.8 Yeast Stress

During wort fermentation, a yeast culture is exposed to a number of stress conditions, and the primary brewing stress factor (not the only one) is the use of high-gravity worts (Pratt-Marshall et al. 2002). In these circumstances, yeast cells are exposed to numerous stresses including high osmotic pressure at the beginning of fermentation due to the presence of high concentrations of wort sugars and ethanol stress at the end of fermentation (Stewart 1999). Other forms of yeast stress are: desiccation (details later), mechanical stress (details later), oxygen and thermal stress (hot and cold) (Fig. 11.6). The yeast is expected to maintain its metabolic activity during stressful conditions by not only surviving these stresses but by rapidly responding to ensure continued acceptable cell viability and vitality (Casey et al. 1985). The question of yeast stress has been discussed in detail in Chap. 11.

12.9 Dried Yeast

Dried yeast has been employed in the baking and distilling industries for over 50 years (Pyke 1958). By comparison its use in brewing is relatively recent (Fels et al. 1998). The reasons for this delay in use are the differences in drying characteristics. Ale yeast strains dry relatively well, whereas lager yeast cultures, following drying, exhibit much lower viabilities and vitalities (Finn and Stewart 2002). The reasons for the drying differences between these two brewing yeast species are still not fully understood, but levels of the storage carbohydrates glycogen and trehalose have been implicated (Gadd et al. 1987; Finn and Stewart 2002). Another reason for the delay in adopting dried yeast in brewing is that this yeast type is often contaminated with various bacteria and wild yeasts (Debourg and Van Nedervelde 1999). However, this problem is not as prevalent as it was 25 years ago!

The use of dried yeast in brewing has several advantages and similarities in comparison to the use of fresh yeast (Fels et al. 1998):

- A dried culture is easier to handle and convenient to store.
- It can replace yeast propagation in breweries.
- In some cases (not all) it can replace the need for wort aeration at pitching for the initial wort fermentation cycle. Recent studies have shown that dried yeast often has characteristics similar to those of its fresh counterpart regarding analytical and flavour profiles, fermentation rate and final attenuation all matching the characteristics of the fresh yeast culture (Debourg and Van Nedervelde 1999; Finn and Stewart 2002).

- The average viability (determined by methylene violet or methylene blue staining) of dried yeast (particularly with lager yeast cultures) is 20–30% lower than that of the same strain freshly propagated. This situation can be accommodated by overpitching according to viable cell number. However, the addition of too many dead yeast cells can influence beer flavour and its foam stability—details later.

During studies with dried yeast samples, differences in flocculation and haze formation characteristics when compared to fresh yeast samples were observed (Finn and Stewart 2002). The flocculation rate with fresh and dried ale cultures was rapid, with most of the yeast sedimenting out of suspension within the first minute of the Helm's Sedimentation Test (Helm et al. 1953), and 80% of the culture eventually flocculated out of suspension. The flocculating differences between fresh lager cultures and dried yeast samples were more pronounced with lager than with ale strains. Virtually no flocculation occurred within the 10 min test period with the dried yeast samples. This test indicated that the lager dried yeast samples (particularly the cell wall structure) were modified in some manner and, as a consequence, exhibited non-flocculent characteristics (details of yeast flocculation are in Chap. 13).

During studies on the flocculation of dried yeast cultures, it was observed that the dried yeast fermentation often left a distinct haze in suspension. Even with an ale yeast fermentation, haze remained in suspension. This may have been due, in part, to the number of dead cells pitched into wort. In addition, with fresh yeast samples (both ale and lager cultures), PrA and other proteinases were released by dried yeast into the wort in much greater quantities than fresh yeast under similar fermentation conditions (Finn and Stewart 2002). PrA release into the wort (already discussed) will have an impact on beer foam characteristics (Cooper et al. 2002; Osmond et al. 1991). The decreased foam stability is due to the hydrolysis of hydrophobic polypeptides by PrA. Hydrophobic polypeptides are known to be a major factor (amongst others) responsible for beer foam stability (Bamforth 2012). Leakage of intracellular proteinase from living brewer's yeast cultures has been demonstrated (Dreyer et al. 1983). This is the case, particularly when they are under stress (Stewart 1999). This phenomenon has been considered in detail in Chap. 11 during discussions of high-gravity brewing. Indeed, the addition of dead cells (as could be the case with a dried culture) would greatly increase the levels of active PrA.

12.10 Cultivation of Distiller's Yeast Cultures

Yeast can be purchased from manufacturers of baking and distilling yeasts, and currently there are a number of specialized strains available depending on the particular fermentation and organoleptic profile desired in the fermented wort (Russell and Stewart 2014). The stock culture is propagated by the yeast manufacturer through a succession of fermentation vessels, during gradual increases in the size of the vessels as a result of scale-up; the goal being rapid cell growth and a culture with good cell viability and vitality (Cheung et al. 2012; Nielsen 2010).

The culture medium usually employs molasses (which predominantly consists of sucrose, vitamins, protein minerals and traces of fat), but whey (containing lactose) is sometimes used and is supplemented with ammonium salts and various other nutrients. The yeast is grown with vigorous aeration (oxygenation) and careful temperature control in order to obtain a product that retains maximum cell viability and vitality. The sugar medium is fed into the fermentation at approximately 0.5% (w/v) in order to maintain the yeast in respiration mode (rather than letting it switch to fermentation—the Crabtree effect again!) (Crabtree 1928). The goal is to accumulate yeast biomass, not produce ethanol!

The yeast biomass can be harvested by a number of methods such as rotary vacuum filtration (compressed yeast), collected and sold as cream yeast (for convenient delivery by tanker trucks and for automated pitching), centrifuged, or dried under a partial vacuum. By using an inert gas, such as nitrogen, for packaging the yeast, the shelf life of the culture is increased and the intracellular glycogen level is maintained.

Manufacturers of yeast for bakeries and distilleries aim for minimal alcohol production and maximum biomass during the production process. They carefully supply the sugar substrate (either sucrose, glucose, fructose, lactose and/or galactose) and the high levels of oxygen required to produce a yeast culture that can survive well in whatever form it will be stored before it is used in a distillery fermentation. Table 12.3 illustrates the difference in the propagation yield between yeast scaled up in the respiration mode compared to propagation in wort with oxygen limitation (Russell and Stewart 2014).

Depending on the method of yeast storage, a liquid reactivation step is usually employed before the yeast is pitched into wort, especially if it has been dried for shipment. This method is carefully detailed by each manufacturer for the specific strain being employed to ensure that rapid fermentation will commence on pitching. This process is called bubbing or livening and is discussed in Chap. 11 (Fig. 11.19).

The ethanol tolerance of four distilling strains commonly used for Scotch whisky fermentations has been assessed under CO_2-induced anaerobic conditions (Cheung et al. 2012). Ethanol tolerance is strain dependent, and this confirms that very high-gravity wort distilling fermentations may require appropriate strain selection to be

Table 12.3 Comparison of oxygen consumption under two propagation modes[a]

Propagation mode	Yield factor	Oxygen consumption
Pure respiration propagation	0.54 g yeast dry solids per g carbohydrate	0.74 oxygen per g yeast dry solids
Propagation in 12°Plato wort	0.10 g yeast dry solids per g carbohydrate	0.12 g oxygen per g yeast dry solids

[a]Adapted from Nielsen (2005, 2010) and Russell and Stewart (2014)

fully successful. Ethanol tolerance is also dependent on the mode in which the strain is supplied. It is suggested (only suggested) that creamed yeast exhibits enhanced ethanol tolerance compared to dried yeast. Reasons for this tolerance are unknown, however, it is worthy of note that the physiological state of the fermentation inoculum (pitching yeast) may influence the capacity of a strain to tolerate fermentation stress and therefore its potential to perform. It is not suggested that creamed yeast offers distinct advantages over other forms of a yeast culture, but changes in yeast supply modes may lead to differences in stress tolerance and/or overall performance (Russell and Stewart 2014).

References

Anderson R (2012) One yeast or two? Pure yeast and top fermentation. Brewery Hist 149:30–38

Bah S, McKee RA (1965) Beer-spoilage bacteria and their control with a phosphoric acid – ammonium persulfate wash. Can J Microbiol 11:309–318

Bamforth CW (2012) Practical guides for beer quality: foam. ASBC Handbook Series. ASBC, St Paul, MN

Bamforth CW, Lentini A (2009) The flavour instability of beer. In: Bamforth CW (ed) Beer: a quality perspective. Academic Press, Burlington, MA, pp 85–109

Boulton CA, Quain DE (1999) A novel system for propagation of brewing yeast. In: Proceedings of 27th congress of the European Brewery Convention, Cannes, pp 647–654

Boulton CA, Quain DE (2001) Brewing yeast and fermentation. Blackwell Science, Oxford

Casey PH, Chen ECH, Ingledew WM (1985) High-gravity brewing: production of high levels of ethanol without excessive concentrations of esters and fusel alcohols. J Am Soc Brew Chem 43:178–182

Chan L-YL, Driscoll D, Kuksin D, Saldi S (2016) Measuring lager and ale yeast viability and vitality using fluorescence-based image cytometry. MBAA TQ 53:49–54

Cheung HC, San Lucas FA, Hicks S, Chang K, Bertuch AA, Ribes-Zamora A (2012) An S/T-Q cluster domain census unveils new putative targets under Tel1/Mec1 control. BMC Genomics 13:664

Chlup PA, Stewart GG (2011) Centrifuges in brewing. MBAA Tech Quart 48:46–50

Chlup PL, Bernard D, Stewart GG (2007a) The disc stack centrifuge and its impact on yeast and beer quality. J Am Soc Brew Chem 65:29–37

Chlup PL, Conery J, Stewart GG (2007b) Detection of mannan from *Saccharomyces cerevisiae* by flow cytometry. J Am Soc Brew Chem 65:151–155

Chlup PL, Bernard D, Stewart GG (2008) Disc stack centrifuge operating parameters and their impact on yeast physiology. J Inst Brew 114:45–61

Cooper DJ, Stewart GG, Bryce JH (2000) Yeast proteolytic activity during high and low gravity wort fermentations and its effect on lead retention. J Inst Brew 106:197–201

Cooper DJ, Husband FA, Mills ENC, Wilde PJ (2002) Role of beer lipid-binding proteins in preventing lipid destabilization of foam. J Agric Food Chem 50:7645–7650

Crabtree HG (1928) The carbohydrate metabolism of certain pathological overgrowths. Biochem J 22:1289–1298

Cunningham S, Stewart GG (2000) Acid washing and serial repitching a brewing ale strain of Saccharomyces cerevisiae in high gravity wort and the role of wort oxygenation conditions. J Inst Brew 106:389–402

Debourg A, Van Nedervelde L (1999) The use of dried yeast in the brewing industry. In: Proceedings of European Brewery Convention Congress, Cannes, pp 751–760

Douglas J (2003) Candida biofilms and their role in infection. Trends Microbiol 11:30–36. ISSN 0966-842X

Dreyer T, Biedermann K, Ottesen M (1983) Yeast proteinase in beer. Carlsberg Res Commun 48:249–253

Fels S, Reckelbus B, Gosselin Y (1998) Why use dried yeast for brewing your beers? Brew Distil Int 29:17–19

Finn DA, Stewart GG (2002) Fermentation characteristics of dried brewer's yeast: effect of drying on flocculation and fermentation. J Am Soc Brew Chem 60:135–139

Gadd GM, Chalmers K, Reed RH (1987) The role of trehalose in dehydration resistance of *Saccharomyces cerevisiae*. FEMS Microbiol Lett 48:249–254

Gosselin Y, Fels S (1998) Fermentation characteristics from dried ale and lager yeasts. Tech Q MBAA 35:129–132

Hansen EC (1883) Undersøgelser over alkoholgjaersvampenes fysiologi og morfologi. II Om askosposedann elsen hos slaegten Saccharomyces. Meddelelser fra Carlsberg Laboratoriet. 2:29–104

Harrison CJ, Hayer-Hartl M, Di Liberto M, Hartl F, Kuriyan J (1997) Crystal structure of the nucleotide exchange factor GrpE bound to the ATPase domain of the molecular chaperone DnaK. Science 276:431–435

Helm E, Nohr B, Thome RSW (1953) The measurement of yeast flocculence and its significance in brewing. Wallerstein Lab Commun 16:315–325

Hill AE, Stewart GG (2009) A brief overview of brewer's yeast. Brewer Distil Int 5:13–15

Holter H, Moller KM (eds) (1976) The Carlsberg laboratory, 1876–1976. International Service and Art, Copenhagen

Jenkins DM (2011) The impact of dehydration and rehydration on brewing yeast. PhD thesis, University of Nottingham

Jenkins D, Powell CD, Smart KA (2010) Dried yeast: impact of dehydration and rehydration on brewing yeast DNA integrity. J Am Soc Brew Chem 68:132–138

Kara BV, Simpson WJ, Hammond JRM (1988) Prediction of the fermentation performance of brewing yeast with the acidification power test. J Inst Brew 94:153–158

Kirsop B, Doyle A (1990) Maintenance of microorganisms and cultured cells. A manual of food practice, 2nd edn. Academic Press, London

Krogerus K, Gibson BR (2013) Influence of valine and other amino acids on total diacetyl and 2,3-pentanedione levels during fermentation of brewer's wort. Appl Microbiol Biotechnol 97:6919–6930

Kunze W (2014) Technology brewing and malting, 5th edn. VLB, Berlin

Miller KJ, Box WG, Jenkins DM, Boulton CA, Linforth R, Smart KA (2013) Does generation number matter? The impact of repitching on wort utilization. J Am Soc Brew Chem 71:233–241

Nielsen O (2005) Control of the yeast propagation process – How to optimize oxygen supply and minimize stress. MBAA Tech Quart 42:128–132

Nielsen O (2010) Status of the yeast propagation process and some aspects of propagation for re-fermentation. Cerevisia 35:71–74

O'Rourke T, Godfrey T, West S (1996) Industrial enzymology, 2nd edn. MacMillan, London

Osmond IHL, Lebor EF, Sharpe FR (1991) Yeast proteolytic enzyme activity during fermentation. In: Proceedings of European Brewery convention congress, Copenhagen, pp 457–464

Pasteur L (1876) Études Sur La Bière. Gauthier-Villars, Paris

Powell CD, Diacetis AN (2007) Long term serial repitching and the genetic stability of brewer's yeast. J Am Soc Brew Chem 113:67–74

Pratt-Marshall PL, Brey SE, de Costa SD, Bryce JH, Stewart GG (2002) High gravity brewing – an inducer of yeast stress. Brew Guard 131:22–26

Pyke M (1958) The technology of yeast. In: Cook AH (ed) The chemistry and biology of yeasts. Academic Press, New York, pp 535–586

Quain DE, Tubb RS (1982) The importance of glycogen in brewing yeasts. MBAA Tech Quart 19:29–33
Quain DE, Thurston PA, Tubb RS (1981) The structural and storage carbohydrates of *Saccharomyces cerevisiae*: changes during fermentation of wort and a role for glycogen catabolism in lipid biosynthesis. J Inst Brew 87:108–111
Russell I, Stewart GG (1981) Liquid nitrogen storage of yeast cultures compared to more traditional storage methods. J Am Soc Brew Chem 39:0019
Russell I, Stewart GG (eds) (2014) Whisky: technology, production and marketing, 2nd edn. Academic Press (Elsevier), Boston
Samp EJ, Sedlin D (2017) Important aspects of controlling sulphur dioxide in brewing. MBAA Tech Quart 54:60–71
Shardlow PJ (1972) The choice and use of cylindrico-conical fermentation vessels. MBAA Tech Quart 9:1–5
Shardlow PJ, Thompson CC (1971) The Nathan system. Conical fermenters. Brew Digest 46 (August):76–80
Simpson WJ, Hammond JRM (1989) The response of brewing yeasts to acid washing. J Inst Brew 95:347–354
Sofie MG, Saerens C, Duong CT, Nevoigt E (2010) Genetic improvement of brewer's yeast: current state, perspectives and limits. Appl Microbiol Biotechnol 86:1195–1212
Speers RA, Stokes S (2009) Effects of vessel geometry, fermenting volume and yeast repitching on fermenting beer. J Inst Brew 115:148–150
Stewart GG (1999) High gravity brewing. Brew Guard 128:31–37
Stewart GG (2004) Chemistry of beer instability. J Chem Educ 81:963–968
Stewart GG (2009) Forty years of brewing research. J Inst Brew 115:3–29
Stewart GG (2015) Chap 2: Yeast quality assessment, management and culture maintenance. In: Hill AE (ed) Brewing microbiology: managing microbes, ensuring quality and valorising waste. Elsevier Woodhead, Oxford, pp 11–29
Stewart GG (2017) Brewer's yeast propagation – the basic principles. MBAA Tech Quart 54:125–131
Stewart GG, Russell I (2009) An introduction to brewing science and technology, series III: Brewer's yeast, 2nd edn. The Institute of Brewing and Distilling, London
Stewart GG, Hall AE, Russell I (2013) 125th Anniversary review: developments in brewing and distilling strains. J Inst Brew 119:202–220
Stoupis T, Stewart G, Stafford RA (2002) Mechanical agitation and rheological considerations of ale yeast slurry. J Am Soc Brew Chem 60:58–62
Stoupis T, Stewart GG, Stafford RA (2003) Hydrodynamic shear damage of brewer's yeast. J Am Soc Brew Chem 61:219–225
Strevey D (2014) Taking control of cone to cone pitching in the Craft Brewery. Dan Strevey, Cellar Manager, Avery Brewing Company, Avery Brewing, 4910 Nautilus Ct, Boulder, CO 80301
Wellman AM, Stewart GG (1973) Storage of brewing yeasts by liquid nitrogen refrigeration. J Appl Microbiol 26:577–583
Young L, Cauvain SP (2007) Technology of breadmaking. Springer, Berlin, p 79
Zheng X, D'Amore T, Russell I, Stewart GG (1994) Factors influencing maltotriose utilization during brewery wort fermentations. J Am Soc Brew Chem 52:41–47

Chapter 13
Harvesting and Cropping Yeast: Flocculation and Centrifugation

13.1 Introduction

The differences and similarities between the brewing and distilling (particularly whisk(e)y) production processes are tabulated in this chapter (Table 13.1). One (not the only) of the principal process differences is that the culture yeast is recycled from one fermentation to another a number of times during the brewing of beer (Stewart et al. 2013). In the distilling of both potable and industrial (fuel) alcohol (details in Chaps. 9 and 11), a culture yeast is invariably (not always) used once and is not recycled (Russell and Stewart 2014). This yeast management difference has implications for both processes and the characteristics of the resulting beer and the distilled spirit (Stewart 2014d). In the later stages of brewing primary fermentation (and also sometimes during maturation), the yeast is harvested from the fermented wort, stored and is usually used in subsequent wort fermentations (Stewart 2015a, b). This harvesting can be accomplished by employing the flocculation characteristics of the yeast culture (Fig. 13.1), alternatively, with a filter or centrifuge using a particular strain's non-flocculation characteristics—details later (Sect. 13.13) (Stewart et al. 2013).

Brewers employ a number of methods to crop their yeast which varies depending on whether one is dealing with traditional ale top-cropping, traditional lager bottom-cropping or cylindroconical fermentation systems [also sometimes called a Nathan fermenter after a Swiss engineer (Leopold Nathan) who designed it] (Fig. 13.2) (Nathan 1930). Here the yeast culture (ale or lager) is recovered from the cone (sometimes repitched cone to cone), a non-flocculent culture where the yeast, still in suspension, is cropped with a centrifuge or a portion of the suspended yeast is blended into the fresh wort of a subsequent brew. With traditional ale top-cropping fermentation systems, although there are many variations to this process, a single, dual or multi-strain yeast strain culture can be employed (Hough 1959), and the timing of the top-cropping process (skimming) can be critical in order to maintain the flocculation characteristics of the cropped culture.

G.G. Stewart, *Brewing and Distilling Yeasts*, The Yeast Handbook,
https://doi.org/10.1007/978-3-319-69126-8_13

Table 13.1 Differences between the production of beer and Scotch Whisky

	Brewing	Scotch Whisky
Barley malt	Yes	Yes
Non-malted carbs	Corn (maize) wheat, rice, sorghum, etc.	Corn and wheat
Hops	Yes	No
Yeast	*S. cerevisiae* *S. pastorianus*	*S. cerevisiae*
Yeast	Recycled	One cycle (generation)
Non-malt enzymes	Occasionally	Never
Fermentation temperature	12–25 °C	28–32 °C
Fermentation time	7–10 days	5–7 days
Maturation time	1 week–6 weeks	Min. 3 years
High-gravity	Extensive	Increasing
Process aids and conditions	Limited	Never

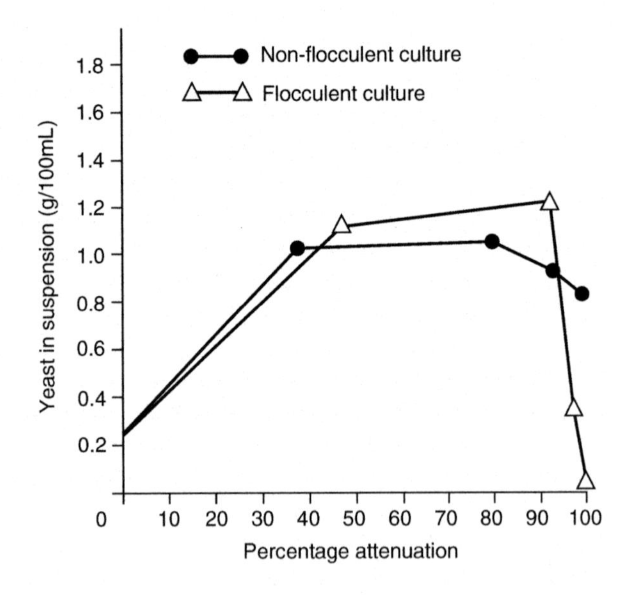

Fig. 13.1 Static fermentation flocculation characteristics of brewer's yeast strains

Traditionally, the first skim or "dirt skim", containing considerable quantities of denatured protein (called trub) and some yeast, is discarded, as is, in most cases, the final (third or fourth) skim. The middle skim (usually the second) is normally maintained for repitching. With the traditional lager bottom-cropping fermentation system, the yeast culture flocculates and descends to the bottom of the vessel at the conclusion of wort fermentation. Yeast cropping, in this situation, is usually non-selective and the yeast often contains entrained trub (denatured protein usually containing polyphenol material). With the cylindroconical fermentation system (now widely adopted worldwide for both ale and lager fermentations), the angle of the cone at the bottom of the tank allows for effective yeast plug removal (Boulton 2011) (Fig. 13.2). The advent and evolution of cylindroconical fermenters

Fig. 13.2 Cylindroconical batch fermenters

have been discussed in detail by Boulton and Quain (2001) and by Speers and Stokes (2009).

Yeast quality is influenced by the way that yeast is cropped and centrifuges currently play an important part in this regard (details later, Sect. 13.13). Yeast quality affects both beer stability and instability which concerns a number of complex reactions involving proteins, carbohydrates, polyphenols, metal ions, thiols and carbonyls. There are many diverse types of beer instability that involves a limited number of different microorganisms, chemical species and reactions (Bamforth 2009). Although our understanding of these reactions has progressed over the past 30 years, a complete comprehension of beer instability reaction systems is still not completely available (Bamforth 2009, 2017).

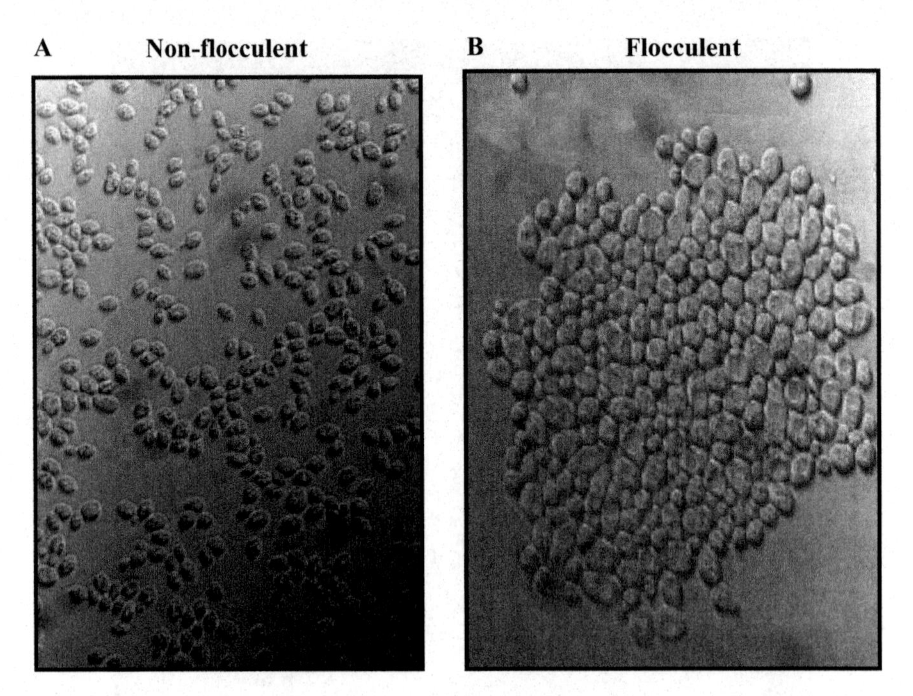

Fig. 13.3 Flocculation characteristics of brewer's yeast strains. (**a**) Non-flocculent culture; (**b**) flocculent culture

Depending on the brewing process, the flocculation characteristics (or lack of flocculation properties) of the yeast being employed will differ (Siero et al. 1994). In contrast, during the fermentation of a wort to be distilled, the yeast should not flocculate (it is non-flocculent) appreciably during fermentation and the yeast culture will be transferred into the still (batch or continuous) along with the fermented wort (Russell and Stewart 2014). Flocculation, as it applies to brewer's yeast, is "the phenomenon wherein yeast cells adhere in clumps (Fig. 13.3) and either sediment from the medium in which they are suspended (predominantly lager yeasts) or rise to the medium's surface (exclusively ale yeasts)" (Stewart 1974). This definition excludes forms of "clumpy growth" and "chain formation" (Guinard and Lewis 1993) (Fig. 13.8). Over the years, there have been a large number of colloquia (Speers 2012), monographs (Stratford 1992a, b; Vidgren and Londesborough 2011), review papers (Soares 2010), peer-reviewed papers (Stewart and Russell 1981; Rose 1984, Calleja 1987; Verstrepen et al. 2001b; Verstrepen and Klis 2006, and many more) and at least two textbooks (Calleja 1984; Speers 2016) specifically focussing on yeast flocculation and related phenomena. Without exception, they have all progressively advanced our knowledge of this complex yeast activity. There are a number of reasons for the advance in our knowledge of yeast flocculation. However, four primary factors (not in any order of priority) have mainly influenced this advance:

- Yeast flocculation (in its various formats) plays an important role in both the brewing and distilling processes. Consequently, detailed information on both its fundamental and applied aspects is critical for an understanding of brewing and distilling fermentation and has stimulated extensive research in industrial, government and university laboratories (Boulton and Quain 2001; Verstrepen et al. 2003; Soares 2010).
- The evolution and development of yeast molecular biology together with gene characterization and function, particularly details of the structure and function of the *FLO* genes (Sprague and Thorner 1992; Zhao and Bai 2009; Soares 2010).
- Development of appropriate equipment, techniques and methods such as electron microscopy (Day et al. 1975; Osumi 2012; Mamvura et al. 2017), confocal imaging (Schlee et al. 2006), flow cytometry (Hutter et al. 2005; Heine et al. 2009; Kuřec et al. 2009), mass spectroscopy (associated with gas chromatography and high-performance liquid chromatography) and visible/NIR spectroscopy. For an overall review of current analyses applied to yeast see Boulton 2012; Stewart and Murray 2011).
- Fundamental studies on the overall structure and function of yeast cell walls, particularly cell-cell interactions (Guo et al. 2000; Chen et al. 2007; Klis et al. 2006; Botstein and Fink 2011; Nayar et al. 2017) (further details in Chap. 5).

Most of this chapter is devoted to a discussion of the nature-nurture (Stewart et al. 1975a, b; Stewart 2014b) characteristics of brewer's and distiller's yeast strains and the factors (geno- and phenotypic) that influence the flocculation properties of cultures during wort fermentation. However, apart from this phenomenon's involvement in wort fermentation, its contribution as a model system in order to advance our knowledge of aspects of the cell-cell interactions of eukaryotic cells in general cannot be overstated (Rose 1984; Teunissen and Steensma 1995; Calleja 1987; Fink and Cookson 2005; Botstein and Fink 2011). As well as flocculation in *Saccharomyces cerevisiae* it also occurs in other yeasts such as *Kluyveromyces marxianus* (Sousa et al. 1992) and *Schizosaccharomyces pombe* (Calleja 1984, 1987).

Studies on the cell-cell interaction of eukaryotic cells using yeast as the model organism have been conducted for many years, by a number of notable research laboratories. One of the exponents of this endeavour was my close friend, colleague and travelling companion, the late Anthony H. Rose (Fig. 13.4) (Rose 1980, 1984). He proposed (Rose and Harrison 1987) that the long-term goal of research on eukaryotic cells should be the complete solution to the molecular biology of *S. cerevisiae*. Although efforts to complete this objective continue, unfortunately Tony passed away in 1993 at the young age of 63. He has been greatly missed by the scientific community ever since!

Yeast flocculation in this context can be defined as nonsexual, homotypic (involving only one type of cell in the interaction) and reversible (flocs can be reversibly dispersed or de-flocculated by washing the culture a number of times with deionized distilled water, EDTA solution or with specific sugars such as mannose) (Speers et al. 2006) (Fig. 13.5). Reflocculation of most yeast cells occurs

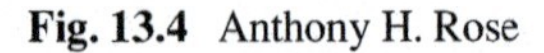

Fig. 13.4 Anthony H. Rose

upon the addition of calcium ions. Yeast clumps are composed of thousands, sometimes even millions, of cells called flocs (Fig. 13.3b). The flocs usually exhibit rapid sedimentation from the medium, in which the cells are suspended, to the bottom of a fermenter (lager yeast) or collect on the surface of the fermentation adhering to CO_2 bubbles to form a top crop (ale yeast) (Stewart et al. 1974).

The word floc derives from the Latin word *floccus*, which means a tuft of wool, while the cells that are not able to form flocs are known as non-flocculent or powdery (Stewart and Russell 1981). *S. cerevisiae* cells can be found aggregated in different ways, which should not be confused with flocculation, such as sexual aggregation (Chap. 5), co-flocculation (Stewart 1972; Stewart and Garrison 1972; Stewart et al. 1973; Zarattini et al. 1993; Rossouw et al. 2015) (Figs. 13.6 and 13.8), biofilm function (Fig. 13.7) (Mamvura et al. 2017) and chain (pseudohyphae) formation (Fig. 13.8) (Kukuruzinska et al. 1987).

Sexual aggregation in haploid strains of *S. cerevisiae* involves complementary mating types (MAT**α** and MAT**a**) which can occur after an exchange of pheromones (an agent secreted by an organism that produces a change in the sexual behaviour of a related organism) **a** and **α**, respectively (Chap. 2) (Sprague and Thorner 1992). This induces the appearance of complementary molecules (proteins) on the surface of cells which facilitates the fusion of haploid cells (Figs. 2.10 and 16.2) (Chen et al. 2007). Co-flocculation (Stewart and Garrison 1972) (Fig. 13.6), mutual agglutination (Eddy 1958), mutual aggregation (Nishihara et al. 2000) and mutual flocculation (White and Kidney 1979) (Sect. 13.11) all describe a heterotypic aggregation process (while single strain flocculation is homotypic) amongst two microbial strains which are usually (but not always)

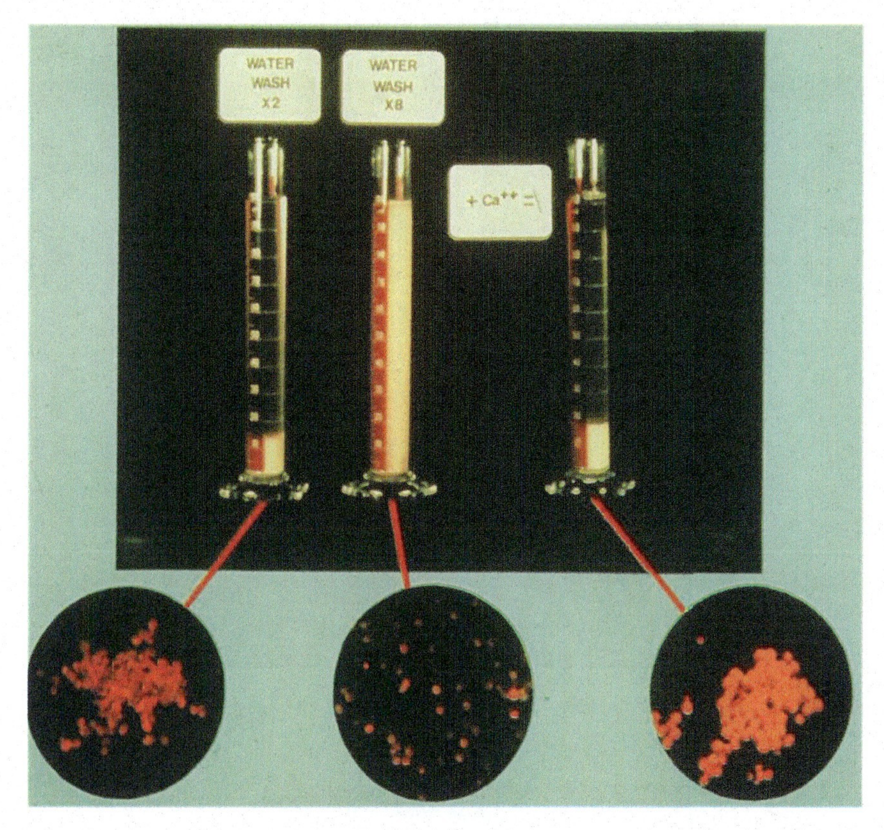

Fig. 13.5 De-flocculation of yeast as a result of repeated water washing and re-flocculation following the addition of the calcium ions

yeast (Peng et al. 2001; Zarattini et al. 1993) (Fig. 13.9). When these two strains (one being non-flocculent and the other very weakly flocculent) are mixed together, in the presence of Ca^{++} ions at pH 4.0–5.5, flocs form which rapidly settle out of suspension or rise to the surface of a fermentation (Stewart 1972, 2009). Chain formation (Fig. 13.8) (also called pseudohyphal growth) occurs in *S. cerevisiae* and the pathogen yeast *Candida albicans* (Kaur et al. 2005; Wang et al. 2009). It results from the failure of a young daughter yeast bud to physically separate from its mother cell. This results in an aggregate composed usually of 30–50 cells (Gimeno et al. 1992; Wang et al. 2009). These cell aggregates are physically attached to each other and consequently, following mechanical dispersion of the cells (e.g. physical disruption, sonication or enzyme treatment), cells will not be able to reaggregate prior to another growth cycle (Vidgren and Londesborough 2011). Other aggregation phenomena in *S. cerevisiae*, and related species, include biofilm (Verstrepen et al. 2004; Mamvura et al. 2017) (Fig. 13.7) and pseudohyphae formation (Fig. 13.8), which grow and invade the agar medium (Gimeno et al. 1992; Song and Kumar 2012) during nitrogen limitation and is a model of cellular responses in

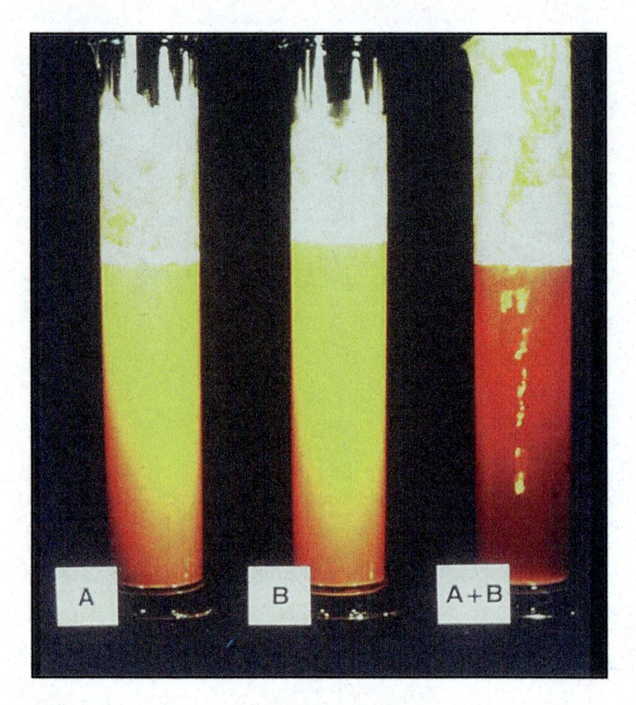

Fig. 13.6 Ale yeast co-flocculation—2 L cylinder wort fermentation test

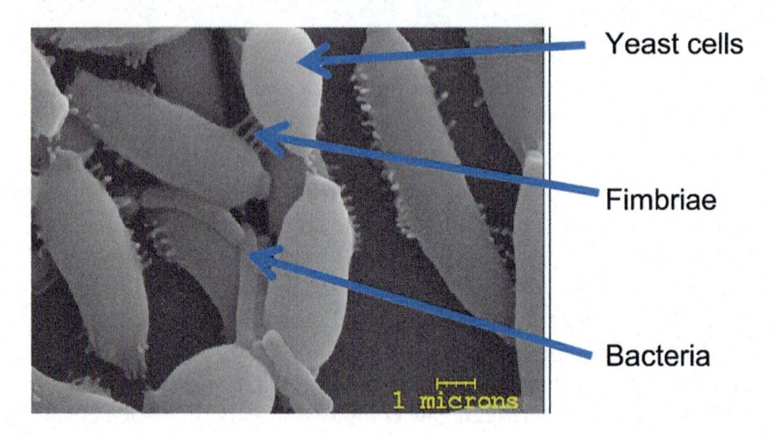

Fig. 13.7 Micrograph of a biofilm on the inner surface of a beer dispensing line

many eukaryotes. The various phenotypic and genotypic effects on yeast flocculation and aggregation will be discussed in greater detail later in this chapter (Sect. 13.3).

It has already been discussed that the ability of some yeast strains to aggregate and flocculate facilitates culture separation particularly during fermentation as a

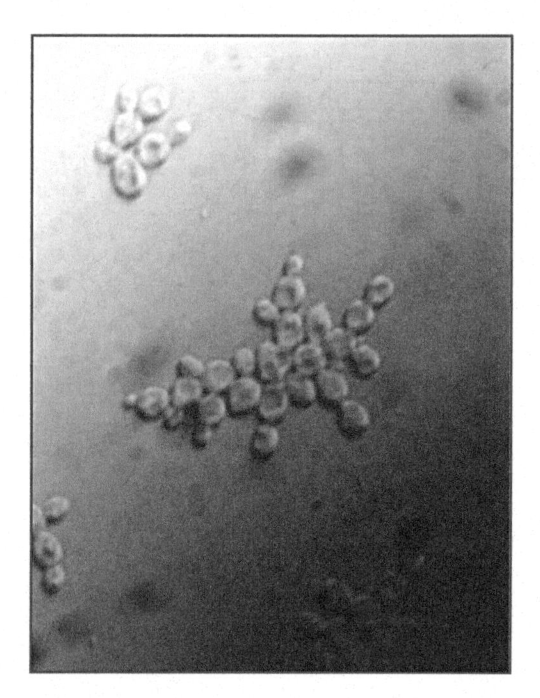

Fig. 13.8 Yeast chains—pseudohyphae

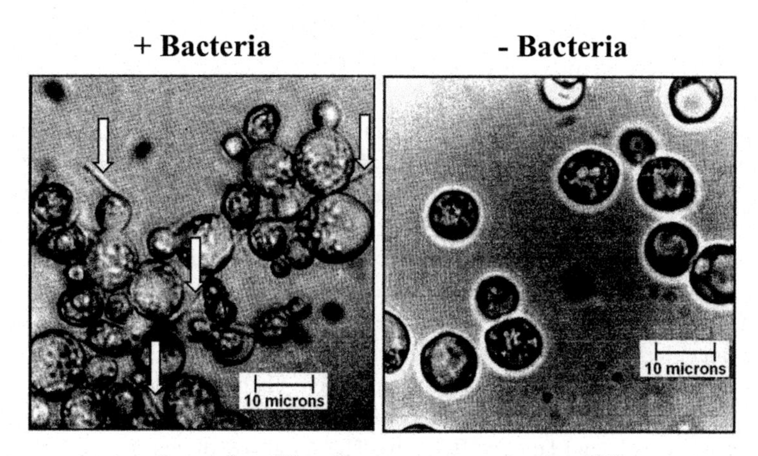

Fig. 13.9 Co-flocculation between an ale yeast strain and a *Lactobacillus* sp. strain—bacterial-induced yeast flocculation

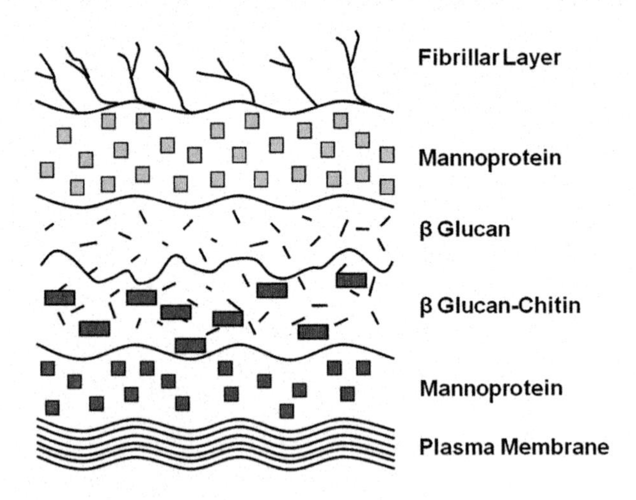

Fig. 13.10 Schematic diagram of the architecture of the yeast cell wall. Components such as mannoproteins occur as a matrix and are distributed throughout the entire wall, and therefore the layered arrangement should show zones of enrichment

part of the brewing process. Flocculation is an off-cost process which does not require significant energy input. However, cooling at the end of fermentation does facilitate yeast separation and this normally requires some energy. Paula and Birrer (2006) consider that flocculation increases process efficiency and reduces the energy consumption associated with cell separation with, for example, the use of a centrifuge (Chlup and Stewart 2011)—details later. Indeed, flocculation in brewing is an example of Lean Manufacturing (Chap. 11) (Dudbridge 2011, Stewart 2014a).

The cell wall structure of *S. cerevisiae* is a critical parameter involved in yeast flocculation (there are others—details later). Although details of cell wall structure have been described elsewhere (Chap. 5), it is worth discussing again that the wall of flocculent cultures consists of an inner layer composed predominantly of β-glucan, chitin and a fibrillar outer layer (Figs. 13.7, 13.10 and 13.11) (Day et al. 1975) constituting primarily of α-mannan associated with mannoproteins (Klis et al. 2006).

Flocculation is a cell surface characteristic (Gilliland 1951; Thorne 1951; Calleja 1984), and some (not all) heat-treated flocculent cells retain their ability to flocculate (Machado et al. 2008), as well as isolated in vitro cell walls prepared from a flocculent culture (Stewart and Russell 1986; Stratford 1992b; Miki et al. 1982a, b). The yeast cell wall has a net negative charge due to the ionization of carboxyl and phosphodiester groups of cell wall proteins and phosphomannans, respectively (Lyons and Hough 1970a, b). The repulsion of charges with the same sign prevents cells from approaching sufficiently close and acts as an effective barrier to cell aggregation (Beavan et al. 1979). Also, a positive correlation between cell surface hydrophobicity (CSH) and flocculation has been reported (Jin et al. 2001). CSH is partially responsible for the triggering of flocculation in brewing yeast strains (Smit et al. 1992; Speers et al. 1993, 2006). Based on these results, an increase in yeast surface hydrophobicity has been described when a number of *FLO* genes (details later) were expressed in the structure of yeast cell walls (Verstrepen et al. 2001a, b; Govender et al. 2008; van Mulders et al. 2009).

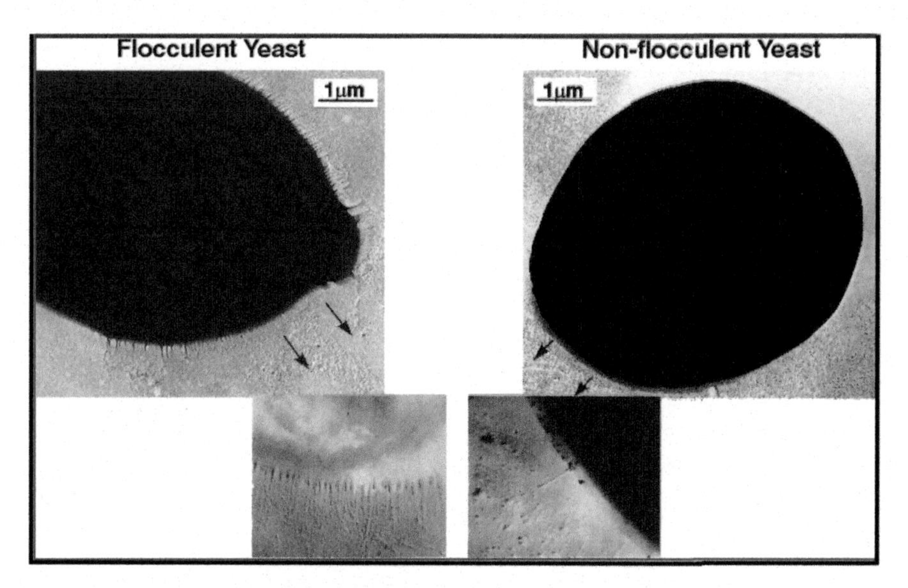

Fig. 13.11 Electron photomicrograph of *Saccharomyces cerevisiae* flocculent and non-flocculent strains shadowcast with tungsten oxide

13.2 Lectin Theory

Lectins (carbohydrate-binding proteins)—are macromolecules that are highly specific for sugar moieties (Halina and Nathan 2007). They have been proposed to be present on the surface of flocculent cells (Nayar et al. 2017). Miki et al. (1982a) have reported that specific lectin-like proteins interact with carbohydrate residues of α-mannans (receptors) on neighbouring yeast cells (Fig. 13.12). Also, calcium ions enable the lectins to achieve their active conformation (Miki et al. 1982a, b; Stratford 1989) (details of calcium effects on yeast flocculation will be discussed later) (Sect. 13.10.1). While flocculation lectins are only present on the surface of flocculent cells, the receptors are present in both flocculent and non-flocculent cells since, as already discussed, the outer layer of the *S. cerevisiae* cell wall is composed (in part) of mannan (Guinard and Lewis 1993). An analysis of the inhibitory action of sugars and the use of mannan synthesis mutants and concanavalin A (a lectin originally extracted from the jack bean) that specifically consists of glycoprotein with mainly α-D-mannosyl and α-D-glucosyl groups (Goldstein and Poretz 1986) supports the theory that flocculation receptors are most likely the nonreducing termini of α(1→3)-linked mannan side branches with two or three mannopyranose residues in length (Stratford and Assinder 1991; Stratford 1992a, b). Besides specific lectin-sugar interactions, other non-specific interactions, such as hydrogen bonds and hydrophobic interactions, can occur (Jin and Speers 2000).

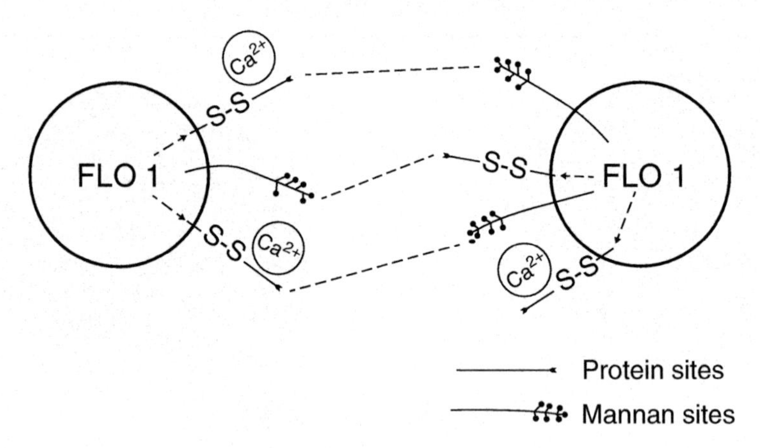

Fig. 13.12 Lectin theory of flocculation. Protein lectins on the yeast cell surface interact with either mannose containing and/or glucose containing carbohydrate determinants on the cell walls of adjacent cells only in the presence of calcium (Miki et al. 1982a)

13.3 Factors Affecting Yeast Flocculation

Several factors affect yeast flocculation (Soares 2010). The first factor is the genetic background of a strain. Flocculin proteins are encoded by members of the *FLO* group of genes (Teunissen and Steensma 1995). The genetic background (nature effects) with regard to *FLO* genes varies greatly amongst various types of brewer's yeast (ale or lager) and other different strains. Different strains contain a variety of *FLO* gene combinations (van Mulders et al. 2010), resulting in a spectrum of flocculation characteristics (Siero et al. 1994; Stratford and Assinder 1991). The current status with respect to *S. cerevisiae FLO* genes will be discussed later (Sect. 13.4).

Second, flocculation is affected by the physiological environment (nurture effects), for example, the pH, availability of metal ions and nutrients during the growth phase (Stewart et al. 1975a, b; Stewart 2014b; Gibson 2011). The pH influence of the cell surface charge will have an effect on the flocculation phenotype. Changes in pH may also modify ionization of functional groups in flocculin proteins which will change their conformation (Jin and Speers 2000). The environment is sensed by yeast cells leading to expression of *FLO* genes, their translation to Flo proteins and their location in the cell wall which is influenced by environmental factors (Verstrepen and Klis 2006; Verstrepen and Fink 2009). Thirdly, the physical environment affects flocculation. The hydrodynamic (pertaining to liquids in motion) conditions must be favourable and promote a sufficient collision rate between cells. However, agitation must not be violent enough to break cell flocs. Also, there must be sufficient cells in suspension to cause the number of collisions to form flocs (van Hamersveld et al. 1997). Factors that increase the hydrophobic character of the yeast cell walls (cell surface hydrophobicity) and factors that decrease the repulsive negative electrostatic charges on the cell walls (cell surface charge) cause stronger flocculation, because they facilitate cell-cell contact (Jin and Speers 2000; Wilcocks and Smart 1995).

Yeast flocculation is not absolutely necessary in order for the yeast cells to gradually sediment out of the green (immature) beer. This is because the size and density of yeast cells can overcome the Brownian motion that would maintain the cells in suspension (Stratford 1992b). The sedimentation rate is slow, especially when the medium is agitated, for example, with gas bubbles formed during fermentation. Also, the sedimentation rate is dependent on particle size. Smaller particles generally settle more slowly than larger particles of similar density, because they are relatively more retarded by friction (viscosity). Consequently, older yeast cells sediment faster than younger, smaller cells. However, the sedimentation of single cells is too slow to be of practical importance during a typical brewery fermentation, where flocculation is required to achieve the sedimentation of most yeast strains during the final quarter of an average wort fermentation cycle (Nayar et al. 2017).

Flocculating yeast cultures have some cells that are not entrapped in flocs which do not necessarily exhibit the ability to flocculate (Speers et al. 2006). There is a dynamic equilibrium between flocs and free cells resulting in continuous exchange. Single cells are constantly set free from flocs and at the same time new cells become entrapped. Loosening of cells occurs from the flocs because of hydrodynamic forces, resulting in a dynamic equilibrium (Jarvis et al. 2005; Stratford and Keenan 1988). Dilution favours an increased proportion of free cells, which is why an appropriate concentration of yeast cells is needed for effective flocculation and why some cells often remain in suspension even after bulk sedimentation of yeast flocs. However, this aspect, as with many aspects of yeast flocculation, is strain dependent.

At the end of fermentation, the following conditions are favourable for the sedimentation of a yeast culture:

- Carbon dioxide production rate is low.
- Wort attenuation is approaching completion. Most of the fermentable sugars have been removed (metabolized) by the yeast culture—glucose, fructose, sucrose, maltose and finally maltotriose (Stewart 2006).
- Flocculation ability is high but not too high.
- The yeast concentration in suspension in the wort is maximal (van Hamersveld et al. 1997).

Several factors influence the rate at which flocs sediment out of the fermented wort:

- The manner that cells pack into flocs.
- The resultant floc size, shape and density.
- Nurture factors such as wort properties including concentration (gravity), viscosity, density and turbulence (Stratford and Keenan 1988; Stratford 1989). In addition, the question of premature yeast flocculation (PYF) (Lake et al. 2008) is a separate, but related, phenomenon which will be discussed later (Sect. 13.9).

- Higher-gravity worts result in green (immature) beers with higher viscosity and density factors will retard sedimentation together with increased osmotic pressure and ethanol prior to dilution (Stewart et al. 1983a; D'Amore et al. 1988; Stewart 2010b; Stewart and Murray 2012; Zhuang et al. 2017).

As settling proceeds, the floc size often decreases because the concentration of yeast cells in suspension continues to decrease and smaller flocs form which will settle out of suspension more slowly (van Hamersveld et al. 1997).

13.4 Yeast Genetics and Flocculation

Genetic studies on brewer's yeast flocculation began 70 years ago (Pomper and Burkholder 1949; Gilliland 1951; Thorne 1951). It was confirmed that this yeast characteristic is an inherited phenomenon and is controlled by dominant genes termed *FLO* genes (Guo et al. 2000). This *FLO* gene from *S. cerevisiae* resembles the *EPA* epithelial adhesion gene family from the opportunistic pathogenic yeast species *Candida glabrata* (Cormack 2004; Kaur et al. 2005; Mundy and Cormack 2009) and the *ALS* genes (agglutinin-like sequence) from *Candida albicans* (Hoyer 2001; Hoyer et al. 2008). *C. glabrata* includes a family of at least 23 *EPA* genes, which encode cell surface proteins capable of mediating adherence to epithelial cells.

The first flocculation gene from *S. cerevisiae* to be studied in detail was *FLO1* (Lewis et al. 1976). However, it has already been discussed (Chaps. 2 and 3) that genetic investigations involving brewer's yeast strains (ale and lager) are fraught with difficulty because of their frequent triploid, polyploid or aneuploid nature and their inability to conveniently sporulate (Panchal et al. 1984a, b, c). Consequently, the early studies focussed on laboratory haploid and diploid flocculent and non-flocculent strains. The flocculent haploid strain studied (coded 169) was a MATα mating type opposite to that of the non-flocculent haploid strain (coded 168-MATa) (Fig. 2.10). These two strains were mated using the micromanipulation techniques discussed in Chap. 2 (Fig. 2.11), and the resulting diploid hybrid (169/168) was found to be flocculent, confirming previous findings that the flocculent character was dominant and stable (Stewart and Russell 1977).

Analysis of spores isolated from asci with tetrad analysis employing micromanipulation techniques (Sherman et al. 1986) (Fig. 16.6) of the 169/168 hybrid revealed that the dominant flocculence character of strain 169 was controlled by a single gene locus (i.e. 2:2 segregation). This gene has been coded *FLO1* (Lewis et al. 1976). The next question to answer was the location of the gene on one of the 16 chromosomes of the *Saccharomyces* genome. A detailed discussion of the chromosome mapping procedures employed is beyond the scope of this chapter (Stewart and Russell 1977; Sherman et al. 1986). Suffice to say, it has been shown that *FLO1* is located on the right-hand side of chromosome I, 33cM from the centromere (Stewart and Russell 1977; Russell and Stewart 1980). Further details of the molecular biology of the *FLO* genes are in Chap. 16.

The mapping of the *FLO1* gene employed traditional gene mapping techniques (mating, sporulation, micromanipulation, tetrad analysis, spore germination, multiple flocculation tests, etc) employed the methods discussed by Sherman et al. 1986. Novel genetic techniques have been developed, the principle of which is the sequencing of the *Saccharomyces* genome (Goffeau et al. 1996; Meaden 1996) with the haploid yeast strain S288C—details later. This has significantly expanded our knowledge regarding the genetic control of flocculation. Taking into account the reversible inhibition of flocculation by sugars, salt, a low pH environment and protease sensitivity (further details later), two main flocculation phenotypes have been distinguished: *FLo1* and *NewFlo* phenotypes (Stratford and Assinder 1991). The *Flo1* phenotype includes strains in which flocculation is specifically inhibited by the monosaccharide mannose and its derivatives. The *NewFlo* phenotype contains the majority of brewing ale strains. In this phenotype, flocculation is reversibly inhibited by mannose, maltose, glucose and sucrose but not galactose. *NewFlo* phenotype strains are more sensitive to inhibition by cations, low pH conditions and digestion by trypsin or proteinase A (Stratford and Assinder 1991). These phenotypes also display different sensitivities to culture conditions such as temperature (Soares et al. 1994), pH (Stratford 1996; Soares and Seynaeve 2000b), ions and nutrient availability (Soares and Mota 1996). Analysis of the N-terminal region of the *Flo1* protein responsible for the *Flo1* phenotype has shown that the domain formed by tryptophan 228 and its neighbouring amino acid residues recognizes the C-2 hydroxyl group of mannose. However, it does not recognize the C-2 hydroxyl group of glucose (Verstrepen et al. 2005). Similar analysis of the *Lg-Flo1p*, responsible for the *NewFlo* phenotype, revealed that the domain formed by leucine 228 and its neighbouring amino acid residues does not recognize the C-2 hydroxyl group of mannose and glucose. On the other hand, threonine 202 most likely interacts with the C-2 hydroxyl group of mannose and glucose, and this permits cell recognition (Kobayashi et al. 1998).

Two further phenotypes have been described: the mannose insensitive (M1) phenotype, composed of strains in which flocculation is not inhibited by sugars, including mannose (Masy et al. 1992) and a phenotype in which flocculation only occurs in the presence of high ethanol concentrations (Dengis et al. 1995). The precise mechanism(s) of strain aggregation belonging to these phenotypes are still not fully understood. However, a lectin mechanism, similar to the *Flo1* and *NewFlo* phenotypes, does not appear to be relevant (Dengis et al. 1995).

It has already been discussed that there are a number of dominant and recessive genes as well as flocculation actuator and suppressor genes, and the genes that encode flocculation lectins as already discussed are called *FLO* genes. *FLO1* is the best known flocculation gene, which has been mapped, cloned, sequenced and characterized by a number of independent research groups (Russell and Stewart 1980; Teunissen et al. 1993; Watari et al. 1994). There are at least nine *FLO* genes—*FLO1*, *FLO5*, *FL08*, *FL09*, *FL010*, *FL011*, *FLONL*, *FLONS* and *Lg-FLO* (Table 13.2)—in *S. cerevisiae* and *S. pastorianus* that encode flocculin proteins.

Table 13.2 Current view of flocculation phenotypes

Genes	Character	Sugars that inhibit flocculation
FLO1, FLO5, FLO9, FLO10	Strong Flo1 phenotype	Only mannose
FL08	Regulation of other FLO genes	Unknown
Lg- FLO1	NewFlo phenotype	Mannose, glucose, sucrose, maltose and maltotriose (not galactose)
FLO11	Chain formation – pseudohyphae	No inhibition by sugars
FLONL, FLONS	Like NewFlo phenotype	Mannose, glucose, sucrose, maltose, maltotriose and galactose
Not known	Mannose-insensitive (MI) flocculation (Ca-independent)	No inhibition by sugars

The current view of flocculation phenotypes is summarized in Table 13.2 (Vidgren and Londesborough 2011). It is characteristic that all of these gene sequences include tandem repeats! The flocculin encoded by *FLO11* differs from the other *FLO* genes in that it is involved in filamentous growth (pseudomycelia), adhesion to solid surfaces and flor formation (Fig. 13.8) rather than flocculation per se (Bayly et al. 2005; Fidalgo et al. 2008; Govender et al. 2008; Halme et al. 2004). Another *FLO* gene, *FL08*, encodes for a transposable factor regulating the expression of other *FLO* genes (Bester et al. 2006). Many yeast strains usually possess several *FLO* genes in their genome. For example, it has already been briefly described that the first completely sequenced laboratory haploid yeast strain—S288C—(Goffeau et al. 1996) contains six *FLO* genes (*FLO1*, *FLO5*, *FL08*, *FLO9*, *FLO10* and *FLO11*) and, in addition, four nonfunctional *FLO* pseudogenes (Teunissen and Steensma 1995; Teunissen et al. 1995; Watari et al. 1999). The amino acid sequences of *Flo5*, *Flo9* and *Flo10* proteins are 96, 94 are 58% identical, respectively, to *Flo10p*. As expected, *Flo11p* is the most distantly related to other flocculins, with only 37% identity to *Flo1p* (Verstrepen and Klis 2006).

The DNA sequences of *FLONL* and *FLONS* are very similar to that of *FLO1*, but compared to *FLO1* they have lost some of the internal tandem repeats. More tandem repeats were lost in *FLONL* than in *FLONS*. Deletion of these repeats appears to have converted the flocculation phenotype from *Flo1* to the *NewFlo* phenotype (Kock et al. 2000). Moreover, the *NewFlo* flocculation phenotype conferred by *FLONS* and *FLONL* is also inhibited by the monosaccharide galactose (Lin et al. 2008), whereas the usual *NewFlo* phenotype is insensitive to galactose. This indicates that the sugar binding properties of the flocculins are dependent upon the number of tandem repeats present in the flocculin gene.

13.5 Flocculation Gene Structure

FLO genes are very long (up to 4.6 kbp), whereas the average yeast gene length is 1.0 kbp, because of the number of tandem repeated DNA sequences of about 100 nucleotides that are repeated 10–20 times in each gene (Verstrepen et al. 2005; Verstrepen and Klis 2006). Although tandem repeated DNA sequences in the *FLO* genes are highly dynamic components of the genome, they change more rapidly than other parts of the genome (van Mulders et al. 2009). They enable rearrangements both between and within flocculin genes. The longer the *Flo* protein, the stronger the yeast strain's flocculation ability (Kihn et al. 1988a, b; Verstrepen et al. 2005; Watari et al. 1994). *FLO1* has been identified as the longest *FLO* gene, with the most repeats and it confers the strongest flocculation phenotype (Verstrepen et al. 2004; Dunn and Sherlock 2008).

Also, all *FLO* genes, except *FLO11*, are located close to telomeres (disposable buffers at the end of chromosomes) (Halme et al. 2004; Ogata et al. 2008). Consequently, they are prone to rearrangement such as deletions, duplications and translocations (Bhattacharyya and Lustig 2006). Indeed, the *Lg-FLO1* gene itself represents a translocation event between *FLO* genes located on different chromosomes. In addition, near-telomeric genes (and also *FLO11*—Halme et al. 2004) can become transcriptionally repressed by an epigenetic process known as telomeric silencing. This is caused by an alteration in chromatin structure near the telomeres leading to the silencing of genes located in that region. This effect can be long lasting. For example, the epigenetic state of *FLO11* is heritable for many generations. The strength of this telomeric silencing varies greatly between different telomeres in yeast (Loney et al. 2009). Significant variation has also been detected between strains. The series of proteins associated with a trithorax-related SET domain protein (COMPASS) complex is involved in telomeric silencing in yeast (Miller et al. 2001). *FLO* and *MAL* (details of the *MAL* genes are in Chap. 7) genes near to telomeres were found to be silenced in some yeast strains whereas in other strains with an inactivated COMPASS complex did not affect expression of these genes. This is consistent with the finding of significant variation in the strength of the silencing effect between different chromosome ends and in different strains. In strains where the COMPASS complex had a strong silencing effect, genetic inactivation of this complex increased the expression of *FLO1*, *FLO5* and *FLO9* genes. Compared to wild-type cells, these mutants displayed enhanced flocculation properties during high-gravity wort fermentation. Flocculation occurred earlier and formed much larger aggregates (Dietvorst and Brandt 2008). Although these findings are of considerable interest concerning the regulation of flocculation, their potential application to high-gravity brewing has already been discussed in Chap. 11.

13.6 Flocculation Instability in Brewer's Yeast Cultures

It has previously been discussed in this text that lager strains are allopolyploid hybrids of *S. cerevisiae* and a *S. bayanus*-like yeast (Dunn and Sherlock 2008; Nakao et al. 2009) that has already been identified as *S. eubayanus* (Chap. 3). A research group from Argentina, Portugal and the United States has published a paper entitled "Microbe domestication and the identification of the wild genetic stock of lager-brewing yeast" (Libkind et al. 2011). This study (amongst others) confirmed that *S. pastorianus* is a domesticated yeast species created by the fusion of *S. cerevisiae* with a previously unknown species that has been designated *S. eubayanus* because of its close relationship to *S. bayanus*. It was also reported that *S. eubayanus* exists in the forests of Patagonia and was not found in Europe until the advent of transatlantic trade between Argentina and Europe. However, *S. eubayanus* has also been isolated in Tibet (Bing et al. 2014). Details of these studies both in Patagonia, Tibet and elsewhere can be found in Chap. 3.

Lager (*S. pastorianus*) strains possess *FLO* genes derived from both parents. Consequently, they have a larger number and more diverse *FLO* genes in their genomes than ale (*S. cerevisiae*) strains. However, as will be discussed later in this chapter, the phenotypic expression of flocculation in ale yeast strains is more complex! Correlation of *FLO* genotypes to *Flo* phenotypes in brewer's yeast strains is only possible to a limited extent because, to date, the *FLO* genotype of a brewer's yeast strain has not been completely characterized. Even though the genome of lager strain *WS34/70* has been completely sequenced (Nakao et al. 2009), its *FLO* genes have not been studied in detail and it is unknown which of them are expressed in a physiologically functional way during wort fermentation.

Studies of both ale and lager strains of *NewFlo* phenotype have revealed that in addition to one or more of the *NewFlo*-type genes, *Lg-FLO1*, *FLONS* and *FLONL*, they also possess *FLO* genes such as *FLO1*, *FL05*, *FL09* and *FLO10* (Damas-Buenrostro et al. 2008; van Mulders et al. 2009, 2010), which are usually linked to a *Flol* phenotype. It is expected that the *Flol* phenotype (flocculation not inhibited by wort sugars) would be dominant over the *Lg-FLO1*-encoded *NewFlo* phenotype (flocculation being inhibited by glucose, maltose and maltotriose). What is the mechanism that gives these yeasts a *NewFlo* phenotype? It has been suggested that genetic or epigenetic mechanisms prevent more than one or more *FLO* genes from being expressed simultaneously in a yeast cell (Verstrepen and Fink 2009). Gene expression analysis confirmed that the tested flocculent lager strains strongly expressed all four flocculin-encoding genes studied, *Lg-FLO1*, *FLO1*, *FLO5* and *FLO9*, simultaneously (whereas non-flocculent yeasts showed nearly no expression) (Heine et al. 2009). One or more flocculin(s) were detected in the cytoplasm and the cell wall fraction of flocculent strains (much less in the non-flocculent strains) by Western analysis (Alwine et al. 1977) (an analytical technique to detect specific proteins) using rabbit antiserum that did not discriminate between the four gene products—*Lg-Flolp*, *Flolp*, *Flo5* and *Flo9p*. Peptide analysis of the excised bands detected a 14 amino acid sequence unique to Lg-Flo1p, but there was no evidence for (or against) the presence of the three other flocculins. Possibly some mechanism ensures that *Lg-Flo1p* is the dominant flocculin even when genes encoding *Flo1* type flocculins are strongly expressed. Perhaps the presence in the cell wall of *NewFlo*-type flocculins, which are blocked by wort sugars, hinders flocculation even when non-blocked *Flo1*-type flocculins are also present. It is possible that sugars bound to *NewFlo*-type receptors might cause steric hindrance of cell-cell linking. It should be emphasized that experimental evidence for this hypothesis has not been published to date.

13.7 Genetic Instability of Flocculation in Brewer's Yeast Strains

The sedimentation performance and characteristics of a brewer's yeast strain very often changes during repeated cropping and repitching in a brewery (Stewart and Russell 1981, 1986; Stewart 2009). In principle, this could be due to either irreversible or reversible genetic change (Watari et al. 1999). Alternatively, it could be due to long-lasting physiological, perhaps epigenetic (the study of cellular and physiological) traits caused by external or environmental factors that switch genes on and affect how cells read genes (Bird 2002) and respond to modifications in yeast handling and fermentation environments [e.g. high-gravity worts (Stewart 2010a, 2014a)]. When a genetic change, conferring a non-flocculent phenotype, occurs in a yeast culture, the culture gradually becomes a mixture of flocculent and non-flocculent cells (Sato et al. 2001). Often within a production lager yeast population, exhibiting moderate flocculent characteristics, a more flocculent variant within the culture can be isolated. An example of this development was when a Canadian brewing company began brewing its lager beer, under contract, in breweries located in the United Kingdom. Most of the contracted UK breweries employed vertical fermenters. However, at that time, the Canadian breweries employed only horizontal tanks (as both fermentation and maturation tanks). This difference in tank geometry influenced the yeast culture's sedimentation characteristics. In vertical fermenters, this yeast culture was too non-flocculent (powdery), with considerable yeast cells remaining in suspension at the end of fermentation (centrifuges were not available in the UK breweries at the time). It was thought that possibly this culture contained a spectrum of isolates that exhibited differing flocculation intensities. Consequently, one of the variants from this strain, with more intensive flocculation characteristics, was successfully employed in the vertical fermenters. The result was less yeast in suspension at the end of fermentation. However, care had to be taken to ensure that the flocculent variant used was not too flocculent became under-fermented wort (unfermented maltose and maltotriose) and residual unwanted beer flavours [(e.g. diacetyl—details of its metabolism in Chap. 15) (Figs. 14.12 and 14.13)] could have been the result (Stewart 2015a, b).

PCR-based methodology (a technique employed to amplify a single copy or a few copies of a piece of DNA in order to generate millions of copies of a particular DNA sequence—details in Chap. 16) has been employed to detect *FLO5* genes (Jibiki et al. 2001). All 48 single colonies isolated from the stock culture of a lager strain exhibited this PCR band. After repeated recycling of this yeast culture in a brewery, single cells were again isolated from batches of the yeast that showed poor sedimentation performance. In one case, most (75%) single cell colonies failed to yield the *FLO5* PCR band, indicating that a genetic mutation (loss of intact *FLO5*) had occurred and spread throughout the population, and it was necessary to replace the brewery yeast with a fresh culture (Stewart 1996). In the other case, nearly all (>90%) of the single colonies still showed the *FLO5* PCR band, indicating that the poor sedimentation behaviour did not have a genetic basis but was due to the yeast culture's physiological condition. In this case, the flocculation ability gradually recovered during repeated use of the yeast in a brewery.

A number of research groups have studied spontaneous changes in flocculation behaviour as a result of repeated repitching in wort during brewery fermentations (Smart and Whisker 1996; Teixeira et al. 1991; Wightman et al. 1996)—stable (Powell and Diacetis 2007) and decreased (Halme et al. 2004; Sato et al. 2001; Watari et al. 1999) flocculation intensity has been observed. Cropped ale yeast, during 30 successive fermentations, has been studied (Smart and Whisker 1996). During the first seven cycles, flocculation intensity increased from 50% to 100% of the original culture. Between the 9th and 23rd cycle, flocculation remained high. Then, between the 24th and 32nd cycles, flocculation ability and cell viability diminished. Sato et al. (2001) reported the flocculation tendency of a lager strain to decrease after serial repitching. A long-term study in a number of breweries showed that flocculation tended to decrease while at the same time other parameters (fermentation rate, ester and higher alcohol production, etc.) usually remained constant. Studies in the author's breweries with high-gravity worts (>16°Plato) do not confirm this fact—it confirmed that many findings of this nature must be yeast strain dependent (Stewart 1988, 2014a, b, c; Stewart and Murray 2012)—further details in Chap. 11.

Cropping methods favour the enrichment of certain cell types. Few modern lager brewers currently recycle their yeast culture more than 20 times (many breweries less than 10 times) (Stewart and Russell 2009) (details in Chaps. 8 and 11). However, as the original wort concentration (gravity) has increased, in the situation that occurs with high-gravity brewing (HGB), the number of yeast cycles has been significantly reduced. Typically, with a 16°Plato wort, the number of cycles is 10 times or less (Stewart 2014a). The results of these studies suggest that it is more likely that a change in flocculation behaviour is due to a modification in process conditions or raw materials [e.g. malt leading to changes in wort composition also increased adjunct (unmalted cereal levels)], especially if the change persists when freshly propagated yeast is introduced into the process.

It has been suggested (Jibiki et al. 2001) that the PCR-based method to detect *FLO5* genes (discussed above) can be used for the early detection of non-flocculent mutants in brewery fermentations. More than 30 different production lager strains with PCR primers designed to detect *FLO5* have been studied. The PCR product varies in size from 4.8 to 2.3 Kb with more flocculent strains showing larger gene products (>4 kb) and less flocculent strains showing smaller products (2.3 kb in a non-flocculent strain). Cell surface hydrophobicity also decreased with reducing size of the PCR product. Thus, the appearance of non-flocculent mutants during a series of brewery fermentations could be tested by PCR analyses. Interestingly, PCR primers designed for *Lg-FLO1*, *FLO1* or the regulatory gene, *FL08*, did not have this predictive power. In strains failing to form a PCR product with the *FLO5* primers, Southern blot hybridization (a method to detect a specific DNA sequence in DNA sample—Southern 1975) after chromosome fingerprinting using a *FLO5* fragment as probe showed that *FLO5* was missing from chromosome VIII, whereas hybridization to chromosome I (the location of *FLO1*) still occurred. The sequences of the successful *FLO5* PCR primers were not revealed. Consequently, we do not know whether they recognized only *FLO5* (which is 96% identical to *FLO1*).

13.8 The Influence of Cell Surface Hydrophobicity (CSH) and Cell Surface Change or Yeast Flocculation

It has been discussed (Sect. 13.1) that there have been a number of published studies that indicate that an increase of CSH and a decrease of cell surface charge occur at the outset of flocculation (Amory et al. 1988a, b; Bayly et al. 2005; Smit et al. 1992; Straver et al. 1993; Speers et al. 2006). CSH increases rapidly as cells pass through exponential growth phase and reach higher and stable levels during stationary phase (Speers et al. 2006). Low CSH in an exponential phase culture is due to the presence of many daughter cells which are significantly less hydrophobic than older (more mature) cells (Powell et al. 2003). It is considered that CSH plays an important role in maintaining the correct conformation of flocculin molecules (Jin et al. 2001), so that the flocculins located in a stationary phase cell are more active (Lange et al. 2001). It has already been discussed that the yeast cell surface has an overall negative charge. As a result, the cell wall phosphate groups and mannoproteins (Lyons and Hough 1970a, b; Rhymes and Smart 2001) are greater when the pH of the fermented medium is higher (Boulton and Quain 2001). The cell surface charge of brewer's yeast strains has been shown to vary during growth (van Holle et al. 2011). However, no clear relationship between cell surface charge and flocculation onset has been observed (Dengis et al. 1995; Smit et al. 1992). A number of environmental factors have been observed to affect CSH. Increased CSH can be stimulated by higher ethanol concentrations (Jin and Speers 2000), lower temperatures (van Lersel et al. 1998) and higher pitching (inoculation) rates (Jin et al. 2001). It has been proposed that hydrophobic oxylipins located at the cell surface of flocculent cells are a cause of increased CSH (Kock et al. 2000; Strauss et al. 2005, 2007). However, a recent publication by Potter et al. (2015) has suggested that the precise rate of 3-OH oxylipin formation is still unclear, and detection methods must be combined with novel techniques to target the cell wall architecture.

Ale strains (*S. cerevisiae*) have been found to be more hydrophobic and less negatively charged than lager (*S. pastorianus*) strains (Amory et al. 1988a). As well as contributing to floc formation, the greater hydrophobicity of ale strains probably explains why the flocs of these strains associate with CO_2 bubbles and rise to the beer surface during traditional top-cropping ale fermentations, whereas the flocs of lager strains sink to the bottom of the fermenter (Dengis et al. 1995). In many modern breweries, ale strains sink to the bottom of fermentations in cylindroconical vessels (CCVs) during the closing stages of a wort fermentation (Stewart et al. 1983a, b, 1984a, b, c; Stewart and Russell 1983; Stewart 1988). It is probable that the greater hydrostatic pressure and the modified geometry in large CCVs restrict CO_2 bubble attachment to yeast flocs. Also, sedimenting mutants of ale strains have been either deliberately or accidentally selected to facilitate a bottom crop during ale production (Stewart and Russell 2009).

13.9 Premature Yeast Flocculation

It has already been discussed that the timing of yeast flocculation during wort fermentation is crucial for the production of beer with the necessary quality characteristics (Sect. 13.1). Occasionally, certain malts can cause premature or heavy flocculation leaving the wort underfermented with sugars still in solution and the alcohol specification not achieved (Axcell et al. 2000) (Fig. 13.13). This phenomenon has been termed premature yeast flocculation (PYF) (Ishimaru et al. 1967) and is detected by way of a fermentability test (Kruger et al. 1982; Lake et al. 2008).

Most of the studies on PYF have not focussed on yeast per se. They have focussed on the malt employed to produce the wort used during the fermentation studies (Axcell et al. 2000). Nevertheless, nature-nurture interactions are critical (Stewart 2014b). Also, the need for future detailed PYF studies with a number of brewer's yeast strains (both ale and lager) will be emphasized later!

Depending on the local beer regulations, often a positive result with the fermentability test does not translate into a problem because this discrepancy is not questioned. However, as will be discussed in detail later (Sect. 13.11), some countries (e.g. Canada) enforce a rigid beer alcohol specification (± 0.2 v/v) which is listed on every bottle, can and keg.

During the past 50 years, more than 30 papers have discussed the phenomenon of PYF. In the last two decades, the research group in South African Breweries in Johannesburg and the research group in Dalhousie University, Nova Scotia have devoted considerable attention to PYF, which has extended our understanding of the problem (Axcell et al. 2000; Jibiki et al. 2006; van Nierop et al. 2004, 2006; Speers 2016). It should be emphasized that PYF is distinct from typical yeast flocculation that is discussed in other sections of this chapter.

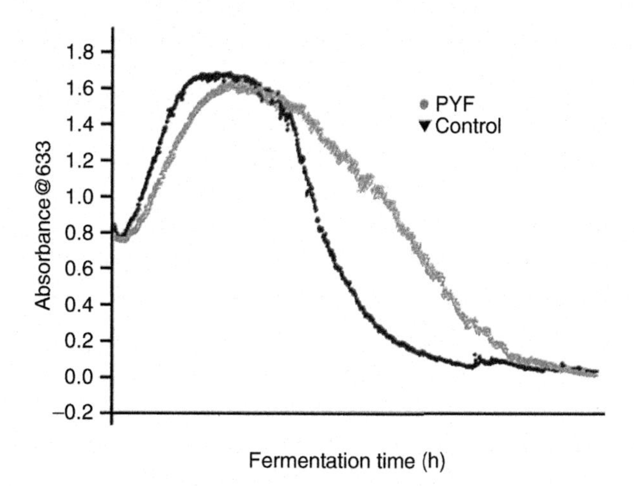

Fig. 13.13 Test tube-sized (15 mL) fermentations of a premature yeast flocculation (PYF) and control malt yeast in suspension continuously measured

What exactly is PYF? (Lake and Speers 2008) The published literature contains a number of different definitions (Koizumi and Ogawa 2005; Inagaki et al. 1994; van Nierop et al. 2004). Although details vary, these PYF definitions are similar. Lake and Speers (2008) have noted that PYF behaviour results in:

- A rapid yeast decline in suspension from peak yeast-in-suspension values occasionally, a high final apparent extract relative to a "normal" malt (depending on the fermentation vessel employed).
- Normal fermentations typically have parabolic yeast-in-suspension trends; PYF yeast-in-suspension curves proceed in a normal and parabolic manner to a peak and then decline in a concave fashion (Fig. 13.13).

Two theories dominate the literature with respect to the development of PYF factors in barley and malt. PYF tends to be a sporadic phenomenon that occurs simultaneously with wet, rainy seasons. In the first theory, it is believed that increased microbial loads during wet seasons lead to the production of PYF factors (van Nierop et al. 2006). The barley husk is the main carrier of microorganisms (Briggs and McGuinness 1992). The microflora consists of bacteria, wild yeasts and filamentous fungi (van Nierop et al. 2006).

The second theory regarding PYF mechanisms is that antimicrobial peptide factors inhibit or negatively affect yeast metabolism. This in turn would initiate flocculation earlier and/or to a greater extent than normal. It has been shown that a PYF-positive malt leads to a minor decrease in wort sugar (particularly maltose and maltotriose) metabolism by yeast (van Nierop et al. 2004). However, it is unclear whether this reduction in sugar metabolism is due to insufficient yeast in suspension (due to PYF) or direct inhibition of yeast metabolism. However, Lake et al. (2008), employing a miniature fermentation assay, did not detect metabolic differences between PYF and a control malt. This course of events does not preclude antimicrobials associating with the yeast cell that would lead to increased flocculation.

It is noteworthy, as would be expected, that yeast strains show different susceptibility to PYF malts (Jibiki et al. 2006; Armstrong and Bendiak 2007). It is unclear whether strain susceptibility is due to differing mechanisms, categories of PYF or whether variability is related to the flocculation potential of various yeast strains. Also, it is unclear whether ale yeast (*S. cerevisiae*) is less susceptible or whether the effects of PYF are masked by the flocculation behaviour of a hydrophobic ale yeast.

Studies by Chinese maltsters have conducted small-scale fermentations with worts produced from seven malts. These fermentations monitored their PYF potential and showed that PYF factors were present in both the malt husk and non-husk portions. Also, it was shown that antimicrobial substances that damage yeast cells were present in the non-husk portion (Ishimaru et al. 1967).

Further research is required (Lake and Speers 2008) on the mechanisms of PYF activity. No one theory supports or challenges any of the currently proposed mechanisms. Most publications have reported on the gross chemical ratio of PYF factors and have avoided discussing or investigating PYF mechanisms.

13.10 Phenotypic Effects on Flocculation

The complexities of yeast flocculation cannot be denied. Besides the genetic characteristics of strains (*FLO* genes, suppressors and activators), a number of nurture parameters affect yeast flocculation (Stewart et al. 1975b; Stewart 2014b).

13.10.1 Cations

Cations have a central role on both ale and lager yeast flocculation. Amongst them, calcium ions are recognized as the most effective ion for the promotion of flocculation (Miki et al. 1982b; Stratford 1989; Bester et al. 2006). This importance has already been discussed (Gilliland 1951) but details of its interaction with the yeast cell surface requires further elaboration. It has already been discussed that calcium can be removed from the yeast cell surface by washing with deionized water, and a deflocculated culture will become flocculent again (Fig. 13.5). Some flocculent yeast strains are not deflocculated by washing with water, the cells need to be treated with a solution of a chelating agent such as EDTA (10 mM usually, Stewart 1973) followed by washing with water to remove the EDTA. This treatment deflocculates these cultures, and the flocculation phenotype is restored upon re-addition of calcium ions. It has been suggested that cell walls isolated from flocculent cultures bind more Ca^{++} ions than walls isolated from non-flocculent cultures. Employing radiolabelled Ca^{45}, studies were conducted to compare the calcium-binding ability of several ale and lager flocculent and non-flocculent brewery yeast cultures (Stewart et al. 1975b). When the final calcium uptake of each culture was analysed, it was clear that no direct correlation existed between the total calcium adsorbed and the flocculation phenotype. There is strain-to-strain variation in calcium binding, and furthermore this variation does not correlate with flocculation and non-flocculation when one strain was compared to another. However, with the knowledge that many flocculent yeast cultures can be deflocculated by washing with deionized water, it was of interest to see if the amount of calcium washed off a yeast culture could be correlated with the visible loss of floc formation. As a result of this study, an improved perspective of calcium-binding behaviour in yeast and its relationship to flocculation might be obtained to test this hypothesis. Aliquots of flocculent and non-flocculent yeast suspensions were taken and incubated with the Ca^{45} solution. The yeast pellet was then washed four times with deionized water, and the activity of each centrifuged supernatant determined with a scintillation counter and the amount of calcium removed with each washing determined. The first wash did not deflocculate the flocculent yeast cultures but did remove adhering calcium around and in the interstitial spaces between the yeast cells. This source of calcium should be relatively the same percentage of total calcium bound for each yeast culture and is in all probability not related to flocculation, since the visible observation of flocculation did not disappear during this first wash.

Subsequent washings gradually dispersed any flocculation characteristics of the yeast culture. The sum of the calcium removed in washings 2–4 were expressed as a percentage of the total calcium removed during washing. When the results were expressed in these terms (Table 13.3) for both flocculent and non-flocculent cultures, the flocculent cultures were found to have bound 28–48% more calcium after four washings than did non-flocculent cultures. As would be expected, there is strain-to-strain variation in calcium adsorption (Stewart et al. 1974). This variation is in all likelihood a reflection of diversities in cell wall structure strain to strain. In addition, this strain-to-strain variation in calcium adsorption per se does not correlate with the flocculation phenotype when one strain is compared to another. The only meaningful measure of calcium behaviour that correlated with flocculation

Table 13.3 Calcium removed from co-flocculent and non-flocculent cultures during de-flocculation washings

Yeast cultures	Flocculation characteristic	Total calcium washed off yeast[a]
Ale (*S. cerevisiae*)	Non-flocculent	18
	Non-flocculent	19
	Flocculent	30
	Flocculent	42
Lager (*S. pastorianus*)	Non-flocculent	12
	Non-flocculent	14
	Flocculent	20
	Flocculent	22

[a]Total calcium washed off yeast—mg/100 mg dry weight of yeast

was the case with which calcium was washed off the cell, and this coincided with the visible loss of flocculation.

Rb^+, Cs^+, Fe^{++}, Co^{++}, Ca^{++}, Ni^{++}, Cd^{++}, Al^{+++} and particularly Zn^{++}, Mg^{++} and Mn^{++} have also been described as inducers of flocculation (Miki et al. 1982a, b; Nishihara et al. 1982; Sousa et al. 1992; Soares and Duarte 2002; Stewart and Goring 1976). Second to calcium, the effect of zinc on flocculation has received the greatest attention and there are a number of publications considering this parameter (Russell et al. 1989; Raspor et al. 1990). In addition, a great deal of research has been conducted studying the effect of Zn^{++} on enzymatic activity and fermentation efficiency (Rees and Stewart 1998). These latter aspects have been discussed in Chap. 7. The flocculation-deflocculation behaviour of *S. cerevisiae* is strongly dependent on the concentration of Zn^{++} ion in the fermentation medium (e.g. wort) and is yeast strain specific. However, *S. pastorianus* is not affected by the presence of Zn^{++}, which suggests another useful method for distinguishing between lager and ale flocculent yeast strains (Raspor et al. 1990).

Cations such as Ba^{++}, Sr^{++} and Ph^{++} competitively inhibit yeast flocculation because of the similarity of their ionic ratio to Ca^{++} (Nishihara et al. 1982; Gouveia and Soares 2004). It is possible that these cations compete for the same "calcium site" of flocculation lectins, but are not able to induce the appropriate conformation of the lectins. At low concentrations, Na^+ and K^+ induce flocculation most likely because of the reduction of the electrostatic repulsive forces the yeasts and/or stimulate the leakage of intracellular Ca^{++} (Nishihara et al. 1982; Stratford 1989; Stewart and Goring 1976). At high concentrations, it seems that these ions provoke distortion of lectins and antagonize calcium-induced flocculation (Stratford 1992a, b, c). Besides the surface action, the presence of cations in the culture medium (viz. Mg^{++} or the Ca^{++}/K ratio) seems to be essential for the expression of flocculation (Smit et al. 1992; Stratford 1992d; Rees and Stewart 1997a, b).

13.10.2 Medium pH

Medium pH can have a profound effect on the yeast flocculation phenotype. With many laboratory and industrial strains, flocculation occurs over a wide pH range (2.5–9.0), while many brewing strains (a subgroup of the *NewFlo* phenotype) only flocculate within a narrow pH range (Stratford 1996; Soares and Seynaeve 2000a, b). In both cases, the optimum pH value takes place between 3.0 and 5.0, according to the yeast strain. Extreme pH values promote a reversible dispersion of flocs. Probably modification of the pH value affects the ionization of lectin amino acids with the consequent change in its conformation (Jin and Speers 2000; Jin et al. 2001).

Most ale strains (*S. cerevisiae*) do not flocculate following growth in a chemically defined medium such as yeast nitrogen base (YNB) (Stewart et al. 1973; Stewart and Russell 1981). It could be that these strains also exhibit a narrow pH range of flocculation (Stratford 1996; Soares and Seynaeve 2000a). However, peptone, certain peptides and wort also play an important role because ale cells grown in a peptone-containing medium or wort exhibit the flocculation phenotype. This does not only occur in the culture medium. Following cell harvesting and washing, the cells were still flocculent in an in vitro flocculation test such as the Helm's flocculation test (Helm et al. 1953; Stewart 1972) (Fig. 13.17).

13.10.3 Temperature

Incubation temperature can act at different levels on the expression of yeast flocculation. The lowering of growth and fermentation temperature leads to a decrease in yeast metabolism and CO_2 production. Consequently, there is a reduction in turbulence which favours yeast sedimentation. During beer fermentation, the agitation (shear force) temperature can also affect yeast flocculation by acting on cell-cell interactions. A rise in temperature to 50–60 °C, for a few minutes, promotes the reversible dispersion of flocs (Taylor and Orton 1978) probably because of denatured flocculation lectins. The incubation of yeast strains at above the optimum temperature (35–37 °C) leads to reduction (Soares et al. 1994; Williams et al. 1992) or impairment of yeast flocculation (Claro et al. 2007). It is probable that this heat stress (<37 °C) acts directly on mitochondrial activity (details in Chap. 14) and indirectly on cell membrane structure affecting the secretion of lectins, with a consequent reduction in flocculation (Stewart 2014c). Details of the structure and influence of both mitochondria and cell membranes and their influence on brewing and distilling fermentation can be found in Chap. 5 together with a discussion regarding the impact of a number of stresses (Chap. 11).

13.10.4 Oxygen

The influence of oxygenation (aeration) in the early stages of both a brewing and a distilling fermentation has been discussed in Chap. 6 (Russell and Stewart 2014). It is well documented that moderate aeration (oxygenation) is beneficial for yeast flocculation (Kida et al. 1989; Soares et al. 1991). However, the principal role of oxygen during the initial stages of wort fermentation is not focussed on yeast flocculation. It is to act as a catalyst for the synthesis of unsaturated fatty acids and sterols whose primary role is in the structure of membranes, particularly the plasma membrane (Lorenz and Parks 1991). This membrane structure is important for the stress protection of yeast cultures during adverse environmental conditions such as high-gravity wort fermentations (details in Chap. 11).

Mannoproteins in the cell wall are differently expressed under aerobic and anaerobic conditions (Abramova et al. 2001). The transition from semi-aerobic to anaerobic conditions, which occurs during typical brewing fermentations, is probably associated with the expression of genes that regulate or encode flocculation lectins. Cells pitched into oxygen-depleted wort flocculated relatively early during the fermentation cycle but to a limited extent. The addition of ergosterol or Tween80 to the same wort restored normal flocculation behaviour of the culture. This study concluded that lack of oxygen inhibited the synthesis of ergosterol and UFAs, consequently, limiting cell growth and resulting in the early onset of stationery phase and cell flocculation (Straver et al. 1993).

13.10.5 Sugars

The presence or absence of fermentable sugars is a major factor influencing flocculation by *NewFlo* phenotypic strains (Speers et al. 2006). As long as glucose, maltose or maltotriose are present in sufficient amounts, flocculation is inhibited because these sugars occupy the flocculins. This inhibits binding to the mannose residues of adjacent cells. Addition of glucose (20 g/L), rapidly dissociated the flocs of a starved ale yeast culture containing the *NewFlo* phenotype (Soares and Duarte 2002). This is a result that might be expected because glucose blocks the mannose-binding sites of *NewFlo*-type flocculins. However, it has been suggested that the loss of flocculation requires energy. Soares and Duarte (2002) examined the effect of nutrients on the loss of flocculation and have presented evidence of six aspects regarding the stimulation of flocculation loss with an ale *NewFlo* phenotype strain, under growing conditions:

- Carbohydrate sources are nutrients that stimulate loss of flocculation in a defined growth medium (e.g. yeast nitrogen base—YNB) (Wickerham 1951).
- All metabolizable carbon sources (e.g. glucose, fructose, galactose, maltose and sucrose) induce the loss of flocculation in YNB, which ethanol does not—details later (Sect. 13.4) (Masy et al. 1992).
- The rate of sugar-induced flocculation appears to be associated with the rate of sugar metabolism (Stewart and Russell 1983).
- The rate of sugar-induced flocculation loss most likely requires energy and this process is blocked by ethanol.
- Growth does not always trigger flocculation loss because cells grown in a medium containing ethanol remained flocculent.
- Glucose-induced loss of flocculation requires de novo protein synthesis—cyclo-heximide addition (an inhibitor of protein synthesis) to glucose-growing cells impairs the loss of flocculation (Baker and Kirsop 1972; Masy et al. 1992).

13.11 Co-flocculation

Co-flocculation, mutual aggregation or mutual flocculation is a heterotypic aggregation process [while flocculation is homotypic (Sect. 13.1)] amongst two separate yeast strains (Mortier and Soares 2007). One strain is non-flocculent and the other strain weakly flocculent. When these strains are mixed together in the presence of Ca^{++} ions, flocs form and the culture rapidly settles out of suspension (Stewart and Garrison 1972) (Figs. 13.16 and 13.17). Heterotypic yeast nonsexual flocculation was first described by Eddy (1958) and Eddy and Rudin (1958) with a number of ale strains. The focus of our research group on co-flocculation began with the top-cropping Labatt ale yeast culture. In the 1960s and early 1970s, consumption of pale ales was popular in Canada; in 1970 it represented 60% of the beer consumed in Ontario and 80% in Quebec (Stewart 2015a, b). However, similar to the situation in Britain, the ale yeast cultures employed (unlike lager strains) were (and many still are) largely uncharacterized. Also, no Labatt employee (or pensioner) could inform us much about the history of their ale culture. The Labatt ale culture possessed classical top-cropping properties. It also exhibited intermittent premature flocculation characteristics, resulting in underfermented worts containing residual sugars (mainly maltotriose—details regarding sugar uptake by brewer's and distiller's yeast strains in Chap. 7). This was a problem in Canada because as previously discussed (13.7) a ±0.2 (v/v) alcohol specification was the legal variation (Federal Health Protection Branch Regulations) and the beer's alcohol composition was (and still is) specified on the label of the bottle, can or keg. This problem was exacerbated during high-gravity brewing trials (details in Chap. 11) of ales. Consequently, it was important to enumerate the number of strains in this ale culture and characterize them.

A suitable method to examine yeast culture's strain composition in the 1960s and 1970s was the giant colony morphology method initially developed by Gilliland (1959) and elaborated by Richards (1967). The Richard's method involved inoculating the yeast culture in question onto wort solid media and examining the colonial morphology that developed after incubating under standard conditions for at least 3 weeks at 18 °C. It had been found that gelatin, as the solidifying matrix, tends to enhance the distinctive features of the colonial morphology to a greater extent than agar and that wort, instead of a synthetic or defined medium, gave distinctive and reproducible results (Fig. 13.14). Also, lager yeast strains do not exhibit distinctive colonial morphologies on wort gelatin media (Fig. 13.15).

Analysis of the Canadian brewery's top-cropping ale yeast culture's strain composition showed that two morphologically different colony types were present (Fig. 13.16) (Stewart 1973). On isolation, both colony types proved to be stable respiratory-sufficient individual yeast strains of the species *S. cerevisiae*, and they were coded LAB A/69 and LAB B/69, with the former strain being ~75% of the ale culture and the latter comprising ~25% of the culture. A production-scale fermentation trial with ale strain LAB A/69 was conducted in a 200 hL (20,000 L) open wood fermenter with a 12°Plato wort at 21 °C. Although the fermentation was

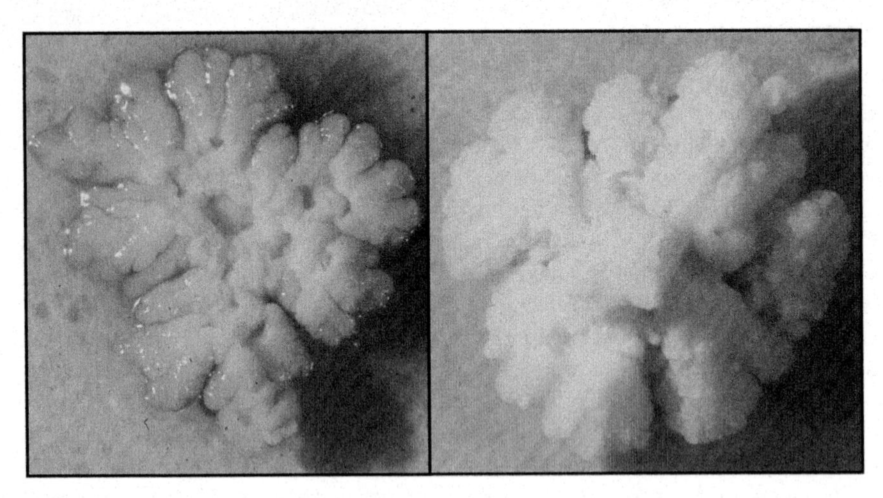

Fig. 13.14 Giant colony morphologies of ale yeast strains that do not exhibit the co-flocculation phenotype

underpitched (under-inoculated), it proceeded rapidly, and all the wort's fermentable sugars were metabolized in less than 96 h. It was then that problems began! A top crop failed to develop on the fermentation's surface (which occurred in the original two-strain ale culture), and most of the yeast culture remained in suspension! As the brewery in question (at that time) did not possess a centrifuge, it was not possible to collect the yeast for reuse in a subsequent fermentation, and the fermented wort (all 200 hL of it) had to be discarded into the sewer and the cost charged to the brewery's effluent budget!

Following the above traumatic brewing-scale trial, laboratory-scale characterization of both ale yeast strains isolated from the Labatt ale cultures was conducted. When the two isolated strains were cultured alone in wort using 2 L glass cylinders, both were non-flocculent during all phases of growth. However, when cultured together in wort in a 1:1 radio, the culture was flocculent in the later stages of fermentation and sedimented out of suspension (Nishihara et al. 2000). A top crop also formed (Fig. 13.6). This type of behaviour, where two yeast strains are non-flocculent alone but flocculent when mixed together (Stewart 1972; Stewart and Garrison 1972), has been termed *co-flocculation*. It has already been discussed that co-flocculation has also been termed *mutual aggregation* and *mutual flocculation* (Eddy 1958). When stationary phase cells of the two strains were mixed together in a 1:1 ratio in the presence of calcium ions at pH 4.5 (Helm et al. 1953) flocs immediately began to appear and a very flocculent culture resulted that sedimented out of suspension (Fig. 13.17).

Protein denaturants (e.g. urea and guanidine) and several sugars (mannose, glucose, fructose and galactose) cause reversible inhibition of co-flocculation in the presence of Ca^{++} ions. Also, the effect of cell treatment with proteolytic enzymes (trypsin and chymotrypsin) and chemical modification of the cell surface protein and carbohydrate components suggests that co-flocculation results from an

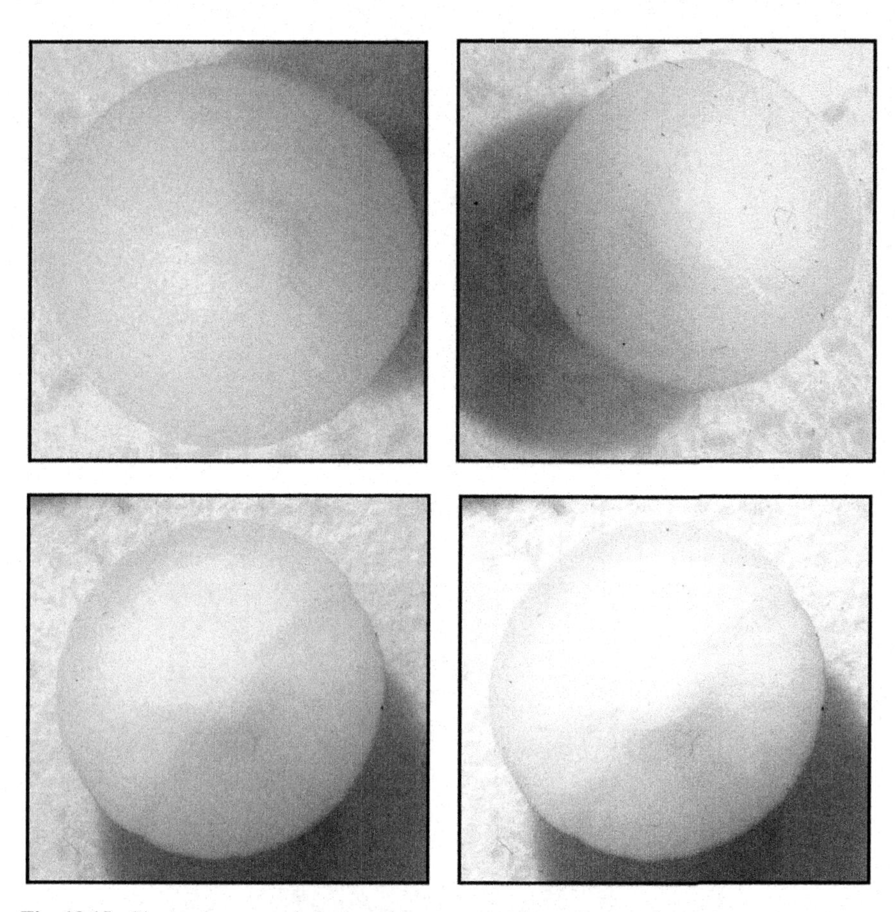

Fig. 13.15 Giant colony morphologies of four separate lager yeast strains

interaction between surface protein cellular components of one of the strains and surface carbohydrate components of the other strains (Nishihara et al. 2000).

To date, co-flocculation has only been observed with ale strains (Stewart et al. 1974; Stewart and Russell 2009). There are no reports (including extensive studies in the Labatt research laboratories) of co-flocculation between non-flocculent lager yeast strains. Another type of co-flocculation reaction that has been described is when an ale yeast strain has the ability to aggregate and co-sediment with contaminating bacteria such as *Hafnia protea* (White and Kidney 1979), *Lactobacillus brevis* (Peng et al. 2001), *Pediococcus* sp. and *Lactobacillus* sp. (Zarattini et al. 1993) (Fig. 13.9). The *Lactobacillus* sp. strain was isolated from a fuel alcohol molasses fermentation in Brazil and its co-flocculation characteristics studied in Canada.

The two-strain composition of the Labatt co-flocculent production ale culture was deemed to be undesirable, particularly because of its tendency for premature

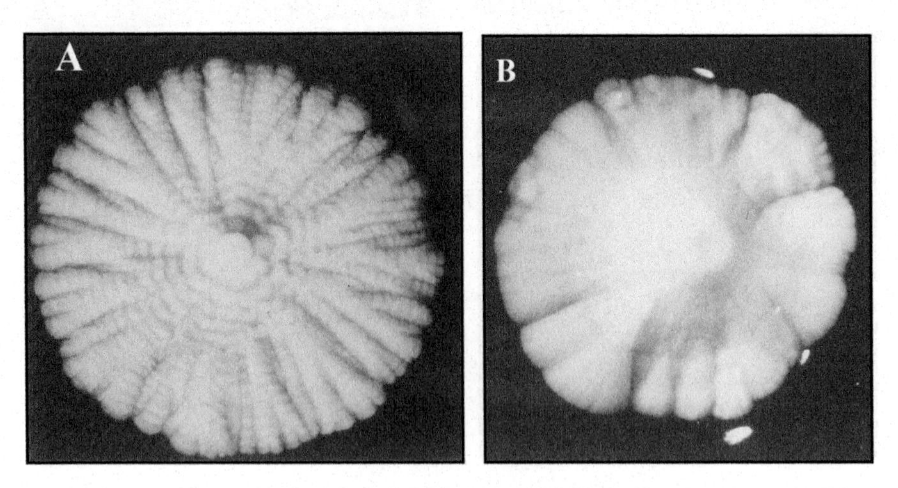

Fig. 13.16 Giant colony morphologies of co-flocculent ale yeast strains

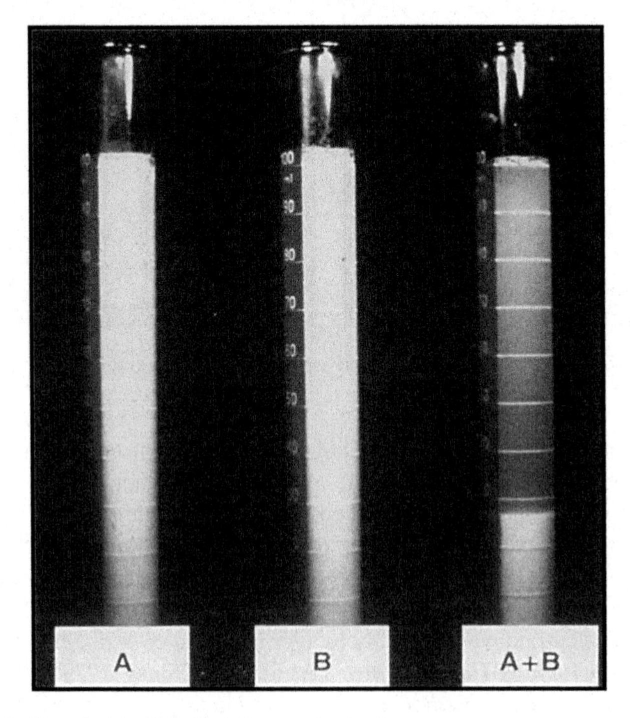

Fig. 13.17 Co-flocculation: Helm's sedimentation in vitro test

flocculation and the consequent wort under-attenuation that occurred. This resulted in the failure to often comply with the beer's alcohol specification (this was prior to the introduction of high-gravity brewing procedures on a production basis, which exacerbated the fermentation problems—details in Chap. 11). Production trials with LAB A/69 ale strain were conducted in both regular (sales) (12°Plato) and higher-gravity (16°Plato) worts. This strain proved to be capable of successfully fermenting both wort gravities but, because of its non-flocculent property, centrifugation was required in order to harvest the culture for yeast removal, beer clarification and yeast collection for reuse. This strain has been employed for ale production by Labatt with high-gravity worts for the past 40 years and was one of the reasons for centrifuge introduction into Labatt plants (further details on the use of centrifuges and their effect on brewer's yeast strains later in this chapter—Sect. 13.13).

Mortier and Soares (2007) have studied co-flocculation from a different perspective. They view co-flocculation as a process to separate non-flocculent yeast cells from a fermentation medium. The possibility of this process being employed for cell separation of different yeast species has been assessed. The fission yeast *Schizosaccharomyces pombe* was used as a control, since these cells are unable to be aggregated by flocculent cells of *S. cerevisiae*, due to lack of compatible receptors with *S. cerevisiae* flocculation lectins (Soares and Mota 1996). However, the yeast *Kluyveromyces marxianus* (details later—Chap. 17) can exhibit co-flocculation and consequently separate from suspension of this culture by settling using flocculent cells of *S. cerevisiae* to enhance it. The various degrees of co-flocculation amongst non-flocculent strains were the consequence of the different composition and structure of the yeast cell wall, particularly subtle variations in its detailed mannan architecture.

Recent studies by Rossouw et al. (2015) have examined the relevance of different Flo proteins in mediating differential interspecies aggregation in a "natural" yeast ecosystem in the wine industry. This paper reports on co-flocculation behaviour in mixed cultures of *S. cerevisiae* and non-*Saccharomyces* yeast cultures. It is suggested that adhesion phenotypes and, in particular, Flo proteins may play roles in system dynamics—beyond the roles assigned to these protein properties previously. It is also suggested that the evolution of these proteins may be driven in response to specific interspecies association within the microbial system. Further elucidation of the molecular mechanisms interpinning different co-flocculation yeast behaviour will be of fundamental importance in order to understand the role of direct cell-cell interactions between different species in a shared environment.

Cell adhesion phenotypes are complex and are influenced, not only by genetic factors which determine the composition and properties of the cell wall, by a number of environmental parameters which impart adhesion behaviour. The observations, under controlled laboratory conditions, therefore presents a simplified view of interspecies co-flocculation has already been discussed in this chapter. The data provides, for the first time, genetic insights into the phenomenon of co-flocculation in *S. cerevisiae*.

Future studies should include additional yeast species and strains under various experimental conditions in order to comprehensively investigate the yeast co-flocculation concept. In addition, the relevance of yeast species adhesion phenotypes to microbial interactions in natural ecosystems needs further detailed investigation (Powell et al. 2003; Powell and Diacetis 2007; Fink and Cookson 2005).

13.12 Adhesion and Biofilm Formation

Yeast cells (including brewing and distilling strains) possess a capacity to adhere to abiotic (nonliving) surfaces, cells and tissues (Verstrepen and Klis 2006). The cell wall serves as a means for yeast cultures (and other microbes) to interact with their environment. It has already been described (Sect. 13.1) that one of the most critical functions of the cell surface is its ability to adhere to other cells (floc formation) and surfaces. Adhesion prevents cells from being washed away when they are present in a nourishing environment and allows them to form biofilms that protect the cells from hazardous and stressful conditions. In addition to industrial environments, pathogenic yeasts exploit their capacity to adhere to abiotic surfaces such as plastic prosthetics in order to gain access to the bloodstream and the internal organs of patients (Ashbee and Bignell 2010).

Yeast cell adhesion is of considerable economic importance for food and beverage processing companies because adherent yeasts (and other fungi) can form highly resistant biofilms. As well as cell-cell adhesion (Sect. 13.1) (flocculation) of brewing (and wine) yeast cultures, biofilm formation is often exploited as a convenient and cost-effective way to separate biomass from fermentation products (e.g. fermented wort and must) (Verstrepen et al. 2003). Fermentation of biofilms appears to be an adaptive mechanism because it usually ensures access to oxygen and permits continued growth on substrates such as nonfermentable ethanol. Biofilm adhering cells have been shown to have elevated and/or modified lipid content and increased surface hydrophobicity (Verstrepen and Klis 2006).

13.13 Centrifuges to Crop Yeast

The incentive to optimize brewing operating costs, while reducing processing times, is imperative for commercial survival of the process. Breweries (and many other manufacturing industries) continuously search for ways to exploit production efficiency (Stewart 2006) and lean technology concepts (Stewart 2014a)— further details in Chap. 11. As a consequence, the disc stack centrifuge has become a popular component of yeast process management systems in order to reduce fermentation, maturation and clarification times and also to control effluent treatment costs (Chlup et al. 2008) (Fig. 13.18).

The use of centrifuges in breweries was initially viewed with misgivings by many brewers (Siebert et al. 1987; Lange et al. 2001). However, the advantages and disadvantages are now much clearer (Chlup and Stewart 2011). The principal advantages are:

Fig. 13.18 A disc stack centrifuge (5 hL hour)

- Rapid and efficient clarification before further filtration stress.
- Most consistent clarification of beer.
- Equipment can be conveniently sterilized.
- Filter aids are not required.
- Space requirements are relatively small.
- Most centrifuges are self-cleaning and operate continuously.
- Lower beer losses occur with minimal oxygen pick-up.

There are also disadvantages:

- High maintenance costs.
- Beer/yeast temperature may increase—details later.
- Mechanical break-up of large particles into finer particles resulting in filtration and beer physical problems.
- Increased mechanical stress and/or increased temperature may influence yeast quality—details later.
- There may be oxygen pick-up with an adverse effect on beer stability (particularly flavour stability). Also, high operating noise levels can negatively affect labour relations.
- Removal of precipitated proteins (cold break) with some yeast at the end of maturation which will adversely affect the quality of the pitching yeast.
- Removal of solid effluent materials from both brewing and packaging operations prior to in-house treatment or by municipal systems.

Modern centrifuges produce gravitational forces in excess of 10,000 times atmospheric, achieving solid separation in seconds with reduced equipment volume. Centrifuges have a number of applications. They can be used with brewing yeast cultures for:

- Cropping of non-flocculent yeast cultures at the end of primary fermentation.
- Reducing the yeast quantity from green beer before the start of secondary fermentation/maturation.
- Beer recovery from cropped yeast (Chlup and Stewart 2011).
- Removal of cold break (precipitated protein, etc.) and yeast at the end of maturation.
- Separation of the hot break after wort boiling.

Yeast management and handling systems are influential in determining the physiological status of yeast, subsequent fermentation performance activity (Chlup et al. 2007a, b, c) and clarification during lagering/maturation (Siebert et al. 1987). It has been documented that beer haze can result from the yeast cell wall releasing mannan as it is processed during agitation in storage and centrifugation (Lewis and Poerwantaro 1991). Hydrodynamic stresses have the potential to inflict damage to the yeast cell wall during beer production and lead to unfilterable beer haze formation. Intermediate haze formation from yeast [there are also other sources of beer haze—a discussion of which is beyond the scope of this chapter (Leiper and Miedl 2009)].

The uses of centrifuges during brewing are diverse (discussed above). This section concentrates on the effect that centrifugation has on brewer's yeast and, particularly, regarding yeast cell wall damage and the resulting impact on beer quality and stability together with yeast viability and vitality. Studies were conducted at Heriot-Watt University during the first decade of this century and focused on a better understanding of the effects of passing brewing yeast cultures through a disc stack centrifuge (Chlup et al. 2007c). In order to confirm that the

Fig. 13.19 The disc stack

effects on yeast was from centrifugation, an extensive number of centrifugation cycles operating at two different G-forces (high and low), was employed.

The passage of yeast through a disc stack centrifuge exposes cells to mechanical and hydrodynamic shear stresses (Siebert et al. 1987). These stresses can result in a decrease in cell viability and vitality, reduced flocculation, cell wall damage, increased yeast extracellular proteinase A (PrA) levels, hazier beers and poorer beer foam stability (Lewis and Poerwantaro 1991; Stoupis et al. 2002). In a more recent study, biological indicators of yeast physiology such as viable and damaged cells, intracellular pH, glycogen and trehalose levels as well as beer physical stability parameters including mannan residues, particle size and beer haze have been employed to quantify the damage which occurs to yeast cells as a function of the number of cycles through a centrifuge operating at high and low G-forces (Chlup et al. 2008; Chlup and Stewart 2011).

A disc stack centrifuge (Fig. 13.18) is typically made up of 50 to 150 discs (Fig. 13.19); the number of discs is determined by the process requirements. The discs are truncated cones and flanged at the inner and outer diameters (Fig. 13.20). The close proximity of the discs reduces the sedimentation distance for yeast cells. Due to centrifugal forces within the centrifuge, numerous reactionary forces take place between liquid, solid and discs including shear stress of the yeast cell surface which may result in beer instability (Fig. 13.21).

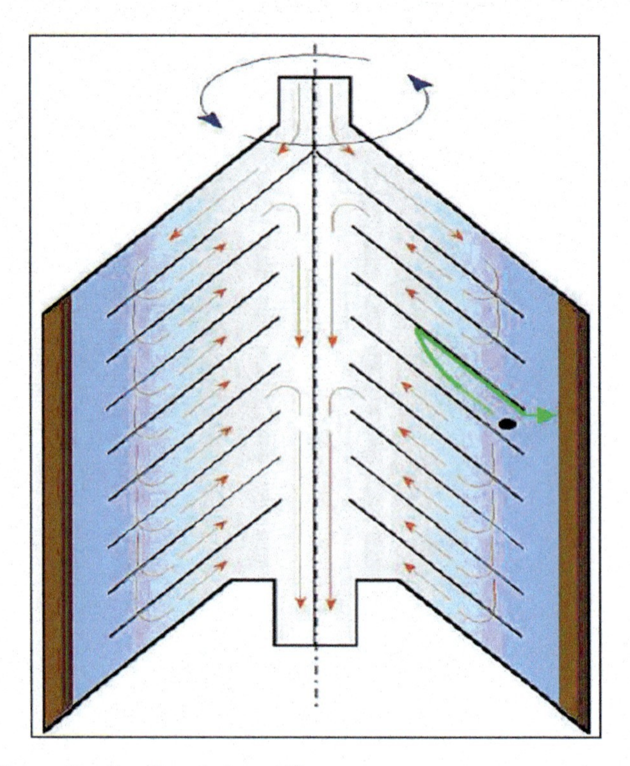

Fig. 13.20 Flow pattern in a disc stack centrifuge

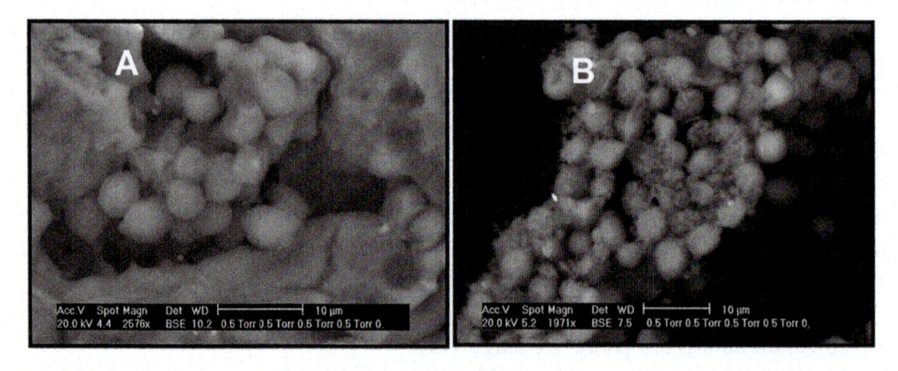

Fig. 13.21 Environmental scanning electron microscopy (ESEM) of ale yeast strain surface damage. (**a**) Cells before passage through a disc centrifuge. (**b**) Cells following passage through a disc centrifuge operating at high G-force

It has already been discussed (Sect. 13.11) that in the late 1970s and early 1980s, Labatt in Canada (Stewart 2015a) (along with a number of other brewing companies) began installing centrifuges in their breweries. This was to harvest ale and lager yeasts at the end of fermentation and, for environmental reasons, defer sewer surcharges. All of a sudden in 1988, one of the Labatt breweries with a centrifuge (operating at 300 hL/h) reported that its ale fermentations were exhibiting slower and incomplete fermentations with a 16°Plato OG wort. Closer study of the fermented wort revealed reduced wort maltose and particularly maltotriose uptake rates (detailed discussion regarding the uptake of both these wort sugars is in Chap. 7), with residual sugars when fermentation ceased. Consequently, the fermented wort's alcohol specification was not achieved. In addition, diacetyl (butterscotch-like flavour) and other vicinal diketone (VDK) levels (Krogerus and Gibson 2013) were elevated at the end of fermentation because of difficulties with VDK reabsorption by yeast at the end of primary fermentation and during maturation (details in Chap. 15). Also, there was yeast autolysis, resulting in reduced foam stability due to excreted intracellular proteinase (details in Chap. 11), elevated unfilterable haze consisting mainly of mannoproteins from disrupted cell walls and autolysed yeast off-flavours (Chlup et al. 2008).

The centrifuged ale yeast exhibited decreasing cell viability (determined with methylene blue staining) during repeated cycles. Also, the same cultures had a higher percentage of respiratory-deficient (RD) petite mutants, determined with the triphenyl tetrazolium chloride overlay method (Ogur and St. John 1956; Ogur et al. 1957) (details in Chap. 14). This increasing RD level and decreasing viability were due to centrifugation when the exit temperature was 30 °C. When the bowl of the centrifuge was cooled and the exit temperature reduced to 20 °C, the cell viability increased, the RD level decreased, the wort fermentation characteristics returned to normal and the beer was drinkable again (Stewart 2010a).

Studies at Heriot-Watt University with a 5hL/hr centrifuge (Fig. 13.18) have quantified the damage that occurs to brewer's yeast strains as a function of cycles through the centrifuge operating at low and high G-forces (Chlup et al. 2007c). Biological indicators of yeast physiology as a result of centrifugation includes non-viable and damaged cells, intracellular pH, glycogen and trehalose levels as well as beer physical stability parameters such as mannan residues (Chlup et al. 2007b) and particle size distribution. Many of these measurements were conducted with a flow cytometer (see Table 13.4) (Chlup et al. 2007c). Flow cytometry possesses technology that performs simultaneous multiparametric analysis of yeast's physical and chemical characterization based on cell size, relative granularity and fluorescence. The particles are transported in a fluid stream to the interrogation point. The extent to which the particle scatters light is dependent on the relative fluorescence. Flow cytometry methods have been developed to measure yeast cell viability, damaged cells, intracellular pH (pHi), extracellular PrA levels, mannose residues and intracellular glycogen and trehalose. When yeast is passed through disc stack centrifuge they exhibit lower cell viabilities when compared with cells that had not been centrifuged. In addition, the effects of G-force upon yeast cells as a function of centrifugation cycles have been studied (Chlup et al. 2008). In order to establish a relationship between a centrifuge G-force and effects on the physiological state of yeast, it was necessary for cells to possess an analogous history and similar physiological states. The yeast strain, *S. cerevisiae* (an ale

Table 13.4 Characteristics of yeast cells recycled nine times in 20°P wort and beer before and after centrifugation at high G-force

Characteristic	Before centrifugation	After centrifugation
Viability (%)	85	42
Extracellular pH	4.2	6.0
Intracellular pH	5.8	5.3
Damaged cells (%)	4	15
Glycogen (ppm)	18	8
Trehalose (ppm)	22	6
Mannan released (counts)	400	1000
Proteinase A (U/mL)	3.1	6.2
Hydrophobic polypeptides (mg/L)	48	25
Beer foam stability (NIBEM)	110	82

strain), employed in this study was subjected to similar fermentation conditions. The yeast culture was exposed to a high G-force (12,000 rpm ≡ 20,000G), and the centrifuged cells were subsequently conditioned for an additional 48 h.

Centrifugation of cultures after nine times recycling through 20°Plato worts demonstrated that the yeast had been damaged in a number of ways. These included a reduction in cell viability, glycogen and trehalose levels. Also, the hydrodynamic shear generated by the centrifuge resulted in cell surface components being released (mannoprotein) (Fig. 13.21), together with a PrA increase resulting in a reduction in wort hydrophobic polypeptides resulting in poor beer foam stability (Cooper et al. 2000) (Table 13.4).

Although centrifugation can exhibit negative effects, the positive effects of controlled centrifugation on beer production and effluent control cannot be overstated. However, yeast is subjected to numerous factors that individually and collectively impose stresses on yeast cells (details in Chap. 11). The effects of environmental conditions and beer production equipment may have been underestimated (even ignored!). A more complete understanding of yeast's biological response to interactions with cell physiology and brewing equipment is an important criterion in maintaining process efficiency and beer quality. It is worthy of note that the advent of flow cytometry (Hutter et al. 2005; Miedl et al. 2005; Chlup et al. 2007b, c) and confocal imaging (Schlee et al. 2006) has introduced analytical methods to yeast research that have broadened the scope of the researcher.

13.14 Summary

Brewers employ a number of methods to crop their yeast cultures which varies depending on whether one is dealing with traditional top-cropping, accepted lager bottom-cropping, and cylindroconical fermentation systems, a non-flocculent culture where the yeast, still in suspension, is cropped with a centrifuge or a portion of the yeast, still in suspension, is blended into the fresh wort of a subsequent fermentation. Yeast quality is influenced by the manner that the yeast is cropped and centrifuges play an important part in this regard. The cell wall structure is a critical parameter involved in yeast flocculation. The wall is composed of an inner layer consisting predominantly of β-glucan and chitin with a fibrillar outer layer containing primarily α-mannan associated with mannoproteins. Yeast management and handling systems, including culture harvesting, are influential in determining the yeast physiological status.

References

Abramova N, Sertil O, Mehta S, Lowry CV (2001) Reciprocal regulation of anaerobic and aerobic cell wall mannoprotein gene expression in *Saccharomyces cerevisiae*. J Bacteriol 183:2881–2887

Alwine JC, Kemp DJ, Stark GR (1977) Methods for detection of specific RNAs in agarose gels by transfer to diazobenzyloxymethyl-paper and hybridization with DNA probes. Proc Natl Acad Sci U S A 74:5350–5354

Amory DE, Genet MJ, Rouxhet PG (1988a) Application of XPS to the surface analysis of yeast cells. Surf Interface Anal 11:478–486

Amory DE, Rouxhet PG, Dufour JP (1988b) Flocculence of brewery yeasts and their surface properties: chemical composition, electrostatic charge and hydrophobicity. J Inst Brew 94:79–84

Armstrong K, Bendiak D (2007) PYF malt: practical brewery observations of fermentability. MBAA Tech Q 44:40–46

Ashbee R, Bignell EM (eds) (2010) Pathogenic yeasts. The yeast handbook. Springer, Berlin

Axcell BC, van Nierop S, Vundla W (2000) Malt induced premature yeast flocculation. MBAA Tech Q 37:501–504

Baker DA, Kirsop BH (1972) Flocculation in *Saccharomyces cerevisiae* as influenced by wort composition and by actidione. J Inst Brew 78:454–458

Bamforth C (2009) Beer: tap into the art and science of brewing. Oxford University Press, Oxford

Bamforth C (2017) Freshness: practical guides to beer quality. American Society of Brewing Chemists, Minneapolis, MN

Bayly JC, Douglas LM, Pretorius IS, Bauer FF, Dranginis AM (2005) Characteristics of Flo11-dependent flocculation in *Saccharomyces cerevisiae*. FEMS Microbiol Lett 5:1151–1156

Beavan MJ, Belki D, Stewart GG, Rose AH (1979) Changes in electrophoretic mobility and lytic enzyme activity associated with developments of flocculating ability in *Saccharomyces cerevisiae*. Can J Microbiol 25:888–895

Bester MC, Pretorius IS, Bauer FF (2006) The regulation of *Saccharomyces cerevisiae FLO* gene expression and Ca^{2+}-dependent flocculation by *Flo8p* and *Mss11p*. Curr Genet 49:375–383

Bhattacharyya MK, Lustig AJ (2006) Telomere dynamics in genome stability. Trends Biochem Sci 31:114–122

Bing J, Han PJ, Liu WQ, Wang QM, Bai FY (2014) Evidence for a Far East Asian origin of lager beer yeast. Curr Biol 24:R380–R381

Bird A (2002) DNA methylation patterns and epigenetic memory. Genes Dev 16:6–21

Botstein D, Fink GR (2011) Yeast: an experimental organism for 21st century biology. Genetics 189:695–704

Boulton C (2011) Yeast handling. Brew Dist Int 7:7–10

Boulton C (2012) Advances in analytical methodology in brewing. J Inst Brew 118:255–263; MBAA Tech Q 50:53–61

Boulton C, Quain E (eds) (2001) Brewing yeast. In: Brewing yeast and fermentation. Blackwell Science, Oxford

Briggs DE, McGuinness G (1992) Microbes on barley grains. J Inst Brew 98:249–255

Calleja GB (1984) Microbial aggregation. CRC Press, Boca Raton, FL

Calleja GB (1987) Cell aggregation. In: Rose AH (ed) The yeasts, vol 2. Academic, London, pp 165–237

Chen EH, Grote E, Mohler W, Vignery A (2007) Cell–cell fusion. FEBS Lett 581:2181–2193

Chlup PH, Stewart GG (2011) Centrifuges in brewing. MBAA Tech Q 48:48–50

Chlup PH, Bernard D, Stewart GG (2007a) The disc stack centrifuge and its impact on yeast and beer quality. J Am Soc Brew Chem 65:29–37

Chlup PH, Conery J, Stewart GG (2007b) Detection of mannan from *Saccharomyces* cerevisiae by flow cytometry. J Am Soc Brew Chem 65:151–155

Chlup PH, Wang T, Lee EG, Stewart GG (2007c) Assessment of the physiological status of yeast during high-and low-gravity wort fermentations determined by flow cytometry. MBAA Tech Q 44:286–295

Chlup PH, Bernard D, Stewart GG (2008) Disc stack centrifuge operating parameters and their impact on yeast physiology. J Inst Brew 114:45–61

Claro FB, Rijsbrack K, Soares EV (2007) Flocculation onset in *Saccharomyces cerevisiae*: effect of ethanol, heat and osmotic stress. J Appl Microbiol 102:693–700

Cooper DJ, Stewart GG, Bryce JH (2000) Yeast proteolytic activity during high and low gravity wort fermentations and its effect on head retention. J Inst Brew 106:197–201

Cormack B (2004) Can you adhere me now? Good. Cell 116:353–354

D'Amore T, Panchal CJ, Stewart GG (1988) Intracellular ethanol accumulation in *Saccharomyces cerevisiae* during fermentation. J Appl Environ Microbiol 54:1471–1510

Damas-Buenrostro LC, Gracia-González G, Hernández-Luna CE, Galán-Wong LJ, Pereyra-Alférez B, Sierra-Benavides JA (2008) Detection of FLO genes in lager and wild yeast strains. J Am Soc Brew Chem 66:184–187

Day AW, Poon NH, Stewart GG (1975) Fungal fimbriae. III. The effect of flocculation in *Saccharomyces*. Can J Microbiol 21:558–564

Dengis PB, Nélissen LR, Rouxhet PG (1995) Mechanisms of yeast flocculation: comparison of top- and bottom-fermenting strains. Appl Environ Microbiol 61:718–728

Dietvorst J, Brandt A (2008) Flocculation in *Saccharomyces cerevisiae* is repressed by the COMPASS methylation complex during high-gravity fermentation. Yeast 25:891–901

Dudbridge M (2011) Handbook of lean manufacturing in the food industry. Wiley, New York, NY

Dunn B, Sherlock G (2008) Reconstruction of the genome origins and evolution of the hybrid lager yeast *Saccharomyces pastorianus*. Genome Res 18:1610–1623

Eddy AA (1958) Composite nature of the flocculation process of top and bottom strains of *Saccharomyces*. J Inst Brew 64:143–151

Eddy AA, Rudin AD (1958) Part of the yeast surface apparently involved in flocculation. J Inst Brew 64:19–21

Fidalgo M, Barrales RR, Jimenez J (2008) Coding repeat instability in the FLO11 gene of *Saccharomyces* yeasts. Yeast 25:879–889

Fink SL, Cookson BT (2005) Apoptosis, pyroptosis, and necrosis: mechanistic description of dead and dying eukaryotic cells. Infect Immun 73:1907–1916

Gibson BR (2011) 125th anniversary review: improvement of higher gravity brewery fermentation via wort enrichment and supplementation. J Inst Brew 117:268–284

Gilliland RB (1951) The flocculation characteristics of brewing yeasts during fermentation. Proceedings of the European Brewery Convention Congress, Brighton, pp 35–58

Gilliland RB (1959) Determination of yeast viability. J Inst Brew 65:424

Gimeno CJ, Ljungdahl PO, Styles CA, Fink GR (1992) Unipolar cell divisions in the yeast *S. cerevisiae* lead to filamentous growth: regulation by starvation and RAS. Cell 68:1077–1090

Goffeau A, Barrell BG, Bussey H, Oliver SG (1996) Life with 6000 genes. Science 274(5287):546, 563–546, 567

Goldstein IJ, Poretz RD (1986) Isolation, physicochemical characterization and carbohydrate-binding specificity of lectins. In: Liener IE, Sharon N, Goldstein IJ (eds) The lectins. Academic, Orlando, FL, p 52

Gouveia C, Soares EV (2004) Pb^{2+} inhibits competitively flocculation of *Saccharomyces cerevisiae*. J Inst Brew 110:141–145

Govender P, Domingo JL, Bester MC, Pretorius IS, Bauer FF (2008) Controlled expression of the dominant flocculation genes *FLO1, FLO5,* and *FLO11* in *Saccharomyces cerevisiae*. Appl Environ Microbiol 74:6041–6052

Guinard J-X, Lewis MJ (1993) Study of the phenomenon by agglomeration in the yeast *Saccharomyces cerevisiae*. J Inst Brew 99:487–503

Guo B, Styles CA, Feng Q, Fink G (2000) A *Saccharomyces* gene family involved in invasive growth, cell–cell adhesion, and mating. Proc Natl Acad Sci U S A 97:12158–12163

Halina S, Nathan L (2007) Lectins. Springer, Netherlands

Halme A, Bumgarner S, Styles C, Fink GR (2004) Genetic and epigenetic regulation of the *FLO* gene family generates cell–surface variation in yeast. Cell 116:405–415

Heine F, Stahl F, Sträuber H, Wiacek C, Benndorf D, Repenning C, Schmidt F, Scheper T, von Bergen M, Harms H, Müller S (2009) Prediction of flocculation ability of brewing yeast inoculates by flow cytometry, proteome analysis and mRNA profiling. Cytometry A 75:140–147

Helm E, Nohr B, Thome RSW (1953) Measurement of yeast flocculation and its significance in brewing. Wallerstein Laboratory Communications 16:315–326

Hough JS (1959) Flocculation characteristics of strains present in some typical British pitching yeasts. J Inst Brew 65:479–482

Hoyer LL (2001) The *ALS* gene family of *Candida albicans*. Trends Microbiol 9:176–180

Hoyer LL, Green CB, Oh SH, Zhao X (2008) Discovering the secrets of the *Candida albicans* agglutinin-like sequence (ALS) gene family—a sticky pursuit. Med Mycol 46:1–15

Hutter K-J, Miedl M, Kushmann B, Nitzsche F, Bryce JH, Stewart GG (2005) Detection of proteinases in *Saccharomyces cerevisiae* by flow cytometry. J Inst Brew 111:26–32

Inagaki H, Yamazumi K, Uehara H, Mochzuki K (1994) Determination of fermentation behaviour-malt evaluation system based on the original small scale fermentation test. Eur Brew Conv 23:111–136

Ishimaru S, Kudo S, Hattan M, Yoshida T, Kataoka J (1967) Selection of small vessels for fermentation tests in the laboratory. Rep Res Lab Kirin Brew Co 10:61–65

Jarvis P, Jefferson B, Parsons SA (2005) Measuring flocstructural characteristics. Environ Sci Biotechnol 4:1–18

Jibiki M, Ishibiki T, Yuuki T, Kagami N (2001) Application of polymerase chain reaction to determine the flocculation properties of brewer's lager yeast. J Am Soc Brew 59:107–110

Jibiki M, Sasaki K, Kagami N, Kawatsura K (2006) Application of a newly developed method for estimating the premature yeast flocculation potential of malt samples. J Am Soc Brew Chem 64:79–85

Jin Y, Speers RA (2000) Effect of environmental conditions on the flocculation of *Saccharomyces cerevisiae*. J Am Soc Brew Chem 58:108–116

Jin Y, Ritcey LL, Speers RA (2001) Effect of cell surface hydrophobicity, charge, and zymolectin density on the flocculation of *Saccharomyces cerevisiae*. J Am Soc Brew Chem 59:1–9

Kaur R, Domergue R, Zupancic M, Cormack BP (2005) A yeast by any other name: *Candida glabrata* and its interaction with the host. Curr Opin Microbiol 8:378–384

Kida K, Yamadaki M, Asno S, Nakata T, Sonoda Y (1989) The effect of aeration on stability of continuous ethanol fermentation by a flocculating yeast. J Ferment Bioeng 68:107–111

Kihn JC, Masy CL, Mestdagh MM (1988a) Yeast flocculation: competition between nonspecific repulsion and specific bonding in cell adhesion. Can J Microbiol 34:773–778

Kihn JC, Masy CL, Mestdagh MM, Rouxhet PG (1988b) Yeast flocculation: factors affecting the measurement of flocculence. Can J Microbiol 34:779–781

Klis FM, Boorsma A, De Groot PWJ (2006) Cell wall construction in *Saccharomyces cerevisiae*. Yeast 23:185–202

Kobayashi O, Hayashi N, Kuroki R, Sone H (1998) Region of Flo1 proteins responsible for sugar recognition. J Bacteriol 180:6503–6510

Kock JLF, Venter P, Smith DP, Van Wyk PWJ, Botes PJ, Coetzee DJ, Pohl CH, Botha A, Ridel K-H, Nigam S (2000) A novel oxylipin-associated 'ghosting' phenomenon in yeast flocculation. Antonie Van Leeuwenhoek 77:401–406

Koizumi H, Ogawa T (2005) Rapid and sensitive method to measure premature yeast flocculation activity in malt. J Am Soc Brew Chem 63:147–150

Krogerus A, Gibson BR (2013) 125th anniversary review: diacetyl and its control during brewery fermentation. J Inst Brew 119:86–97

Kruger L, Ryder DS, Alcock C, Murray JP (1982) Malt quality: prediction of malt fermentability. Part I. Tech Q Master Brew Assoc Am 19:45–51

Kukuruzinska MA, Bergh MLE, Jackson BJ (1987) Protein glycosylation in yeast. Annu Rev Biochem 56:915–944

Kuřec M, Baszczyňski M, Lehnert R, Brányik T (2009) Flow cytometry for age assessment of a yeast population and its application in beer fermentations. J Inst Brew 115:253–258

Lake JC, Speers RA (2008) A discussion of malt-induced premature yeast flocculation. MBAA Tech Q 4:253–262

Lake JC, Speers RA, Porter AV, Gill TA (2008) Miniaturizing the fermentation assay: effect of fermentor size and fermentation kinetics on detection of premature yeast flocculation. J Am Soc Brew Chem 66:94–102

Lange C, Nett JH, Trumpower BL, Hunte C (2001) Specific roles of protein-phospholipid interactions in the yeast cytochrome bc1 complex structure. EMBO J 20:6591–6600

Leiper KA, Miedl M (2009) Colloidal stability of beer. In: Bamforth CW, Russell I, Stewart GG (eds) Beer: a quality perspective. Boston, MA, Elsevier, pp 111–161

Lewis MJ, Poerwantaro WM (1991) Release of haze material from the cell walls of agitated yeast. J Am Soc Brew Chem 49:43–46

Lewis CW, Johnston JR, Martin PA (1976) The genetics of yeast flocculation. J Inst Brew 82:158–160

Libkind D, Hittinger CT, Valério E, Gonçalves C, Dover J, Johnston M, Gonçalves P, Sampaio JP (2011) Microbe domestication and the identification of the wild genetic stock of lager-brewing yeast. Proc Natl Acad Sci U S A 108:14539–14544

Lin CH, MacGurn JA, Chu T, Stefan CJ, Emr SD (2008) Arrestin-related ubiquitin-ligase adaptors regulate endocytosis and protein turnover at the cell surface. Cell 135:714–725

Loney ER, Inglis PW, Sharp S, Pryde FE, Kent NA, Mellor J, Louis EJ (2009) Repressive and non-repressive chromatin at native telomeres in *Saccharomyces cerevisiae*. Epigenetics Chromatin 2:18

Lorenz RT, Parks LW (1991) Involvement of heme components in sterol metabolism of *Saccharomyces cerevisiae*. Lipids 26:598–603

Lyons TP, Hough JS (1970a) Flocculation of brewer's yeast. J Inst Brew 76:564–571

Lyons TP, Hough JS (1970b) The role of yeast cell walls in brewing. Brew Digest 45:52–60

Machado MD, Santos MSF, Gouveia C, Soares HMVM, Soares EV (2008) Removal of heavy metals using a brewer's yeast strain of *Saccharomyces cerevisiae:* the flocculation as a separation process. Bioresour Technol 99:2107–2115

Mamvura TA, Paterson AE, Fanucchi D (2017) The impact of pipe geometry variations on hygiene and success of orbital welding of brewing industry equipment. J Inst Brew 123:81–97

Masy CL, Henquinet A, Mestdagh MM (1992) Flocculation of *Saccharomyces cerevisiae*: inhibition by sugars. Can J Microbiol 38:1298–1306

Meaden PG (1996) DNA fingerprinting of brewer's yeast. Ferment:9267–9272

Miedl M, Stewart GG, Bryce JH, Kuchman B, Hutter K-H (2005) A novel procedure for the determining yeast pitching rates employing flow cytometry. In: Proceedings of the 29th European Brewery Convention Congress, Prague, CD Paper No 33

Miki BLA, Poon NH, James AP, Seligy VL (1982a) Possible mechanism for flocculation interactions governed by gene *FLO1* in *Saccharomyces cerevisiae*. J Bacteriol 150:878–889

Miki BLA, Poon NH, Seligy VL (1982b) Repression and induction of flocculation interactions in *Saccharomyces cerevisiae*. J Bacteriol 150:890–899

Miller T, Krogan NJ, Dover J (2001) COMPASS: A complex of proteins associated with a trithorax-related SET domain protein. Proc Natl Acad Sci U S A 98:12902–12907

Mortier A, Soares EV (2007) Separation of yeasts by addition of flocculent cells of *Saccharomyces cerevisiae*. World J Microbiol Biotechnol 23:1401–1407

Mundy RD, Cormack B (2009) Expression of *Candida glabrata* adhesions following exposure to chemical preservatives. J Infect Dis 199:1891–1898

Nakao Y, Kanamori T, Itoh T, Kodama Y, Rainieri S, Nakamura N, Shimonaga T, Hattori M, Ashikari T (2009) Genome sequence of the lager brewing yeast, an interspecies hybrid. DNA Res 16:115–129

Nathan L (1930) Improvements in the fermentation and maturation of beers. J Inst Brew 36:538–544

Nayar A, Walker G, Wardrob F, Adya A (2017) Flocculation in industrial strains of *Saccharomyces cerevisiae*: role of cell wall polysaccharides and lectin-like receptors. J Inst Brew 123:211–218

Nishihara H, Toraya T, Fukui S (1982) Flocculation of cell-walls of brewers-yeast and effects of metal-ions, protein-denaturants and enzyme treatments. Arch Microbiol 131:112–115

Nishihara H, Kio K, Imamura M (2000) Possible mechanism of co-flocculation between non-flocculent yeasts. J Inst Brew 106:7–10

Ogata T, Izumikawa M, Kohno K, Shibata K (2008) Chromosomal location of Lg-*FLO1* in bottom-fermenting yeast and the *FLO5* locus of industrial yeast. J Appl Microbiol 105:1186–1198

Ogur M, St. John R (1956) A differential and diagnostic plating method for population studies of respiration deficiency in yeast. J Bacteriol 72:500–504

Ogur M, St. John R, Nagai S (1957) Tetrazolium overlay technique for population studies of respiration deficiency in yeast. Science 125:928–929

Osumi M (2012) Visualization of yeast cells by electron microscopy. J Electron Microsc 61:343–365

Panchal CJ, Whitney GK, Stewart GG (1984a) Susceptibility of *Saccharomyces* spp. and *Schwanniomyces* spp. to the aminoglycoside antibiotic G418. Appl Environ Microbiol 47:1164–1166

Panchal CJ, Russell I, Sills AM, Stewart GG (1984b) Fermentation ethanol production – application of the new genetics to an ancient art. In: Proceedings of the 11th energy technology conference, Washington, DC, pp 1270–1273

Panchal CJ, Russell I, Sills AM, Stewart GG (1984c) Genetic manipulation of brewing and related yeast strains. Food Technol 111:99–106

Paula L, Birrer F (2006) Including public perspectives in industrial biotechnology and the biobased economy. J Agric Environ Ethics 19:253–267

Peng X, Sun J, Iserentant D, Michiels C, Verachtert H (2001) Flocculation and coflocculation of bacteria by yeasts. Appl Microbiol Biotechnol 55:777–781

Pomper S, Burkholder PR (1949) Studies on the biochemical genetics of yeast. Proc Natl Acad Sci U S A 35:456–464

Potter G, Budge SM, Speers RA (2015) Flocculation, cell surface hydrophobicity and 3-OH oxylipins in the SMA strain of *Saccharomyces pastorianus*. J Inst Brew 121:31–37

Powell CD, Diacetis AN (2007) Long term serial repitching and the genetic and phenotypic stability of brewer's yeast. J Inst Brew 113:67–74

Powell CD, Quain DE, Smart KA (2003) The impact of brewing yeast cell age on fermentation performance, attenuation and flocculation. FEMS Yeast Res 3:149–157

Raspor P, Russell I, Stewart GG (1990) An update of zinc ion as an effector of flocculation in brewer's yeast strains. J Inst Brew 96:303–305

Rees EMR, Stewart GG (1997a) The effects of divalent ions magnesium and calcium on yeast fermentation performance in conventional (12°P) and high (20°P) gravity worts in both static and shaking fermentations. In: Proceedings of the 26th congress – European Brewery Convention, Maestricht, The Netherlands, pp 461–468

Rees EMR, Stewart GG (1997b) The effects of increased magnesium and calcium concentrations on yeast fermentation performance in high gravity worts. J Inst Brew 103:287–291

Rees EMR, Stewart GG (1998) Strain specific response of brewer's yeast strains to zinc concentrations in conventional and high gravity worts. J Inst Brew 104:255–264

Rhymes MR, Smart KA (2001) Effect of storage conditions on the flocculation and cell wall characteristics of an ale brewing yeast strain. J Am Soc Brew Chem 59:32–38

Richards M (1967) The use of giant-colony morphology for the differentiation of brewing yeasts. J Inst Brew 73:162–166

Rose AH (1980) *Saccharomyces cerevisiae* as a model eukaryote. In: Stewart GG, Russell I (eds) Current developments in yeast research. Pergamon, Toronto, pp 645–652

Rose AH (1984) Physiology of cell aggregation: flocculation by *Saccharomyces cerevisiae* as a model system. In: Marshall CK (ed) Physiology of cell aggregation. Springer, New York, pp 323–335

Rose AH, Harrison JF (1987) The yeasts, vol 1–5. Academic, London

Rossouw D, Bagheri B, Setati ME, Bauer FF (2015) Co-flocculation of yeast species, a new mechanism to govern population dynamics in microbial ecosystems. PLoS One 10(8): e0136249

Russell I, Stewart GG (1980) Revised nomenclature of genes that control yeast flocculation. J Inst Brew 86:120–121

Russell I, Stewart GG (eds) (2014) Whisky: technology, production and marketing, 2nd edn. Academic (Elsevier), Boston, MA

Russell I, Dowhanick T, Raspor P, Stewart GG (1989) Yeast flocculation – the influence of divalent ions. In: Proceedings of the 22nd congress – European Brewery Convention, Zurich. IRL Press, Oxford, pp 529–536

Sato M, Watari J, Shinotsuka K (2001) Genetic instability in flocculation of bottom-fermenting yeast. J Am Soc Brew Chem 59:130–134

Schlee C, Miedl M, Leiper KA, Stewart GG (2006) The potential of confocal imaging for measuring physiological changes in brewer's yeast. J Inst Brew 112:134–147

Sherman F, Fink GR, Hicks JB (1986) Methods in yeast genetics. Cold Spring Harbor Laboratory Press, New York

Siebert KJ, Stenroos LE, Reid DS, Grabowski D (1987) Filtration difficulties resulting from damage to yeast during centrifugation. MBAA Tech Q 24:1–8

Siero C, Reboredo NM, Villa TH (1994) Flocculation of industrial and laboratory strains of *Saccharomyces cerevisiae*. J Ind Microbiol 14:461–466

Smart KA, Whisker S (1996) Effect of serial repitching on the fermentation properties and condition of brewing yeast. J Am Soc Brew Chem 54:41–44

Smit G, Straver MH, Lugtenberg BJJ, Kijne JW (1992) Flocculence of *Saccharomyces cerevisiae* cells is induced by nutrient limitation, with cell surface hydrophobicity as a major determinant. Appl Environ Microbiol 58:3709–3714

Soares EV (2010) Flocculation in *Saccharomyces cerevisiae*: a review. Appl Microbiol 110:1–18

Soares EV, Duarte AA (2002) Addition of nutrients induce a fast loss of flocculation in starved cells of *Saccharomyces cerevisiae*. Biotechnol Lett 24:1957–1960

Soares EV, Mota M (1996) Flocculation onset, growth phase, and genealogical age in *Saccharomyces cerevisiae*. Can J Microbiol 42:539–547

Soares EV, Seynaeve J (2000a) The use of succinic acid, as a pH buffer, expands the potentialities of utilisation of a chemically defined medium in *Saccharomyces cerevisiae* flocculation studies. Biotechnol Lett 22:859–863

Soares EV, Seynaeve J (2000b) Induction of flocculation of brewer's yeast strains of *Saccharomyces cerevisiae* by changing the calcium concentration and pH of culture medium. Biotechnol Lett 22:1827–1832

Soares E, Teixeira JA, Mota M (1991) Influence of aeration and glucose concentration in the flocculation of *Saccharomyces cerevisiae*. Biotechnol Lett 13:207–212

Soares EV, Teixeira JA, Mota M (1994) Effect of cultural and nutritional conditions on the control of flocculation expression in *Saccharomyces cerevisiae*. Can J Microbiol 40:851–857

Song Q, Kumar A (2012) An overview of autophagy and yeast pseudohyphal growth: integration of signaling pathways during nitrogen stress. Cells 1:263–283

Sousa MJ, Teixeira JA, Mota M (1992) Differences in the flocculation mechanism of *Kluyveromyces marxianus* and *Saccharomyces cerevisiae*. Biotechnol Lett 14:213–218

Southern EM (1975) Detection of specific sequences among DNA fragments separated by gel electrophoresis. J Mol Biol 98:503–517

Speers RA (2012) A review of yeast flocculation. In: Speers RA (ed) Proceedings of the 2nd international brewers symposium: yeast flocculation, vitality and viability. MBAA, St Paul, MN

Speers RA (2016) Brewing fundamentals, Part 3: Yeast settling and flocculation. MBAA Tech Q 53:17–22

Speers RA, Stokes S (2009) Effects of vessel geometry, fermenting volume and yeast repitching on fermenting beer. J Inst Brew 115:148–150

Speers RA, Durance TD, Odense P, Owen S, Tung MA (1993) Physical properties of commercial brewing yeast suspensions. J Inst Brew 99:159–164

Speers RA, Wan Y-Q, Jin Y, Stewart RJ (2006) Effects of fermentation parameters and cell wall properties on yeast flocculation. J Inst Brew 112:246–254

Sprague GF Jr, Thorner JW (1992) Pheromone response and signal transduction during the mating process of *Saccharomyces cerevisiae*. In: Jones EW, Pringle JR, Broach JR (eds) The molecular and cellular biology of the yeast saccharomyces: gene expression. Cold Spring Harbor Laboratory Press, Cold Spring Harbor, NY, pp 657–744

Stewart GG (1972) Co-flocculation of brewer's yeast. MBAA Tech Q 9:25

Stewart GG (1973) Recent developments in the characterization of brewery yeast strains. MBAA Tech Q 9:183–191

Stewart GG (1974) Some thoughts on the microbiological aspects of brewing and other industries utilizing yeast. Adv Appl Microbiol 17:233–262

Stewart GG (1988) Twenty-five years of yeast research. Dev Ind Microbiol 29:1–21. SIM Charles Thom Award Lecture

Stewart GG (1996) Yeast performance and management. The Brewer 82:211–215

Stewart GG (2006) Studies on the uptake and metabolism of wort sugars during brewing fermentations. MBAA Tech Q 43:265–269

Stewart GG (2009) The IBD horace brown medal lecture – forty years of brewing research. J Inst Brew 115:3–29

Stewart GG (2010a) MBAA award of merit lecture. A love affair with yeast. MBAA Tech Q 47:4–11

Stewart GG (2010b) The ASBC award of distinction lecture – high gravity brewing and distilling – past experiences and future prospects. J Am Soc Brew Chem 68:1–9

Stewart GG (2014a) Brewing intensification. American Society for Brewing Chemists, St. Paul, MN

Stewart GG (2014b) The concept of nature – nurture applied to brewer's yeast and wort fermentations. MBAA Tech Q 51:69–80

Stewart GG (2014c) Yeast mitochondria – their influence on brewer's yeast fermentation and medical research. MBAA Tech Q 51:3–11

Stewart GG (2014d) *Saccharomyces*. In: Catt C, Tortorello ML (eds) Encyclopedia of food microbiology, vol 3, 2nd edn. Elsevier, Oxford, pp 297–315
Stewart GG (2015a) Seduced by yeast. J Am Soc Brew Chem 73:1–21
Stewart GG (2015b) Yeast quality assessment, management and culture maintenance, Chap. 2. In: Hill AE (ed) Brewing microbiology: managing microbes, ensuring quality and valorising waste. Elsevier, Oxford, pp 11–29
Stewart GG, Garrison IF (1972) Some observations on co-flocculation in *Saccharomyces cerevisiae*. Am Soc Brew Chem Proc:118–131
Stewart GG, Goring TE (1976) Effect of some monovalent and divalent metal ions on the flocculation of brewer's yeast strains. J Inst Brew 82:341–342
Stewart GG, Murray JP (2011) Using brewing science to make good beer. MBAA Tech Q 48:13–19
Stewart GG, Murray JP (2012) Brewing intensification – successes and failures. MBAA Tech Q 49:111–120
Stewart GG, Russell I (1977) The identification, characterization, and mapping of a gene for flocculation in *Saccharomyces* sp. Can J Microbiol 23:441–447
Stewart GG, Russell I (1981) Yeast flocculation. In: Pollock JAR (ed) Brewing science, food science and technology. Academic, New York, pp 61–92
Stewart GG, Russell I (1983) Aspects of the biochemistry and genetics of sugar and carbohydrate uptake by yeasts. In: Spencer JFT, Spencer DM, Smith ARW (eds) Yeast genetics: fundamental and applied aspects. Springer, New York, pp 461–484
Stewart GG, Russell I (1986) The relevance of the flocculation properties of yeast in today's brewing industry. In: European Brewing Convention – Symposium on 'Brewers' yeast, Vuoranta, Helsinki, Finland, pp 24–25, 53–68
Stewart GG, Russell I (2009) An introduction to brewing science and technology, Series lll, Brewer's yeast, 2nd edn. The Institute of Brewing and Distilling, London
Stewart GG, Russell I, Garrison IF (1973) Further studies on flocculation and co-flocculation in brewer's yeast strains. Am Soc Brew Chem Proc 31:100–106
Stewart GG, Russell I, Garrison IF (1974) Factors influencing the flocculation of brewers' yeast strains. MBAA Tech Q II:xiii
Stewart GG, Russell I, Garrison IF (1975a) Some considerations of the flocculation characteristics of ale and lager yeast strains. J Inst Brew 81:248–257
Stewart GG, Russell I, Goring IF (1975b) Nature-nurture anomalies – further studies in yeast flocculation. Am Soc Brew Chem Proc 33:137–147
Stewart GG, Goring TE, Russell I (1983a) (issued October 11, 1983) Yeast strain for fermenting high plato value wort. US Patent 4,409,246
Stewart GG, Panchal CJ, Russell I (1983b) Current developments in the genetic manipulation of brewing yeast strains – a review. J Inst Brew 89:170–188
Stewart GG, Murray CR, Panchal CJ, Russell I, Sills AM (1984a) The selection and modification of brewer's yeast strains. Food Microbiol 1:289–302
Stewart GG, Panchal CJ, Russell I, Sills AM (1984b) Biology of ethanol producing microorganisms. CRC Crit Rev Biotechnol 1:161–188
Stewart GG, Russell I, Panchal CJ (1984c) Genetically stable allopolyploid somatic fusion product useful in the production of fuel alcohols. Australian Patent: 570,260 (issued August 15, 1984)
Stewart GG, Hill A, Russell I (2013) 125th anniversary review - developments in brewing and distilling yeast strains. J Inst Brew 119:202–220
Stoupis T, Stewart GG, Stafford RA (2002) Mechanical agitation and rheological considerations of ale yeast slurry. J Am Soc Brew Chem 60:58–62
Stratford M (1989) Yeast flocculation: calcium specificity. Yeast 5:487–496
Stratford M (1992a) Lectin-mediated aggregation of yeasts- yeast flocculation. Biotechnol Genet Eng Rev 10:283–341
Stratford M (1992b) Yeast flocculation – a new perspective. Adv Microb Physiol 33:1–71

Stratford M (1992c) Yeast flocculation – receptor definition by *mnn* mutants and concanavalin-A. Yeast 8:635–645
Stratford M (1992d) Yeast flocculation: calcium specificity. Yeast 5:487–496
Stratford M (1996) Induction of flocculation in brewing yeasts by change in pH value. FEMS Microbiol Lett 136:13–18
Stratford M, Assinder S (1991) Yeast flocculation: Flo1 and new Flo phenotypes and receptor structure. Yeast 7:559–574
Stratford M, Keenan MMJ (1988) Yeast flocculation: Quantification. Yeast 4:107–115
Strauss CJ, Kock JLF, van Wyk PWJ, Lodolo EJ, Pohl CH, Botes PJ (2005) Bioactive oxylipins in *Saccharomyces cerevisiae*. J Inst Brew 111:304–308
Strauss CJ, van Wyk PWJ, Lodolo EJ, Botes PJ, Pohl CH, Nigam S, Kock JLF (2007) Mitochondrial associated yeast flocculation – the effect of acetylsalicylic acid. J Inst Brew 113:42–47
Straver MH, Aar PCVD, Smit G, Kijne JW (1993) Determinants of flocculence of brewer's yeast during fermentation in wort. Yeast 9:527–532
Taylor NW, Orton WI (1978) Aromatic compounds and sugars in flocculation of *Saccharomyces cerevisiae*. J Inst Brew 84:113–114
Teixeira JM, Teixeira JA, Mota M, Manuela M, Guerra B, Machado Cruz JM, S'Almeida AM (1991) The influence of cell wall composition of a brewer's flocculant lager yeast on sedimentation during successive industrial fermentations. In: Proceedings of the European Brewery Convention Congress, Lisbon, pp 241–248
Teunissen AWRH, Steensma HY (1995) Review: the dominant flocculation genes of *Saccharomyces cerevisiae* constitute a new subtelomeric gene family. Yeast 11:1001–1013
Teunissen AWRH, Holub E, Van Der Hucht J, Van Den Berg JA, Steensma HY (1993) Sequence of the open reading frame of the *FLO1* gene from *Saccharomyces cerevisiae*. Yeast 9:423–427
Teunissen AWRH, Van Den Berg JA, Teunissen SHY (1995) Transcriptional regulation of flocculation genes in *Saccharomyces cerevisiae*. Yeast 11:435–446
Thorne RSW (1951) Some aspects of yeast flocculence. In: Proceedings of the European Brewery Convention Congress, Brighton, pp 21–34
van Hamersveld EH, van der Lans RG, Luyben KC (1997) Quantification of brewers' yeast flocculation in a stirred tank: effect of physical parameters on flocculation. Biotechnol Bioeng 56:190–200
van Holle A, Machado MD, Soares EV (2011) Flocculation in ale brewing strains of *Saccharomyces cerevisiae*: re-evaluation of the role of cell surface charge and hydrophobicity. Appl Microbiol Biotechnol 93:1221–1229
Van Lersel MFM, Meersman E, Arntz M, Rombouts FM, Abee T (1998) Effect of environmental conditions on flocculation and immobilisation of brewer's yeast during production of alcohol-free beer. J Inst Brew 104:131–136
Van Mulders SE, Christianen E, Saerens SM, Daenen L, Verbelen PJ, Willaert R, Verstrepen KJ, Delvaux FR (2009) Phenotypic diversity of Flo protein family-mediated adhesion in *Saccharomyces cerevisiae*. FEMS Yeast Res 9:178–190
Van Mulders SE, Ghequire M, Daenen L, Verbelen PJ, Verstrepen KJ, Delvaux FR (2010) Flocculation gene variability in industrial brewer's yeast strains. Appl Microbiol Biotechnol 88:1321–1331
Van Nierop SNE, Cameron-Clarke A, Axcell BC (2004) Enzymatic generation of factors malt responsible for premature yeast flocculation. J Am Soc Brew Chem 62:108–116
Van Nierop SNE, Rautenbach M, Axcell BC, Cantrell IC (2006) The impact of microorganisms on barley and malt quality – a review. J Am Soc Brew Chem 62:69–79
Verstrepen KJ, Fink GR (2009) Genetic and epigenetic mechanisms underlying cell-surface variability in protozoa and fungi. Annu Rev Genet 43:1–24
Verstrepen KJ, Klis FM (2006) Flocculation, adhesion and biofilm formation in yeasts. Mol Microbiol 60:5–15

Verstrepen KJ, Bauer FF, Winderickx J, Derdelinckx G, Dufour JP, Thevelein JM, Pretorius IS, Delvaux FR (2001a) Genetic modification of *Saccharomyces cerevisiae*: fitting the modern brewer's needs. Cerevisia 26:89–97

Verstrepen KJ, Derdelinckx G, Delvaux FR, Winderickx J, Thevelein JM, Bauer FF, Pretorius IS (2001b) Late fermentation expression of *FLO1* in *Saccharomyces cerevisiae*. J Am Soc Brew Chem 59:69–76

Verstrepen KJ, Van Laere SD, Vanderhaegen BM, Derdelinckx G, Dufour JP, Pretorius IS, Winderickx J, Thevelein JM, Delvaux FR (2003) Expression levels of the yeast alcohol acetyltransferase genes ATF1, Lg-ATF1, and ATF2 control the formation of a broad range of volatile esters. Appl Environ Microbiol 69:5228–5237

Verstrepen KJ, Reynolds TB, Fink GR (2004) Origins of variation in the fungal cell surface. Nat Rev Microbiol 2:533–540

Verstrepen KJ, Jansen A, Lewitter F, Fink GR (2005) Intragenic tandem repeats generate functional variability. Nat Genet 37:986–990

Vidgren V, Londesborough J (2011) 125th anniversary review: yeast flocculation and sedimentation in Brewing. J Inst Brew 117:475–487

Wang A, Raniga PP, Lane S, Lu Y, Liu H (2009) Hyphal chain formation in *Candida albicans*: Cdc28-Hgc1 phosphorylation of Efg1 represses cell separation genes. Mol Cell Biol 29:4406–4416

Watari J, Takata Y, Ogawa M, Sahara H, Koshino S, Onnela M, Airaksinen U, Jaatinen R (1994) Molecular cloning and analysis of the yeast flocculation gene *FlO1*. Yeast 10:211–225

Watari J, Sato M, Ogawa M, Shinotsuka K (1999) Genetic and physiological instability of brewing yeast. Eur Brew Conv Monogr 28:148–160

White FH, Kidney E (1979) The influence of yeast strain on beer spoilage bacteria. Proceedings of the 17th European Brewery Convention Congress, Berlin, DSW, Dordrecht, The Netherlands, pp 801–815

Wickerham LJ (1951) Taxonomy of yeasts. Tech Bull 27.8. Dep Agric no 1029

Wightman P, Quain DE, Meaden PG (1996) Analysis of production brewing strains of yeast by DNA fingerprinting. Lett Appl Microbiol 22:90–94

Wilcocks KL, Smart KA (1995) The importance of surface charge and hydrophobicity for the flocculation of chain-forming brewing yeast strains and resistance of these parameters to acid washing. FEMS Microbiol Lett 15:293–297

Williams LJ, Barnett GR, Ristow JL, Pitkin J, Perriere M, Davis RH (1992) Ornithine decarboxylase gene of Neurospora crassa: isolation, sequence and polyamine-mediated regulation of its mRNA. Mol Cell Biol 12:347–359

Zarattini RA, Williams JW, Ernandes JR, Stewart GG (1993) Bacterial-induced flocculation in selected brewing strains of *Saccharomyces*. Cerevisia Biotechnol 18:65–70

Zhao XQ, Bai FW (2009) Yeast flocculation: new story in fuel ethanol production. Biotechnol Adv 27:849–856

Zhuang S, Smart K, Powell C (2017) Impact of extracellular osmolarity on *Saccharomyces* yeast population during brewing fermentation. J Am Soc Brew Chem 2017:244–254

Chapter 14
Yeast Mitochondria and the Petite Mutation

14.1 Introduction

The structure and function of mitochondria from yeast and a plethora of other eukaryotes have been the subject of intensive investigations since the end of the World War II. Studies have long forged the concept that mitochondria are the "energy powerhouse of the cell" (McBride et al. 2006). These studies, combined with the unique evolutionary origin of mitochondria, led the way to decades of research focusing on this organelle as an essential, yet independent, functional component of the cell.

A mitochondrion (plural mitochondria) is a membrane-enclosed organelle found in most eukaryotic cells including yeast (Henze and Martin 2003). These organelles range from 0.5 to 1.0 μm in diameter (Fig. 14.1a). Mitochondria are described as the "energy powerhouse" because they generate most of the cell's ATP supply. However, mitochondria are also involved in a number of related tasks (McBride et al. 2006), including control of the cell cycle and cell growth (Seo et al. 2010). In animals, including humans, mitochondria have been implicated in a number of diseases (Schapira 2006). The part that yeast mitochondria have played in fundamental studies on eukaryotes that have assisted research into a number of diseases will be discussed later. Indeed, studies on other intracellular organelles and structures in brewer's and distiller's yeast have also been extensive. For example, cell walls, various membranes, vacuoles, nucleus and endoplasmic reticulum (Fig. 1.2), as well as mitochondria, have all contributed considerably to our basic knowledge regarding eukaryote organelles and are discussed in this text (Chap. 5).

Mitochondria are readily recognizable in electron micrographs (Fig. 14.1b) of aerobically grown yeast cells as spherical or rod-shaped structures surrounded by a double membrane. They contain cristae, which are formed by the folding of the inner membrane (Fig. 14.1a). A considerable amount of research has been carried out on the structure of mitochondria (Chinnery and Schon 2003) and the distribution of many mitochondrial enzymes in the matrix of this organelle. Most of the

G.G. Stewart, *Brewing and Distilling Yeasts*, The Yeast Handbook,
https://doi.org/10.1007/978-3-319-69126-8_14

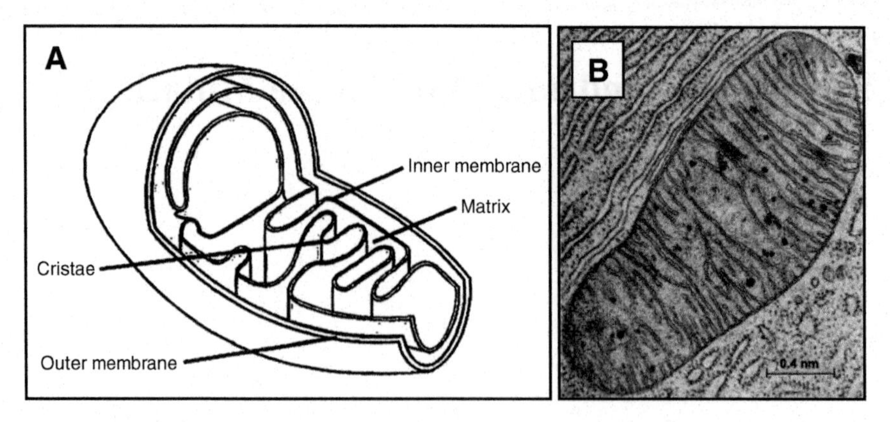

Fig. 14.1 Structure of the mitochondrion. (**a**) Diagram of the overall structure of the mitochondrion and (**b**) electron micrograph of a mitochondrion

Tricarboxylic Acid Cycle enzymes (Krebs cycle) (Fig. 6.22) are presented in the matrix (details in Chap. 6), whereas the enzymes participating in electron transport and oxidative phosphorylation are associated with the mitochondrion's inner membrane, including the cristae.

At one time it was considered that mitochondria were absent from anaerobically grown yeast (Stewart and Russell 2009) since they could not be detected and because such cells lacked many of the enzymes associated with mitochondria. However, it is now clear that yeast cells grown anaerobically in the absence of lipids have simple mitochondria (sometimes called pro-mitochondria), consisting of an outer double membrane but lacking cristae. The addition of lipids and sterols to a yeast culture results in the development of cristae. The development of mitochondria is influenced by the lack of oxygen, the presence of lipids, at the level of glucose typically present in the medium (Stewart et al. 2013). Consequently, there is a change in the structure of mitochondria upon transfer from anaerobic to aerobic conditions but no de novo generation of mitochondria. There is recent evidence (Samp 2012; Samp and Sedin 2017) supporting a role for yeast mitochondria in sulphur dioxide production during wort fermentation and stress resistance within the cone of large vertical fermenters (Lawrence et al. 2012).

14.2 Spontaneous Mutation

In yeast strains, a number of spontaneous mutations can occur in both brewing and distilling fermentations (Stewart and Russell 2009; Gibson et al. 2007), but the most frequently identified spontaneous mutation is the respiratory-deficient (RD) or cytoplasmic "petite" mutation, also known as the rho-mutation (Silhankova et al. 1970a, b). The reason this type of mutation is called "petite" is because colonies

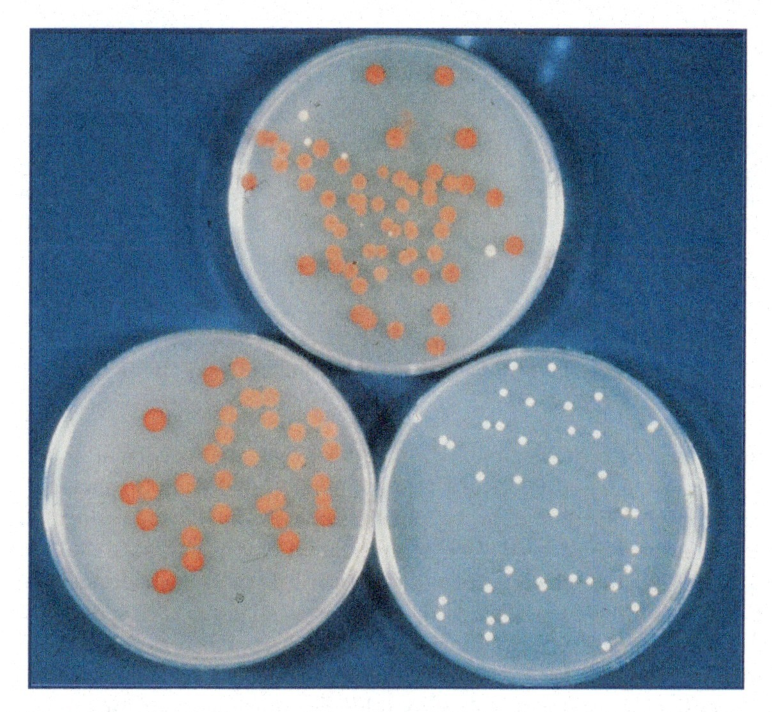

Fig. 14.2 Respiratory-sufficient (RS) and respiratory-deficient (RD) mutants—triphenyl tetrazolium chloride overlay. RS colonies dark and RD colonies white

(not individual cells) of this mutant are usually much smaller (white colonies) than the unmutated wild-type respiratory-sufficient (RS) colonies (also called "grande") (dark or red colonies) (Fig. 14.2). The reason for the French terminology is that this mutation was first identified and described by two Polish scientists—Ephrussi and Slonimski—working in the Pasteur Institute in Paris (Ephrussi et al. 1949).

One of the unique features of yeast mitochondria is that they contain their own DNA—separate from the nucleus. This is termed mitochondrial (mt)DNA. The mtDNA genome ranges in size from 65 to 80 Kp and makes up less than 0.2% of the total cellular DNA with the rest of the 99% plus DNA located within the nucleus. Deletions and amplification of the mtDNA occurs spontaneously in yeast cells at an incidence of approximately 1%. Most often, such mutations are the cause of respiratory deficiency (Gibson et al. 2008a). The frequency of this type of mutation can be increased by stressing the cells, for example, with heat (Lu et al. 2008) or ethanol (Lawrence et al. 2012) and centrifugation plus heat (details in Chap. 13) (Stewart 2010; Chlup and Stewart 2011). In addition, various mutagens such as ethidium bromide (Slonimski et al. 1968) and methyl methane sulfonate (Lawrence et al. 1985) can be used to induce RD mutants in *Saccharomyces* spp. (Piskur et al. 1998).

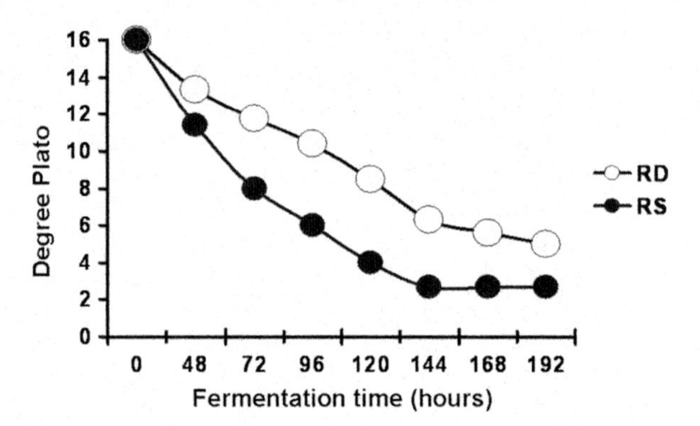

Fig. 14.3 Specific gravity decrease of a 16°Plato wort by a lager brewing strain RS (filled circle) and its spontaneously generated respiratory generated RD (open circle) mutant. Fermentation was conducted in 30 L static fermentation at 15 °C

In brewing yeast strains, the RD mutation normally occurs at frequencies between 0.5 and 5.0% (Silhankova et al. 1970a). However, RD mutation levels up to 50% have been reported for the same strain depending on the temperature and culture conditions (Good et al. 1993). Information on physiological and morphological differences between RS strains and their RD mutants is important in order to aid the understanding of mitochondrial gene expression in yeast. In addition, the use of mutants with modified physiological characteristics, such as overall wort fermentation rate (Fig. 14.3), sugar uptake (Fig. 14.4), metabolite production, flocculation (details later) and reduced stress tolerance are important for brewing and distilling yeast strains (Ernandes et al. 1993; Stewart 2014).

As already discussed, deficiencies in mitochondrial function result in diminished ability to function aerobically, and, consequently, these yeasts are unable to metabolize non-fermentable carbon sources such as lactate, glycerol or ethanol but are able to metabolize fermentable sugars such as glucose and maltose but less efficiently than wild-type cultures (Fig. 14.5). Further to the physiological effects already discussed, RD mutants are much more difficult to store in a culture collection compared to RS cultures (particularly as a result of freeze drying). However, liquid nitrogen storage at −196 °C or refrigeration at −70 °C have been found to be effective storage procedures for RD cultures—details in Chap. 4 (Russell and Stewart 1981).

Although studies with RD mutants from both brewing and distilling strains have been reported since the early 1970s (Silhankova et al. 1970b), they are still not fully understood, particularly with respect to the involvement of mtDNA deletion with a yeast strain's overall growth and fermentation characteristics.

The following aspects of brewing yeast strains RD mutants have been considered:

- Effects on the uptake of wort sugars—particularly maltose and maltotriose and the production of beer flavour congeners (Ernandes et al. 1993)
- Effects of RD mutants on yeast flocculation and sedimentation characteristics
- Influence of stress conditions (e.g. centrifugation and heat) on the formation of RD mutants—details in Chap. 13 (Good et al. 1993)

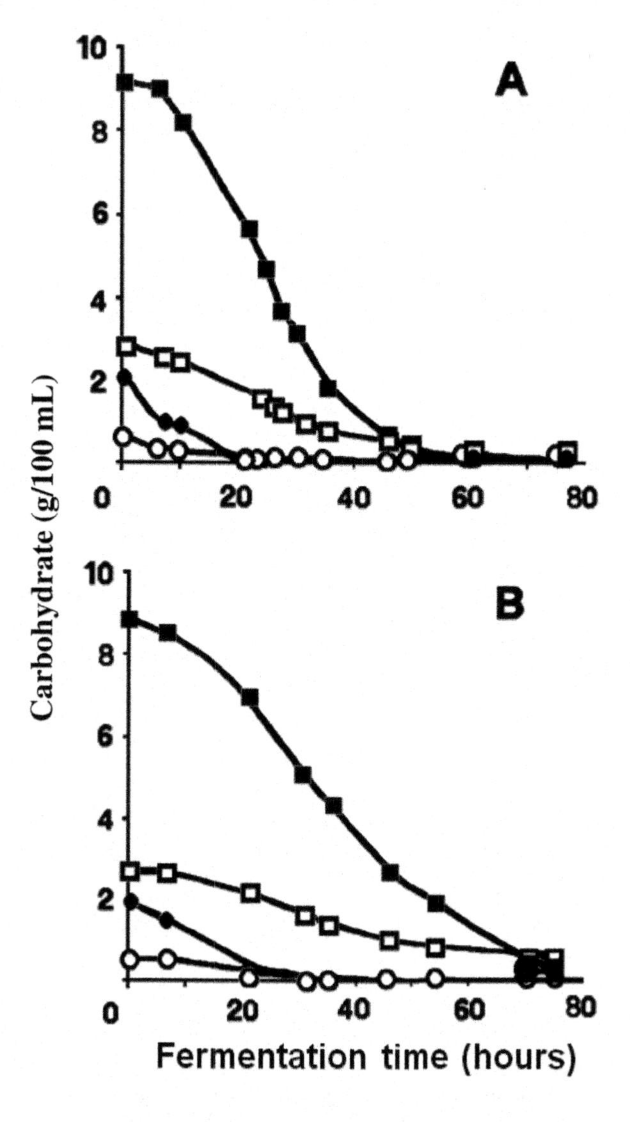

Fig. 14.4 Maltose (filled square), maltotriose (open square), glucose (filled circle) and fructose (open circle) uptake during low-cell-density fermentation of 16°Plato wort by brewing lager strain 3021 RS (**a**) and its spontaneously generated respiratory-deficient (RD) mutant (**b**). Fermentations were conducted in 30 L static fermenters at 15 °C

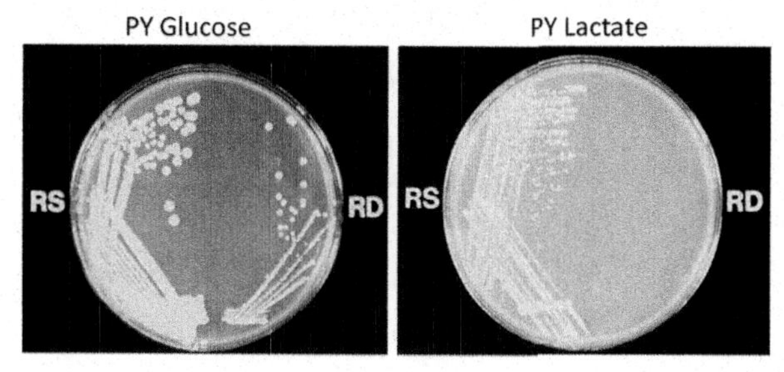

Fig. 14.5 Growth of respiratory-sufficient (RS) and respiratory-deficient (RD) cultures on fermentable (glucose) and non-fermentable (lactate) carbon sources. *PY* peptone-yeast extract

At the conclusion of this chapter, current research on diseases caused by dysfunctional mitochondria, often as a result of mtDNA mutation, will be discussed and related to fundamental studies on yeast mitochondria.

14.3 Wort Sugars Uptake

The uptake of wort sugars by brewer's and distiller's yeast has been discussed in detail elsewhere in this text (Chap. 7). However, the influence of mitochondria on the uptake of wort sugars should also be discussed here. It has already been discussed that wort contains the sugars sucrose, fructose, glucose, maltose and maltotriose together with unfermentable dextrin material—this is particularly true of brewer's wort fermentation. In the normal situation, brewing yeast strains (ale and lager strains) are capable of utilizing sucrose, glucose, fructose, maltose and maltotriose in this approximate sequence (or priority) (Fig. 14.4a) although some overlap does occur. The majority of brewing strains (not all) leave maltotetraose (G4) and other dextrins unfermented (details in Chap. 7).

The initial step in the utilization of any sugar by yeast is usually either its passage intact across the cell (plasma) membrane or the sugar is hydrolysed outside the cell membrane followed by entry into the cell by some or all of the hydrolysis products (Fig. 14.6). Maltose and maltotriose (Fig. 14.7) are examples of sugars that pass intact across the cell membrane. Sucrose (Fig. 14.7) is hydrolysed by the extracellular enzyme invertase, and the hydrolysed products (glucose and fructose) are taken into the cell. Maltose and maltotriose are the major sugars in brewer's wort (Table 14.1) and, as a consequence, a brewer's

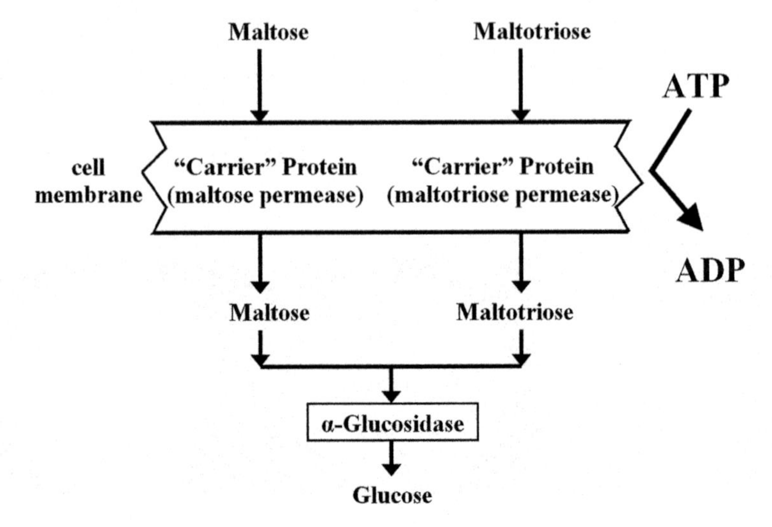

Fig. 14.6 Uptake and metabolism of maltose and maltotriose by yeast

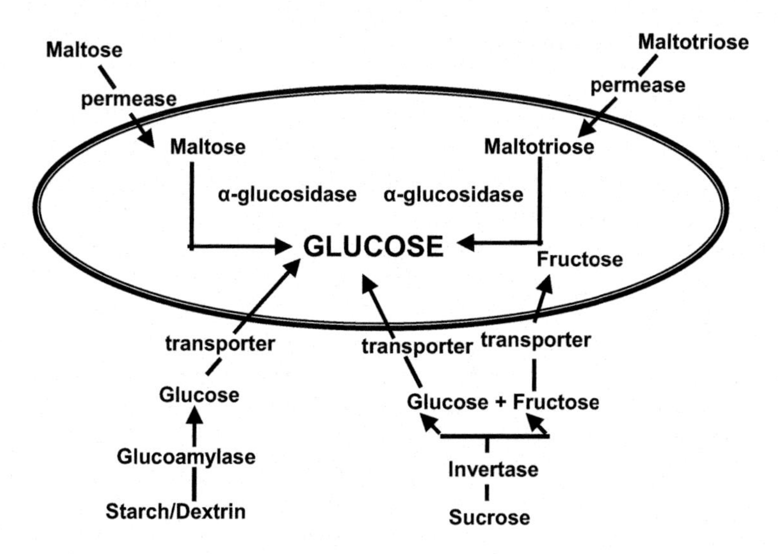

Fig. 14.7 Uptake of sugars by the yeast cell

Table 14.1 Typical sugar composition of wort

	Percent composition (%)
Glucose	10–15
Fructose	1–2
Sucrose	1–2
Maltose	50–60
Maltotriose	15–20
Dextrins	20–30

Stewart GG (2006) Studies on the uptake and metabolism of wort sugars during brewing fermentations. MBAA Tech Quart 43:265–269

yeast's ability to use these two sugars is critical and depends on the correct genetic complement. It is probable that brewer's yeast possesses independent uptake mechanisms (maltose and maltotriose permease) in order to transport these two sugars into the cell. The uptake of both these sugars is by active transport. This is a process that requires the utilization of energy from ATP (Gibson et al. 2008b). It has already been discussed (Chap. 6) that RD cells have reduced mitochondrial efficiency, and therefore depleted overall intracellular energy and, as a consequence, the uptake of maltose and maltotriose will be compromised. On the other hand, glucose and fructose are taken up by passive transport—a process that does not require energy, and their uptake into RD cells is not seriously compromised (Stewart 2009).

14.4 Wort Fermentations

Fermentations with a 16°Plato wort (30% corn adjunct) have been conducted at 15 °C with an RS lager yeast and its RD spontaneous mutant. Both the RS and RD isolates were isolated from a wild-type culture (Stewart 2014). The RD mutant was identified by the TTC overlay method (Fig. 14.2) (Ogur and St. John 1956) and characterized by its inability to grow on lactate; it was however able to grow on glucose (Fig. 14.5). Also, the difference in colony size between the wild-type culture (RS—grande) and its RD mutant (petite) was a factor confirming their identity (Fig. 14.5—PY glucose plate). At specified times during the wort fermentations, a sample of the cell suspension was withdrawn and centrifuged and the supernatant stored at −20 °C prior to subsequent analysis. In addition, samples were taken for determination of the yeast in suspension concentration as previously discussed (Ernandes et al. 1993) and the RS and RD characteristics of the cultures confirmed.

The uptake of wort sugars was assessed by HPLC analysis together with gravity assessment in degrees Plato. Also, the accumulation of biomass and yeast in suspension in both the RS and RD cultures is shown in Fig. 14.8. The RS culture used the total wort sugar more rapidly and accumulated more biomass than the RD mutant. The uptake of the four major fermentable wort sugars: glucose, fructose, maltose and maltotriose of an RS lager strain (a) and its RD mutant (b) is shown in Fig. 14.4. The uptake pattern of glucose in the two fermentations (RS and RD) was similar (Fig. 14.9), but the uptake of both maltose and maltotriose was more rapid in the RS fermentation compared to the RD fermentation (Figs. 14.10 and 14.11). This confirms the differences between sugars that require an active transport process (maltose and maltotriose) to be taken into the cell and those employing passive transport systems (e.g. glucose). RD cultures contain reduced concentrations of metabolic energy (mostly as ATP) compared to their wild-type RS cultures. As a result, maltose and maltotriose uptake is inhibited in RD cells, whereas glucose uptake in both sets of fermentation (RS and RD) was very similar (Figs. 14.9, 14.10 and 14.11).

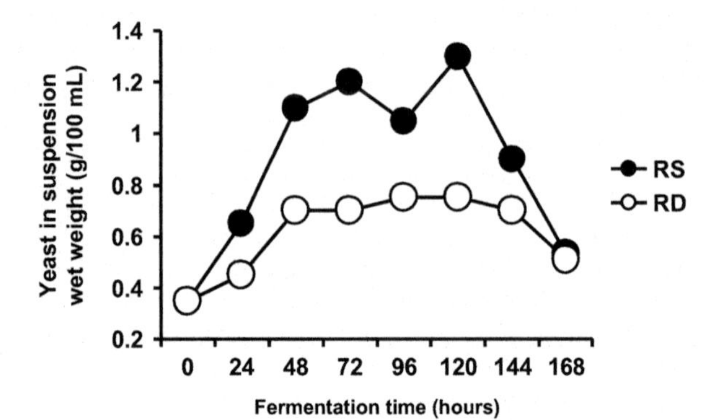

Fig. 14.8 Yeast in suspension during fermentation of a 16°Plato wort by a lager brewing strain RS (filled circle) and its spontaneously generated respiratory generated RD (open circle) mutant. Fermentation was conducted in 30 L static fermentation at 15 °C

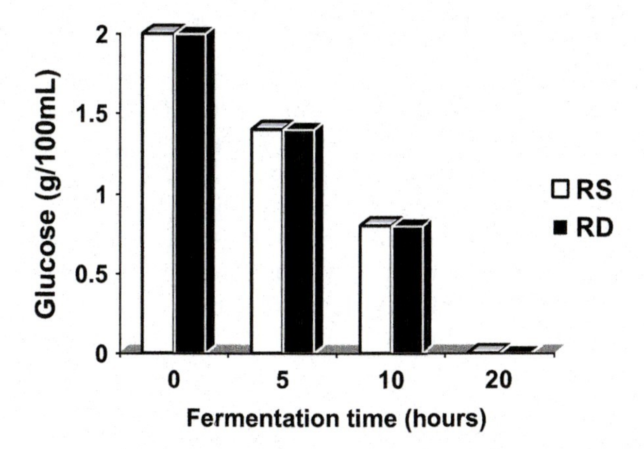

Fig. 14.9 Glucose uptake from a 16°Plato wort by a lager yeast strain RS (open square) and its RD (filled square) mutant

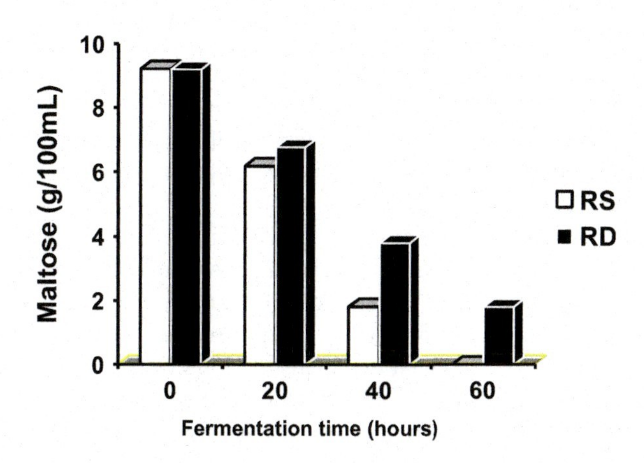

Fig. 14.10 Maltose uptake from a 16°Plato wort by a lager yeast strain RS (open square) and its RD (filled square) mutant

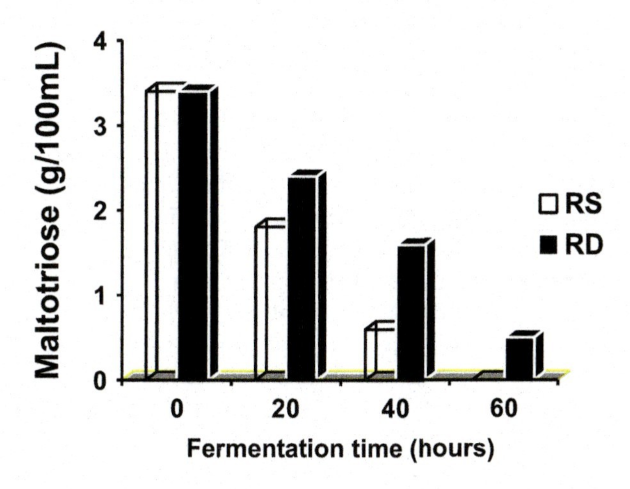

Fig. 14.11 Maltotriose uptake from a 16°Plato wort by a lager yeast strain RS (open square) and its RD (filled square) mutant

14.5 RD Mutants and Beer Flavour

It will be discussed elsewhere in this text (Chap. 16) that many factors contribute to beer flavour (Bamforth 2016) and of paramount importance amongst these is yeast performance and stability (Bamforth and Stewart 2011). Spontaneously generated RD yeasts have been reported to adversely affect beer flavour (Meilgaard 1975a, b). It has been documented (Stewart 2009) that appropriate amounts of esters, fusel oils and carbonyls are critical compounds in determining beer flavour. Esters, in particular, constitute an important group amongst beer volatiles because of their strong, penetrating fruit flavours (Stewart 2008, 2010; Stewart and Russell 2009; Xu et al. 2017).

The concentration of carbonyls (acetaldehyde, diacetyl, etc.), esters and fusel oils in lager beer produced by RS and RD cultures as a result of fermentation of a 16°Plato brewer's wort have been compared. The RD cultures produced considerably higher amounts of fusel oils (a consolidation of propanol, isobutanol and isoamyl alcohol concentrations) compared to the RS (control) culture. However, lower amounts of aliphatic esters (ethyl acetate and isoamyl acetate) were detected in beer produced by the RD mutant (Silhankova et al. 1970b). This was also the case with acetaldehyde, which can give a characteristic grassy or "green apple" immature flavour (Meilgaard 1975a).

Another important group of flavour-active carbonyls are the vicinal diketones (VDK), particularly diacetyl and 2,3-pentanedione. Both compounds impart to beer a "butterscotch" or stale milk flavour and aroma. A detailed discussion of VDKs in beer is in Chap. 15. However, they are relevant in the complexity of respiratory-deficient yeast mutant metabolism during wort fermentation.

Quantitatively, diacetyl in beer is the most important VDK since its threshold is 0.1 mg/L for most tasters. Diacetyl and 2,3-pentanedione arise in beer as by-products of the metabolic pathways leading to the amino acids valine and isoleucine (Fig. 14.12). The α-acetodroxy acids, which are intermediates in these biosyntheses, are excreted into the fermenting wort. They subsequently undergo spontaneous oxidative decarboxylation, giving rise to the VDKs. Diacetyl is then reduced by yeast to acetoin and ultimately 2,3 butanediol (Fig. 14.13). The flavour threshold concentrations of the diols are relatively high and considerably higher than diacetyl. For this oxidative decarboxylation to occur in either the later stages of fermentation or during maturation, it is important that viable and vital yeast in suspension is present (Krogerus and Gibson 2013).

The diacetyl metabolic pattern has been compared with the RS and RD cultures of typical lager yeast strain in a 16°Plato wort. Figure 14.14 illustrates the production of free diacetyl and total diacetyl (free diacetyl plus α-acetohydroxylactate) is higher with the RD mutant throughout wort fermentation compared to the RS wild type. In addition to producing higher levels of diacetyl, the RD mutant, unlike the RS wild type, lacks the ability to reduce it to the same extent, late in the fermentation or during maturation. This is due to a metabolic

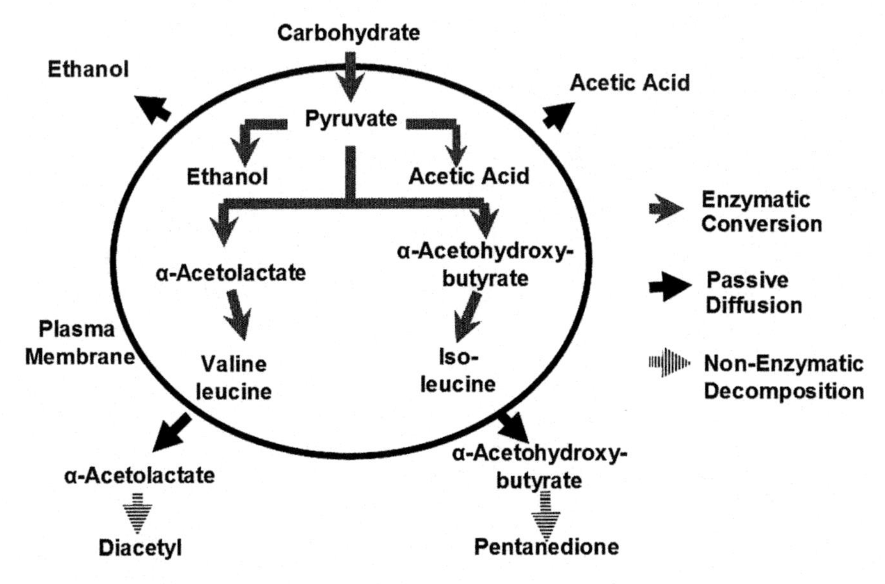

Fig. 14.12 Formation of diacetyl and 2,3-pentanedione

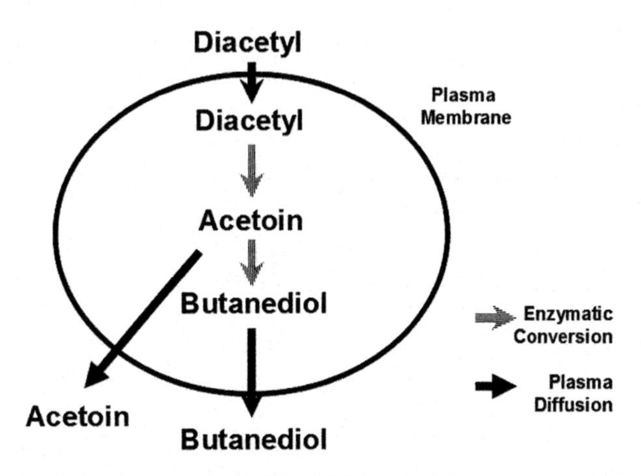

Fig. 14.13 Reduction of diacetyl to acetoin and 2,3-butanediol

imbalance caused by a malfunctioning Tricarboxylic Acid Cycle (Fig. 6.22) and defects in the enzymes of the isoleucine-valine pathways, which affect the production of diketones (predominantly diacetyl) during wort fermentation. The effect of RD mutants producing elevated levels of VDKs has previously been reviewed (Krogerus and Gibson 2013; Silhankova et al. 1970a, b)—further details of VDK metabolism are in Chap. 16.

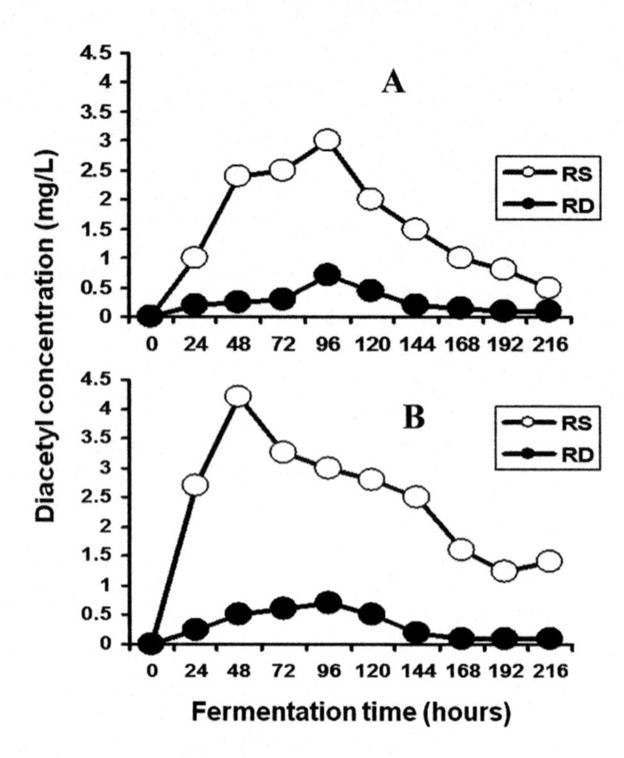

Fig. 14.14 Production of free diacetyl (**a**) and total diacetyl (diacetyl plus α-acetolactate) (**b**) during fermentation of 16°Plato wort by a lager strain RS (filled circle) and its spontaneously generated RD mutant (open circle). Fermentation was conducted in 30 L static fermentation at 15 °C

14.6 Effects of RD Mutants on Yeast Flocculation and Sedimentation Characteristics

The flocculation properties of brewer's (and distiller's) yeast strains or, conversely, lack of flocculation of a particular strain is one of their most important properties (Stewart 2009). This important property has been discussed in detail in Chap. 13. Of particular importance, in the context of this chapter, is that mitochondria of yeast cells have been found to have a controlling influence on yeast cell wall structure and their flocculation characteristics (Ernandes et al. 1993).

The absence of distinct mtDNA regions leads to a loss of flocculation characteristics (Good et al. 1993). In this study, the wild-type (RS) lager brewing strain exhibited strong flocculation, whereas its spontaneously arising RD mutant was non-flocculent. During fermentations of 16°Plato wort, the yeast in suspension was determined throughout the fermentation cycle for both the RS and RD cultures (Fig. 14.8). The RS culture grew considerably faster than the RD culture, and towards the end of fermentation, it sedimented out of suspension more rapidly than the RD culture. This demonstrated that the RD culture's cell wall structure had been modified as a result of mutation to the mtDNA (Stewart 2014). This resulted in loss of this mutant's flocculation properties.

14.7 Mitochondrial Diseases

Damage and subsequent dysfunction of mitochondria is an important factor in a range of human diseases due to their influence on cellular metabolism. Mitochondrial malfunction in mammals (including humans) often presents itself as a neurological disorder but also as myopathy, diabetes, liver disease, heart disease and a variety of other manifestations (DiMauro and Davidzon 2005). The recent Charlie Gard baby, who suffered with mitochondrial depletion syndrome, is an example of such a disability! Diseases induced by mitochondrial disorders may initially be considered by many to be unrelated to the influence of yeast mitochondria on brewing yeast fermentations. However, fundamental research on mitochondrial structure and function of brewer's (and baker's) yeast strains has been extrapolated to an understanding of mammalian mitochondrial disorders.

Diseases caused by mutation in the mtDNA include Kearns-Sayre syndrome (negative effects on the retina and eye lipids), MELAS syndrome (muscle weakness and pain, recurrent headaches, loss of appetite, vomiting and seizures) and Leber's hereditary optic neuropathy (loss of colour vision). These maladies are only transmitted through the mother primarily due to mtDNA mutations, and only the egg (not the sperm) contributes mitochondria to the embryo. Mitochondria-mediated oxidative stress plays a role in cardiomyopathy with Type 2 diabetic patients.

In summary, mitochondrial induced disease results in crippling conditions that can lead to serious disabilities and even death. It is passed from mothers to their children and often manifests itself at a very young age. It is caused by faulty mitochondria that, until recently, could not be treated or prevented. However, new approaches to three-person in vitro fertilization (IVF), pioneered by British scientists (Tachibana et al. 2009), now promise to give families affected by mitochondrial disease the opportunity to have healthy children. In most countries, the law currently does not permit these novel techniques to be employed. However, in 2015, changes in the appropriate legislation (promoted by the Human Fertilization and Embryology Authority [HFEA]) were approved by both houses of the British Parliament. There are still many opponents to this legislation (including church leaders, pro-life groups and even 55 Italian MPs). They warn "that the change has been brought about too hastily and marked the start of a 'slippery slope' towards designer babies and eugenics". There is no doubt that their objections will not succeed, and it is expected that the first baby conceived with this treatment to be born shortly.

Specifically, the genetic material from the two parents is injected into a healthy donor egg with normal mitochondria from the donor that has had its nucleus removed. The resulting embryo carries nuclear DNA from the parents, and the mitochondrial DNA comes from the donor. Children conceived in this way would inherit DNA from their parents with 99.8% of total DNA coming from the mother and father and only 0.2% from the donor (mtDNA) (Fig. 14.15).

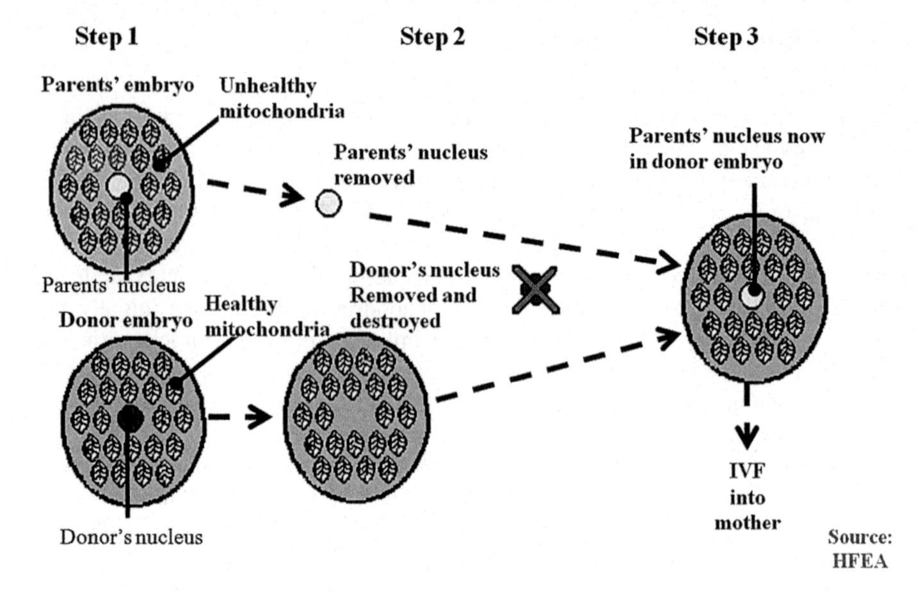

Fig. 14.15 Mitochondrial exchange

A paper (Reinhardt et al. 2013) entitled "Mitochondrial Replacement, Evolution, and the Clinic" cautions the application of this technique and advises as follows: ". . . although there has been increased government interest, especially in the United Kingdom, for using this approach to treat patients, there are reasons to believe that it is premature to move this technology into the clinic at this stage". The HFCA immediately published the following statement as a reaction to Reinhardt and his colleagues' paper (2013): "There are still hurdles to overcome before these techniques could be used clinically and it won't be a suitable treatment option for everyone at risk of having a child with a mitochondrial disease. As in every area of medicine, moving from research into clinical practice always involves a degree of uncertainty. Experts should be satisfied that the results of further safety checks are reassuring and long term follow-up studies are critical. Even then patients will need to carefully weigh the risk and benefits for them".

References

Bamforth CW (2016) Flavor some components of beer. In: Bamforth CW (ed) Brewing materials and processes. Academic, London, pp 151–155

Bamforth CW, Stewart CC (2011) Brewing – its evolution from a craft into a technology. Biologist 57:138–147

Chinnery PF, Schon EA (2003) Mitochondria. J Neurol Neurosurg Psychiatry 74:1188–1199

Chlup PH, Stewart GG (2011) Centrifuges in brewing. MBAA Tech Q 48:46–50

Dimauro S, Davidzon G (2005) Mitochondrial DNA and disease. Ann Med 37:222–232
Ephrussi B, Hottinguer H, Chimenes AM (1949) Action de l'acriflavine sur les levures. I. La mutation "petite colonie". Ann Inst Pasteur 76:351–367
Ernandes JR, Williams JW, Russell I, Stewart GG (1993) Respiratory deficiency in brewing yeast strains – effects on fermentation, flocculation, and beer flavor components. J Am Soc Brew Chem 151:16–20
Gibson BR, Lawrence SJ, Leclaire JP, Powell CD, Smart KA (2007) Yeast responses to stresses associated with industrial brewery handling. FEMS Microbiol Rev 31:535–569
Gibson BR, Prescott KA, Smart KA (2008a) Petite mutation in aged and oxidatively stressed ale and lager brewing yeast. Lett Appl Microbiol 46:636–642
Gibson BR, Boulton CA, Box WG, Graham NS, Lawrence SJ, Linforth RS, Smart KA (2008b) Carbohydrate utilization and the lager yeast transcriptome during brewery fermentation. Yeast 25:549–562
Good L, Dowhanick TM, Ernandes JE, Russell I, Stewart GG (1993) Rho – mitochondrial genomes and their influence on adaptation to nutrient stress in lager yeast strains. J Am Soc Brew Chem 51:35–39
Henze K, Martin W (2003) Evolutionary biology: essence of mitochondria. Nature 426 (6963):127–128
Krogerus K, Gibson BR (2013) Influence of valine and other amino acids on total diacetyl and 2,3-pentanedione levels during fermentation of brewer's wort. Appl Microbiol Biotechnol 97:6919–6930
Lawrence CW, Krauss BR, Christensen RB (1985) New mutations affecting induced mutagenesis in yeast. Mutat Res 150:211–216
Lawrence SJ, Wimalasena TT, Nicholls SM, Box WG, Boulton C, Smart KA (2012) Incidence and characterization of petites isolated from lager brewing yeast *Saccharomyces cerevisiae* populations. J Am Soc Brew Chem 70:268–274
Lu B, Yadav S, Shah PG, Liu T, Tian B, Pukszta S, Villaluna N, Kutejová E, Newlon CS, Santos JH, Suzuki CK (2008) Roles for the human ATP-dependent Lon protease in mitochondrial DNA maintenance. J Biol Chem 282:17363–17374
McBride HM, Neuspiel M, Wasiak S (2006) Mitochondria: more than just a powerhouse. Curr Biol 16:R551–R560
Meilgaard MC (1975a) Flavour chemistry of beer. Part I. Flavour interaction between principal volatiles. MBAA Tech Q 12:107–117
Meilgaard MC (1975b) Flavor chemistry of beer. Part II. Flavour and threshold of 239 aroma volatiles. MBAA Tech Q 12:151–168
Ogur M, St. John R (1956) A differential and diagnostic plating method for population studies of respiration deficiency in yeast. J Bacteriol 72:500–504
Piskur J, Smole S, Groth C, Petersen RF, Pedersen MB (1998) Structure and genetic stability of mitochondrial genome vary among yeasts of the genus *Saccharomyces*. Int J Syst Bacteriol 48:1015–1024
Reinhardt K, Dowling DK, Morrow EH (2013) Medicine. Mitochondrial replacement, evolution, and the clinic. Science 341(6152):1345–1346
Russell I, Stewart GG (1981) Liquid nitrogen storage of yeast cultures compared to more traditional storage methods. J Am Soc Brew Chem 39:19–24
Samp EJ (2012) Possible roles of the mitochondria in sulfur dioxide production by lager yeast. J Am Soc Brew Chem 70:219–229
Samp EJ, Sedin D (2017) Important aspects of controlling sulfur dioxide in brewing. MBAA Tech Q 54:60–71
Schapira AH (2006) Mitochondrial disease. Lancet 368(9529):70–82
Seo AY, Joseph A-M, Dutta D, Hwang JCY, Aris JP, Leeuwenburgh C (2010) New insights into the role of mitochondria in aging: mitochondrial dynamics and more. J Cell Sci 123:2533–2542
Silhankova C, Savel J, Mostek J (1970a) Respiratory deficient mutants of bottom brewer's yeast. I. Frequencies and types of mutants in various strains. J Inst Brew 76:280–288

Silhankova C, Mostek J, Savel J (1970b) Respiratory deficient mutants of bottom brewer's yeast. II. Technological properties of some RD mutants. J Inst Brew 76:289–295

Slonimski PP, Perrodin G, Croft JH (1968) Ethidium bromide induced mutation of yeast mitochondria: complete transformation of cells into respiratory deficient non-chromosomal "petites". Biochem Biophys Res Commun 30:232–239

Stewart GG (2008) Esters – the most important group of flavour-active compounds in alcoholic beverages. In: Bryce JH, Piggott JR, Stewart GG (eds) Distilled spirits. Production, technology and innovation. Nottingham University Press, Nottingham, pp 243–250

Stewart GG (2009) Forty years of brewing research. J Inst Brew 115:3–29

Stewart GG (2010) High gravity brewing and distilling – past experiences and future prospects. J Am Soc Brew Chem 68:1–9

Stewart GG (2014) Yeast mitochondria – their influence on brewer's yeast fermentation and medical research. MBAA Tech Q 51:3–11

Stewart GG, Russell I (2009) An introduction to brewing science and technology, Series lll, Brewer's yeast, 2nd edn. The Institute of Brewing and Distilling, London

Stewart GG, Hill AE, Russell I (2013) 125th anniversary review – developments in brewing and distilling yeast strains. J Inst Brew 119:202–220

Tachibana M, Sparman M, Sritanaudomchai H, Ma H, Clepper L, Woodward J, Li Y, Ramsey C, Kolotushkina O, Mitalipov S (2009) Mitochondrial gene replacement in primate offspring and embryonic stem cells. Nature 461:367–372

Xu Y, Wang D, Li H, Hao J-Q, Jiang W, Liu Z, Qin Q (2017) Flavour contribution of esters in lager beers and an analysis of their flavour thresholds. J Am Soc Brew Chem 75:201–206

Chapter 15
Flavour Production by Yeast

15.1 Introduction

The final/aroma of beer (and spirits—details later) is the sum of several hundreds of flavour-active compounds produced in every stage of the brewing process (Meilgaard 1975a, b). The great majority (not all) of these substances are yeast metabolites. They are produced during wort fermentation and consist of fermentation intermediates or by-products (metabolites). Higher alcohols (also called fusel oils), esters, vicinal diketones (VDKs), other carbonyls and sulphur compounds are the key (not all) flavour elements produced by yeast. These compounds (plus hop constituents) determine a beer's final quality, particularly when it is fresh (Bamforth 2011). Higher alcohols and esters are desirable volatile beer constituents, with a few exceptions (Inoue 2008). Together with these compounds, yeast metabolism contributes to the biosynthesis of three other groups of beer flavour-active compounds: organic acids, sulphur compounds (both organic and inorganic) and aldehydes (Molina et al. 2007; Ekberg et al. 2013). This is an area of yeast activity that is currently termed metabolomics (Nicholson and Lindon 2008). It is the study of chemical processes involving metabolites. Metabolomics is the "systematic study of the unique chemical fingerprints that specific cellular processes leave behind", the study of the small-molecule metabolite profiles (Daviss 2005). In the context of ester formation by brewer's and distiller's yeast (Stewart 2008), the phenomena of both proteomes (the study of proteins, particularly their structure and function) (Anderson and Anderson 1998) and genomics (genetics that applies to DNA sequence, assembly, function and structure) (Culver and Labow 2002; Smart 2007) are also considered later (details in Chap. 16).

Obviously, flavour-active compounds must be maintained within certain limits. Otherwise, a single compound or group of compounds (e.g. VDKs) may predominate and prejudice a beer's flavour balance. Furthermore, flavour compounds such as esters often act in synergy with other components to affect beer flavour in concentrations well below their individual threshold values (Kluba et al. 1993).

G.G. Stewart, *Brewing and Distilling Yeasts*, The Yeast Handbook,
https://doi.org/10.1007/978-3-319-69126-8_15

Although ethanol, carbon dioxide and glycerol are the major products produced by yeast during wort fermentation (details in Chaps. 6 and 7) (Fig. 6.16), they have minimal impact on the flavour of the final beer (Stewart and Russell 2009). It is the type and concentration of the other excretion products, already discussed, that primarily determines the beer flavour balance. There are many factors that can modify this balance. Yeast strain, malt variety fermentation temperature, malt type and variety, adjunct type and percentage in the grist, fermentation design and geometry, wort pH, buffering capacity, wort gravity, etc. are all influencing factors (Boulton and Quain 2001). In addition, hop variety and hopping procedures are important factors (Roberts and Falconer 2018).

Each type of beer has its own particular aroma triggered largely (but not exclusively) either by the yeast strain employed during the fermentation (Peddie 1990; Ramos-Jeunehomme et al. 1981; Rossouw et al. 2008) or by a plethora of process parameters used during fermentation (Lodolo et al. 2008; Saerens et al. 2008a, b; Bravi et al. 2009; Verbelen et al. 2009a, b; Blasco et al. 2011; Berner and Arneborg 2012; Dekoninck et al. 2012; Hiralal et al. 2014; Pires et al. 2014). While only isoamyl acetate (banana-like aroma) concentrations are often above the threshold level in most lager beers, ales normally have ethyl acetate (solvent-like aroma) and ethyl hexanoate (apple-like aroma) as supplementary flavouring compounds with levels often above their taste threshold (Alvarez et al. 1994). However, compounds such as diacetyl and other VDKs (with certain exceptions) should usually be below flavour threshold values. Diacetyl contributes negatively to a buttery (stale milk) flavour in most beers. Table 15.1 lists threshold values of the main esters and higher alcohols present in ale and lager beers.

The biosynthesis of higher alcohols and esters during wort fermentation has been extensively reviewed by Pires and colleagues (2014), and the following discussion has extensively employed this publication (with permission) (Table 15.2). This also includes the effects of wort maltose concentrations in high-gravity circumstances on ester formation which has been discussed in detail in Chap. 11 (Younis and Stewart 1998, 1999). Reduced ester formation in a maltose synthetic medium compared to a glucose synthetic medium has also been documented (Table 15.3). In addition, material compiled in *An Introduction to Brewery Science and Technology* published by the Institute of Brewing and Distilling has been employed here (Stewart and Russell 2009) together with *Whisky: Technology, Production and Marketing* (ed. by Russell and Stewart 2014)—with permission.

15.2 Higher Alcohols

In flavour terms, the higher alcohols that occur in beer and many spirits are: *n*-propanol, isobutanol, 2-methyl-1-butanol and 3-methyl-1-butanol. However, more than 40 other alcohols have been identified. Regulation of higher alcohol biosynthesis is complex since they may be produced as by-products of amino acid metabolism or via pyruvate produced from carbohydrate metabolism (Fig. 15.1).

Table 15.1 Major esters and higher (fusel) alcohols in beer[a]

	Organoleptic threshold (mg/L)
Esters	
Ethyl acetate (fruity, solvent-like)	3.0
Isoamyl acetate (fruity, banana aroma)	1.2
Isobutyl acetate (pineapple)	0.7
Ethyl caproate (sour apple)	0.22
Ethyl caprylate (sour fruit)	0.9
Phenyl ethyl acetate (roses, honey, fruity)	0.4
Higher alcohols	
Amyl alcohol (Alcohol)	65
Isobutanol (Solvent)	70
Propanol (Solvent)	200
Methyl butanol (Banana, medicinal)	65
Phenyl alcohol (Roses, sweet, perfume)	125

[a]Adapted from: Olaniran AO, Hiralal L, Mokoena MP, Pillay B. Flavour-active volatile compounds in beer: production, regulation and control. J Inst Brew 2017 123:13–23

Table 15.2 Effects of process variables on beer ester and higher alcohol levels

Variable	Possible results
Higher fermentation temperature (Hiralal et al. 2014)	Increased esters and alcohols
Trub present in the fermenter (Stewart and Martin 2004)	Reduced esters of increased alcohols
Yeast nutrients added to wort (Hiralal et al. 2014)	Increased esters and alcohols
Increased wort oxygen at pitching (Kucharczyk and Tuszyński 2015)	Reduced esters
Pressure fermentation (Landaud et al. 2001)	Reduced esters and alcohols
Higher wort gravity (Pires et al. 2014)	Increased esters and alcohols
Increased wort gravity and maltose (Stewart et al. 1999)	Decreased esters
Increased wort lipids (Klug and Daum 2014)	Decreased esters

Table 15.3 Ethyl acetate and isoamyl acetate produced by brewing yeast strains during fermentation of synthetic media

	Ethyl acetate (mg/L)		Isoamyl acetate (mg/L)	
	Glucose	Maltose	Glucose	Maltose
Ale 1	4.13	2.79	0.14	0.14
Ale 2	2.97	2.59	0.06	0.04
Ale 3	3.13	2.71	0.05	0.03
Lager 1	6.00	5.22	0.22	0.21
Lager 2	3.75	3.28	0.26	0.22
Lager 3	4.13	3.51	0.23	0.17

Peptone—yeast extract—4% sugar medium

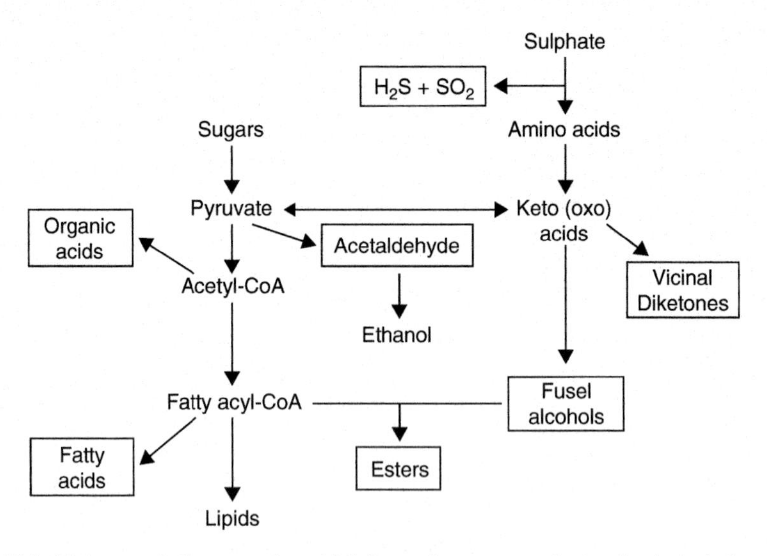

Fig. 15.1 Major metabolic routes by which brewer's yeasts synthesize fusel alcohols, esters, sulphur compounds, VDKs, acetaldehydes and ethanol

Both brewer's and distiller's yeast strains (similar to wort sugars) absorb wort amino acids in a distinct order (Jones and Pierce 1964; Lekkas et al. 2007) (uptake details in Chap. 7) from which they take the amino group so it can be incorporated into its own structures. What remains from the amino acid are α-keto acids which enter into an irreversible chain reaction that will ultimately form higher alcohols. This metabolic pathway was originally suggested by the German physician Paul Ehrlich (1907) (Fig. 15.2) who was intrigued with the structural molecular similarities between amyl alcohol and isoleucine and isoamyl alcohol with leucine (Figs. 15.3 and 15.4). Ehrlich was awarded the 1908 Nobel Prize in Physiology or Medicine for his contribution to immunology. His observations led to an investigation of the relationship between these amino acids and higher alcohol synthesis. When he supplemented a fermenting medium with these amino acids, increased higher alcohol production was observed. Ehrlich also proposed that amino acids were enzymatically hydrolysed to form the corresponding higher alcohol, along with ammonia and carbon dioxide. As ammonia was not detected in the medium, it was assumed that it was incorporated into yeast proteins. Subsequently, Neubauer and Fromherg (1911) proposed a few intermediate steps to the Ehrlich pathway, completing a metabolic pathway that is currently accepted. However, a detailed enzymatic chain reaction was demonstrated several decades after Ehrlich's original studies (Sentheshanmuganathan and Elsden 1958; Sentheshanmuganathan 1960) when the enzymatic sequence for the Ehrlich pathway—transaminase, decarboxylase and alcohol dehydrogenase. Although this pathway is the most studied in this context, higher alcohols are also formed during the upstream (anabolic pathway) biosynthesis of amino acids (Chen 1978;

Fig. 15.2 Paul Ehrlich

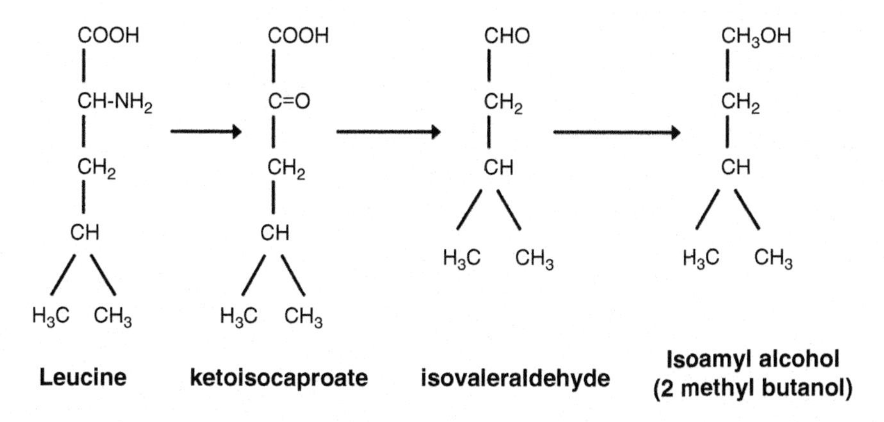

Fig. 15.3 Ehrlich pathway—relationship between higher (fusel) alcohols and amino acids

Dickinson and Norte 1993). The most important specific pathway in the brewing context is the *de novo* synthesis of branched-chain amino acids (BCAAs) through the isoleucine-leucine-valine (ILV) pathway (Dickinson and Norte 1993) (Fig. 15.3).

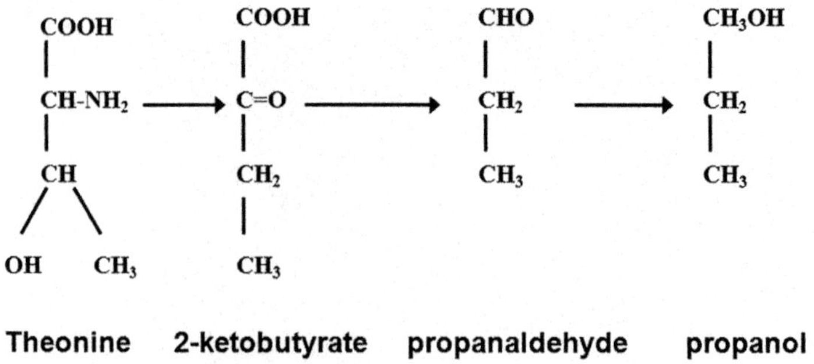

$$HOOC{-}CH(NH_2){-}CH(OH)CH_3 \longrightarrow HOOC{-}C(=O){-}CH_2{-}CH_3 \longrightarrow CHO{-}CH_2{-}CH_3 \longrightarrow CH_3OH{-}CH_2{-}CH_3$$

Theonine **2-ketobutyrate** **propanaldehyde** **propanol**

Fig. 15.4 Anabolic route for the synthesis of fusel oils

Ethyl acetate is the most common ester produced by yeast :

$$CH_3CH_2OH + CH_3COSCoA \rightarrow CH_3CH_2COOCH_3 + CoASH$$

Ethanol **acetyl CoA** **ethyl acetate**

Fig. 15.5 Production of an ester during fermentation

15.3 Esters

Esters are predominantly formed during the vigorous phase of primary fermentation by the enzymatic condensation of organic acids with alcohols. Volatile esters in beer can be divided into two major groups: the acetate esters and the medium-chain fatty acid (MCFA) ethyl esters (Fig. 15.5). The acetate esters are synthesized from acetic acid (acetate) with ethanol or a high alcohol. In the ethyl ester's family, ethanol will form the alcohol radical, and the acid side is an MCFA (Malcorps and Dufour 1992). Although dozens of different esters can be formed during the fermentation stage of any beer or spirit (Engan 1981; Meilgaard 1975b), six of them are of major importance as aromatic constituents: ethyl acetate, (solvent-like aroma), isoamyl acetate (banana aroma), isobutyl acetate (fruity aroma), phenylethyl acetate (roses and honey aroma), ethyl hexanoate (sweet apple aroma) and ethyl octanoate (sour apple aroma) (Table 15.1).

Esters are synthesized in the cytoplasm of the yeast cell, but they readily leave the cell as they are lipophilic. However, while small-chain acetate esters easily diffuse through the plasma membrane whereas MCFA ethyl esters may have their passage hindered (Nykänen and Nykänen 1977; Nykanen et al. 1977; Dufour and Malcorps 1994).

Ester synthesis involves organic acids linked to a coenzyme A in order to form an acyl-CoA molecule. Acyl-CoAs are highly energetic entities, which in the presence of oxygen can be β-oxidized (spliced) into smaller units (acetyl-CoA) in the mitochondria. This will occur unless the organic acid involved is acetic acid, where it will be converted to acetyl-CoA. The great majority of acetyl-CoA produced by a yeast cell comes from the oxidative decarboxylation of pyruvate (details in Chap. 6). Aerobic conditions inside mitochondria will induce acetyl-CoA to enter the TCA Cycle (Krebs cycle) (Fig. 15.4). In the absence of oxygen, acetyl-CoA will be enzymatically esterified with an alcohol to form acetate esters (Table 15.5). Also, MCFA ethyl esters are formed from longer chains of acyl-CoA with ethanol. Figure 15.1 summarizes the major metabolic routes by which brewer's yeast strains synthesize both higher alcohols and esters.

15.3.1 Influence of Yeast Strains on Ester Formation

Brewers and distillers have a large number of yeast strains from which to select—distillers, ales and lagers. The use of alternative yeast strains offers the prospect of more efficient fermentations, diverse characteristics together with beer and spirit flavour manipulation. The characteristics of strain variation have been extensively reported, but a basic understanding of the reasons for these differences is still necessary (Stewart 2008). To illustrate yeast strain diversity, ester production under standard conditions has been extensively studied. Fermentations were conducted in 1L static fermenters with 16°Plato all-malt wort at 20 °C. When the fermentations were complete, ethyl acetate and isoamyl acetate were determined in the fermented wort samples (Table 15.4). These results of differences regarding strain variations concerning the production of esters during wort fermentation illustrate strain differentiation within the two brewing yeast species *Saccharomyces cerevisiae* and *S. pastorianus*. Although the *S. cerevisiae* and the *S. pastorianus* genomes have now been sequenced (details in Chap. 16) (Goffeau et al. 1996; Krogerus et al. 2015), and this has enabled an intense study of the genetic control of the species (Lyness et al. 1997; Zhang et al. 2013), an understanding of the reasons for ester production differences is still largely unclear!

Table 15.4 Ethyl acetate and isoamyl acid (mg/L) produced by distillers and brewer's yeast strains

	Ethyl acetate	Isoamyl acetate
Distillers	46.2	1.3
Distillers	36.2	0.9
Ale	20.6	0.6
Ale	36.2	0.8
Ale	25.6	3.5
Lager	60.2	4.6
Lager	32.6	1.6

15.3.2 Influence of Wort Clarity on Ester Formation

The level of solids carried over from cereal mashing into wort, also solids resulting from grape crushing, their effect on yeast fermentation performance in the production of beer, whisky and wine has been the subject of considerable study (Ancin et al. 1996; Merritt 1967; O'Connor-Cox et al. 1996a, b; Thomas et al. 1994; Stewart and Martin 2004; Stewart 2007).

Wort solids (trub) influence:

- The removal of CO_2 from solution during fermentation by acting as nucleation sites. Removal of CO_2 from solution can occur with no suspended particles, but it requires a great deal of energy (Merritt 1967).
- Wort solids confer nutritive value to the fermentation medium. The rate of fermentation is faster in the presence of insoluble material. Wort solids are generally associated with higher levels of fatty acids (Thomas et al. 1994).
- In some instances, yeast cells are able to attach themselves to wort solids and display enhanced growth because of the concentration of nutrients at the particle surface (Stewart and Martin 2004; Yonezawa and Stewart 2004).

To assess the influence of wort clarity on ester formation, 15°Plato wort, produced in the ICBD 2hL pilot brewery (Fig. 15.6) and a number of wort types were fermented in 1L volume batches employing 2L tall tubes at 27 °C with a *S. cerevisiae* distiller's strain. Carbon dioxide evolution rates during fermentation and ethyl acetate and isoamyl acetate concentrations at the end of fermentation were determined. The following wort conditions were studied:

Fig. 15.6 The 2hL brewing pilot plant at Heriot-Watt University, Edinburgh

Table 15.5 Concentration of CO_2 (g/L) present during fermentation of 15°Plato worts

	24 h	48 h
Cloudy wort	2	4
Clear wort	5	8
Clear wort plus DE	2	4
Clear wort plus bentonite	5	8

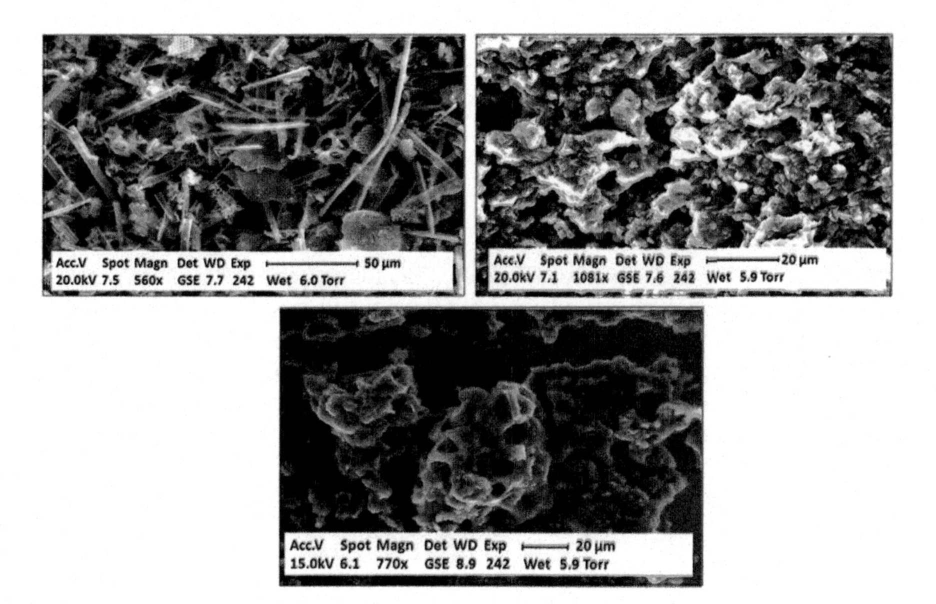

Fig. 15.7 Environmental scanning electron microscopy (SEM) of different wort solid materials

- Cloudy wort.
- Clear wort.
- Clear wort plus 0.2 g/L diatomaceous earth (DE).
- Clear wort plus DE and 5.5 mg/L C16:1 fatty acid.
- Clear wort plus 0.2 g/L bentonite.

The concentration of CO_2 during fermentation of the wort types was determined (Table 15.5) (Stewart and Martin 2004; Stewart 2008). Cloudy wort, containing trub, and wort containing DE acted as nucleation sites and increased CO_2 evolution out of the wort. Clear wort and wort plus bentonite did not function as nucleation sites, and consequently CO_2 remained in the fermentation medium to a much greater extent. Why was there a difference between trub, DE and bentonite? In order to visualize each type of particle and obtain data on their surface characteristics, environmental scanning electron microscopy (ESEM) was conducted (Fig. 15.7). In DE, there was a heterogeneous mix of shapes and sizes of particles. The surface of most of the particles revealed an extremely porous structure, as would be expected. The micrograph of the cloudy wort solids showed a mix of

Table 15.6 Ethyl acetate concentration (mg/L) following 160 h fermentation of different 15°Plato wort types

15°Plato wort type	mg/L
Cloudy wort	30
Clear wort	16
Clear wort plus DE and C16:1 fatty acid	25

Table 15.7 Isoamyl acetate concentration (mg/L) following 160 h fermentation of different 15°Plato wort types

15°Plato wort type	mg/L
Cloudy wort	0.95
Clear wort	0.55
Clear wort plus DE and C16:1 fatty acid	0.75

different structures, again porous in nature. Bentonite had a more homogenous structure. In addition, it possessed a much different surface topology, but it did not appear to possess the same porous nature as DE or wort solids.

Ester levels are also influenced by the wort particle concentration and type. Ethyl acetate and isoamyl acetate concentrations are high in cloudy wort and DE containing C16:1 fatty acid (Tables 15.6 and 15.7). This reflects the fact that unsaturated fatty acids that would be absorbed to wort trub induced the synthesis of esters (Äyräpää and Lindström 1973; O'Connor-Cox et al. 1996a, b; Stewart 2007).

15.3.3 Biosynthesis of Acetate Esters

Research on enzymatic synthesis during ester production dates from the early 1960s (Nordström 1962), and the critical enzyme was purified and named alcohol acetyltransferase (AATases) in the early 1980s by Yoshioka and Hashimoto (1981). The most studied and comprehensively characterized enzymes responsible for ester synthesis are AATases I and II, which are encoded by the genes *ATF1* and *ATF2* (Yoshioka and Hashimoto 1981; Malcorps and Dufour 1992; Fujii et al. 1994; Lyness et al. 1997; Nagasawa et al. 1998; Yoshimoto et al. 1998; Verstrepen et al. 2003; Molina et al. 2007; Dekoninck et al. 2012; Zhang et al. 2013). It was also found that lager yeast strains have an additional *ATF1* homologous gene—*Lg-ATF1* (Fujii et al. 1996a, b) which encodes an AATase very similar to that encoded from the original *ATF1* gene. This additional gene expression on lager yeast strains compared to ale strains enhances acetate ester production and ultimately a beer's aroma profile.

A brewer's yeast strain was designed to increase the ester/higher alcohol ratio by overexpressing *ATF1* and decreasing the expression of a gene related to higher alcohol synthesis (Zhang et al. 2013). Ester production by this genetically modified strain was considerably higher than the parental strain. It has also been shown that

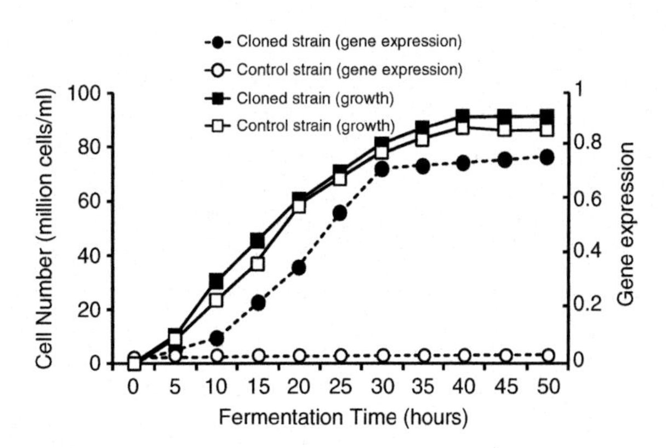

Fig. 15.8 Expression of the ATF1 during fermentation. Cells cultured (with shaking) in yeast extract—peptone—glucose medium at 21 °C. Control strain (growth) (open square); cloned strain (growth) (filled circle); control strain (gene expression) (open circle); cloned strain (gene expression) (filled square)

the level of *ATF* genes has an impact on acetate ester production (Fujii et al. 1994, 1996b; Nagasawa et al. 1998). Overexpression of *ATF1* strains may have up to a 180-fold increased isoamyl acetate production and a 30-fold increased ethyl acetate production when compared to wild-type yeast strains. Analysis has also revealed that *ATF1*-encrypted ATTases appear to be responsible for the majority of acetate ester production. When *ATF1* and *ATF2* were deleted, no acetate esters originated from alcohols with more than five carbon atoms, such as isoamyl acetate and phenylethyl acetate, were formed. This means that the desirable beer banana aroma (isoamyl acetate) depends on *ATF1*- and *ATF2*-encoded enzymes. Subsequently, Saerens et al. (2008b) confirmed that the maximum expression levels of *ATF1* and *ATF2* are directly correlated to the final concentration of acetate esters. It is possible that there might be more ATTases involved in acetate ester production, but current published knowledge is unclear. Figure 15.8 illustrates the reactions involved in the production of the major acetate esters and the relevant ATF genes (Stewart et al. 1999).

Pires et al. (2014) have discussed the importance of acetate esters in alcohol-free beers (AFBs). AFBs can either be produced as a result of physical removal of ethanol from finished beer or control of the biological process involved in beer fermentation (Brányik et al. 2012). The lack of reduced ethanol greatly affects the retention of volatile aroma-active compounds (Perpete and Collin 2000). Strejc et al. (2013) have isolated a brewing yeast mutant capable of overproducing isoamyl acetate and isoamyl alcohol. The resulting sweet banana aroma of the resulting beer could be employed to overcome the undesirable worty off-flavour of an AFB. As a consequence, sensory analyses showed that the enhanced level of isoamyl acetate ester had a positive effect on the fruity palate fullness and aroma intensity of the AFB.

15.3.4 Regulation of Ethyl Ester(s) Production

The net rate of ester production by yeast depends on both the availability of precursor substrates (Saerens et al. 2006; Hiralal et al. 2014) and on the enzymatic balance during the synthesis by AATases (Yoshimoto et al. 1998; Mason and Dufour 2000; Verstrepen et al. 2003; Saerens et al. 2006; Zhang et al. 2013) and the breakdown of esters by esterases (Lilly et al. 2006). Esterases are a group of hydrolysing enzymes that catalyse the cleavage and/or avoid the formation of ester bonds. Instead of enhancing AATase activity, Fukuda et al. (1998) have increased the amount of isoamyl acetate production by a saké strain of *S. cerevisiae*. Instead of enhancing AATase activity, isoamyl acetate cleavage was avoided by deleting the acetate-hydrolysing esterase gene (*IAHI*). The mutant-deficient strain produced approximately 19 times higher amounts of isoamyl acetate when compared with the parent strain. Fukuda et al. (1998) have subsequently demonstrated the important balance between AATase and esterase activity for the overall rate of ester synthesis by *S. cerevisiae*. Further evidence regarding the *IAHI*-encoded esterase on the breakdown of esters was reported by Lilly et al. (2006). In addition to isoamyl acetate, a decreased production of ethyl acetate, phenylethyl acetate and hexyl acetate by the overexpressing *IAHI* mutant strain has been reported.

15.3.5 Ester Metabolism During Beer Ageing (Maturation)

During beer ageing (maturing) after primary fermentation, a beer's flavour will usually change. This change will occur as a result of the yeast culture (refermentation in the cask, keg or bottle) (Vanderhaegen et al. 2003) or by spontaneous chemical condensation of organic acids with ethanol (Vanderhaegen et al. 2006; Saison et al. 2009a; Rodrigues et al. 2011). Some esters such as isoamyl acetate are known to be hydrolysed during beer ageing (Neven et al. 1997). Chemical hydrolysis and esterification are acid-catalysed (Vanderhaegen et al. 2006), but the remaining esterases from yeast autolysis can also play a role in unpasteurized beers (Neven et al. 1997). Other ethyl esters such as ethyl nicotinate (medicinal, solvent), ethyl pyruvate (peas, freshly cut grass) and ethyl lactate (fruity, buttery) are also formed in ageing beers (Saison et al. 2009b).

15.4 Carbonyls

Over 200 carbonyl compounds are reported to contribute to the flavour of beer and other alcoholic beverages (Stewart and Russell 2009). Those compounds influencing beer flavour, produced as a result of yeast metabolism during fermentation, are various aldehydes and vicinal diketones, notably diacetyl. Also, carbonyl compounds exert a significant influence on beer flavour stability. Excessive concentrations of carbonyl compounds are known to cause stale flavour in beer. The effects of aldehydes on flavour stability are reported to develop grassy notes (propanol, 2-methyl butanol, pentanol and a papery taste trans-2-nonenal, furfural) and acetaldehyde.

15.4.1 Acetaldehyde

Quantitatively, acetaldehyde is the carbonyl present in beer at the highest concentration (10–15 mg/L). It is produced by yeast as a result of the decarboxylation of pyruvate (Figs. 6.16 and 15.1) and is an intermediate in the metabolic formation of ethanol (glycolysis). It can be present in beer at concentrations above its flavour threshold (approximately 10 mg/L), and it imparts an undesirable "grassy" or "green apple" character. Acetaldehyde accumulates during the period of active growth. As with higher alcohols and esters, the extent of acetaldehyde accumulation is determined by the yeast strain employed and the fermentation conditions. Although the yeast strain is of primary importance (details later), elevated wort oxygen concentration, pitching rate and fermentation temperatures all influence acetaldehyde accumulation. In addition, premature flocculation of yeast from suspension in fermenting wort (Lake and Speers 2008; Lake et al. 2008) does not allow the reutilization of excreted acetaldehyde associated with later fermentation stages (further details in Chaps. 8 and 13).

Acetaldehyde has long been known as a product of alcoholic fermentation by yeast. The concentration of acetaldehyde in various wine musts after fermentations lasting from 5 to 15 days varied from 20 to 200 mg/L in the fermented medium. Many years later, Romano et al. (1994) studied in detail acetaldehyde production in *S. cerevisiae* wine yeasts. Eighty-six strains were investigated for their ability to produce acetaldehyde in a synthetic fermentation medium and in grape must. Acetaldehyde production characteristics did not differ significantly between the two media, ranging from a few mg/L to 60 mg/L and was found to be a yeast strain characteristic. A 30 °C fermentation temperature increased the acetaldehyde produced. This study allowed the strains to be assigned into three different phenotypes: low, medium and high acetaldehyde producers. The low and high phenotypes also differed considerably in the production of acetic acid, acetoin and higher alcohols and can be useful for studying acetaldehyde production in *S. cerevisiae* from both technological and genetic aspects.

The metabolism of acetaldehyde, ethanol and acetate in *S. cerevisiae* is complex (Aranda and del Olmo 2003) because several enzymes are involved, and there is a dependence on the cell's redox balance. The enzymes that are ultimately responsible for their metabolic pathways are alcohol dehydrogenase (ADH) and aldehyde dehydrogenase (ALDH), each of which consists of several isoenzymes. The interconversion between ethanol and acetaldehyde is catalysed by aldehyde dehydrogenase (Lutstorf and Megnet 1968; Beier et al. 1985). Acetaldehyde can also be metabolized by its oxidation to acetate, which is carried out by aldehyde dehydrogenases (ALDH) (Collin et al. 1991).

The presence of acetaldehyde is one of the many stress conditions that yeast cells may encounter (Aranda et al. 2002; Aranda and del Olmo 2003). Several heat-shock proteins (HSP) genes are induced that are also involved in the response to other forms of stress (e.g. ethanol) (Piper et al. 1994). ALDHs play an important role in yeast acetaldehyde metabolism when the cells are growing in ethanol. Under several growth conditions, further addition of acetaldehyde (or ethanol) to wine yeasts induced the expression of some *ALD* genes and led to an increase in ALDH activity. This result is consistent with their need to obtain energy from ethanol during the biological ageing processes. Also, under certain conditions, acetaldehyde functions as a mutagen with respiratory-deficient (petite) mutants as the primary result (Ernandes et al. 1993) (Chap. 14).

15.4.2 Vicinal Diketones (VDKs): Diacetyl and 2,3-Pentanedione

Diacetyl and 2,3-pentanedione are produced during wort fermentation as by-products of specific amino acid synthesis (valine and isoleucine, respectively) by ale and lager yeast strains. Both of these VDKs (especially diacetyl) will have a significant effect on the flavour, aroma and drinkability of beer. Diacetyl imparts a butterscotch-like stale milk flavour with a threshold usually reported to be around 0.1–0.2 mg/L in lagers and 0.1–0.4 mg/L in ales (Meilgaard 1975a, b; Wainwright 1973; Krogerus and Gibson 2013a). However, diacetyl flavour thresholds as low as 1.7 mg/L (Saison et al. 2009b) and 1.4–6.1 mg/L (Kluba et al. 1993) have been reported. There is no doubt that the flavour threshold of diacetyl varies with a taster's geographical background, ethnicity and diet (Inoue 2008). 2,3-Pentanedione has a similar flavour to diacetyl, although often described as more toffee-like but with a much higher flavour aroma threshold—1 mg/L. The presence of VDKs above their flavour threshold in beer is generally regarded as a defect because their flavour is undesirable in many styles of beer and can also (not always) indicate microbial contamination. Diacetyl, at detectable concentrations, is acceptable in some beer styles such as Czech Pilsners and some (only some) English ales. Diacetyl metabolism during brewing has been the subject of significant polemic over the years, but a comprehensive review by Krogerus and Gibson (2013a) has discussed the history and current situation regarding diacetyl biosynthesis and subsequent metabolism during wort fermentation.

Although microbial contamination can be responsible for diacetyl formation (Inoue 2008), yeast metabolism is the major metabolic route and this is why it is discussed in detail here. Diacetyl and 2,3-pentanedione are formed indirectly as a result of valine and isoleucine biosynthesis (Fig. 14.12). They arise from the spontaneous non-enzymatic oxidative decarboxylation of α-acetohydroxy acids that are intermediates in the valine and isoleucine biosynthesis pathways (Radhakrishnan et al. 1960) (Fig. 15.9). Valine and isoleucine synthesis is localized in the mitochondria (Ryan and Kohihaw 1974). In the former pathway, the reaction between α-acetolactate and 2,3-dihydroxy-isovalerate is rate-limiting, and thus during fermentation and yeast growth, α-acetolactate is secreted through the cell plasma membrane into the wort (Dillemans et al. 1987). Recovery and mechanisms for α-acetolactate secretion are not fully understood but may involve protecting the yeast from carbonyl stress (van Bergen et al. 2006). The α-acetolactate is then spontaneously decarboxylated, oxidatively or non-oxidatively, forming diacetyl or

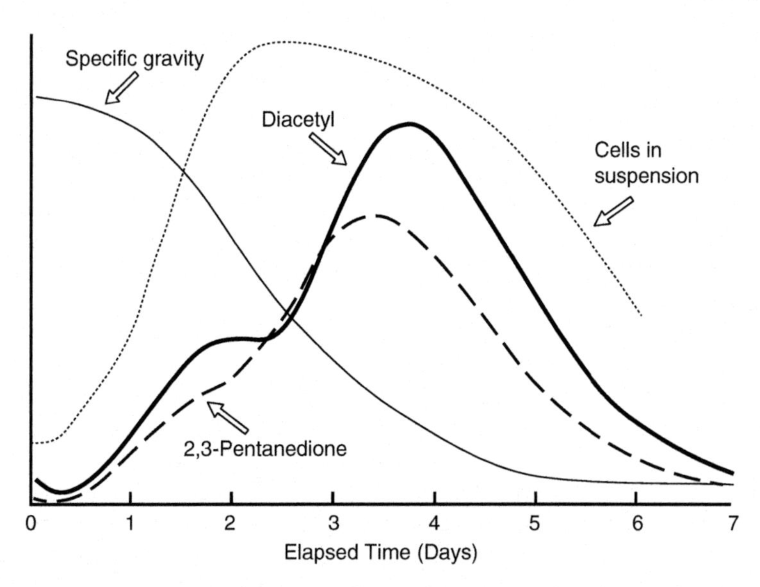

Fig. 15.9 Metabolism of VDKs during wort fermentation. The figure shows the relationship between diacetyl and 2,3-pentanedione, yeast cell growth and specific wort gravity

acetoin, respectively, and in both cases releasing carbon dioxide. The non-oxidative decarboxylation of acetoin can be enhanced by heating under anaerobic conditions and by maintaining a low redox wort potential (Kobayashi et al. 2005). Diacetyl production is enhanced with increasing valine biosynthesis, which depends on the cell's need for, and access to, valine and other amino acids. Fermentation conditions favouring rapid yeast growth can give rise to increased diacetyl production if the wort's free amino nitrogen content is insufficient. The diacetyl and "total diacetyl" concentrations during a typical lager fermentation are depicted in Fig. 15.9. During fermentation, the concentration of free diacetyl in wort is usually low, whereas α-acetolactate constitutes the majority of the "total diacetyl" present (Landaud et al. 1998). As a result, diacetyl concentrations are often expressed as "total diacetyl" concentrations, that is, the sum of free diacetyl and α-acetolactate ("potential diacetyl"), during analysis in order to highlight potential diacetyl concentration in fermenting wort. Details of the uptake of valine (along with other amino acids) by brewer's and distiller's yeast during wort fermentation are discussed in Chap. 7.

15.4.3 Process Conditions and Diacetyl Formation

It has already been discussed (Sect. 15.4.2) that diacetyl metabolism is directly linked to valine biosynthesis. The intracellular valine concentration affects the amount of diacetyl generated during fermentation (Wainwright 1973). Valine inhibits acetohydroxyacid synthase (AHAS)—the enzyme responsible for catalysing the formation of α-acetolactate from pyruvate (Fig. 14.12). Consequently, the more valine present in the yeast cell, the less α-acetolactate will be synthesized. As the catalysing enzyme is inhibited, less diacetyl will be formed.

The effect of valine and isoleucine addition to fermenting wort on the production of diacetyl has found that increased wort valine concentrations significantly reduce the amount of diacetyl produced during fermentation (Nakatani et al. 1984). During fermentation trials with lager yeast strains involving wort with a spectrum of original gravities, FAN and valine concentrations, Petersen et al. (2004) observed that low levels of valine in wort resulted in the formation of double-peak diacetyl profiles (most likely as a result of valine depletion towards the end of fermentation). However, high concentrations of wort valine resulted in single-peak diacetyl profiles with lower maximum diacetyl levels compared to the wort with low valine concentrations. Wort valine concentrations influence the amount of diacetyl formed but as the wort employed in these studies varied in specific gravity and FAN, no definite conclusions regarding a relationship between wort valine and diacetyl concentrations can be drawn. Cyr et al. (2007) conducted trials with two lager yeast strains (dried) and observed that diacetyl concentrations in the fermenting wort were constant or decreased when valine uptake increased, while diacetyl concentrations increased when valine uptake decreased. Krogerus and Gibson (2013b) showed that direct supplementation of wort with valine (100–300 mg/L) and consequently greater uptake of valine by yeast cultures resulted in less diacetyl being formed during fermentation. However, other fermentation parameters such as fermentation rate and yeast growth were unaffected. During the same series of fermentation trials, Krogerus and Gibson (2013a) showed that when wort FAN levels were decreased, diacetyl production also decreased. This effect is presumably due to faster absorption of preferred amino acids (Jones and Pierce 1964), resulting in an earlier and enhanced demand for valine and its increased uptake due to less competition for amino acid permease interactions. Increased background levels of initial wort amino acids (maintaining the valine concentration constant) resulted in greater production of diacetyl. Pugh et al. (1997) also reported that the maximum diacetyl concentration during fermentation decreased as the initial FAN concentrations were increased from 122 to 144 mg/L. Verbelen et al. (2009a) has reported lower diacetyl production rates and increased valine uptake during fermentation of 18°Plato adjunct wort (FAN content approximately 150–210 mg/L) compared with an all-malt 18°Plato wort (FAN content approximately 300 mg/L). The conflicting diacetyl results are presumably owing to differences in the valine uptake rate and extent.

Yeast pitching rate and cell density also affects the amount of diacetyl present in wort at the end of fermentation, as it has been observed that the diacetyl concentration increases with higher cell pitching rates. In trials with a number of lager yeast strains, Verbelen et al. (2008, 2009b) observed over 10-fold increases in beer diacetyl concentrations when the pitching rate was increased severalfold. This can be explained because more α-acetolactate was presumably produced, and fermentation times were shorter as a result of higher pitching rates. This reduced the amount of α-acetolactate spontaneously decarboxylated to diacetyl outside the cell during active fermentation during the diacetyl rate-limiting step of diacetyl removal (Fig. 14.12). This has led to increased post-fermentation α-acetolactate and eventually diacetyl concentrations. Ekberg et al. (2013) observed increased concentrations of total VDK in wort fermented for 72 h with a stress-tolerant yeast strain, compared with a wort fermented for 193 h with the control (less tolerant) strain. Sigler et al. (2009) observed enhanced diacetyl production with increased wort osmolarity [adjunct with sorbitol—a sugar that is not metabolized by *Saccharomyces* spp. (Panchal et al. 1982)], most likely as a result of decreased yeast vitality.

15.4.4 Process Conditions and Diacetyl Reduction

The pattern of diacetyl formation (Fig. 14.12) and subsequent breakdown in relation to yeast growth and gravity during a lager fermentation is shown in Fig. 15.9. Diacetyl is reduced to acetoin and ultimately 2,3-butanediol (Fig. 14.13) and 2,3-pentanedione to its corresponding diol. The flavour threshold concentrations of these diols are relatively high—details later. Therefore, the final reductive stages of VDK metabolism are critical in order to obtain a beer with acceptable organoleptic properties.

The reduction of diacetyl takes place during the latter stages of fermentation when active yeast growth has ceased. In terms of practical fermentation management, the need to achieve a desired diacetyl specification may be the factor which determines when a beer may be moved to its conditioning phase, diluted, filtered and/or centrifuged (depending on the processing procedures). Thus, diacetyl metabolism is an important determinant of overall vessel residence time, which already affects the efficiency of plant utilization (Stewart 2014a, b; Stewart et al. 2016).

The removal of diacetyl during the later stages of fermentation is not as well understood and documented, as is the formation of this carbonyl. Diacetyl removal during the closing stages of wort fermentation is usually rapid (Boulton and Box 2003; Inoue and Yamamoto 1970). It has been suggested that diacetyl reduction by yeast is not the rate-limiting step but rather the spontaneous decarboxylation of α-acetolactate to diacetyl is rate-limiting. The exact mechanisms of diacetyl uptake by yeast are unknown, but the diacetyl uptake rate into yeast cells, and consequently the diacetyl removal rate has been shown to be affected by a number of fermentation parameters influencing yeast membrane composition (such as temperature and oxygenation of yeast or wort), while any effects will be amplified by phenomena such as yeast flocculation and sedimentation characteristics (details in Chap. 13) (Boulton and Box 2003).

Wort pH and the fermentation temperature also influence the amount and rate of diacetyl formed and subsequent reduction as both parameters affect yeast growth rate, the reaction rate of the spontaneous decarboxylation of α-acetolactate into diacetyl, and the activities of the enzymes responsible for reducing diacetyl to acetoin and 2,3-butanediol (Garcia et al. 1994). Increased fermentation temperatures lead to higher initial diacetyl production rates as a consequence of increased yeast growth but also produce more yeast mass to reduce the diacetyl to 2,3-butanediol and acetoin and also increase the reaction rate of the oxidative decarboxylation of α-acetolactate to diacetyl. This suggests that the rate-limiting conversion of α-acetolactate to diacetyl is expedited at higher temperatures leading to sharper diacetyl concentration peaks during fermentation and consequently faster diacetyl reduction rates (Wainwright 1973; Garcia et al. 1994; Saerens et al. 2008a).

The increased decarboxylation rate of α-acetolactate to diacetyl at higher temperatures can be exploited during fermentation in a so-called diacetyl rest (also called a "free rise" by some brewers), where temperatures are increased towards the conclusion of fermentation in order to more rapidly decrease wort α-acetolactate concentrations and shorten the maturation period (Quain and Smith 2009). Bamforth and Kanauchi (2004) have reported an optimum pH of 3.5 for an acetoin dehydrogenase isolated from a commercial lager yeast strain, suggesting that diacetyl reduction rates are higher at lower wort pH values (towards the end of fermentation). Several sources (Garcia et al. 1994; Rondags et al. 1999; Kobayashi et al. 2005) have reported an increased reaction rate for the oxidative decarboxylation of α-acetolactate to diacetyl at lower pH values, which also suggests that the rate-limiting conversion of α-acetolactate to diacetyl is more rapid at more acidic wort conditions. Consequently, the maturation time needed for diacetyl reduction could be reduced at lower pH values as long as the pH within the preferred range is suitable for a palatable beer. Beer is typically in the pH range of 4.0–4.5 but can vary from 3.6 to 5.0 depending on beer style, raw materials and the brewing process employed (Stewart and Russell 2009).

A process has been developed utilizing immobilized yeast cells for the accelerated maturation of beer—primarily (but not exclusively) for diacetyl management (Pajunen 1995). The Finnish biotechnology company Cultor, working in association with the Sinebrychoff and Bavaria brewing companies from Finland and the Netherlands, respectively, and the German engineering company Tuchenhagen have developed a process utilizing immobilized cells for the accelerated maturation of beer. The immobilization of the yeast cells on the carrier (DEAE cellulose particles—Spezyme®) was accomplished by surface adsorption in a downflow packed bed continuous bioreactor through which a yeast slurry was recirculated. The main advantage of this technology is its high volumetric production with corresponding residence times of only a few hours. The system developed by Cultor and their associates is industrially available and has been employed on an industrial scale. Figure 15.10 depicts a schematic of the Cultor system. In order to achieve rapid diacetyl reduction of "green beer" (immature beer), the remaining freely suspended yeast cells are completely centrifuged out of the primary fermentation and the resulting "clear" beer is subjected to heat treatment (65–90 °C with a holding time of 7–20 min). The non-enzymatic conversion of the α-acetolactate (the diacetyl precursor) to acetoin is enhanced in this step. After cooling, the beer is then introduced into the packed bed column containing yeast cells immobilized on DEAE cellulose particles. In this final stage, the yeast cells complete the conversion of the remaining diacetyl into acetoin, while other aspects of flavour maturation also occur. The application of immobilized cell technology in brewing (and to a much less extent in distilling) has been discussed in detail in Chap. 12.

Although research on the metabolism of diacetyl (and other VDKs) has been ongoing since the 1960s (Inoue et al. 1968), a novel approach has been undertaken

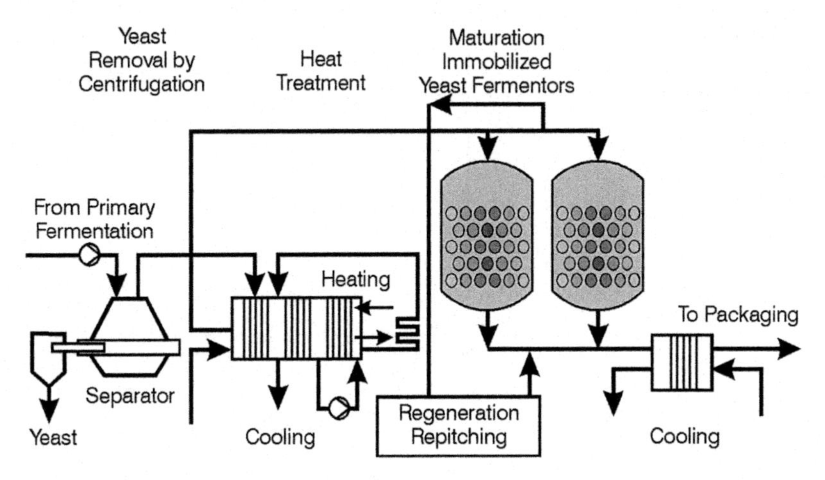

Fig. 15.10 Cultor's 2 h continuous maturation system

since the 1990s. The enzyme α-acetolactate decarboxylase (ALDC) catalyses the following reaction without the formation of diacetyl:

$$\alpha\text{-acetolactate} \rightarrow \text{acetoin} + \text{carbon dioxide}$$

ALDC is not produced by brewer's yeast strains or any other yeast. However, it is produced by some other bacteria that are *g*enerally *r*egarded *a*s *s*afe (GRAS) (e.g. the Gram-negative bacterium *Acetobacter aceti* which converts ethanol to acetic acid during the production of vinegar) (Godtfredsen and Ottesen 1982). ALDC has been isolated, purified and added to a brewer's wort fermentation and the total diacetyl concentration throughout the fermentation cycle determined (Fig. 15.11) (Suihko et al. 1989). Compared with an untreated control, little diacetyl was produced, and its concentration was inversely related to the concentration of ALDC added to the fermentation. However, the ALDC must be added at the start of the fermentation. ALDC does not reduce diacetyl levels if it is added once the fermentation has commenced (no one is sure why). In addition, wort fermentation rates with and without added ALDC were similar (Fig. 15.12). It is worthy of note that commercial quantities of ALDC (under the brand name Maturex®) are produced by a genetically modified strain of *Bacillus subtilis* that has received the genetic coding for ALDC from a strain of *B. brevis*.

The genetic code for ALDC has been cloned into brewing yeast strains (Yamano et al. 1995). Wort fermentation trials with the ALDC yeast were conducted and compared with the same uncloned yeast strain as the control. The diacetyl production and reduction profiles were profoundly different when compared with the uncloned control culture (Fig. 15.13). Due to the pressure of ALDC, the acetolactate was not spontaneously converted to diacetyl but to acetoin instead. However, acetoin does not exhibit the same flavour impact as the butterscotch/stale milk aroma of diacetyl. The overall fermentation performance of some cloned yeast strains can be adversely affected when compared with the uncloned strain

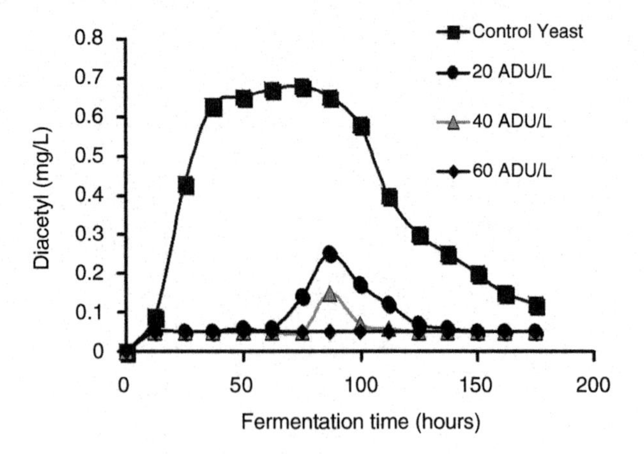

Fig. 15.11 Effect of α-acetolactate decarboxylase (ALDC) on diacetyl metabolism in fermenting wort

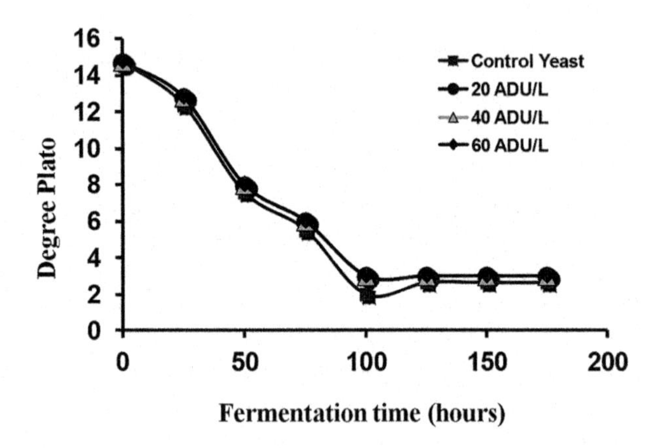

Fig. 15.12 Effect of α-acetolactate decarboxylase (ALDC) on wort fermentation rate and extent

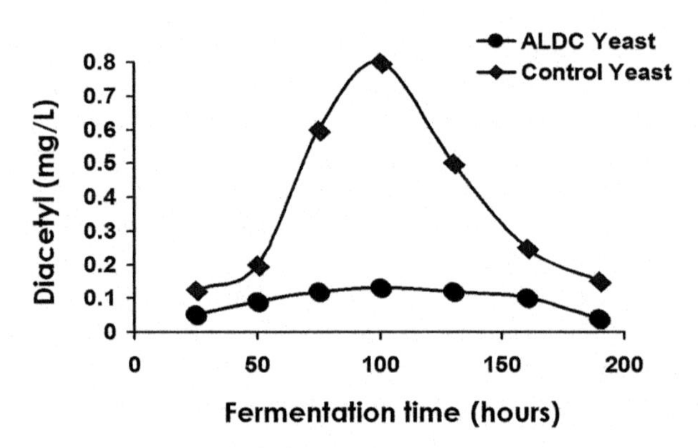

Fig. 15.13 Effect of α-acetolactate decarboxylase (ALDC) in a brewing yeast strain on diacetyl metabolism during wort fermentation

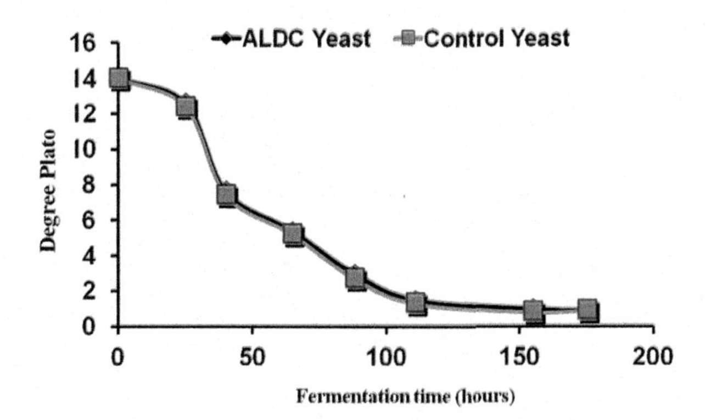

Fig. 15.14 Effect of α-acetolactate decarboxylase (ALDC) expression in a brewing yeast strain on overall fermentation rate during wort fermentation

(Yamano et al. 1995) although this was not the case with the ALDC yeast (Fig. 15.14). However, as far as the author is aware, this yeast is not currently being employed for the production of beer on an industrial scale.

15.5 Sulphur Compounds

Sulphur compounds make a significant contribution to beer flavour and to fresh spirit prior to being matured in oak casks. Although small amounts of sulphur compounds can be acceptable or even desirable in beer, in excess they give rise to unpleasant off-flavours and special process, such as purging with CO_2, use of a copper electrode (Pfisterer et al. 2004) and prolonged maturation times are necessary to remove them (Munroe 2006). Some of the sulphur compounds present in beer are not directly associated with fermentation and are derived from the raw materials (malt, hops, etc.) employed. However, the concentrations of hydrogen sulphide (rotten egg aroma) and sulphur dioxide (burnt match aroma) are dependent on yeast activity. Failure to manage fermentation properly can result in unacceptably high levels of these compounds in finished beer.

The concentration of hydrogen sulphide and sulphur dioxide (Samp and Sedin 2017) formed during fermentation is primarily determined by the yeast strain used, although wort composition and fermentation conditions are important factors, particularly where levels of these compounds are abnormally high. Both compounds arise as by-products of the synthesis of the sulphur-containing amino acids cysteine and methionine from sulphate (Fig. 15.15). These syntheses are influenced by wort composition in that the yeast will preferentially assimilate sulphur-containing amino acids such as methionine (Fig. 7.3) compared to other amino acids. It is only when wort is depleted of wort amino acids does the biosynthetic route(s) of amino acids come into operation (Lekkas et al. 2007).

The peak of hydrogen sulphide and sulphur dioxide production occurs in the second to the fourth days of fermentation (depending on the yeast strain, the wort gravity and the incubation temperature) (Nagami et al. 1980; Duan et al. 2004). Presumably, by this time the sulphur-containing amino acids in wort will have been

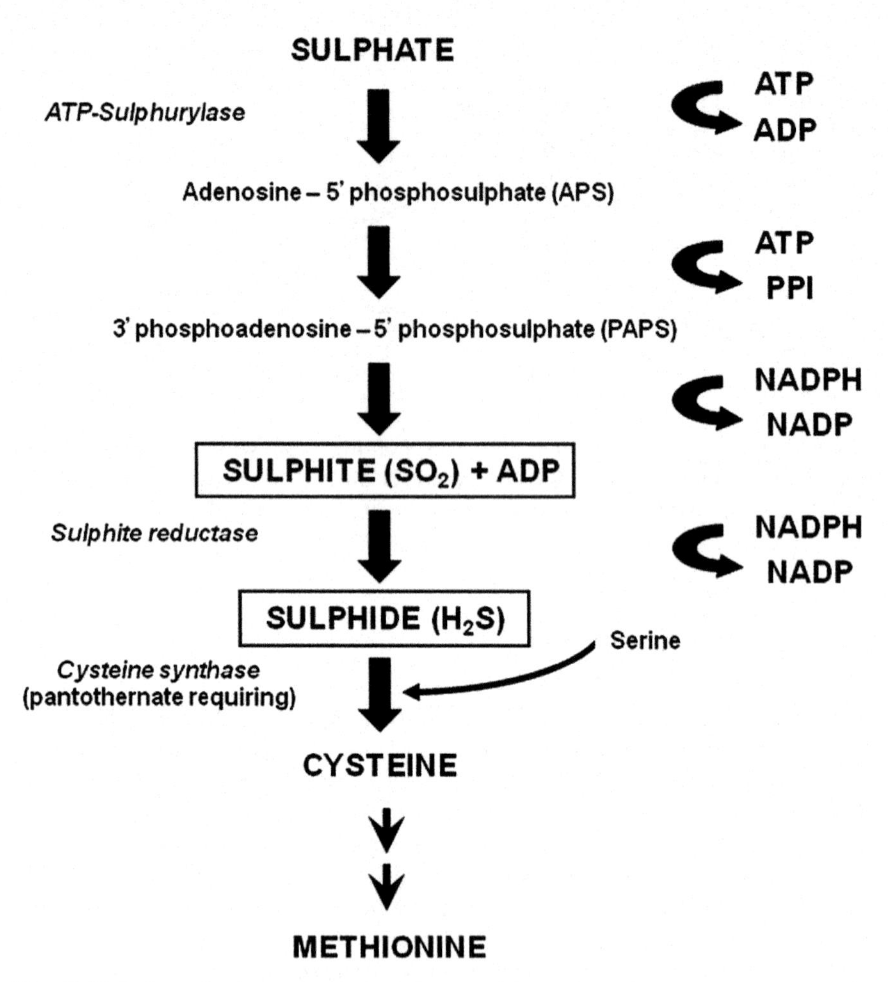

Fig. 15.15 Pathway for the synthesis of sulphur-containing amino acids

utilized. Yeast growth during wort fermentation is approximately synchronous with cell division occurring at the same time, and hydrogen sulphide evolution seems to occur in a number of peaks which correspond to the phase of the yeast cell just prior to the onset of budding (Ogata 2013).

The formation of excessive levels of hydrogen sulphide and sulphur dioxide during fermentation is, therefore, associated with conditions that restrict yeast growth. In this regard, the provision of adequate oxygen at the time of pitching is a critical factor. Since both hydrogen sulphide and sulphur dioxide are volatile, it follows that vigorous fermentation will promote their removal via carbon dioxide purging. The type and geometry of the fermentation vessel is also influential. Sulphur dioxide, as well as contributing to beer flavour (burnt match), has a number of other functions in beer (and other alcoholic beverages). It can act as an antimicrobial agent and antioxidant (Barker et al. 1983; Collin et al. 1991), and it retards the development of the beer staling characters. Bisulphite's antimicrobial activity

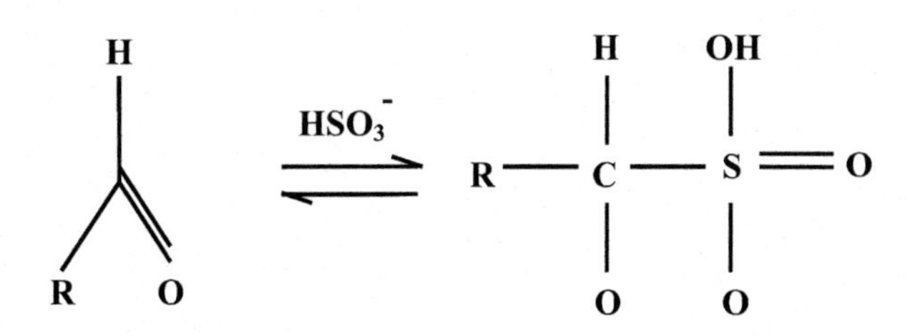

Fig. 15.16 Binding of bisulphate to carbonyls

can only occur at concentrations in excess of 50 mg/L which is well above the permitted limit in beer for most countries (in the United States it is 10 mg/L, Canada less than 10 mg/L and the United Kingdom 25 mg/L). Due to the adverse effects of oxygen on the flavour stability of finished beer, some brewers (not all) add bisulphites or other antioxidants, such as ascorbic acid (vitamin C), to beer prior to packaging in order to provide protection against oxygen pickup. However, because of bisulphite's allergenic properties, its use is questionable and is decreasing.

The effectiveness of bisulphite, besides its antioxidant properties, is also its ability to bind carbonyl compounds into flavour neutral complexes (Barker et al. 1983). This can improve beer flavour stability. The efficiency of bisulphite binding to carbonyl compounds does vary. The reaction is reversible, so that an excess of bisulphite will increase the yield of the flavour inactive adjunct (Fig. 15.16). The addition of bisulphites to fresh beer minimizes increases in acetaldehyde concentrations during ageing. In addition, when added to stale beer, bisulphite lowers the concentration of free aldehyde and affects the removal of the cardboard flavour (Barker et al. 1983). Over time, the bisulphite will be oxidized, by oxygen in the beer, to sulphate, thus increasing the concentration of free aldehydes (Narziss 1986). Research at the Carlsberg Technical Centre in Copenhagen (Chap. 2) has developed a genetically manipulated yeast strain that hyper-produces sulphur dioxide. Results would indicate that beer produced with it has enhanced flavour stability (Hansen and Kielland-Brandt 1996). More recently, the development of a *S. cerevisiae* mutant which produces higher levels of sulphur dioxide and glutathione but a lower level of hydrogen sulphide with improved beer flavour stability has been reported (Chen et al. 2012). This mutant has been developed by nongenetic engineering means and employed UV mutagenesis together with specific plate screening methods. The antioxidizing ability of the mutant was significantly improved and would be safe for public use.

The process of maturation (ageing or lagering) is a complex process (one aspect—the management of VDK has already been discussed in this chapter). Maturation also manages the levels of unwanted sulphur compounds (e.g. H_2S, SO_2 and thiols) during primary fermentation. The reduction of unwanted volatile sulphur compounds occurs principally during secondary fermentation (lagering). For this purpose, some yeast must be present during secondary fermentation. The yeast employed for this role is usually the same yeast strain as that used for primary fermentation. However, sometimes a speciality lagering yeast strain is employed (Munroe 2006), and the conditions that the yeast encounters during maturation are different to those during primary fermentation—lower pH, higher alcohol and

Table 15.8 Content of hydrogen sulphide (H_2S) and copper in maturing beer samples

	Before electrolysis		After electrolysis	
	Copper (μg/L)	H_2S (μg/L)	Copper (μg/L)	H_2S (μg/L)
Beer A	32	4	69	Traces
Beer B	29	3	68	Traces

carbon dioxide concentrations, CO_2 pressure, reduced osmotic and temperature pressure, etc. Copper ions in fermented wort have a positive effect on the reduction of these sulphur compounds (Boulton and Quain 2001; Pfisterer et al. 2004). The positive effects of copper in this regard can also be illustrated by its importance in the distilling industry (Bringhurst and Brosnan 2014).

The use of copper in the construction of pot stills for the distillation of malt spirit (Nicol 2014) and for incorporation in the plates of continuous stills during grain whisky distillation (Murray 2014) is invaluable for the removal of unwanted sulphur compounds. The expanding application of stainless steel for the manufacture of brewing equipment, replacing copper, has reduced the beer copper concentration in many breweries since World War II. Copper ions can precipitate hydrogen sulphide (and other sulphur-containing compounds) as insoluble copper sulphide. The resultant CuS subsequently can be removed by filtration of the maturing beer or adsorbed by the filter bed. The traditional use of copper vessels, pipes and plates during wort and beer production does not permit precise copper ion addition. Consequently, deliberate and careful treatment of beer with copper is advisable. Copper electrolysis of maturing beer is a potential alternative (Pfisterer et al. 2004). It should be noted that Cu^{++} ions can have a negative effect on beer flavour stability. For example, Irwin and his colleagues reported (Irwin et al. 1991) that the rate of beer staling increased in the presence of small amounts of copper ions. As a transition metal ion, copper catalyses the activation of molecular oxygen, which in turn can oxidize primary alcohols to beer-staling aldehydes. Nevertheless, the application of a copper electrode during maturation cannot be ignored. In a typical application (Pfisterer et al. 2004), a copper electrolysis system reduced H_2S levels in beer from 4 μg/L to undetectable levels and beer copper levels increased from 32 to 68 μg/L (Table 15.8). Many years ago, in an attempt to reduce maturation times, copper sulphate was added to boiling wort or to the initial stages of fermentation to remove compounds such as copper sulphide (Boulton and Quain 2001). Although, copper's negative effect on beer flavour stability was unknown at the time, copper sulphate addition to wort has not been employed during the brewing process for a long time!

15.6 Summary

The final flavour of beer and spirits is the sum of several hundreds of flavour-active compounds produced in every stage of the brewing process. The great majority (not all) of these substances are yeast metabolites. They are produced during wort fermentation and consist of fermentation intermediates or by-products. Higher alcohols (also called fusel oils), esters, vicinal diketones (VDKs), other carbonyls and sulphur compounds (inorganic and organic) are key elements produced by yeast. The constituent compounds determine a beer's final quality, particularly when it is fresh. Although ethanol, carbon dioxide and glycerol are the major products produced by yeast during wort fermentation, they have little impact on the flavour of the final beer. It is the type and concentration of the excretion products already discussed that primarily determines beer flavour balance. There are many factors that can modify this balance including: the yeast strain, malt variety, fermentation temperature, adjunct type and level, fermenter design and geometry, wort pH and clarity, media buffering capacity, wort gravity, etc.

References

Alvarez P, Sampedro M, Molina M, Nombela C (1994) A new system for the release of heterologous proteins from yeast based on mutant strains deficient in cell integrity. J Biotechnol 38:81–88

Ancin H, Roysam B, Dufresne TE, Chestnut MM, Ridder GM, Szarowski DH, Turner JN (1996) Advances in automated 3D image analyses of cell populations imaged by confocal microscopy. Cytometry 25:221–234

Anderson NL, Anderson NG (1998) Proteome and proteomics; new technologies, new concepts and new words. Electrophoresis 19:1853–1861

Aranda A, del Olmo M (2003) Response to acetaldehyde stress in the yeast *Saccharomyces cerevisiae* involves a strain-dependent regulation of several *ALD* genes and is mediated by the general stress response pathway. Yeast 20:747–759

Aranda A, Querol A, Olmo d (2002) Correlation between acetaldehyde and ethanol resistance and expression of *HSP* genes in yeast strains isolated during the biological ageing of sherry wines. Arch Microbiol 177:304–312

Äyräpää T, Lindström I (1973) Influence of long-chain fatty acids on the formation of esters by brewer's yeast. Proc Congr Eur Brew Conv 14:271–283

Bamforth CW (2011) 125th anniversary review: the non-biological instability of beer. J Inst Brew 117:488–497

Bamforth CW, Kanauchi M (2004) Enzymology of vicinal diketone reduction in brewer's yeast. J Inst Brew 110:83–93

Barker RL, Gracey DEF, Irwin AJ, Pipasts P, Leiska E (1983) Liberation of staling aldehydes during storage of beer. J Inst Brew 89:411–415

Beier DR, Sledziewski A, Young ET (1985) Deletion analysis identifies a region, upstream of the *ADH2* gene of *S. cerevisiae*, which is required for *ADR1*-mediated derepression. Mol Cell Biol 5:1743–1749

Berner TS, Arneborg N (2012) The role of lager beer yeast in oxidative stability of model beer. Lett Appl Microbiol 54:225–232

Blasco L, Vinas M, Villa TG (2011) Proteins influencing foam formation in wine and beer: the role of yeast. Int Microbiol 14:61–71

Boulton C, Box W (2003) Formation and disappearance of diacetyl during lager fermentation. In: Smart K (ed) Brewing yeast fermentation performance. Blackwell Science, Oxford, pp 183–195

Boulton C, Quain D (2001) Chap. 3: The biochemistry of fermentation. Blackwell Science, Oxford, pp 69–142

Brányik T, Silva DP, Baszczyňski M, Lehnert R, João B Almeida e Silva RJB (2012) A review of methods of low alcohol and alcohol-free beer production. J Food Eng 108:493–506

Bravi E, Perretti G, Buzzini P, Della Sera R, Fantozzi P (2009) Technological steps and yeast biomass as factors affecting the lipid content of beer during the brewing process. J Agric Food Chem 57:6279–6284

Bringhurst TA, Brosnan J (2014) Scotch whisky: raw material selection and processing. In: Russell I, Stewart G (eds) Whisky: technology, production and marketing, 2nd edn. Elsevier, Oxford, pp 49–122

Chen E-H (1978) Relative contribution of Ehrlich and biosynthetic pathways to the formation of fusel alcohols. J Am Soc Brew Chem 36:39–43

Chen Y, Siewers V, Nielsen J (2012) Profiling of cytosolic and peroxisomal acetyl-coa metabolism in *Saccharomyces cerevisiae*. PLoS One 7(8):e42475

Collin S, Montesinos M, Meersman E, Swinkles W, Dufour JP (1991) Yeast dehydrogenase activities in relation to carbonyl compounds removal from wort and beer. Proc Congr Eur Conv 23:409–416

Culver KW, Labow MA (2002) Genomics. In: Robinson R (ed) Genetics. Macmillan Science Library. Macmillan Reference USA, New York

Cyr N, Blanchette M, Price S, Sheppard J (2007) Vicinal diketone production and amino acid uptake by two active dry lager yeasts during beer fermentation. J Am Soc Brew Chem 65:138–144

Daviss B (2005) Growing pains for metabolomics. The Scientist 19:25–28

Dekoninck T, Verbelen PJ, Delvaux F, Van Mulders SE, Delvaux F (2012) The importance of wort composition for yeast metabolism during accelerated brewery fermentations. J Am Soc Brew Chem 70:195–204

Dickinson JR, Norte V (1993) A study of branched-chain amino acid aminotransferase and isolation of mutations affecting the catabolism of branched-chain amino acids in *Saccharomyces cerevisiae*. FEBS Lett 326:29–32

Dillemans M, Goossens E, Goffin O, Masschelein C (1987) The amplification effect of the *ILV5* gene on the production of vicinal diketones in *Saccharomyces cerevisiae*. J Am Soc Brew Chem 45:81–84

Duan W, Roddick F, Higgins V, Rogers P (2004) A parallel analysis of H_2S and SO_2 formation by brewing yeast in response to sulfur-containing amino acids and ammonium ions. J Am Soc Brew Chem 62:35–41

Dufour J-P, Malcorps P (1994) Ester synthesis during fermentation: enzymes characterization and modulation mechanism. In: Campbell I, Priest FG (eds) Proceedings of the 4th aviemore conference on malting, brewing and distilling. The Institute of Brewing, London, pp 137–151

Ehrlich F (1907) Über die Bedingungen der Fuselölbildung und über ihren Zusammenhang mit dem Eiweissaufbau der Hefe. Ber Dtsch Chem Ges 40:1027–1047

Ekberg J, Rautio J, Mattinen L, Vidgren V, Londesborough J, Gibson BR (2013) Adaptive evolution of the lager brewing yeast *Saccharomyces pastorianus* for improved growth under hyperosmotic conditions and its influence on fermentation performance. FEMS Yeast Res 13:335–349

Engan S (1981) Beer composition: volatile substances. In: Pollick J (ed) Brewing science, vol 2. Academic, London, pp 7–67

Ernandes JR, Williams JW, Russell I, Stewart GG (1993) Respiratory deficiency in brewing yeast strains – effects on fermentation, flocculation, and beer flavor components. J Am Soc Brew Chem 51:16–20

Fujii T, Nagasawa N, Iwamatsu A, Bogaki T, Tamai Y, Hamachi M (1994) Molecular cloning, sequence analysis and expression of the yeast alcohol acetyltransferase gene. Appl Environ Microbiol 60:2786–2792

Fujii T, Yoshimoto H, Nagasawa N, Bogaki T, Tamai Y, Hamachi M (1996a) Nucleotide sequence of alcohol acetyltransferase genes from lager brewing yeast. *Saccharomyces carlsbergensis*. Yeast 12:593–598

Fujii T, Yoshimoto H, Tamai T (1996b) Acetate ester production by *Saccharomyces cerevisiae* lacking the ATF1 gene encoding the alcohol acetyltransferase. J Ferment Bioeng 81:538–542

Fukuda K, Yamamoto N, Kiyokawa Y, Yanagiuchi T, Wakai Y, Kitamoto K, Inoue Y, Kimura A (1998) Balance of activities of alcohol acetyltransferase and esterase in *Saccharomyces* cerevisiae is important for production of isoamyl acetate. Appl Environ Microbiol 64:4076–4078

Garcia A, Garcia L, Diaz M (1994) Modelling of diacetyl production during beer fermentation. J Inst Brew 100:179–183

Godtfredsen SE, Ottesen M (1982) Maturation of beer with alpha-acetolactate decarboxilase. Carlsb Res Commun 47:93–102

Goffeau A, Barrell BG, Bussey H, Davis RW, Dujon B, Feldmann H, Galibert F, Hoheisel JD, Jacq C, Johnston M, Louis EJ, Mewes HW, Murakami Y, Philippsen P, Tettelin H, Oliver SG (1996) Life with 6000 genes. Yeast 274:546–567

Hansen J, Kielland-Brandt MC (1996) Inactivation of MET2 in brewer's yeast increase the level of sulfite in beer. J Biotechnol 50:75–87

Hiralal L, Olaniran AO, Pillay B (2014) Aroma-active ester profile of ale beer produced under different fermentation and nutritional conditions. J Biosci Bioeng 117:57–64

Inoue T (2008) Diacetyl in fermented foods and beverages. St Paul, MN, American Society of Brewing Chemists

Inoue T, Yamamoto Y (1970) Report of the Research Laboratories of the Kirin Brewery Co. 13:79

Inoue T, Masuyama K, Yamamoto Y, Okada K (1968) Report of the Research Laboratories of the Kirin Brewery Co. 11:1

Irwin AJ, Barker RL, Pipasts P (1991) The role of copper, oxygen, and polyphenols in beer flavor instability. J Am Soc Brew Chem 49:140–149

Jones M, Pierce J (1964) Absorption of amino acids from wort by yeasts. J Inst Brew 70:307–315

Kluba R, de Banchs N, Fraga A, Jansen G, Langstaff S, Meilgaard M, Nonaka R, Thompson S, Verhagen L, Word K, Crumplen R (1993) Sensory threshold determination of added substances in beer. J Am Soc Brew Chem 51:181–183

Klug L, Daum G (2014) Yeast lipid metabolism at a glance. FEMS Yeast Res 14:369–388

Kobayashi K, Kusaka K, Takahashi T, Sato K (2005) Method for the simultaneous assay of diacetyl and acetoin in the presence of α-acetolactate: application in determining the kinetic parameters for the decomposition of α-acetolactate. J Biosci Bioeng 99:502–507

Krogerus K, Gibson BR (2013a) Influence of valine and other amino acids on total diacetyl and 2,3-pentanedione levels during fermentation of brewer's wort. Appl Microbiol Biotechnol 97:6919–6930

Krogerus K, Gibson BR (2013b) 125th anniversary review: diacetyl and its control during brewery fermentation. J Inst Brew 119:86–97

Krogerus K, Magalhães F, Vidgren V, Gibson BR (2015) New lager strains generated by interspecific hybridization. J Ind Microbiol Biotechnol 42:769–778

Kucharczyk K, Tuszyński T (2015) The effect of pitching rate on fermentation, maturation and flavour compounds of beer produced on an industrial scale. J Inst Brew 121:349–355

Lake JC, Speers RA (2008) A discussion of malt-induced premature yeast flocculation. MBAA Tech Quart 4:253–262

Lake JC, Speers RA, Porter AV, Gill TA (2008) Miniaturizing the fermentation assay: effects of fermentor size and fermentation kinetics on detection of premature yeast flocculation. J Am Soc Brew Chem 66:94–102

Landaud S, Lieben P, Picque D (1998) Quantitative analysis of diacetyl, pentanedione and their precursors during beer fermentation by an accurate gc/ms method. J Inst Brew 104:93–99

Landaud S, Latrille E, Corrieu G (2001) Top pressure and temperature control the fusel alcohol/ester ratio through yeast growth in beer fermentation. J Inst Brew 107:107–117

Lekkas C, Stewart GG, Hill A, Taidi B, Hodgson J (2007) Elucidation of the role of nitrogenous wort components in wort fermentation. J Inst Brew 113:183–191

Lilly M, Bauer FF, Styger G, Lambrechts MG, Pretorius IS (2006) The effect of increased branched-chain amino acid transaminase activity in yeast on the production of higher alcohols and on the flavour profiles of wine and distillates. FEMS Yeast Res 6:726–743
Lodolo EJ, Koc k JL, Axcell BC, Brooks M (2008) The yeast *Saccharomyces cerevisiae* – the main character in beer brewing. FEMS Yeast Res 8:1018–1036
Lutstorf U, Megnet R (1968) Multiple forms of alcohol dehydrogenase in *S. cerevisiae.* I. Physiological control of ADH-2 and properties of ADH-2 and ADH-4. Arch Biochem Biophys 126:933–944
Lyness CA, Steele GM, Stewart GG (1997) Investigating ester metabolism: characterisation of the ATF1 gene in *Saccharomyces cerevisiae*. J Am Soc Brew Chem 55:141–146
Malcorps P, Dufour JP (1992) Short-chain and medium-chain aliphatic-ester synthesis in *Saccharomyces cerevisiae*. Eur J Biochem 210:1015–1022
Mason B, Dufour J-P (2000) Alcohol acetyltransferases and the significance of ester synthesis in yeast. Yeast 16:1287–1298
Meilgaard M (1975a) Flavour chemistry of beer. Part I flavour interaction between principal volatiles. MBAA Tech Q 12:107–117
Meilgaard M (1975b) Flavour chemistry of beer. Part II flavour and threshold of 235 aroma volatiles. MBAA Tech Q 12:151–168
Merritt NR (1967) The effect of suspended solids on the fermentation of distiller's malt wort. J Inst Brew 73:484–488
Molina AM, Swiegers JH, Varela C, Pretorius IS, Agosin E (2007) Influence of wine fermentation temperature on the synthesis of yeast-derived volatile aroma compounds. Appl Microbiol Biotechnol 77:675–687
Munroe JH (2006) Fermentation. In: Priest FG, Stewart GG (eds) Handbook of brewing, 2nd edn. Taylor & Francis, Boca Raton, FL, pp 487–524
Murray D (2014) Grain whisky distillation. In: Russell I, Stewart G (eds) Whisky: technology, production and marketing, 2nd edn. Elsevier, Oxford, pp 179–198
Nagami K, Takahashi T, Nakatani K, Kumada J (1980) Hydrogen sulphide in brewing. MBAA Tech Q 17:64–68
Nagasawa N, Bogaki T, Iwamatsu A, Hamachi M, Kumagai C (1998) Cloning and nucleotide sequence of the alcohol acetyltransferase II gene (ATF2) from *Saccharomyces cerevisiae* Kyokai No 7. Biosci Biotechnol Biochem 62:1852–1857
Nakatani K, Takahashi T, Nagami K, Kumada J (1984) Kinetic study of vicinal diketones in brewing II: theoretical aspect for the formation of total vicinal diketones. MBAA Tech Q 21:175–183
Narziss L (1986) Centenary review. Technological factors of flavour stability. J Inst Brew 92:346–353
Neubauer O, Fromherg K (1911) Über den Abbau der Aminosäuren bei der Hefegärung.–De Gruyter. Hoppe-Seyler's. Z Physiol Chem 70:326–350
Neven H, Delvaux F, Derdelinck G (1997) Flavor evaluation of top fermented beers. MBAA Tech Q 34:115–118
Nicholson JK, Lindon JC (2008) Systems biology: metabonomics. Nature 455(7216):1054–1056
Nicol DA (2014) Batch distillation. In: Russell I, Stewart G (eds) Whisky: technology, production and marketing, 2nd edn. Elsevier, Oxford, pp 155–178
Nordström K (1962) Formation of ethyl acetate in fermentation with brewer's yeast. J Inst Brew 67:173–181
Nykänen L, Nykänen I (1977) Production of esters by different yeast strains in sugar fermentations. J Inst Brew 83:30–31
Nykänen L, Nykänen I, Suomalainen H (1977) Distribution of esters produced during sugar fermentation between the yeast cell and the medium. J Inst Brew 83:32–34
O'Connor-Cox ESC, Lodolo EJ, Axcell BC (1996a) Mitochondrial relevance to yeast fermentative performance: a review. J Inst Brew 102:19–25

O'Connor-Cox ESC, Lodolo EJ, Steyn GJ, Axcell BC (1996b) High-gravity wort clarity and its effect on brewing yeast performance. MBAA Tech Q 33:20–29

Ogata T (2013) Hydrogen sulphide production by bottom-fermenting yeast is related to nitrogen starvation signalling. J Inst Brew 119:228–236

Pajunen E (1995) Immobilised yeast lager beer maturation: DEAE-cellulose at Sinebrychoff. Proc Eur Brew Conv 24:23–34

Panchal C, Peacock L, Stewart GG (1982) Increased osmotolerance of genetically modified ethanol producing strains of *Saccharomyces* sp. Biotechnol Lett 4:639–644

Peddie HAB (1990) Ester formation in brewery fermentations. J Inst Brew 96:327–331

Perpete P, Collin S (2000) Influence of beer ethanol content on the wort flavour perception. Food Chem 71:379–385

Petersen E, Margaritis A, Stewart R, Pilkington P, Mensour N (2004) The effects of wort valine concentration on the total diacetyl profile and levels late in batch fermentations with brewing yeast *Saccharomyces carlsbergensis*. J Am Soc Brew Chem 62:131–139

Pfisterer E, Richardson I, Soti A (2004) Control of hydrogen sulfide in beer with a copper electrolysis system. MBAA Tech Q 41:50–52

Piper PW, Talreja K, Panaretou B, Moradas-Ferreira P, Byrne K, Praekelt UM, Meacock P, Recnacq M, Boucherie H (1994) Induction of major heat-shock proteins of *Saccharomyces cerevisiae*, including plasma membrane Hsp30, by ethanol levels above a critical threshold. Microbiology 140:3031–3038

Pires EJ, Teixeira JA, Brányik T, Vicente AA (2014) Yeast: the soul of beer's aroma-a review of flavour-active esters and higher alcohols produced by the brewing yeast. Appl Microbiol Biotechnol 98:1937–1949

Pugh T, Maurer J, Pringle A (1997) The impact of wort nitrogen limitation on yeast fermentation performance and diacetyl. MBAA Tech Q 34:185–189

Quain D, Smith I (2009) The long and short of maturation. Brew Guard 138:56–61

Radhakrishnan AN, Wagner RP, Snell EE (1960) Biosynthesis of valine and isoleucine III α-keto-β-hydroxy acid reductase and α-hydroxy-β-keto acid reductoisomerase. J Biol Chem 235:2232–2242

Ramos-Jeunehomme C, Laub R, Masschelein CA (1981) Why is ester formation in brewery fermentations yeast strain dependent? Proc Cong Eur Brew Conv 23:257–264

Roberts T, Falconer F (2018) Hops. In: Stewart GG, Anstruther A, Russell I (eds) Handbook of brewing, 3rd edn. Taylor and French, Boca Raton, FL, pp 148–224

Rodrigues D, Rocha-Santos TAP, Pereira CI, Gomes AM, Malcata FX, Freitas AC (2011) The potential effect of FOS and inulin upon probiotic bacterium performance in curdled milk matrices. LWT – Food Sci Technol 44:100–108

Romano P, Suzzi G, Turbanti L, Polsinelli M (1994) Acetaldehyde production in *Saccharomyces cerevisiae* wine yeasts. FEMS Microbiol Lett 118:213–218

Rondags E, Germain P, Marc I (1999) Quantification of extracellular α-acetolactate oxidative decarboxylation in diacetyl production by an α-acetolactate overproducing strain of *Lactococcus lactis* sp. *lactis* bv. *Diacetylactis*. Biotechnol Lett 21:303–307

Rossouw D, Naes T, Bauer FF (2008) Linking gene regulation and the exo-metabolome: a comparative transcriptomics approach to identify genes that impact on the production of volatile aroma compounds in yeast. BMC Genomics 9:530

Russell I, Stewart GG (2014) Whisky: technology, production and marketing. Elsevier, Boston, MA

Ryan ED, Kohlhaw GB (1974) Subcellular localization of isoleucine-valine biosynthetic enzymes in yeast. J Bacteriol 120:631–637

Saerens SMG, Verstrepen KJ, Laere V, Voet SDM, Van Dijck ARD, Delvaux P (2006) The *Saccharomyces cerevisiae* EHT1 and EEB1 genes encode novel enzymes with medium-chain fatty acid ethyl ester synthesis and hydrolysis capacity. J Biol Chem 281:4446–4456

Saerens SM, Delvaux F, Verstrepen KJ, Van Dijck P, Thevelein JM, Delvaux FR (2008a) Parameters affecting ethyl ester production by *Saccharomyces cerevisiae* during fermentation. Appl Environ Microbiol 74:454–461

Saerens SM, Verbelen PJ, Vanbeneden N, Thevelein JM, Delvaux FR (2008b) Monitoring the influence of high-gravity brewing and fermentation temperature on flavour formation by analysis of gene expression levels in brewing yeast. Appl Microbiol Biotechnol 80:1039–1051

Saison D, De Schutter DP, Delvaux F, Delvaux FR (2009a) Determination of carbonyl compounds in beer by derivatisation and headspace solid-phase microextraction in combination with gas chromatography and mass spectrometry. J Chromatogr A 1216:5061–5068

Saison D, De Schutter DP, Uyttenhove B, Delvaux F, Delvaux FR (2009b) Contribution of staling compounds to the aged flavour of lager beer by studying their flavour thresholds. Food Chem:114:1206–114:1215

Samp EJ, Sedin D (2017) Important aspects of controlling sulfur dioxide in brewing. MBAA Tech Q 54:60–71

Sentheshanmuganathan S (1960) The purification and properties of the tyrosine-2-oxoglutarate transaminase of *Saccharomyces cerevisiae*. Biochem J 74:568–576

Sentheshanmuganathan S, Elsden SR (1958) The mechanism of the formation of tyrosol by *Saccharomyces cerevisiae*. Biochem J 69:210–218

Sigler K, Matoulková D, Dienstbier M, Gabriel V (2009) Net effect of wort osmotic pressure on fermentation course, yeast vitality, beer flavor, and haze. Appl Microbiol Biotechnol 82:1027–1035

Smart KA (2007) Brewing yeast genomes and genome-wide expression and proteome profiling during fermentation. Yeast 24:993–1013

Stewart GG (2007) The influence of high gravity wort on the stress characteristics of brewer's yeast and related strains. Cerevisia 32:37–48

Stewart GG (2008) Esters – the most important group of flavour-active compounds in alcoholic beverages. In: Bryce JH, Piggott JR, Stewart GG (eds) Distilled spirits. Production, technology and innovation. Nottingham University Press, Nottingham, pp 243–250

Stewart GG (2014a) Saccharomyces. In: Catt C, Tortorello M-L (eds) Encyclopedia of food microbiology, vol 3, 2nd edn. Elsevier, Oxford, pp 297–315

Stewart GG (2014b) Brewing intensification. American Society for Brewing Chemists, St. Paul, MN

Stewart GG, Martin SA (2004) Wort clarity: effects on fermentation. MBAA Tech Q 41:18–26

Stewart GG, Russell I (2009) An introduction to brewing science and technology, Series lll, Brewer's yeast, 2nd edn. The Institute of Brewing and Distilling, London

Stewart GG, Lyness CA, Younis OS (1999) The control of ester synthesis during wort fermentation. MBAA Tech Q 36:61–66

Stewart GG, Maskell DL, Speers A (2016) Brewing fundamentals – fermentation. MBAA Tech Q 53:2–22

Strejc J, Siříšťová L, Karabín M, e Silva JBA, Brányik T (2013) Production of alcohol-free beer with elevated amounts of flavouring compounds using lager yeast mutants. J Inst Brew 119:149–155

Suihko M-L, Penttila M, Sone H, Home S, Blomqvist K, Tanaka J, Inoue T, Knowles J (1989) Pilot brewing with alpha-acetolactate decarboxylase active yeasts. In: Proceedings of the 22nd congress of the European brewery convention, Zurich, IRL Press, Oxford, p 483

Thomas C, Hynes SH, Ingledew WM (1994) Effects of particulate materials and osmoprotectants on very high gravity ethanolic fermentation by *Saccharomyces cerevisiae*. Appl Environ Microbiol 60:1519–1524

van Bergen B, Strasser R, Cyr N, Sheppard JD, Jardim A (2006) α,β-dicarbonyl reduction by *Saccharomyces* D-arabinose dehydrogenase. Biochim et Biophys Acta (BBA) – General Subjects 1760:1636–1645

Vanderhaegen V, Neven H, Coghe S, Verstrepen KJ, Verachtert H, Derdelinckx G (2003) Evolution of chemical and sensory properties during aging of top-fermented beer. J Agric Food Chem 51:6782–6790

Vanderhaegen V, Neven H, Verachtert H, Derdelinckx G (2006) The chemistry of beer aging – a critical review. Food Chem 95:357–381

Verbelen P, van Mulders S, Saison D, van Laere S, Delvaux F, Delvaux FR (2008) Characteristics of high cell density fermentations with different lager yeast strains. J Inst Brew 114:127–133

Verbelen P, Dekoninck T, Saerens S, van Mulders S, Thevelein J, Delvaux F (2009a) Impact of pitching rate on yeast fermentation performance and beer flavour. Appl Microbiol Biotechnol 82:155–167

Verbelen PJ, Dekoninck TM, Van Mulders SE, Saerens SM, Delvaux F, Delvaux FR (2009b) Stability of high cell density brewery fermentations during serial repitching. Biotechnol Lett 31:1729–1737

Verstrepen KJ, Van Laere SDM, Vanderhaegen BMP, Derdelinckx G, Dufour JP, Pretorius IS, Winderickx J, Thevelein JM, Delvaux FR (2003) Expression levels of the yeast alcohol acetyltransferase genes ATF1, Lg-ATF1, and ATF2 control the formation of a broad range of volatile esters. Appl Environ Microbiol 69:5228–5237

Wainwright T (1973) Diacetyl – a review. J Inst Brew 79:451–470

Yamano S, Tomizuka K, Sone H, Imura M, Takeuchi S, Inoue T (1995) Brewing performance of a brewer's yeast having alpha-acetolactate decarboxylase from *Acetobacter aceti* spp. *xylinum* in brewer's yeast. J Biotechnol 39:21–26

Yonezawa T, Stewart GG (2004) Monitoring and controlling whisky fermentation. In: Bryce JH, Stewart GG (eds) Distilled spirits. Tradition and innovation. Nottinghan University Press, Nottingham, pp 103–111

Yoshimoto H, Fujiwara D, Momma T, Ito C, Sone H, Kaneko Y, Tamai Y (1998) Characterization of the ATF1 and Lg-ATF1 genes encoding alcohol acetyltransferase in the bottom fermenting yeast *Saccharomyces pastorianus*. J Ferment Bioeng 86:15–20

Yoshioka K, Hashimoto N (1981) Ester formation by alcohol acetyl transferase from brewers' yeast. Agric Biol Chem 45:2183–2190

Younis OS, Stewart GG (1998) Sugar uptake and subsequent ester and alcohol production in *Saccharomyces cerevisiae*. J Inst Brew 104:255–264

Younis OS, Stewart GG (1999) The effect of malt wort, very high gravity malt wort and very high gravity adjunct wort on volatile production in *Saccharomyces cerevisiae*. J Am Soc Brew Chem 52:38–45

Zhang CY, Liu YL, Qi YN, Zhang JW, Dai LH, Lin X, Xiao DG (2013) Increased esters and decreased alcohols production by brewer's yeast strains. Eur Food Res Technol 236:1009–1014

Chapter 16
Yeast Genetic Manipulation

16.1 Introduction

The importance of the molecular biology of *Saccharomyces cerevisiae* and its closely related species has recently been clearly described by Pretorius (2016). He discusses that the remarkable academic and industrial track record of this microorganism that is highlighted by many world-firsts. This yeast was the first microorganism to be domesticated for the production of fermented food and beverages, and it was the first microbe to be observed under the microscope and described it as a living biochemical agent for biological transformations (details in Chap. 2) (Stewart 2002). In more recent times, *S. cerevisiae* has been the host for the production of the first recombinant vaccine (against hepatitis B) and the first recombinant food enzyme [the milk coagulation enzyme chymosin, for cheese making (Pretorius et al. 2003; Jacob et al. 2011)].

In 1996, the complete genome of a *S. cerevisiae* haploid strain (S288c) became the first eukaryote genome to be fully sequenced. Its 16 chromosomes encode approximately 6000 genes, and approximately 5000 of them are individually non-essential to the yeast cell (Goffeau et al. 1996; Oliver 1996a). This publicly available genome sequence has prepared the way to build the first systematic collection of deletion mutants, which enables high-throughput functional genetics experiments (Winzeler et al. 1999). In 2014, *S. cerevisiae* became the first eukaryote to be equipped with a functional synthetic chromosome (Peris et al. 2014; Gibson and Liti 2015).

S. cerevisiae is a budding yeast species (Figs. 1.1 and 16.1) (details in Chaps. 1 and 3). There are three basic cell mating types—a, α and a/α (Fig. 16.2). These cells are easy and relatively inexpensive to grow in the laboratory under optimal nutritional and cultural conditions. They double their mass approximately every 90 min (sometimes less depending on the growth medium). Through an asexual mitotic budding process, heterothallic strains can multiply as stable a or α haploids, while both heterothallic and homothallic strains can reproduce in the a/α diploid state or

G.G. Stewart, *Brewing and Distilling Yeasts*, The Yeast Handbook,
https://doi.org/10.1007/978-3-319-69126-8_16

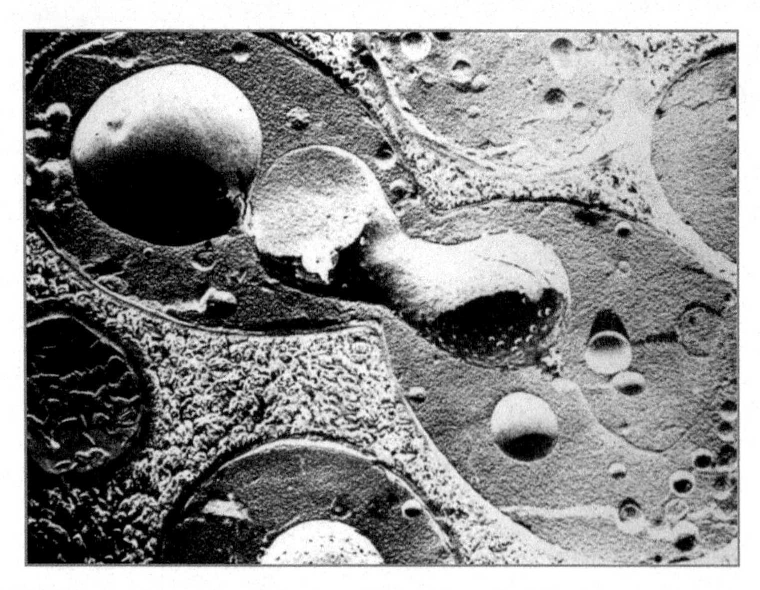

Fig. 16.1 Electron micrograph of a freeze-etched budding cell of *Saccharomyces cerevisiae*. Bar represents 1 μm. Photograph courtesy of the late C.F. Robinow, University of Western Ontario, London, Ontario, Canada

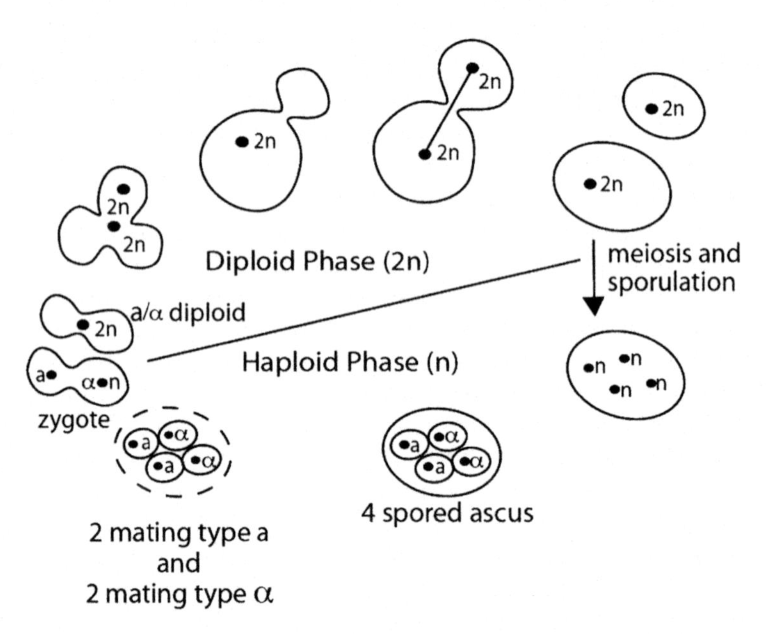

Fig. 16.2 Haploid/diploid life cycle of *Saccharomyces* sp.

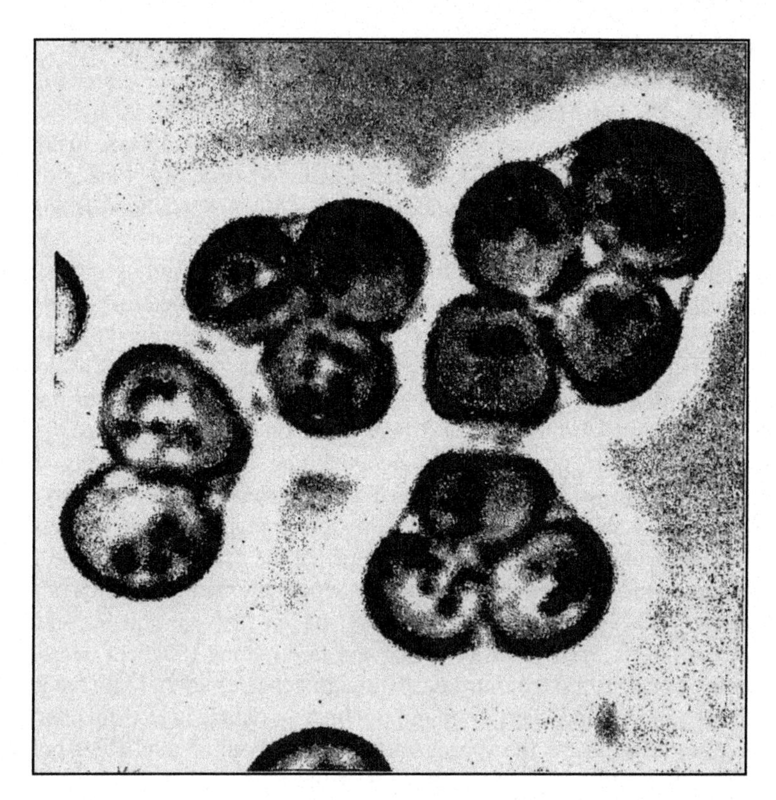

Fig. 16.3 Asci and ascospores of *Saccharomyces cerevisiae*. Spores stained with malachite green, vegetative cells with safranin. Photograph courtesy of the late C.F. Robinow, University of Western Ontario, London, Ontario, Canada

in a state of higher ploidy (Fig. 16.2). Most laboratory strains are haploid or diploid, whereas industrial yeast strains (brewing, distilling, baking and wine) are predominantly diploid, aneuploid and polyploid (Stewart et al. 1983).

It would appear that the widespread use of aneuploid/polyploid brewing yeast cultures is no accident. Due to their multiple gene structure, polyploids are genetically more stable and less susceptible to mutational forces than either haploid or diploid yeast strains (Stewart and Russell 1977), thus enabling such strains to be used by brewers with a high degree of confidence.

Two haploid cells of opposite mating type, a and α, can mate and produce a/α diploid mating types, which can undergo meiosis to generate four haploid ascospores, two of each mating type according to Mendelian segregation laws (Figs. 16.2 and 16.3). Homothallic haploids can switch their mating type from a to α and vice versa through the HO-controlled MATa, MATα, HML and HMR loci on chromosome III and conjugate with cells of the same single colony.

The mating type alleles (abbreviated *mt*) control a number of events in the life cycle of yeast (Crandall et al. 1977). One of them is the secretion of diffusible agglutination factors involved in the recognition and cell contact of opposite mating types and the repulsion of cells with the same mating type. However, mating efficiency can be variable, particularly with derivatives of many brewing yeast strains (Anderson and Martin 1975). Consequently, alternatives to hybridization have been sought and techniques such as spheroplast (protoplast) fusion (van Solingen and van der Plaat 1977) and rare mating (Conde and Fink 1976) and latterly transformation of DNA plasmids (Hinnen et al. 1978; Kielland-Brandt et al. 1979) have been studied—details later.

Sporulation in *S. cerevisiae* strains encompasses meioses and ascopore formation (Esposito and Klapholz 1981). Cells of diploid strains, heterozygous for the mating type locus (*MATa/MATα*), transferred from a presporulation medium (PSm) to the sporulation medium (SPm), undergo the two meiotic nuclear divisions and differentiate into asci containing four haploid cells (Fig. 16.3). Cells of polyploid (aneuploid) brewing strains—ale and lager—(*S. cerevisiae* and *S. pastorianus*, respectively) sporulate poorly, and asci containing four spores rarely develop (Anderson and Martin 1975; Gjermansen and Sigsgaard 1981; Stewart 1981). Moreover, spore viabilities are low and those spores that are viable often lack the ability to mate (Johnston 1965). This has impeded the cross-breeding and genetic characterization of brewing yeast. As a consequence, alternative methods of genetic manipulation such as protoplast (spheroplast) fusion (Russell and Stewart 1979; Skatrud et al. 1980; Spencer et al. 1980) and rare mating (Spencer and Spencer 1977; Stewart et al. 1985) that introduces foreign genetic material into the genome have been required to facilitate strain improvement—details later (Molzahn 1977).

A programme to assess the sporogenic ability of both ale and lager polyploid yeast strains has been conducted (Bilinski et al. 1986, 1987). Final sporulation percentages of five colony isolates were compared employing several agar media and the most sporogenic of these isolates selected for further study. Cultivation in liquid, rather than agar, media improved ascus production substantially. An analysis of the effects on ascosporogenesis temperature and of presporulation growth conditions, with various carbon sources, led to identification of culture conditions for enhanced ascus formation. Sporulation at 21 °C, instead of the usual 27 °C, gave significant increases in ascus yield. Substitution of glucose with acetate as the presporulation carbon source increased the ascus yield further. A comparison of the effects of fermentable versus non-fermentable carbon sources suggested that tetrad production depends closely on presporulation growth conditions inducing complete carbon catabolite derepression. Although spore viabilities were low, the increases in sporulation obtained by manipulation of the culture conditions permitted rapid isolation of an array of segregants. These included both a and α isolates for use during hybridization and genetic characterization (Gjermansen and Sigsgaard 1981).

Studies with lager yeast strains (Bilinski et al. 1987) have examined the effects on sporulation of temperature and of presporulation growth conditions on several carbon sources. Experiments at 21 °C in which the sporogenic abilities were monitored at various stages in the growth cycle demonstrated maximum ascus production in acetate-grown cells harvested from stationary rather than exponential phase. Increasing the potassium acetate concentration in the sporulation medium gave further increases in total sporulation together with a marked induction of

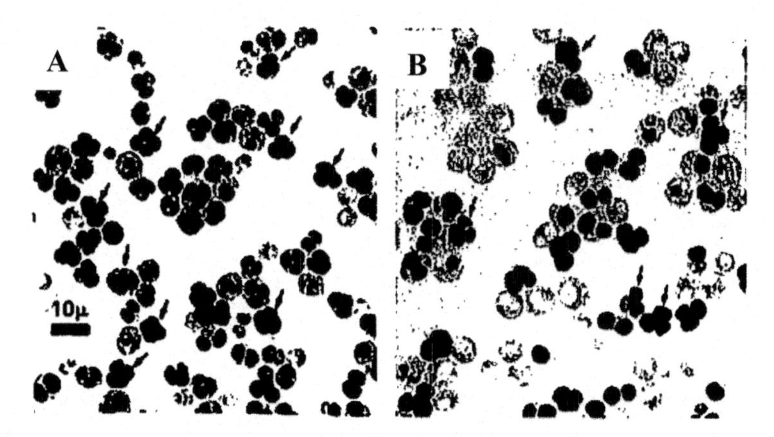

Fig. 16.4 Morphology of ascospores produced in (**a**) ale *versus* (**b**) lager production yeast strains. Both photographs are the same magnification as given by the bar scale in (**a**). Arrows indicate tetrads. Asci were stained with malachite green and counterstained with safranin 0 to permit observation of ascospores

tetrad production. Also, asci produced by induction of sporulation in acetate-grown media exhibited several morphological differences between ale and lager production yeast strains. The ale asci (Fig. 16.4a) were small and uniform in size, whereas lager spores in asci (Fig. 16.4b) tended to be more variable in size and shape. Unlike ale strains, it is not usually possible to observe lager spores in asci by direct microscopic examination without enzymatic digestion of ascus walls or the use of a specific spore staining procedure. This was found to be due to extensive vacuolar fragmentation which was initiated in stationary phase acetate-grown cells that persisted in mature asci. These developments have presented an opportunity to construct novel brewing yeast strains by direct intraspecific and interspecific hybridization, and this will be discussed later.

Saccharomyces diastaticus is a subspecies of *S. cerevisiae*. It differs from the parental species by producing the extracellular enzyme glucoamylase (α-1,4-glucan glucohydrolase) (Andrews and Gilliland 1952; Gilliland 1966) (Fig. 7.1). This enzyme removes successive glucose units from the nonreducing ends of dextrins (partially hydrolysed starch) (Figs. 7.11 and 16.5). Using hybridization techniques, three non-allelic genes have been identified and characterized from strains of *S. diastaticus* (Erratt and Stewart 1978). Two *DEX* genes (*DEX1* and *DEX2*) have been identified that code for the production of *S. diastaticus* glucoamylase, with glucose being the sole hydrolysis product. A third dextrinase gene (*DEX3*) was found to be allelic to *STA3* (Tamaki 1978), a gene also reported to control starch fermentation in *Saccharomyces* sp. (Erratt and Stewart 1981a, b).

Using hybridization techniques (Sherman and Hicks 1991), employing a micromanipulator (Fig. 2.11), a number of diploid strains have been developed, each being either heterozygous or homozygous for the individual *DEX* genes. Wort fermentations with the *DEX*-containing cultures showed that, as well as the wort sugars (glucose, fructose, sucrose, maltose and maltotriose), the dextrins (G4-G15 polymers) were

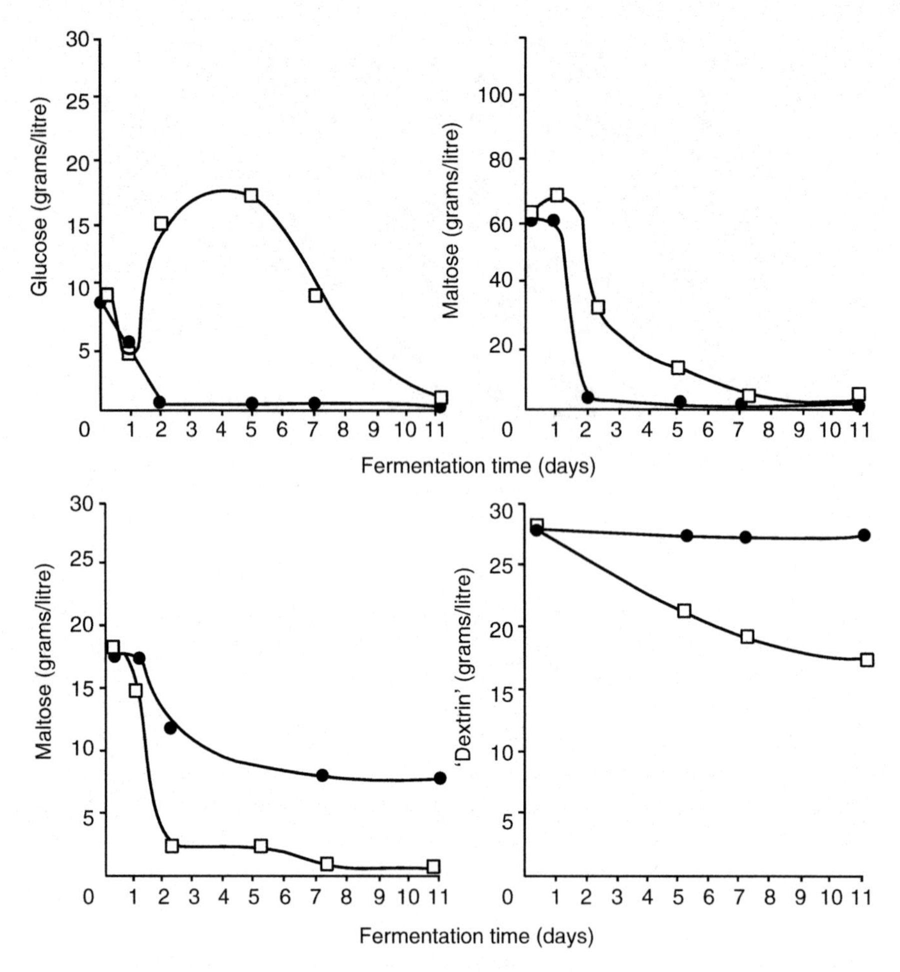

Fig. 16.5 Carbohydrate fermentation pattern of an 11.3°P wort static 40 L fermentation. 0.3% (w/v) pitching rate at 21 °C by polyploidy *Saccharomyces cerevisiae* strain (filled circle) and diploid DEX2/DEX2 Saccharomyces diastaticus strain (open square)

partially fermented (Fig. 7.15). There was residual dextrin remaining in the fermented worts because the glucoamylase from *S. diastaticus* strains was unable to hydrolyse the α(1-6) branch points, only the α(1-4) linkages (Fig. 16.6).

Beer produced with *DEX*-containing yeast strains has been studied. Glucose concentrations in the pasteurized beer increased during storage for 3 months at room temperature—this indicates that the glucoamylase is still active not heat sensitive (Erratt and Stewart 1981a). The reasons for this heat insensitivity are not at all clear. The fact that glucoamylase is heavily glycosylated could be the reason (Russell and Stewart 1981). Also, overall beer flavour produced by these *DEX*-containing yeast strains was unacceptable because they contained 4-vinylguaiacol (clove-like) (due to decarboxylation of ferulic acid) (Fig. 2.12);

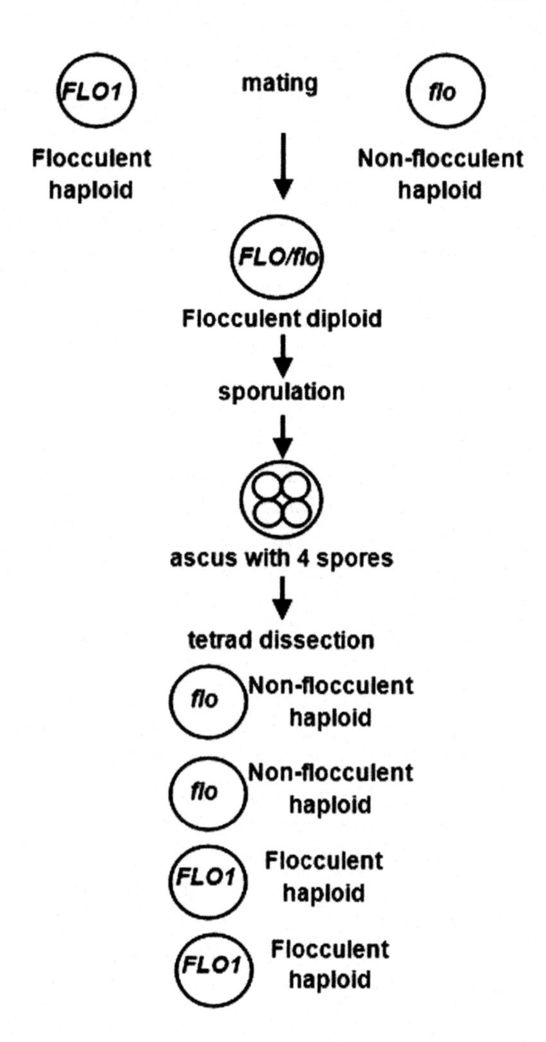

Fig. 16.6 Haploid/diploid life cycle of *Saccharomyces* sp.—tetrad analysis of the *FLO1* gene

because of the presence of the *POF* gene in the yeast genome, the beer was estery, sulphury and musty (Erratt and Stewart 1981b).

Nevertheless, the brewing industry has continued to be interested in the glucoamylases produced by *S. diastaticus* strains. Consequently, recombinant DNA (rDNA), plasmids and transformation techniques have been employed to clone the *DEX* genes into brewing strains (Hammond 1996). In order to enhance the activity of the glucoamylase, the glucoamylase enzyme from *Aspergillus niger* was also cloned into a brewing yeast along with the *STA2* gene. The *A. niger* gene had the advantage of possessing debranching activity and being heat sensitive, unlike the yeast enzyme, so that it can degrade more of the wort dextrin material (Hammond 2012).

The cloned amylolytic yeast strains have received approval by the UK Agriculture and Health Ministries, who concluded that there were no food safety reasons why this yeast should not be used in brewing (Ministry of Agriculture 1993). The reduced dextrin produced with this cloned yeast was called Nutfield Lyte (Fig. 7.16) and was the first beer in the world produced from a genetically modified yeast to receive approval. After considerable deliberation, it was decided that specific labelling was unnecessary. However, a back label of Nutfield Lyte bottles explained the process by which it had been produced and the advantages of doing so. Reaction to this beer from both the general public was neutral or positive. Media reports were largely factual and supportive. There was some criticism from the environmental lobby. The criticism has not diminished over the years, and the production of modified foodstuffs (including alcoholic beverages) in Europe (not North America) has been, and still is, severely restricted (Hammond 2012).

As well as employing hybridization to develop new hybrid strains with novel characteristics, it has also been used to map the location of a gene on one of the 16 chromosomes of the *Saccharomyces* sp. genome (Sherman and Hicks 1991; Sherman 1998). An example of this technique is the mapping of a flocculation gene. Flocculation, as it applies to brewer's yeast, is: "the phenomenon wherein yeast cells adhere in clumps and either sediment from the medium in which they are suspended or rise to the medium's surface" (Stewart and Russell 2009) (Fig. 13.3) (details in Chap. 13). This definition excludes other forms of aggregation, particularly that of "clumpy growth" and "chain formation" (Stewart et al. 2016) (Figs. 5.4 and 13.7) where daughter cells are physically attached to mother cells. However, distilling yeast cultures do not flocculate and are only used for one cycle (Russell and Stewart 2014) (Table 16.1).

It has already been discussed (Chap. 13) that genetic studies on yeast flocculation began over 60 years ago. The first publication on this subject was by Pomper and Burkholder (1949) who reported crossing a haploid culture that possessed a "dispersed" character (non-flocculent) with a haploid culture of opposite mating type that possessed a "non-disperse" character (flocculent). The "disperse" character was reported to be dominant over the non-disperse character. In the early 1950s, Gilliland and Thorne independently conducted extensive studies on yeast flocculation genetics. Gilliland (1951) studied two non-brewing strains of *S. cerevisiae* which differed only in their flocculation properties. It was proposed that a single gene was responsible for the flocculation phenotype. Thorne (1951) confirmed Gilliland's studies and demonstrated that flocculation was an inherited characteristic and that flocculation was dominant over non-flocculation.

It has also already been discussed that genetic studies involving brewer's yeast strains are fraught with difficulties because of their frequent triploid, polyploid or aneuploid nature (Stewart 1973). Consequently, studies have focussed on haploid and diploid flocculent and non-flocculent strains. The flocculation characteristic studied was reversible and induced by calcium ions (Stewart 1975). The flocculent haploid strain (coded 169) was of opposite mating type to the non-flocculent haploid strain (coded 168). These two strains were mated using micromanipulation (Fig. 2.11) techniques, and the resulting diploid hybrid (169/168) was found to be flocculent, confirming previous findings that the flocculent character was dominant and stable (Stewart and Russell 1977).

Table 16.1 Improvement factors that influence the characteristics of brewer's/distiller's yeast strains

Process improvements	Raw material flexibility	Novel products
Fermentation rate	Hydrolysis of starch/dextrins	Low carbohydrate
Flocculation characteristics	Hydrolysis of cellulose	Low/high alcohol
Optimum temperature	Metabolism of cellobiose	Specific flavours
Osmotic tolerance—high-gravity brewing	Lactose utilisation	
Ethanol tolerance—high-gravity brewing	Maltose/maltotriose utilisation	
Growth rate and amount	Greater use of adjuncts	
Infection proofing—zymocidal (killer) yeasts		
Protein/polypeptide hydrolysis—beer stability		
Diacetyl management		
Control of ester, higher alcohol and sulphur compounds production—flavour management		

Tetrad analysis of spores isolated from asci of the 169/168 hybrid revealed that the dominant flocculence of strain 169 was controlled by a single gene locus (i.e. 2:2 Mendelian segregate). This gene has been coded *FLO1*. The next question was the location of this gene on one of the 16 chromosomes of the *Saccharomyces* genome. A detailed discussion of chromosome mapping procedures employed in this study is in Chap. 13. Suffice to say, it has been shown that FLO1 is located on chromosome 1, 33 cM from the centromere on the right-hand side of the chromosome (Russell and Stewart 1980) (Fig. 16.6).

However, the mapping of the *FLO1* gene employed traditional gene mapping techniques (mating, sporulation, micromanipulation, tetrad analysis, etc.) (Sherman 1998). Novel genetic techniques have been (and are still being) developed, the principle of which is the sequencing of the *Saccharomyces* genome (Meaden 1996; Goffeau et al. 1996; Swinnen et al. 2012). Our knowledge of the genetic control of flocculation has expanded. At least nine genes (*FLO1, FLO5, FLO8, FLO9, FLO10, FLO11, FLONL, FLONS* and *Lg-FLO*) have been identified in both *S. cerevisiae* (ale) and *S. pastorianus* (lager) that encode flocculation proteins (Table 13.2). Although *FLO1* is well known and the best characterized flocculation gene (Teunissen and Steensma 1995), the flocculin encoded by *FLO11* differs from other flocculation gene products. It is involved in filamentous growth adhesion to solid surfaces (chain formation, Fig. 13.7) and flocculation *per se* (Halme et al. 2004).

16.2 Spontaneous Mutations in Yeast

Spontaneous yeast mutations commonly occur throughout the growth and fermentation cycle. They are usually recessive due to loss of a single gene (Meaden 1996). Because of the aneuploid or polyploid nature of most industrial yeast strains, the dominant gene functions adequately in a strain, as it is phenotypically normal. Only if the mutation takes place in all complementary genes is the recessive character expressed. However, if the mutation weakens the yeast (which is usually the case), the mutated strain is unable to compete and is soon outgrown (not always) by the nonmutated yeast population (Belton et al. 2012).

It has already been discussed in Chap. 14 that the respiratory-deficient (RD) mutation is the most frequent spontaneous mutant found in brewing yeast strains (Silhankova et al. 1970; Stewart 2014a), but a discussion of RDs is also appropriate in this chapter. This mutant arises spontaneously when a segment of the mitochondrial DNA (mtDNA) becomes defective and a flawed mitochondrial genome is formed. The mitochondria are then unable to synthesize certain proteins. This type of mutation is also called the "petite" mutation because colonies of such a mutant are much smaller than respiratory-sufficient (RS) cultures (also called "grande") (Fig. 14.2). The reason for the French terminology is that this mutation was first identified and described in the Pasteur Institute in Paris (Ephrussi et al. 1966), although the principal investigators (Ephrussi and Slonimski 1955) were Polish refugees living in Paris. The RD mutation usually occurs at frequencies between 0.5 and 5% of a yeast's population, but in some strains, levels as high as 50% have been reported (Stewart 2009). RD mutants can also occur as a result of deficiencies in nuclear DNA, but these are much rarer.

Yeasts with mitochondrial functional deficiencies are unable to metabolize non-fermentable carbon sources such as lactate, glycerol and ethanol but are able to metabolize fermentable sugars such as glucose (Fig. 14.5) and, as will be discussed shortly, maltose. Further, cultures of RD mutants are much more difficult to store in a culture collection compared to the RS culture. However, liquid nitrogen at −196 °C and refrigeration at −70 °C have been found to be effective storage procedures for both RS and RD cultures (Russell and Stewart 1981) (details in Chap. 4).

Studies with RD mutants of brewing and related yeast strains have been ongoing for over 50 years. However, the effects of this mutation on yeast activity are still not fully understood, particularly with respect to the involvement of mtDNA deletion with a yeast strain's overall growth and metabolic characteristics. Specifically, four areas have been considered (Stewart 2014a):

- Effects on the uptake of wort sugars—particularly maltose and maltotriose—and on the development of beer flavour congeners. Although the uptake pattern of glucose in the two fermentations (RS and RD) was similar, the uptake of both maltose and maltotriose was more rapid in the RS fermentation compared to the RD fermentation (Ernandes et al. 1993a, b, c). This confirms the differences between sugars where an active transport process is employed (maltose and maltotriose) to be taken up into the cell and those that use passive transport systems (e.g. glucose). RD cultures contain reduced concentrations of metabolic energy (mostly as ATP) compared to their wild-type RS cultures. As a result, maltose and maltotriose uptake is inhibited in RD cells, whereas glucose uptake rates and extent in the two sets of fermentations (RS and RD) were very similar (Fig. 14.4).

- Many factors contribute to beer flavour (Bamforth 2016), and of paramount importance is yeast performance and stability. Spontaneously generated RD yeasts have been reported to adversely affect beer flavour. The RD cultures produced considerably higher amounts of fusel oils (a consolidation of propanol, isobutanol and isoamyl alcohol concentrations) compared to the same strain's RS culture (further details in Chap. 14). However, reduced amounts of esters (ethyl acetate and isoamyl acetate) were detected in beer produced by the RD mutant. This was also the case with acetaldehyde which gives beer a characteristic grassy or "green apple" flavour (Meilgaard 1975a, b; Stewart 2014b). Regarding the effects of diacetyl (and other VDKs), the RD mutant produces higher amounts of this beer flavour compared to the RS culture. Also, the RD mutant, unlike the RS culture, lacks the ability to reduce diacetyl late in the fermentation. This is due to a metabolic imbalance caused by a malfunctioning Tricarboxylic Acid Cycle (Chap. 6) and defects in the enzymes of the isoleucine—valine pathways, which affects the production of diketones (predominantly diacetyl) during wort fermentation (Krogerus and Gibson 2013).
- RD mutants also affect yeast sedimentation characteristics. The genetic control of yeast flocculation has already been discussed in this text. The wild-type (RS) lager brewing strain used in this study (Stewart 2014a) exhibited strong flocculation, whereas its spontaneous RD mutant was non-flocculent. Also, the RS culture grew considerably faster than the RD culture from the same yeast strain, and towards the end of wort fermentation, it sedimented more rapidly out of suspension than the RD culture. This illustrated that the RD culture's cell wall structure had been modified as a result of changes to the mtDNA. This resulted in loss of this RD mutant's flocculation properties (Ernandes et al. 1993b).
- Stress conditions affect the formation of RD mutants. These conditions include osmotic pressure, ethanol, temperature, yeast centrifugation (Chlup and Stewart 2011) and particularly during high-gravity brewing (Stewart 2014a). All of these stresses affect yeast cultures, the formation of RD mutants is one result, and dual stresses can exacerbate these effects. One example of dual stresses upon yeast is centrifugation with a flow through bowl centrifuge at an elevated temperature (>30 °C). This caused mutation of mtDNA increasing the level of RD mutants. As already discussed (Chap. 14), these mutants were able to ferment wort efficiently, and the resulting beer had poor overall quality, stability and drinkability (Stewart 2010).

Finally, fundamental research on the molecular biology and function of yeast mitochondria has enhanced and assisted our knowledge of human mitochondrial function and disease (a typical eukaryote) (Dimauro and Davidzon 2005). Current research involves removing the mtDNA from ovaries of a diseased patient and replacing it with unmutated (healthy) mtDNA from a donor to produce a non-diseased zygote that can be inserted by IVF into a female recipient. Although caution in the clinical application of this technique has recently been expressed (Reinhardt et al. 2013), details of yeast mtDNA have greatly assisted research on this endeavour Stewart (2014c) (Chap. 14).

16.3 Spheroplast (Protoplast) Fusion

Problems with yeast hybridization, particularly inefficient a/α mating, can be overcome by the use of somatic or spheroplast (protoplast) fusion. Somatic fusion is a technique initially proposed in response to the question: "Can we achieve gene recombination in cells other than from the germ tract?" (Pontecorvo and Kafer 1958). A spheroplast (Fig. 5.1) is a structure from which the cell wall has been almost completely removed by the action of specific enzymes (glucanases—details later). The name comes from the fact that after the yeast's cell wall has been digested, membrane tension causes the cell to acquire a characteristic spherical shape. Spheroplasts are osmotically fragile and will only remain intact "if suspended in a hypertonic solution" (e.g. 0.8–1.0 mM sorbitol or 0.7 M sodium chloride). Spheroplasts will lyse if transferred to a hypotonic solution. Protoplasts are structures where the cell wall has been completely removed from the rest of the intact cell. In this author's opinion, it is doubtful if they really exist with vestigial cell wall material probably always remaining. Consequently, they are in reality spheroplasts not protoplasts.

Fusion of cells that have their walls removed and are suspended in a hypertonic solution has been studied, with the initial studies being conducted on plants (Kao and Michayluk 1974). In addition, polyethylene glycol (PEG) has been introduced as an aid to increase fusion frequency of plant spheroplasts (protoplasts) (Anné and Peberdy 1976; Ferenczy et al. 1976).

Van Solingen and van der Plaat (1977) adapted the reported studies on plant spheroplast fusion to be applied to *S. cerevisiae*. Haploid mutants (both *a* mating type) of *S. cerevisiae* strains were spheroplasted, with zymolyase containing β-mercaptoethanol and 0.8M sorbitol included in the incubation medium to maintain spheroplast integrity. Spheroplasts were washed with the sorbitol solution to remove the spheroplasting medium and centrifuged. The pellets (both strains) were mixed together in PEG 4000 solution containing 10 mM $CaCl_2$ for 15 to 30 min. For reversion to vegetative cells, the spheroplast mixture was mixed with melted hypotonic regeneration agar [3% (w/v) – (3% not the usual 2 w/v)] medium and poured gently into a petri dish. After 2–4 days incubation at approximately 25 °C, colonies of regenerated cells appeared in the petri dish medium. The results obtained presented evidence that not only fusion of spheroplasts, together with cell wall regeneration, has taken place but that the fusion process is accompanied by nuclear fusion leading to a stable diploid with an *a* mating type.

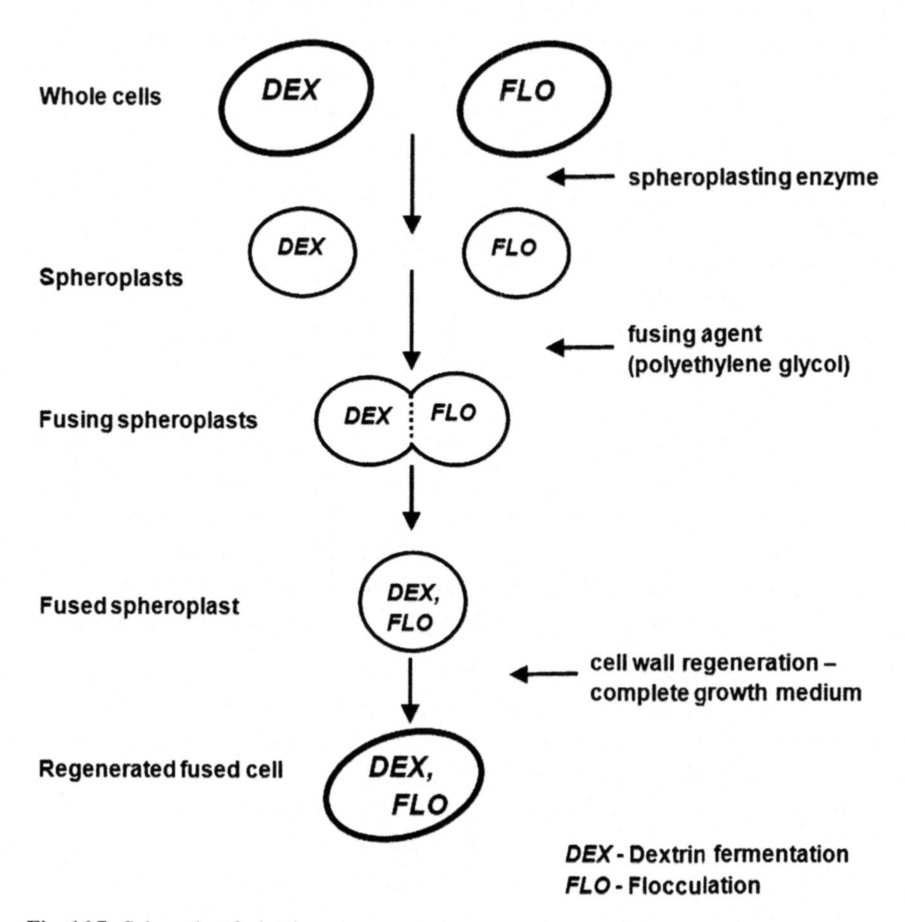

Fig. 16.7 Spheroplast fusion in yeast

We (Russell and Stewart 1979) applied fusion methodology to brewing yeast strains that were either polyploid or aneuploid and consequently did not possess a mating type. A dextrin-positive strain was fused with a flocculent culture (*FLO1*—containing haploid strain that possessed a number of nutritional growth characteristics) (Fig. 16.7). In order to aid the selection of the fusion recombinant, both fusion strains were respiratory deficient or "petite" (i.e. unable to grow on a medium containing a non-fermentable carbon source such as lactate, glycerol or acetate). The recombinant resulting from the fusion was dextrin positive and flocculent (it contained *FLO1*), had no nutritional deficiencies, contained all the characteristics of the parent lager strain and was respiratory sufficient or "grand" (capable of growth on a non-fermentable carbon source) (Ernandes et al. 1993c). In the intervening years, there have been many examples of spheroplast fusion with *S. cerevisiae* and other yeast species (e.g. Sipiczki and Ferenczy 1977a, b; Svoboda 1978; Sipiczki et al. 1985; Curran and Bugeja 1996; Crumplen et al. 1990; Brueggemann and Zou 2004).

Using yeast spheroplasts, whole genomes from bacteria have now been transferred to yeast using whole bacterial cells (Karas et al. 2013, 2014). Over 88% of bacterial genomes transferred in this way are complete, on the basis of both structural and functional tests. Excluding the time required for preparing starting cultures and for the incubation of cells to regenerate and form final colonies, the fusion protocol can be complete in 3 h!

Spheroplast fusion has been employed to study many aspects of the genetics and physiological characters of yeast strains—particularly aneuploid and polyploid (Anné and Peberdy 1976; Peberdy et al. 1977). This technique affords a means of genetically analysing such strains in order to identify the genes that code for important brewing properties such as maltose and maltotriose metabolism, dextrin utilization, flocculation and stress tolerance (osmotic pressure, ethanol and temperature) (Stewart and Russell 1979). Also, fusion affords a means in conjunction with other genetic manipulation methods such as hybridization (Stewart and Russell 1977), rare mating (Conde and Fink 1976; Hinnen et al. 1978) and mutation (Molzahn 1977) of genetically improving strains for production purposes. Spheroplast fusion has been employed to fuse strains constructed by hybridization with brewing strains in order to introduce the novel capabilities of the hybridized strain into the brewing strain while still maintaining all the characteristics of the latter (Russell and Stewart 1985; Stewart et al. 1986).

Although spheroplast fusion is an efficient genetic manipulation technique, it relies mainly on trial and error. It is not specific enough to modify strains in a predictable manner. The fusion product is nearly always very different from both original fusion partners because the genome of both donors becomes integrated. Consequently, it is difficult to selectively introduce a single trait such as flocculation or maltose/maltotriose uptake into a particular yeast strain employing the spheroplast fusion technique (Stewart 1981).

16.4 Rare Mating and the Killer (Zymocidal) Toxin(s)

Rare mating in *S. cerevisiae* is a modification of traditional classical mating in yeasts that has already been described here. It was critical to study the switching of mating types in a yeast species (Gunge and Nakatomi 1972). Later, rare mating was adapted to include investigations with industrial yeasts (brewing, distilling, wine, baking, etc.) that have both low mating frequencies and spore viabilities or that do not mate or sporulate at all (Spencer and Spencer 1977, 1996). Rare mating depends on the fact that occasional mating type switching occurs in industrial yeasts, which are normally diploid or of higher ploidy and may also display aneuploidy (Fig. 10.2). This results in the occurrence of low frequency mating cells that can then conjugate with a laboratory mating strain of either a or α mating type (also aa or $\alpha\alpha$ mating types) and whose progeny subsequently carry the characteristics present in the laboratory strain. In addition, the ability to sporulate in the progeny is often enhanced. Consequently, genetic analysis can be carried out on this strain and some information on the nature of the genome of the industrial strain obtained.

A mutation (*kar 1-1*), which causes a defect in nuclear fusion, has been described (Conde and Fink 1976). The existence of such a mutation suggests that nuclear fusion is an autonomous morphogenetic process, dissectible by genetic analysis. Tetrad analysis of the mutation for this defect has shown that it segregates in a Mendelian manner. The defect in *kar 1-1* appears to be nuclear limited. Cytological and genetic evidence showed that, in this mutant, the events associated with zygote formation are normal until the point of nuclear fusion. The consequence of this defect is the formation of a multinucleate zygote which, during subsequent divisions, can segregate heterokaryons and haploid heteroplasmons. These heteroplasmons allow an investigation of cytoplasmic mixing without nuclear fusion. This alternative to the usual life cycle provides a potential approach to problems of cytoplasmic inheritance.

The *kar* mutation has been employed to manipulate the killer yeast system in *S. cerevisiae* and related strains. Killer yeast strains secrete toxins that are lethal to sensitive strains of the same or related yeast species (details in Chap. 10). All the known killer toxins produced by killer yeasts are proteins that kill sensitive cells via a two-step mode of action (Liu et al. 2015), while antibiotics are bioactive substances that are produced by any organism and possess activity against selective fungi, bacteria, viruses and cancer cells. To date, it is known that under competitive conditions, the killer phenomenon offers considerable advantages to these yeast strains against other sensitive microbial cells in their ecological niches (Wang et al. 2012).

Killer yeasts and their toxins have several applications. For example, killer yeasts have been used to combat contaminating yeasts in food and to control pathogenic fungi in plants (Schmitt and Breining 2002). In the medical field, killer yeasts have been used for the development of novel antimycotics, used in treatment of human and animal fungal infections and in the biotyping of pathogenic yeasts and yeast-like fungi (Magliani et al. 2008). Moreover, killer yeasts have been used to control contaminating wild yeasts in the winemaking, brewing and other fermentation industries (Schmitt and Breining 2002).

The killer factor has been renamed a zymocide in order to indicate that it is only lethal towards yeast, other fungi and a few bacteria and not higher organisms. Zymocidal yeasts have been recognized to be a problem in both batch and continuous fermentation systems. An "infection" of as little as 1% of the cell population can completely eliminate all of the yeast from the fermenters (Russell and Stewart 1985). Consequently, a brewer can protect the process from this occurring in one of two ways:

- Maintain vigorous standards of hygiene in order to ensure that contamination with a wild yeast possessing zymocidal activity is prevented.
- Genetically modify the brewery yeast so that it is not susceptible to the zymocidal toxin.

Although the first method has been employed for many years in order to protect the brewing process, genetic manipulation is now available to produce a brewing yeast strain that is less vulnerable to destruction by zymocidal yeast infection.

The killer character of *Saccharomyces* sp. is determined by the presence of two cytoplasmically located double-stranded (ds) RNAs (Woods and Bevan 1968). The M-dsRNA (killer plasmid), which is killer strain specific, codes for a killer toxin and also for a protein or proteins that render the host immune to the toxin (Hammond and Eckersley 1984). The L-dsRNA, which is also present in many non-killer yeast strains, specifies a capsid protein that encapsulates batch forms of dsRNA, thereby yielding virus-like particles. Although the killer plasmid is contained within these virus-like particles, the killer genome is not naturally transmitted from cell to cell by any infection process. The killer plasmid behaves as a true cytoplasmic element and requires at least 29 different chromosomal genes (*mak* for its maintenance in the cell). In addition, three other chromosomal genes (*kex1*, *kex2* and *rex*) are required for toxin production and resistance to the toxin (Woods and Bevan 1968).

Rare mating has also been employed to produce hybrids with enhanced fermentation ability (e.g. Stewart et al. 1986). The *kar* mutation has been successfully employed for this purpose (Conde and Fink 1976). As already described in this chapter, this mutation prevents nuclear fusion, and hybrids, from such a rare mutation, can be selected that contain only the brewing strain's nucleus (Fink and Styles 1972). Such hybrids contain the cytoplasm of both parental cells. This permits the introduction of cytoplasmically transmitted characteristics such as killer toxin production into brewing strains, without modifying the brewing strain's nucleus. A brewing polyploid lager strain and several rare mating products have been isolated (Fig. 10.2).

Biochemical tests have characterized the rare mating products. Also, agarose gel electrophoresis demonstrated that some rare mating products contained the 2 μm plasmid (from the parental brewing lager strain) and the L- and M-dsRNA plasmids (from the haploid partner), which code for killer toxin production (Fig. 10.3).

The effect of zymocidal lager strains on a typical brewery fermentation was studied by mixing it at a concentration of 10% with an ale brewing strain (Fig. 10.4). The yeast culture was sampled throughout the fermentation and viable cells determined by plating onto nutrient agar plates. The plates were incubated at 37 °C (a temperature that inhibits the growth of lager yeast but permits the growth of ale strains—Stewart et al. 2013). Within 10 h after inoculation, the killer lager strain had almost totally eliminated the ale strain. When the concentration of killer was reduced to 1%, within 24 h incubation, the ale yeast was again eliminated (Fig. 10.4).

The speed at which death of the brewing yeast culture occurred made the brewing community apprehensive about employing killer strains during fermentation, particularly in a brewery where several yeast strains are employed for the production of different beers (e.g. an ale, lager and two licensed beers). An error by an operator in maintaining lines and yeast tanks separate could result in serious consequences. In a brewery where only one yeast strain is employed (increasingly the current situation in lager breweries but not in operations serving the craft brewing sector), this would not be a cause for concern!

An alternative to employing a killer strain would be to produce a yeast strain that does not kill but is killer resistant. It has to receive the genetic complement that renders a brewing yeast strain immune to zymocidal activity. The constitution of such a strain would be an interesting compromise because it does not itself kill. This allays the brewer's fear that this yeast might kill all other production strains in a plant, and, at the same time, it is not itself killed by a contaminating yeast possessing killer activity. The initial protein product from translation of the M-dsRNA is called a preprotoxin, which is targeted to the yeast secretory pathway. The preprotoxin is processed and cleaved to produce an α/β dimer, which is the active form of the toxin and is released into the environment (Bussey 1991).

Although they were first described in *S. cerevisiae*, toxin-producing yeasts have now been identified in other yeast genera, such as *Candida*, *Cryptococcus*, *Debaryomyces*, *Hanseniaspora*, *Hansenula*, *Kluyveromyces*, *Metschnikowia*, *Pichia*, *Ustilago*, *Torulopsis*, *Williopsis*, *Zygosaccharomyces*, *Aureobasidium*, *Zygowilliopsis* and *Mrakia* (details of the yeast species in Chap. 17), indicating that the killer phenomena are widespread amongst yeasts (Liu et al. 2015; Buzdar et al. 2011; Naumov et al. 2011; Naumov and Li 2011; Wang et al. 2008)—further details of killer (Zymocidal) yeasts are in Chap. 10.

In conclusion, rare mating to manipulate the killer toxin phenomenon (and other traits) has many potential applications in fermentation (food and beverages), taxonomy, medicine and the marine culture industry. These applications have been thoroughly reviewed (e.g. Chi et al. 2010; Wickner and Edskes 2015). Killer toxins kill sensitive cells by the inhibition of DNA replication, induction of membrane permeability changes and the arrest of the cell cycle. In some cases, a toxin can interfere with cell wall synthesis by inhibiting β-1,3-glucan synthase or by hydrolysing the major cell wall components, β-1,3 glucans and β-1,6 glucans. It is still unknown what the receptors of many other killer toxins on the sensitive cells are and how the killer toxins kill sensitive cells. In addition, little is known about the relationship between the structure of killer toxins, their killer activity and binding to targets on sensitive cells (Liu et al. 2015).

16.5 Transformation and Recombinant DNA in Yeast

The introduction of genetic engineering in 1973 (Cohen et al. 1973) laid the foundations for the current biotechnology industry which is based on using microorganisms on cell cultures for the production of proteins that can serve as pharmaceuticals or as novel enzymes (Walsh 2010). A few years later, research at the biotech research company Genentech cloned the genes for human insulin and growth hormone and expressed them in *Escherichia coli* (Lacroix and Citovsky 2016). In 1982, this led to the marketing of the first biopharmaceutical, human insulin, by Eli Lilly, who licensed the technology from Genentech. In 1985,

Genentech received FDA approval to market their own first product, Protropin®, the human growth hormone for children with growth hormone deficiency. Protropin® was followed by tissue plasminogen activator (t-PA, Activase®), another Genentech product. This is an enzyme that can resolve blood clots in patients with acute myocardial infarction.

In 1987, Novo (now Novo Nordisk), a major insulin-producing company launched human insulin produced by *S. cerevisiae* as a replacement for their human insulin enzymatically derived from porcine insulin. Following these early developments, many other products were launched, and currently, there are more than 300 biopharmaceutical proteins and antibodies on the market produced by *S. cerevisiae* (Martinez et al. 2002; Hou et al. 2012; Kim et al. 2012).

The advantages of using yeast as a cell factory for the production of biopharmaceuticals are that this eukaryote model system enables production and proper folding of many human proteins. Also, the proteins can be secreted out of the yeast cell, and this enables subsequent purification. *S. cerevisiae* is also widely used as a eukaryote model organism (Petranovic et al. 2010), and much information is available about this organism through high-throughput studies (Kim et al. 2012), databases and sequenced genomics, and an extensive toolbox for molecular modification provides researchers with an extensive knowledge base for further engineering of this and other organisms. One of the limitations with the use of yeast, however, is that products are heavily glycosylated with a mannose polymer (Nielsen 2013). The result of this is a short half-life of the modified *in vivo* protein that can exhibit a reduced efficiency for therapeutic use (Wildt and Gerngross 2005). Considerable studies have been conducted on engineering yeasts, both *S. cerevisiae* and *Pichia pastoris*, in order for both yeast species to carry out human-like N-glycosylation patterns. This has ensured wider use of both yeast species as cell factories for the production of biopharmaceuticals, and there is much interest to further engineer yeast in order to ensure efficient production of recombinant proteins. *P. pastoris* is a species of the methylotrophic yeast species *Pichia*. It is widely used for protein production using rDNA techniques (Daly and Hearn 2005). A number of properties make *P. pastoris* suited for this task (Cregg et al. 2009):

- A high growth rate.
- Ability to grow on a simple, inexpensive medium.
- Can grow in both shake flasks or in a fermenter.
- Suitable for both small- and large-scale production.

P. pastoris has two advantages when compared to *S. cerevisiae* in both laboratory and industrial settlings:

- It can grow to very high cell densities to the point where the cell suspension is practically a paste. The protein yield from expression in this yeast is approximately equal to the number of cells.
- *Pichia* is a methylotroph (it can grow with methanol as its energy source that would kill most other microorganisms—a system that is inexpensive to set up and maintain).

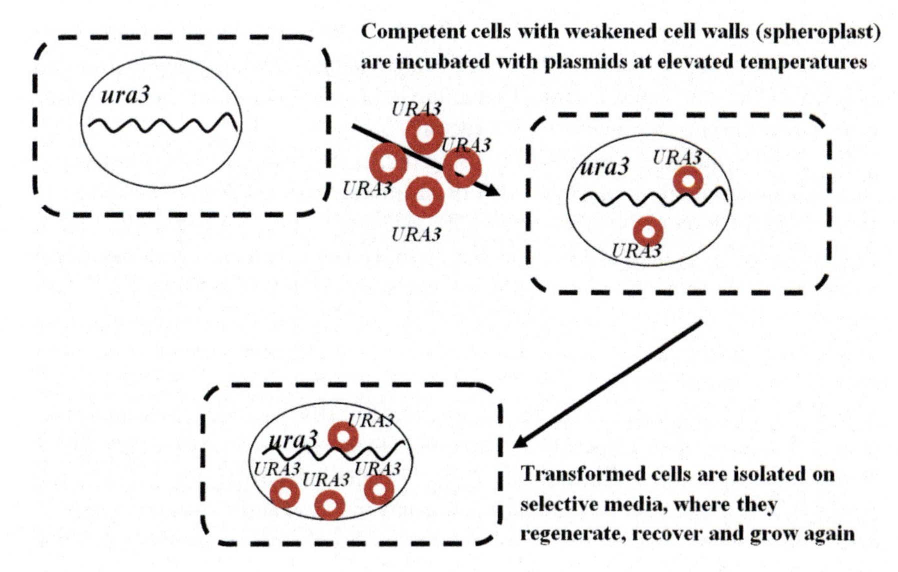

Fig. 16.8 A simplified protocol for plasmid DNA transformation in *Saccharomyces* sp.

- Although *S. cerevisiae* is well studied, *P. pastoris* has similar growth conditions and tolerances. It can be readily adopted by laboratories without specialist equipment (Daly and Hearn 2005).

Unlike the other techniques already discussed, genetic engineering (recombinant DNA—rDNA) affords the possibility of cloning a specific genetic character, thereby eliminating the possibility of introducing additional factors that could be detrimental to the host. In addition, genetic engineering methods permit the transfer of genetic information between completely unrelated organisms. Consequently, the recipient organism becomes able to produce heterologous proteins and peptides that are not produced by their mated constituents. This provides considerable scope for the transfer of new constituents into industrial yeast strains (Gietz and Woods 2001).

Considerable literature concerning transformation and recombinant DNA strategies in yeast is extensive and varied. For example, Pretorius (2016) has recently discussed the current situation regarding molecular biology, systems biology, regeneration biology and synthetic biology as they apply to industrial yeast strains (particularly brewing and wine yeasts). As a consequence, only the basics of rDNA and transformation, as they apply to brewing and distilling yeasts, are discussed here.

S. cerevisiae has become the most sophisticated model for recombinant DNA (rDNA) technology. When transformed into yeast cells, plasmids are inserted into yeast chromosomes generally by homologous recombination (Fig. 16.8). rDNA molecules are DNA molecules formed *in vitro* (by laboratory methods) of genetic recombination (such as molecular cloning) to bring together genetic material from multiple sources, creating sequences that would not otherwise be found in the yeast (or other eukaryote) genome. rDNA is possible because DNA molecules from all organisms share the same chemical structure. They differ only in the nucleotide sequence within the identical overall structure.

The DNA sequences used in the construction of recombinant DNA molecules can originate from any living species. For example, yeast DNA may be joined to human DNA or insect DNA (Hayama et al. 2002), or plant DNA to bacterial DNA (Kawai et al. 2010; Gietz et al. 1995), etc. Also, DNA sequences that do not occur anywhere in nature may be created by the chemical synthesis of DNA and incorporated into recombinant molecules. Using recombinant DNA technology uses palindromic (reads the same backward and forward) sequences to the production of sticky and blunt ends.

Proteins that can occur from the expression of rDNA within yeast and other living cells are termed recombinant proteins. However, when rDNA encoding a protein is introduced into *Saccharomyces* sp., or other host organism, the recombinant protein is not necessarily produced (Rosano and Ceccarelli 2014). Expression of foreign proteins requires the use of specialized expression vectors and often necessitates significant restructuring by foreign coding sequences.

Pretorius (2016) has succinctly outlined the history of molecular biology and its application to a yeast culture. The discovery of the double-helical structure of DNA (Fig. 16.9), more than 60 years ago by James Watson and Francis Crick (Fig. 16.10), with the collaboration of Maurice Wilkinson and Rosalind Franklin, has captivated imaginations and continues to formulate our understanding of life's genetic basis (Watson and Crick 1953a, b). The late 1950s saw the uncovering of the mysteries of the universal genetic code, which was followed in the 1960s with the unravelling of the workings of regulatory genetic circuits and genetic expression. The 1970s were marked by the development of recombinant DNA (rDNA) and molecular cloning techniques. With the improvement of transformation procedures and the arrival of polymerase chain reaction (PCR) techniques developed by Kary Mullis of Cetus (Fig. 16.11) in the 1980s, genetic engineering became entrenched in mainstream molecular biology research (Mullis et al. 1987). This was followed by the advent of automated DNA sequencing, high-throughput "omics" technologies, bioinformatics and sophisticated computational tools which permitted the decoding of complete genomes in the 1990s and beyond (Pray 2008).

Developments in molecular biology have facilitated yeast strain improvement (particularly brewing strains) by metabolic engineering (Saerens et al. 2010). Although concerns about public health and environmental safety of genetically modified organisms (GMO) represent a hurdle for commercialization (details later), an array of desired properties has been successfully introduced into brewer's and other *Saccharomyces* sp. strains by this strategy. Rational metabolic engineering differs from traditional genetic approaches as it allows the direct modification of genetic information, strongly reducing the risk of losing positive traits present in the host strain or accumulating negative strains (Walters 2001).

Based on the assumption that a phenotypic trait is primarily controlled by proteins, rational metabolic engineering refers to the targeted engineering of

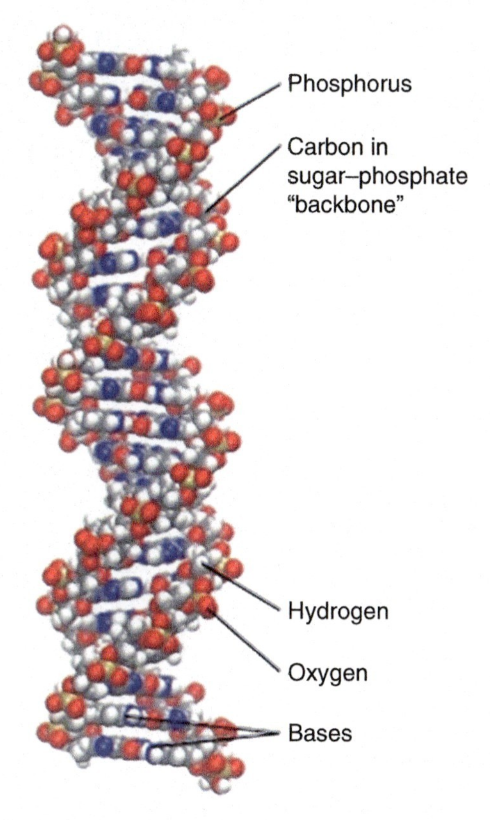

Fig. 16.9 Double helical structure of DNA

Fig. 16.10 James Watson and Francis Crick

Fig. 16.11 Kary Mullis

enzymes, transporters or regulatory proteins. Amongst existing possibilities for engineering protein activity, examples are:

- The level and regulation of gene expression.
- *In vivo* protein/enzyme activity.
- Protein subcellular localization.

It is often necessary to modify several enzymes/transport proteins of a pathway for improved metabolic fluxes. To introduce useful enzyme activities which are not naturally present in the host yeast strain, heterologous genes have been introduced from other organisms. The content of a homologous or heterologous gene in a cell can be increased by introducing single- or multicopy plasmid vectors or via (multiple) genomic integration(s). Also, gene deletions have been applied in order to decrease the level of homologous proteins. Native *S. cerevisiae* promoters of different strengths are useful to adjust protein levels. However, a gene expression customized in its level and induction time point may require finer adjustments (Alper et al. 2005; Nevoigt et al. 2007). This fine tuning can be important, for example, when considering the introduction of the flocculation gene into a yeast strain (Watari et al. 1991).

It has already been discussed that strong flocculation of a yeast culture is usually not required before the latter part of wort fermentation (Chap. 13). Consequently, specific promoters have been employed to ensure yeast culture flocculation at the onset of the stationary growth phase. Overexpressing FLO genes has enhanced yeast flocculent intensity (Van Mulders et al. 2009), but many of these cloned cultures did not flocculate at the correct time in the wort fermentation cycle (Speers 2016). Indeed, premature flocculation leads to sluggish and incomplete fermentations, and often off-flavours occur in the resulting beer. The major challenge for correctly timed flocculation has been the lack of promoters that can be induced under production conditions at the appropriate time in wort fermentation. A major advance in the control of flocculation by *FLO* genes is the replacement of its promoter—the *HSP30* promoter. Under laboratory conditions, it has been shown (Verstrepen et al. 2001) that flocculation occurs towards the end of wort fermentation. The *HSP30* promoter is a gene that promotes heat shock. During the past few decades, many useful strategies have been developed in order to optimize brewer's yeast, and other yeast strains, by genetic manipulation and engineering (Tyo et al. 2010; Hammond 2012).

16.6 The Development and Application of Bioengineered Yeast Strains

It has already been described that a laboratory strain of *S. cerevisiae* was the first to be transformed in the late 1970s (Hinnen et al. 1978). Since then, numerous genes have been cloned into a number of strains, including industrial yeast strains (Gibson et al. 2017). During the past three decades or so, a number of genetically engineered yeast strains have been developed for a number of purposes. In the biotechnology-based industries, genes encoding enzymes, interferon, insulin, growth hormones and vaccines have been expressed in yeast. Patents have been filed and several by-products generated by genetically modified (GM) yeast strains (with either *S. cerevisiae* or *P. pastorianus*) resulting in varying degrees of commercial success (Zimmermann and Stokell 2010; Annibali 2011). The fiercest resistance regarding this development by the general public was aimed against food products containing ingredients produced by genetically modified organisms (GMOs), such as the yeast-derived milk-clotting enzyme, chymosin. However, much less resistance has been encountered with the commercialization of the yeast-derived recombinant vaccine against hepatitis B also insulin produced by yeast.

The advent of synthetic biology has led to GM technology becoming much more powerful, precise and sophisticated (Steensels et al. 2014; Bae et al. 2016). In addition to the production of yeast-derived ethanol in the fermentation beverage and biofuel industries, this organism with GRAS status is increasingly being employed as "cell factories" for the production of high-value, low-volume products. These could improve health, wellness and nutrition (Kavšček et al. 2015). Firms use a combination of genetic and genome engineering and analyse specific *Saccharomyces* strains by employing three major strategies: rational engineering, inverse metabolic engineering and evolutionary strategies (Shi et al. 2009; Tao et al. 2012; Stephanopoulos 2012). There is a long list of products generated by bioengineered *S. cerevisiae* cell factories. This list of products includes aliphatic alcohols, β-amyrin, β-amyrin, β-carotene, casbene, cinnamoyl anthranilates, cabebol, propanediol, ribitol and acetic, adipic, artemisinic and eicosapentaenoic acids. A company such as Evolva (which stands for less sugar in food—discovering, producing and supplying innovative and sustainable ingredients for health, wellness and nutrition) specializes in the use of yeast cell factories for the production of high-value, low-volume products (Nielsen 2015). Currently, three approaches are being employed: re(design), re(construct) and analyse specific *Saccharomyces* strains by employing three major strategies: rational engineering, inverse metabolic engineering and evolutionary strategies for the production of a variety of drugs for pain management and palliative care (Galanie et al. 2015). Also, *Escherichia coli* is being employed for this purpose (Nakagawe et al. 2016). The most common approach currently is to re-engineer the metabolism of existing cells and to retrofit existing strains. Efforts are underway to create synthetic cells that could be used to generate valuable bioproducts.

A synthetically engineered yeast cell factory with a significant commercial profile is a high-yielding *S. cerevisiae* strain capable of producing artemisinic acid, a precursor of artemisinin and a potent anti-malarial compound. Artemisinin is a sesquiterpene endoperoxide naturally produced by the sweet wormwood plant *Artemisia annua*. Artemisinin and its endoperoxide derivatives have been used for the treatment of *Plasmodium falciparum* infections, but low bioavailability, poor pharmocokinetic properties and high cost of the drugs are major drawbacks of their use (Soares 2010). There is an unstable supply of plant-derived artemisinin resulting from regular shortages and fluctuations for the manufacturers of "artemisinin-based combination therapies (ACTs) for "uncomplicated" malaria, caused by the protozoan parasite *P. falciparum*. Commercialization of a more reliable source of semi-synthetic artemisinin was achieved by engineering the complete biosynthetic artemisinic acid pathway into *S. cerevisiae* (Ro et al. 2006; Li et al. 2016).

Methods based on the clustered regulatory interspaced short palindromic repeats (CRISPR/Cas) system (gene editing) have rapidly gained popularity for genome editing and transcriptional regulation in many organisms, including a number of yeast species (Stovicek et al. 2017).

CRISPR/Cas9 technology has advantages over conventional marker-based genome editing in several aspects. It enables rapid strain engineering of wild-type and industrial yeast strains. It also allows performing multiple genome editing simultaneously and is independent of marker cassette integration. However, to enable the wide adaptation of CRISPR, the following limitations need to be addressed:

- Design of efficient and specific targeting for different yeast species.
- Elimination of a necessity for cloning.
- Large-scale multiplexing can occur.
- Intellectual property (IP) issues must be resolved!

The uncertainty regarding the ownership of CRISPR technology is delaying its adoption for industrial biotechnology and pharmaceutical applications. This must be resolved as soon as possible so that the technology can reveal its true potential.

What will genetic engineering mean for our future progress and development? What will our planet look like in 2050? There are many other questions, such as:

- Will synthetic chloroplasts and artificial leaves enable the development of green, self-powered buildings?
- Will synthetic yeast cell factories provide the next-generation biofuels, food enzymes, antibiotics, vaccines and personalized medicines?
- Will synthetic genomics accelerate the discovery of beneficial gene therapies and cures for cancers and dementia?

These questions cannot be answered today with any certainty. However, it is sure that synthetic biology will impact upon our future world and the unimaginable will be made possible by genome engineering. A new scientific frontier is now available with unknown opportunities and perils full of hopes and fears. Close collaboration will be necessary by researchers representing multiple disciplines, regulations and society leaders. The constructive dialogue that is now ongoing concerning the potential impact of synthetic genomics and the advent of novel yeast strains and their practices and products must continue (Oliver 1996a, b)!

16.7 Summary

The importance of the molecular biology of *S. cerevisiae* and its closely related species is well documented. This yeast was the first microorganism to be domesticated for the production of fermented food and beverages and to be described as a living biochemical agent for biological transformations. In 1986, the complete genome of an *S. cerevisiae* haploid strain became the first eukaryote genome to be fully sequenced. Its 16 chromosomes encode approximately 6000 genes with approximately 5000 of them being individually non-essential to the yeast cell. Knowledge of this genome sequence has prepared the way to build the first collection of deletion mutants, enabling high-throughput functional genetic experiments.

As well as hybridization with a and α mating type cells, manipulation with techniques such as spheroplast (protoplast) fusion and rare mating that introduces foreign genetic material into the genome have been employed to facilitate strain improvement. Unlike other manipulative techniques recombinant DNA (rDNA) affords the possibility of introducing additional factors. It permits the transfer of genetic information between completely unrelated organisms. Consequently, the recipient organism becomes able to produce heterologous proteins or peptides that are not produced by their mated constituents. This method provides considerable scope for the transfer of new constituents and functions into industrial yeast strains.

References

Alper H, Fischer C, Nevoigt E, Stephanopoulos G (2005) Tuning genetic control through promoter engineering. PNAS 102:12678–12683

Anderson JE, Martin PA (1975) The sporulation and mating of brewing yeasts. J Inst Brew 81:242–247

Andrews J, Gilliland RB (1952) Super-attenuation of beer: a study of three organisms capable of causing abnormal attenuations. J Inst Brew 58:189–196

Anné J, Peberdy JF (1976) Induced fusion of fungal protoplasts following treatment with polyethylene glycol. J Gen Microbiol 92:413–417

Annibali N (2011) Process for obtaining aspart insulin using a Pichia pastoris yeast strain. Patent No. US20110117600A1

Bae SJ, Kim S, Hahn JS (2016) Efficient production of acetoin in *Saccharomyces cerevisiae* by disruption of 2,3-butanediol dehydrogenase and expression of NADH oxidase. Sci Rep 6:27667

Bamforth CW (2016) Brewing materials and processes. Elsevier, Boston, pp 151–156

Belton JM, McCord RP, Gibcus JH, Naumova N, Zhan Y, Dekker J (2012) Hi-C: comprehensive technique to capture the conformation of genomes. Methods 58:268–276

Bilinski CA, Russell I, Stewart GG (1986) Analysis of sporulation in brewer's yeast: induction of tetrad formation. J Inst Brew 92:594–598

Bilinski CA, Russell I, Stewart GG (1987) Physiological requirements for induction of sporulation in lager yeast. J Inst Brew 93:21

Brueggemann M, Zou X (2004) Spheroplast fusion. Patent No. WO2004101802 A2

Bussey H (1991) K1 killer toxin, a pore-forming protein from yeast. Mol Microbiol 5:2339–2343

Buzdar MA, Chi Z, Wang Q, Hua MX, Chi ZM (2011) Production, purification, and characterization of a novel killer toxin from *Kluyveromyces siamensis* against a pathogenic yeast in crab. Appl Microbiol Biotechnol 91:1571–1579

Chi ZM, Liu GM, Zhao SF, Li J, Peng Y (2010) Marine yeasts as biocontrol agents and producers of bio-products. Appl Microbiol Biotechnol 86:1227–1241

Chlup PH, Stewart GG (2011) Centrifuges in brewing. MBAA Tech Q 48:46–50

Class III Ministry of Agriculture, Fisheries and Food and Intervention Board 1993 to 1994 – executive agency: supply estimates https://www.google.co.uk/#q=Ministry+of+agriculture+1993

Cohen SN, Chang AC, Boyer HW, Helling RB (1973) Construction of biologically functional bacterial plasmids *in vitro*. Proc Natl Acad Sci U S A 70:3240–3244

Conde J, Fink GR (1976) A mutant of *Saccharomyces cerevisiae* defective for nuclear fusion. Proc Natl Acad Sci U S A 73:3651–3655

Crandall MA, Egel R, Mackay VL (1977) Physiology of mating in three yeasts. Adv Microb Physiol 15:307–398

Cregg JM, Tolstorukov I, Kusari A, Sunga J, Madden K, Chappell T (2009) Expression in the yeast *Pichia pastoris*. Methods Enzymol 463:169–189

Crumplen RM, D'Amore T, Russell I, Stewart GG (1990) The use of spheroplast fusion to improve yeast osmotolerance. J Am Soc Brew Chem 48:58–61

Curran BPG, Bugeja VC (1996) Protoplast fusion in *Saccharomyces cerevisiae*. Yeast Protoc Methods Cell Mol Biol 53:45–49

Daly R, Hearn MT (2005) Expression of heterologous proteins in *Pichia pastoris*: a useful experimental tool in protein engineering and production. J Mol Recogn 18:119–138

Dimauro S, Davidzon G (2005) Mitochondrial DNA and disease. Ann Med 37:222–232

Ephrussi B, Slonimski PP (1955) Subcellular units involved in the synthesis of respiratory enzymes in yeast. *Nature* 176(4495):1207–1208

Ephrussi B, Jakob H, Grandchamp S (1966) Etudes sur la suppressivite des mutants a deficience respiratoire de la levure. 11. Etapesde la mutation grande en petite provoquee par le facteur suppressif. Genetics 54:1–29

Ernandes JR, D'Amore T, Russell I, Stewart GG (1993a) Regulation of glucose and maltose transport in strains of *Saccharomyces*. J lndust Microbiol 9:127–130

Ernandes JR, Williams JW, Russell I, Stewart GG (1993b) Effect of yeast adaptation to maltose utilization on sugar uptake during the fermentation of brewer's wort. J Inst Brew 99:67–71
Ernandes JR, Williams JW, Russell I, Stewart GG (1993c) Respiratory deficiency in brewing yeast strains – effects on fermentation, flocculation, and beer flavor components. J Am Soc Brew Chem 51:16–20
Erratt JA, Stewart GG (1978) Genetic and biochemical studies on yeast strains able to utilize dextrins. J Am Soc Brew Chem 36:151–161
Erratt JA, Stewart GG (1981a) Fermentation studies using *Saccharomyces diastaticus* yeast strains. Dev Ind Microbiol 22:577–586
Erratt JA, Stewart GG (1981b) Genetic and biochemical studies on glucoamylase from *Saccharomyces diastaticus*. In: Stewart G, Russell I (eds) Advances in biotechnology. Pergamon Press, Toronto, pp 177–183
Esposito RE, Klapholz S (1981) Meiosis and ascospore development. In: Strathern JN, Jones EW, Broach JR (eds) The molecular biology of the yeast *Saccharomyces*: life cycle and inheritance. Cold Spring Harbor Laboratory Press, Cold Spring Harbor, NY, pp 211–287
Ferenczy L, Kevei F, Szegedi M, Franko A, Rojik I (1976) Factors affecting high frequency fungal protoplast fusion. Experientia 32:1156–1158
Fink GR, Styles CA (1972) Curing of a killer factor in *Saccharomyces cerevisiae*. Proc Natl Acad Sci U S A 69:2846–2849
Galanie S, Thodey K, Trenchard IJ, Interrante MF, Smolke CD (2015) Complete biosynthesis of opioids in yeast. Science 349(6252):1095–1100
Gibson B, Liti G (2015) *Saccharomyces pastorianus*: genomic insights inspiring innovation for industry. Yeast 32:17–27
Gibson B, Geertman JMA, Hittinger CT, Krogerus K, Libkind D, Louis EJ, Magalhães F, Sampaio JP (2017) New yeasts – new brews: modern approaches to brewing yeast design and development. FEMS Yeast Res 17:216–230
Gietz RD, Woods RA (2001) Genetic transformation of yeast. Bio Techniques 30:816–831
Gietz RD, Schiestl RH, Willems AR, Woods RA (1995) Studies on the transformation of intact yeast cells by the LiAc/SS-DNA/PEG procedure. Yeast 11:355–360
Gilliland RB (1951) The flocculation characteristics of brewing yeasts during fermentation. In: Proceedings of the European Brewery Convention Congress, Brighton, pp 35–58
Gilliland RB (1966) *Saccharomyces diastaticus* – a starch fermenting yeast. Wellerstein Lab Commun 17:165–176
Gjermansen C, Sigsgaard P (1981) Construction of a hybrid brewing strain of *Saccharomyces carlsbergensis* by mating of meiotic segregants. Carlsb Res Commun 46:1–11
Goffeau A, Barrell BG, Bussey H, Davis RW, Dujon B, Feldmann H, Galibert F, Hoheisel JD, Jacq C, Johnston M, Louis EJ, Mewes HW, Murakami Y, Philippsen P, Tettelin H, Oliver SG (1996) Life with 6000 genes. Science 274(5287):546, 563–7
Gunge N, Nakatomi Y (1972) Genetic mechanisms of rare matings of the yeast *Saccharomyces cerevisiae* heterozygous for mating type. Genetics 70:41–58
Halme A, Bumgarner S, Styles C, Fink GR (2004) Genetic and epigenetic regulation of the *FLO* gene family generates cell-surface variation in yeast. Cell 116:405–415
Hammond JRM (1996) Yeast genetics. In: Priest FG, Campbell I (eds) Brewing microbiology. Chapman and Hall, London, pp 45–82
Hammond J (2012) Brewing with genetically modified amylolytic yeast, Chap 7. In: Harlander S, Roller S (eds) Genetic modification in the food industry: a strategy for food quality improvement. Springer Science & Business Media, pp 129–157
Hammond JRM, Eckersley KW (1984) Fermentation properties of brewing yeast with killer character. J Inst Brew 90:167–177
Hayama Y, Fukuda Y, Kawai S, Hashimoto W, Murata K (2002) Extremely simple, rapid and highly efficient method for the yeast *Saccharomyces cerevisiae* using glutathione and early lag phase cells. J Biosci Bioeng 94:166–171
Hinnen A, Hicks JB, Fink GR (1978) Transformation of yeast. Proc Natl Acad Sci U S A 75:1929–1933
Hou J, Tyo KE, Liu Z, Petranovic D, Nielsen J (2012) Metabolic engineering of recombinant protein secretion by *Saccharomyces cerevisiae*. FEMS Yeast Res 12:491–510

Jacob M, Jaros D, Rohm H (2011) Recent advances in milk clotting enzymes. Int J Dairy Technol 64:14–33
Johnston JR (1965) Breeding yeasts for brewing. I. Isolation of breeding strains. J Inst Brew 71:130–135
Kao KN, Michayluk MR (1974) A method for high-frequency intergeneric fusion of plant protoplasts. Planta 115:355–367
Karas BJ, Jablanovic J, Sun L, Ma L, Goldgof JM, Stam J, Ramon A, Manary MJ, Winzeler EA, Venter JC, Weyman PD, Gibson DG, Glass JI, Hutchison CA III, Smith HO, Suzuki Y (2013) Direct transfer of whole genomes from bacteria to yeast. Nat Methods 10:410–412
Karas BJ, Jablanovic J, Irvine E, Sun L, Ma L, Weyman PD, Gibson DG, Glass JI, Venter JC, Hutchison CA III, Smith HO, Suzuki Y (2014) Transferring whole genomes from bacteria to yeast spheroplasts using entire bacterial cells to reduce DNA shearing. Nat Protoc 9:743–750
Kavšček M, Stražar M, Curk T, Natter K, Petrovič U (2015) Yeast as a cell factory: current state and perspectives. Microb Cell Factor 94:201514
Kawai S, Hashimoto W, Murata K (2010) Transformation of *Saccharomyces cerevisiae* from other fungi: methods and possible underlying mechanisms. Bioeng Bugs 1:395–403
Kielland-Brandt MC, Nilsson-Tillgren T, Holmberg S, Petersen JGL, Svenningsen BA (1979) Transformation of yeast without the use of foreign DNA. Carlsb Res Commun 44:77–87
Kim IK, Roldão A, Siewers V, Nielsen J (2012) A systems-level approach for metabolic engineering of yeast cell factories. FEMS Yeast Res 12:228–248
Krogerus K, Gibson BR (2013) Influence of valine and other amino acids on total diacetyl and 2,3-pentanedione levels during fermentation of brewer's wort. Appl Microbiol Biotechnol 97:6919–6930
Lacroix B, Citovsky V (2016) Transfer of DNA from bacteria to eukaryotes. MBio 7:e 00863-16
Li C, Li J, Wang G, Li X (2016) Heterologous biosynthesis of artemisinic acid in *Saccharomyces cerevisiae*. J Appl Microbiol 120:1466–1478
Liu GL, Zhe C, Wang GY, Wang ZP, Li Y, Chi ZM (2015) Yeast killer toxins, molecular mechanisms of their action and their applications. Crit Rev Biotechnol 35:222–234
Magliani W, Conti S, Travassos LR, Polonelli L (2008) From yeast killer toxins to antibiobodies and beyond. FEMS Microbiol Lett 288:1–8
Martinez LA, Naguibneva I, Lehrmann H, Vervisch A, Tchénio T, Lozano G, Harel-Bellan A (2002) Synthetic small inhibiting RNAs: efficient tools to inactivate oncogenic mutations and restore pathways. In: Vogt PK (ed) Proceedings of the National Academy of Sciences of the USA. The Scripps Research Institute, La Jolla, CA, p 53
Meaden PG (1996) Yeast genome now completely sequenced. Ferment 9:213–214
Meilgaard MC (1975a) Aroma volatiles in beer: purification, flavour, threshold and interaction. In: Drawert F (ed) Geruch und Geschmacksstoffe. Verlag Hans Carl, Nurnberg, pp 211–254
Meilgaard MC (1975b) Flavor chemistry of beer. Part I. Flavor interaction between principal volatiles. MBAA Tech Q 12:107–117
Molzahn SW (1977) A new approach to the application of genetics to brewing yeast. J Am Soc Brew Chem 35:54–59
Mullis KB, Erlich HA, Arnheim N, Horn GT, Saiki RK, Scharf SJ (1987) Process for amplifying, detecting, and/or cloning nucleic acid sequences. Patent No. US4683195 A
Nakagawe A, Matsumura E, Koyanagi T, Katayama T, Kawano N, Yoshimatsu K, Yamamoto K, Kumagai H, Sato F, Minami H (2016) Total biosynthesis of opiates by stepwise fermentation using engineered *Escherichia coli*. Nat Commun 7:10390–10396
Naumov GI, Li C-F (2011) Species specificity of the action of *Zygowilliopsis californica* killer toxins on *Saccharomyces* yeasts: investigation of the Taiwanese populations. Mikol Fitopatol 45:332–336
Naumov GI, Kondratieva VI, Naumova ES, Chen G-Y, Li C-F (2011) Polymorphism and species specificity of killer activity formation in the yeast *Zygowilliopsis californica*. Biotekhnologiya 3:29–33
Nevoigt E, Fischer C, Mucha O, Matthaus F, Stahl U, Stephanopoulos G (2007) Engineering promoter regulation. Biotechnol Bioeng 96:550–558

Nielsen J (2013) Production of biopharmaceutical proteins by yeast: advances through metabolic engineering. Bioengineered 4:207–211
Nielsen J (2015) Yeast cell factories on the horizon. Science 349(6252):1050–1051
Oliver SG (1996a) From DNA sequence to biological function. Nature 379:597–600
Oliver SG (1996b) A network approach to the systematic analysis of yeast gene function. Trends Genet 12:241–242
Peberdy JF, Eyssen H, Anné J (1977) Interspecific hybridisation between *Penicillium chrysogenum* and *Pencillium cyaneo-fulvum* following protoplast fusion. Mol Gen Genet 157:281–284
Peris D, Sylvester K, Libkind D, Gonçalves P, Sampaio JP, Alexander WG, Hittinger CT (2014) Population structure and reticulate evolution of *Saccharomyces eubayanus* and its lager-brewing hybrids. Mol Ecol 23:2031–2045
Petranovic D, Tyo K, Vemuri GN, Nielsen J (2010) Prospects of yeast systems biology for human health: integrating lipid, protein and energy metabolism. FEMS Yeast Res 10:1046–1059
Pomper S, Burkholder PR (1949) Studies on the biochemical genetics of yeast. Proc Natl Acad Sci U S A 35:456–464
Pontecorvo G, Kafer E (1958) Genetic analysis based on mitotic recombination. Adv Genet 9:71–104
Pray LA (2008) Discovery of DNA structure and function: Watson and Crick. Nat Educ 1:100
Pretorius IS (2016) Synthetic genome engineering forging new frontiers for wine yeast. Crit Rev Biotechnol 37:112–136
Pretorius IS, du Toit MD, van Rensburg P (2003) Designer yeasts for the fermentation industry of the 21st century. Food Technol Biotechnol 41:3–10
Reinhardt K, Dowling DK, Morrow EH (2013) Medicine. Mitochondrial replacement, evolution, and the clinic. Science 341:1345–1346
Ro DK, Paradise EM, Ouellet M, Fisher KJ, Newman KL, Ndungu JM, Ho KA, Eachus RA, Ham TS, Kirby J, Chang MCY, Withers ST, Shiba Y, Sarpong R, Keasling JD (2006) Production of the antimalarial drug precursor artemisinic acid in engineered yeast. Nature 440:940–943
Rosano GL, Ceccarelli EA (2014) Recombinant protein expression in *Escherichia coli*: advances and challenges. Front Microbiol 5:172
Russell I, Stewart GG (1979) Spheroplast fusion of brewer's yeast strains. J Inst Brew 85:95–98
Russell I, Stewart GG (1980) Transformation of maltotriose uptake ability into a haploid strain of *Saccharomyces* spp. J Inst Brew 86:55–59
Russell I, Stewart GG (1981) Liquid nitrogen storage of yeast cultures compared to more traditional storage methods. J Am Soc Brew Chem 39:19–24
Russell I, Stewart GG (1985) Valuable techniques in the genetic manipulation of industrial yeast strains. J Am Soc Brew Chem 43:84–90
Russell I, Stewart GG (2014) Whisky: technology, production and marketing. Academic (Elsevier), Boston, MA
Saerens SM, Duong CT, Nevoigt E (2010) Genetic improvement of brewer's yeast: current state, perspectives and limits. Appl Microbiol Biotechnol 86:1195–1212
Schmitt MJ, Breining F (2002) The viral killer system in yeast: from molecular biology to application. FEMS Microbiol Rev 26:257–276
Sherman F (1998) An introduction to the genetics and molecular biology of the yeast *Saccharomyces cerevisiae*. http://dbburmcrochester.edu/labs/sherman_f/yeast/
Sherman F, Hicks J (1991) Micromanipulation and dissection of asci. Methods Enzymol 194:21–37
Shi S, Chen T, Zhang Z, Chen X, Zhao X (2009) Transcriptome analysis guided metabolic engineering of Bacillus subtilis for riboflavin production. Metab Eng 11:243–252
Silhankova L, Mostek J, Savel J, Solinova H (1970) Respiratory deficient mutants of bottom brewer's yeast II technological properties of some RD mutants. J Inst Brew 76:289–295
Sipiczki M, Ferenczy L (1977a) Protoplast fusion of *Schizosaccharomyces pombe* auxotrophic mutants of identical mating-type. Mol Gen Genet 151:7781
Sipiczki M, Ferenczy L (1977b) Fusion of *Rhodosporidiuin (Rhodotorula)* protoplasts. FEMS Microbiol Lett 2:203–205

Sipiczki M, Heyer WD, Kohli J (1985) Preparation and regeneration of protoplasts and spheroplasts for fusion and transformation of *Schizosaccharomyces pombe*. Curr Microbiol 12:169
Skatrud PL, Jaeck DM, Kot EJ, Helbert JR (1980) Fusion of *Saccharomyces uvarum* with *Saccharomyces cerevisiae*: genetic manipulation and reconstruction of a brewers yeast. J Am Soc Brew Chem 38:49–53
Soares EV (2010) Flocculation in *Saccharomyces*: a review. J Appl Microbiol 110:1–18
Speers A (2016) Brewing fundamentals. Part 3: Yeast settling—flocculation. MBAA Tech Q 53:17–22
Spencer JFT, Spencer DM (1977) Hybridization of non-sporulating and weakly sporulating strains of brewer's yeasts. J Inst Brew 83:287–289
Spencer JF, Spencer DM (1996) Rare-mating and cytoduction in *Saccharomyces cerevisiae*. Methods Mol Biol 53:39–44
Spencer JFT, Laud P, Spencer DM (1980) The use of mitochondrial mutant sin the isolation of hybrids involving industrial yeast strains. Mol Gen Genet 178:651–654
Steensels J, Snoek T, Meersman E, Nicolino MP, Voordeckers K, Verstrepen KJ (2014) Improving industrial yeast strains: exploiting natural and artificial diversity. FEMS Microbiol Rev 38:947–995
Stephanopoulos G (2012) Synthetic biology and metabolic engineering. ACS Synth Biol 1:514–525
Stewart GG (1973) Recent developments in the characterization of brewery yeast strains. MBAA Tech Q 9:183–191
Stewart GG (1975) Yeast flocculation – practical implications and experimental findings. Brew Dig 50:42–62
Stewart GG (1981) The genetic manipulation of industrial yeast strains. Can J Microbiol 27:973–990 (Canadian Society for Microbiology, Hotpack Award Lecture)
Stewart GG (2002) Yeast: the most important microorganism in use. Food Essent 1:2–4
Stewart GG (2009) The IBD Horace Brown Medal Lecture – forty years of brewing research. J Inst Brew 115:3–29
Stewart GG (2010) MBAA Award of Merit Lecture. A love affair with yeast. MBAA Tech Q 47:4–11
Stewart GG (2014a) Yeast mitochondria – their influence on brewer's yeast fermentation and medical research. MBAA Tech Q 51:3–11
Stewart GG (2014b) Brewing intensification. The American Society for Brewing Chemists, St. Paul, MN
Stewart GG (2014c) The concept of nature – nurture applied to brewer's yeast and wort fermentations. MBAA, Tech Q 51:69–80
Stewart GG, Russell I (1977) The identification, characterization, and mapping of a gene for flocculation in *Saccharomyces* sp. Can J Microbiol 23:441–447
Stewart GG, Russell I (1979) Current use of the "new" genetics in research and development of brewer's yeast strains. In: Proceedings of the 17th European Brewery Convention Congress, Berlin (West) EBC/DSW, Dordrecht, pp 475–490
Stewart GG, Russell I (2009) An introduction to brewing science and technology, Series lll, Brewer's yeast, 2nd edn. The Institute of Brewing and Distilling, London. isbn:0900498-13-8
Stewart GG, Panchal CJ, Russell I (1983) Current developments in the genetic manipulation of brewing yeast strains – a review. J Inst Brew 89:170–188
Stewart GG, Jones RM, Russell I (1985) The use of derepressed yeast mutants in the fermentation of brewery wort. In: Proceedings of the 20th European Brewery Convention Congress, Helsinki. IRL Press, Oxford, pp 243–250
Stewart GG, Russell I, Panchal CJ (1986) Genetically stable allopolyploid somatic fusion product useful in the production of fuel alcohols. Canadian Patent 1,199,593
Stewart GG, Jones RM, Russell I (2013) 125th anniversary review – developments in brewing and distilling yeast strains. J Inst Brew 119:202–220
Stewart GG, Maskell DL, Speers A (2016) Brewing fundamentals – fermentation. MBAA Tech Q 53:2–22

Stovicek V, Holkenbrink C, Borodina I (2017) CRISPR/Cas system for yeast genome engineering: advances and applications. FEMS Yeast Res 17:1–16

Svoboda A (1978) Fusion of yeast protoplasts induced by polyethylene glycol. J Gen Microbiol 109:169–175

Swinnen S, Thevelein JM, Nevoigt E (2012) Genetic mapping of quantitative phenotypic traits in *Saccharomyces cerevisiae*. FEMS Yeast Res 12:215–227

Tamaki H (1978) Genetic studies of the ability to ferment starch in *Saccharomyces*: gene polymorphism. Mol Gen Genet 164:205–209

Tao X, Zheng D, Liu T, Wang P, Zhao W, Zhu M, Jiang X, Zhao Y, Wu X (2012) A novel strategy to construct yeast *Saccharomyces cerevisiae* strains for very high gravity fermentation. PLoS One 7(2):e31235

Teunissen AWRH, Steensma HY (1995) Review: The dominant flocculation genes of *Saccharomyces cerevisiae* constitute a new subtelomeric gene family. Yeast 11:1001–1013

Thorne RSW (1951) Some aspects of yeast flocculence. In: Proceedings of the European Brewery Convention Congress, Brighton, pp 21–30

Tyo KE, Kocharin K, Nielsen J (2010) Toward design-based engineering of industrial microbes. Curr Opin Microbiol 13:255–262

Van Mulders SE, Christianen E, Saerens SMG, Daenen L, Verbelen PJ, Willaert R, Verstrepen KJ, Delvaux FR (2009) Phenotypic diversity of Flo protein family-mediated adhesion in *Saccharomyces cerevisiae*. FEMS Yeast Res 9:178–190

van Solingen P, van der Plaat JB (1977) Fusion of yeast spheroplasts. J Bacteriol 130:946–947

Verstrepen KJ, Derdelinckx G, Delvaux FR, Winderickx J, Thevelein JM, Bauer FF, Pretorius IS (2001) Late fermentation expression of *FLO1* in *Saccharomyces cerevisiae*. J Am Soc Brew Chem 59:69–76

Walsh G (2010) Biopharmaceutical benchmarks. Nat Biotechnol 28:917–924

Walters VL (2001) Conjugation between bacterial and mammalian cells. Nat Genet 29:375–376

Wang L, Yue L, Chi ZM, Wang X (2008) Marine killer yeasts active against a yeast strain pathogenic to crab *Portunus trituberculatus*. Dis Aquat Org 80:211–218

Wang XX, Chi ZM, Peng Y (2012) Purification, characterization and gene cloning of the killer toxin produced by the marine-derived yeast *Williopsis saturnus* WC91-2. Microbiol Res 167:558–563

Watari J, Takata Y, Ogawa M, Murakami J, Shohei K (1991) Breeding of flocculent industrial *Saccharomyces cerevisiae* strains by introducing the flocculation gene FLO1. Agric Biol Chem 55:1547–1552

Watson JD, Crick FHC (1953a) A structure for deoxyribose nucleic acid. Nature 171:737–738

Watson JD, Crick FHC (1953b) Genetical implications of the structure of Deoxyribonucleic Acid. Nature 171:964–967

Wickner RB, Edskes HK (2015) Yeast killer elements hold their hosts hostage. PLoS Genet 11: e1005139

Wildt S, Gerngross TU (2005) The humanization of N-glycosylation pathways in yeast. Nat Rev Microbiol 3:119–128

Winzeler E, Shoemaker DD, Astromoff A, Liang H, Anderson K, Andre B, Bangham R, Benito R, Boeke JD, Bussey H, Chu AM, Connelly C, Davis K, Dietrich F, Dow SW, El Bakkoury M, Foury F, Friend SH, Gentalen E, Giaever G, Hegemann JH, Jones T, Laub M, Liao H, Liebundguth N, Lockhart DJ, Lucau-Danila A, Lussier M, M'Rabet N, Menard P, Mittmann M, Pai C, Rebischung C, Revuelta JL, Riles L, Roberts CJ, Ross-MacDonald P, Scherens B, Snyder M, Sookhai-Mahadeo S, Storms RK, Véronneau S, Voet M, Volckaert G, Ward TR, Wysocki R, Yen GS, Yu K, Zimmermann K, Philippsen P, Johnston M, Davis RW (1999) Functional characterization of the *S. cerevisiae* genome by gene deletion and parallel analysis. Science 285:901–906

Woods DR, Bevan EA (1968) Studies on the nature of the killer factor produced by *Saccharomyces cerevisiae*. J Gen Microbiol 51:15–126

Zimmermann RE, Stokell DJ (2010) Insulin production methods and pro-insulin constructs. US Patent No. 7790677 B2

Chapter 17
Non-*Saccharomyces* (and Bacteria) Yeasts That Produce Ethanol

17.1 Introduction

In excess of a thousand unique yeast strains have been identified, and many of them have been characterized (to a lesser or greater extent) (Boekhout 2005; Sicard and Legras 2011). Ninety percent (and more) of the fermentation ethanol produced globally employs species of the genus *Saccharomyces* (predominantly *S. cerevisiae*, *S. pastorianus* and related species). However, there are a number of non-*Saccharomyces* yeast species that can produce ethanol [also called nonconventional yeast species (Radecka et al. 2015)]. An excellent monograph entitled *Nonconventional Yeasts in Biotechnology* (edited by Wolf 1996) summarizes this subject in considerable detail. Although it may be out of date to a degree! Nevertheless, only non-*Saccharomyces* yeasts that produce ethanol will be discussed here. Also, *Zymomonis mobilis*, a Gram-negative ethanol-producing bacteria, is considered (Agrawal et al. 2011)

Nonconventional yeasts are a large, and barely exploited, resource of yeast biodiversity. Many of these nonconventional yeast species exhibit industrially relevant traits such as an ability to utilize complex substrates and nutrients and tolerate stresses and fermentation inhibitors (Steensels and Verstrepen 2014). The evolution of most of these species was independent of *Saccharomyces* sp. (Forsburg and Rhind 2006; Souciet et al. 2009), and it is widely speculated that many of them possess novel and unique mechanisms that are not present in *Saccharomyces* yeasts. Most of them have individually been characterized as spoilage yeasts due to their isolation from contaminated foods and beverages (Dujon 2010). However, novel sequencing technology and advanced molecular engineering tools offer the possibility to reveal the underlying molecular basis of the superior stress tolerance of some of these nonconventional yeast species. In this chapter, the phenotype characteristics of some of these nonconventional yeast species are discussed. Particular emphasis is placed on species that are extremely tolerant to stresses commonly encountered during first- and second-generation

G.G. Stewart, *Brewing and Distilling Yeasts*, The Yeast Handbook,
https://doi.org/10.1007/978-3-319-69126-8_17

bioethanol production such as osmotic, ethanol and thermal stresses together with a plethora of different fermentation inhibitor stresses. Also, an overview of the potential industrial applications of these nonconventional species is provided. It should be emphasized that the nonconventional yeast species that are discussed are not an exhaustive listing. In the author's opinion, they are the most interesting and potentially industrially valuable species for the future (Dujon 2010).

17.2 *Schizosaccharomyces pombe*

Schizosaccharomyces pombe, also called "fission yeast", is a yeast species used in traditional brewing (originally isolated from African millet beer). It is also employed as a model organism in molecular and cell biology. It has a distinguished research history in studies of the cell cycle and mitosis, chromosome dynamics and epigenetics. It has also been used to address other cell biology questions such as DNA repair, meiosis, cytokinesis and mRNA processing. Paul Nurse (Hagen et al. 2016), a fission yeast researcher (Fig. 17.1), successfully merged fission yeast genetics and cell cycle research. Together with Lee Hartwell and Tim Hunt,

Fig. 17.1 Paul Nurse

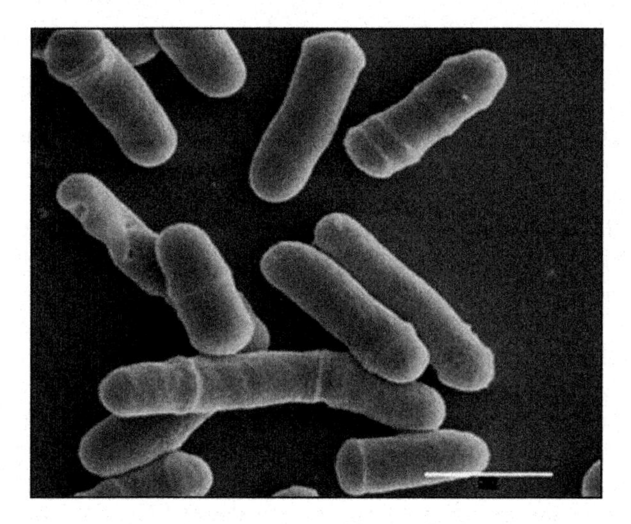

Fig. 17.2 *Schizosaccharomyces pombe*

Nurse was awarded the 2001 Nobel Prize in Physiology or Medicine for their studies on cell cycle regulation. (http://www.nobelprize.org/nobel_prizes/medicine/laureates/2001/illpres/index.html).

S. pombe is a unicellular fungus (a eukaryote) that is rod shaped. Cells of this yeast (Fig. 17.2) measure 2–3 microns in diameter and 7–14 microns in length. Its genome, which consists of approximately 14.1×10^6 base pairs, is estimated to contain 4970 protein-coding genes and at least 450 non-coding RNAs (Wilhelm et al. 2008). The cells maintain their shape by growing exclusively through the cell tips and divide through the cell tips by medial fission to produce two daughter cells of equal sizes (Fig. 17.2), which make them a powerful tool for cell cycle research (Novak and Mitchison 1990a, b; Hagen et al. 2016).

Fission yeast was first isolated in 1893 by Paul Linden from East African millet beer. Linden was working in a Brewery Association Laboratory in Germany. He was examining a sediment found in millet beer imported from East Africa that gave it an acidic taste. The term "schizo", meaning "split" or "fission", had previously been used to describe other Schizosaccharomycetes. The addition of the word pombe was due to its isolation from East African beer, as pombe means "beer" in Swahili. The standard *S. pombe* strains were isolated by Urs Leupold in 1946 and 1947 from a culture he obtained from the Delft yeast culture collection in the Netherlands (Leupold 1950). It was deposited there by A. Osterwalder under the name *S. pombe var. liquefaciens*, following isolation in 1924 from French wine at a research station in Wädenswil, Switzerland. Urs Leupold studied its genetics (Leupold 1993), and Murdoch Mitchison, working at the University of Edinburgh, studied the cell cycle (Mitchison 1957).

The DNA sequence of the *S. pombe* genome was published in 2002, by a consortium led by researchers at the Sanger Institute (Wood et al. 2002). It became the sixth model eukaryote organism whose genome has been fully sequenced (the first eukaryote organism to be sequenced was *S. cerevisiae*) (Goffeau et al. 1996; Goffeau 1996).

S. pombe belongs to the phylum Ascomycota, which represents the largest and most diverse group of fungi. Amongst the ascomycetous yeast genera, the fission yeast *Schizosaccharomyces* is unique because of the deposition of wall α-(1,3)-glucan in addition to better known β-glucans and a lack of chitin (Calleja et al. 1984). Specimens of this genus also differ in mannan composition because they contain β-galactose sugars in their mannan side chains. *S. pombe* undergoes aerobic fermentation in the presence of excess sugar (Lin and Li 2011). Also, *S. pombe* can degrade L malic acid (a dominant organic acid in wine) which makes them diverse amongst *Saccharomyces* strains.

It has already been described that *S. pombe* is a single-celled fungus with a fully characterized genome and a rapid growth rate. It has been used in brewing, baking and molecular genetics. *S. pombe* is a rod-shaped cell, approximately 3 μm in diameter, that grows entirely by elongation at the cell ends. Following mitosis, division occurs by the formation of a septum or cell plate that clears the cell at its midpoint. If the cells have a mutation that prevents the normal progression of the cell cycle, they become elongated. This occurs because the cells can continue to grow, but they are unable to divide. These mutants are called *cdc* mutants, for cell division cycle (Fantes and Nurse 1977).

S. pombe is chemoorganotrophic, and it uses organic compounds as a source of energy and does not require light. These yeasts can grow under both aerobic and anaerobic conditions. Fission yeasts are facultatively fermentative and exhibit aerobic fermentation in the presence of excess sugar (Novak and Mitchison 1990a, b). Alcohol dehydrogenase (ADH) catalyses the reduction of acetaldehyde to ethanol in the last step of glycolysis (details in Chap. 6). This reduction is coupled with the oxidation of NADH and provides the NAD^+ essential for glyceraldehyde-3-phosphate oxidation in glycolysis. Ethanol production is important in order to maintain the redox balance in the cytoplasm. For a while (Crichton et al. 2007), it has been assumed that *S. pombe* and *Kluyveromyces lactis* (details later) do not contain a mitochondrial ADH isoenzyme and, consequently, does not have ethanol-dependent respiratory activity in their mitochondria. Ethanol-dependent respiratory activity is generally attributed to the presence of mitochondrial ADH isoenzymes. However, it has recently been shown that knockout strains of *S. pombe* do not exhibit mitochondrial ADH activity, but the physiological function of *S. pombe* mitochondrial ADH enzymes (unlike *S. cerevisiae* ADH enzymes—Saliola et al. 1990) is unclear.

As already discussed, *S. pombe* is found in alcoholic beverages. It is one of the yeasts that play an important role in the ecology of Kombucha fermentation (Teoh et al. 2004). Kombucha is a traditional fermentation of sweetened tea, involving a symbiosis of yeast species and acetic acid bacteria. The study of yeast ecology in Kombucha fermentation has revealed that *S. pombe* operates in conjunction with other yeast species such as *Brettanomyces bruxellansis*, *Candida stellata*, *Torulaspora delbrueckii* and *Zygosaccharomyces bailii*. *S. pombe* and other yeasts ferment the sugar in the tea to ethanol, and then the ethanol is oxidized by the acetic acid bacteria to acetic acid. *S. pombe* is suited to the Kombucha environment due to its ability to tolerate a high-sugar environment (Teoh et al. 2004; Martorell et al. 2007). Isolates of *S. pombe*, *T. delbrueckii* and *Z. bailii* can exhibit tolerance up to 60% glucose concentration and are commonly associated with alcoholic fermentation for wine and champagne production. As the fermentation progressed, species with low acid tolerance decreased in population. Species such as *S. pombe*, with moderate tolerance to acidic conditions, become non-viable after day 10. Kombucha fermentation is initiated by osmotolerant species of yeast which are capable of growing in the presence of high concentrations of sugar. The process is ultimately dominated by acid-tolerant species (Basso et al. 2016).

Currently, *S. pombe* is not employed in any modern biotechnology application. However, this yeast species remains an important eukaryote used in research on cell division in model genetics and cell biology (Forsburg and Rhind 2006).

17.3 Kluyveromyces

Kluyveromyces is a genus of ascomycetous yeasts in the family *Saccharomycetaceae*. Some of the species, for example, *K. marxianus*, are the teleomorphs of *Candida* species [a teleomorph is the sexual reproductive stage (Pecota et al. 2007)]. This yeast genus has been reviewed in a comprehensive discussion of nonconventional yeast species (Radecka et al. 2015).

Kluyveromyces marxianus

Kluyveromyces marxianus (Fig. 17.3) is a yeast species in the genus *Kluyveromyces* and is the teleomorph of *Candida kefyr*. It has a number of commercial applications (details later) (Fuson et al. 1987). Its current primary use is to produce lactase (β-galactosidase) similar to the use of other fungi such as the genus *Aspergillus* (Seyis and Aksoz 2004).

K. marxianus is well known for its extreme thermotolerance. It has been reported to grow at 47 °C (Limtong et al. 2007), 49 °C and even as high as 52 °C and to produce ethanol above 40 °C (Nonklang et al. 2008). *K. marxianus* is not only thermotolerant but also offers additional benefits including a high growth rate and the ability to utilize a wide variety of industrially relevant substrates such as sugar cane, corn silage juice, molasses and whey (Prazeres et al. 2012). It has also been used for recombinant protein (Nonklang et al. 2008) and industrial enzyme production, such as inulinase (Rouwenhorst et al. 1988) and β-galactosidase (Martinez et al. 2002).

K. marxianus was first described in 1888 by E.C. Hansen (Fig. 2.20) working in the Copenhagen Carlsberg Laboratory (discussed by Phaff 1981), and he named it *Saccharomyces marxianus*. Numerous strains of this species have been isolated, mostly from cheese and other dairy products. Strains of the *Kluyveromyces* genus have the ability to mate and produce fertile hybrids, both intraspecies and interspecies

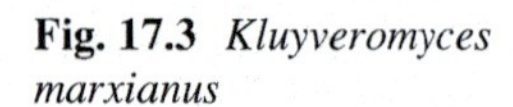

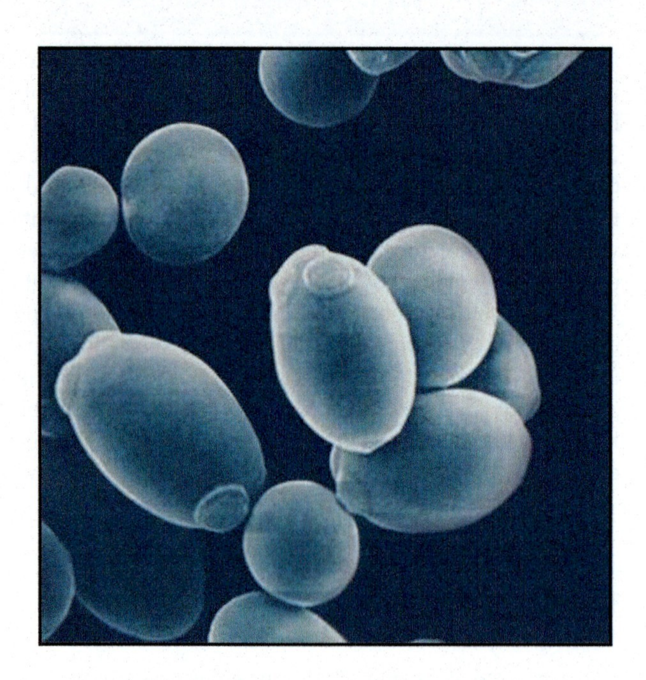

Fig. 17.3 *Kluyveromyces marxianus*

hybrids. Frequent isolation of these hybrids leads to difficulties in identifying a distinct species. This problem has been addressed by DNA reassociation studies (Fuson et al. 1987; Martini and Martini 1987; Llorente et al. 2000). *K. marxianus* shows a high intraspecies polymorphism with a common species-species pattern (Belloch et al. 1998a, b). The strain *K. marxianus* CBS 6556 has been sequenced, and a genome size of 10.9 Mb has been estimated (Jeong et al. 2012).

The metabolism of *K. marxianus* has been described as respirofermentative. Interestingly, *K. marxianus*, along with its sister species *K. lactis* (details later), is traditionally classified as a "Crabtree-negative" yeast. It has been suggested that such a contradiction might be due to highly divergent phenotypes amongst isolates (Lane et al. 2011). In contrast to *S. cerevisiae*, *K. marxianus* is able to utilize xylose, xylitol, cellobiose, lactose and arabinose both on solid and in liquid media (Nonklang et al. 2008). It also has the ability to ferment glucose at between 30 and 45 °C. At 30 °C, it achieved similar levels of ethanol and glucose consumption as *S. cerevisiae* will achieve at 45 °C. *S. cerevisiae* (unlike *K. marxianus*) was unable to ferment at all (Nonklang et al. 2008)! Previous reports described the temperature tolerance range of several *K. marxianus* isolates, with most being able to grow at 42 °C and only a few isolates to 48 °C (Nonklang et al. 2008; Abdel-Banat et al. 2010; Lane et al. 2011). The mechanism(s) behind this extreme temperature tolerance are currently unknown. During the last decade, the availability of tools for genetic modification of *K. marxianus* is increasing, which will introduce opportunities to uncover the molecular basics of this unique yeast genus (Kegel et al. 2006; Pecota et al. 2007; Nonklang et al. 2008; Abdel-Banat et al. 2010; Lee et al. 2013; Yarimizu et al. 2013; Hoshida et al. 2014).

Fig. 17.4 Albert Jan Kluyver

Several different biotechnological applications have been investigated with strains of *K. marxianus* (Fonseca et al. 2008). This includes the production of enzymes (e.g. β-galactosidase, β-glucosidase, inulinase and polygalacturonases), single-cell protein, aroma components and ethanol (including both at high temperature and simultaneous fermentation processes). In addition, reduction of the lactose content of food products, production of bioingredients from cheese whey, bioremediation, an anticholesterolemic agent and a host for heterologous protein products are also potential uses for strains of *K. marxianus*. This yeast species can envisage a great biotechnological future because of properties already discussed, namely, a broad substrate spectrum, thermotolerance, high growth rates and a reduced tendency to ferment when exposed to excess sugar (Prazeres et al. 2012).

The so-called milk yeast *K. lactis* is a close relative of *K. marxianus*. It has been isolated from milk and constitutes the predominant eukaryote during cheese production, which decreases whey pH and paves the way for lactic acid bacteria to thrive. *K. lactis* has been named in honour of the Dutch microbiologist Albert Jan Kluyver (1888–1956) (Fig. 17.4). Honouring this name, *K. lactis* displays the Kluyver effect—it grows on sugars like galactose, sucrose and maltose only under aerobic, but not oxygen-limiting conditions.

The Kluyver effect is associated with specific combinations of yeast species and sugars. For example, *Debaryomyces robertsiae* exhibits the Kluyver effect positive on galactose and is Kluyver effect negative on maltose, whereas *Candida kruisli* and *Pichia heimii* are contrary, with the Kluyver effect negative on galactose and positive on maltose. There are many more examples that could be cited (Barnett 1992; Fukuhara 2003).

This Kluyver effect has been explained by the lack of sugar transport under anaerobic conditions (Fukuhara 2003). It does not grow on any sugar under strict anaerobiosis, not even on glucose, or in the presence of growth factors such as ergosterol and unsaturated fatty acids (Snoek and Steensma 2006). However, respiration can be inhibited by the addition of antimycin A, in most *K. lactis* strains, which continue to ferment glucose (RAG^+ phenotype). Mutants are unable to grow

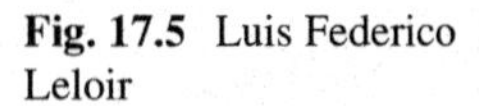
Fig. 17.5 Luis Federico Leloir

if respiration is inhibited [*rag*$^-$ are mainly affected in hexose transport and glycolysis and/or its regulation (Breunig et al. 2000; Hnatova et al. 2008)].

With regard to lactose utilization, this disaccharide is first transported through a permease (encoded by *LAC12*), which also transports galactose with low affinity (Baruffini et al. 2006; Rigamonte et al. 2011). Lactose is then fed into glycolysis through the Leloir (Fig. 17.5) pathway (Fig. 17.6) (reviewed in Breunig et al. 2000). External galactose can also be imported by the high-affinity carrier *Hgt1*, whose gene expression escapes glucose repression but requires induction by galactose (Boze et al. 1987a, b; Baruffini et al. 2006).

Due to its traditional employment in cheese production, *K. lactis* has GRAS (generally regarded as safe) status and consequently is an attractive production organism in the food industry (Coenen et al. 2000). It is also the natural choice for the production of β-galactosidase, which is used to remove lactose from milk for patients with lactose intolerance. Also, the enzyme is applied in the production of cheese and yoghurt and ice cream (Lomer et al. 2008; Panesar et al. 2010). To improve β-galactosidase production, vector systems, genetically modified yeast strains and adjustment of growth parameters have been intensively investigated (Becerra et al. 2004a, b; Panesar et al. 2010).

Intact cells of *K. lactis* have been suggested to possess probiotic potential. Its ability to use lactose is also of growing importance for waste disposal in the dairy industry because whey constitutes a major environmental problem due to its high organic load (Prazeres et al. 2012; Rives et al. 2011). Also, cheese whey represents an inexpensive carbon source for yeast growth. The price of molasses (the traditionally used source of carbon) had increased considerably during the past two to three decades. However, *S. cerevisiae* cannot use lactose; consequently, the genes

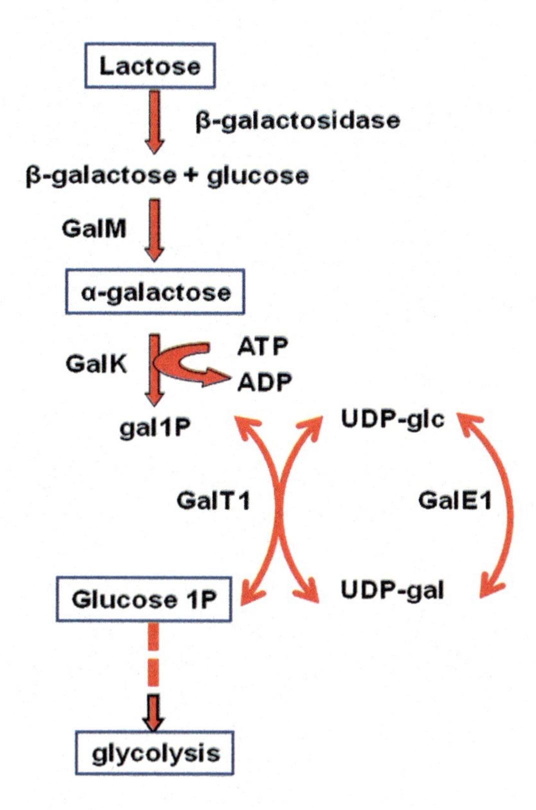

Fig. 17.6 The Leloir pathway for lactose and galactose metabolism

encoding the lactose permease (*LAC12*) and the β-galactosidase (*LAC4*) from *K. lactis* have been heterologously expressed (Domingues et al. 2010).

K. lactis has also been employed as a host for heterologous protein production and the most prominent is chymosin, which was traditionally obtained from calf thymus (van den Berg et al. 1990). Many other enzymes have been obtained from heterologous expression systems in *K. lactis* (van Ooyen et al. 2006). A commercially available cloning vector directs heterologous gene expression from a modified (*LAC4*) promoter, with the option of secreting the product by coupling it to the signal peptide from various proteins (Colussi et al. 2005; Colussi and Taron 2005; Krijger et al. 2012).

S. cerevisiae is of invaluable importance as a unicellular and genetically tangible model system for most eukaryotic cells in all aspects of basic cell function (Clapp et al. 2012; Winderickx et al. 2008). However, the redundancy of genes encoding different isoforms of the same protein, primarily caused by the whole-genome duplication event, frequently complicates the analysis of protein function. This is one of the reasons why alternative yeast models, especially those that have not undergone genome duplication, are gaining interest. In this context, *K. lactis*, which has a similar life cycle and ease of manipulation, has been employed in most of the methods established for *S. cerevisiae*. However, *K. lactis* suffers from a few drawbacks:

- In vivo recombination methods can be considerably improved by using a few genetic tools. Currently, its efficiency is at least an order of magnitude lower than that in *S. cerevisiae*).
- A central collection of auxotrophic mutants deleted for open reading frames still needs to be established and will not cover the entire genome at any time soon (Yarimisu et al. 2013).
- Instead of working in an isogenic background, several *K. lactis* strains with considerable physiological differences are employed in different laboratories.

The above situation will hopefully change as a result of the growing interest of new research groups working on *K. lactis* and the availability of a congenic strain series and a corresponding set of molecular tools (Heinisch et al. 2010).

17.4 *Schwanniomyces*: Starch-Degrading Yeast Genes

The hydrolysis of starch en route to the production of fermentation ethanol can be achieved by two basic processes:

- The starch is first hydrolysed by a spectrum of amylases (mashing and similar processes) by malted cereals (usually barley malt) or by amylases isolated from specific bacteria and fungi. The sugars obtained as a result of starch hydrolysis are subsequently fermented by yeast, usually *S. cerevisiae*.
- Alternatively, the starch is hydrolysed and subsequently fermented to ethanol by one (or more) strain microorganism. One such yeast belongs to the genus *Schwanniomyces* which is a member of the family Saccharomycetaceae (order Endomycetalis). It belongs to the same subfamily (Saccharomycodeae) as well as the genera *Saccharomyces*, *Kluyveromyces* and *Hansenula* (Lodder 1984).

Schwanniomyces occidentalis (syn. *Debaryomyces occidentalis*) is a yeast species with large enzymatic potential. It is capable of utilizing a broad spectrum of carbon sources, including starch and inulin (a fructose polymer). Four species of Schwanniomyces have been described—*Schw. alluvius*, *Schw. castellii*, *Schw. peroonie* and *Schw. occidentalis* (Phaff 1970). A subsequent study of DNA reassociation between species of *Schwanniomyces* (Price et al. 1978) has led to single-species designation—*S. occidentalis*. Also, *Lipomyces kononenkae* converts starch into fermentable sugars (Spencer-Martins and van Uden 1979, 1982). The amylase activity of this yeast species has been cloned into *S. cerevisiae* (Eksteen et al. 2003a, b).

Cells of *Schw. occidentalis* are ovoid, or egg-shaped, occasionally elongated or cylindrical, and they reproduce vegetatively by multilateral budding. Cells are haploid with a transient diploid stage immediately preceding sporulation. Before sporulation, the nucleus divides. This process is followed by fusion and the formation of an abortive daughter bud, called the meiosis bud, which remains a distinct feature of the ascus. Usually only one, but more rarely two, ascospore is formed, with the remaining meiotic products disintegrating (Ferreira and Phaff 1959). *Schwanniomyces* yeasts can be distinguished from other ascomycetous yeasts (e.g. *Saccharomyces*) by the unique shape of the ascospores (Phaff et al. 1966).

The number of research groups studying the properties of *Schw. occidentalis* has increased significantly over the past 20 years as it has become clear that this organism is equipped with an extremely powerful amylolytic system. Because starch-containing raw materials are inexpensive and widely available, investigators have increased their efforts to characterize the amylolytic system of *Schw. occidentalis*, as well as the yeast itself (Rossi and Clementi 1985).

Schw. occidentalis is capable of utilizing a wide range of organic compounds as carbon sources: glucose, fructose, galactose, D-xylose, sucrose, raffinose (partially), cellobiose, trehalose, lactose (by some strains), succinate, citrate, ethanol and n-alkanes, as well as maltose, iso-maltose, pullulan and dextrin together with soluble and raw starch. Melibiose cannot be used because strains of this species do not produce the enzyme α-galactosidase. *Schw. occidentalis* (and related species) is Crabtree-negative and shows a strong Pasteur effect (Fig. 6.1). This means that oxidative metabolism is not repressed by high sugar concentration, and there is little fermentation under aerobic conditions (Ingledew 1987; Poinsot et al. 1987). (Details of both the Crabtree and Pasteur effects in yeast can be found in Chaps. 6 and 7.) The most important property of *Schw. occidentalis* is its ability to efficiently degrade starch as a result of the activity of extracellular α-amylase and glucoamylase. The expression of both enzymes is both inducible and repressed by glucose depending on the formation conditions (Sills et al. 1984a).

It has been an objective for a number of decades to generate *S. cerevisiae* strains that can degrade starch and dextrins and produce ethanol. A large number of yeast strains have been described that are able to grow on starch (Lodder 1984; Fuji and Kawamura 1985). However, only a few degrade starch with high efficiency as a result of the combined action of a α-amylase, glucoamylase and debranching activity—the hydrolysis of α-1,6 glycosidic bonds (Fig. 17.6) (Erratt and Stewart 1978, 1981a; Spencer-Martins and van Uden 1979; Clementi and Rossi 1986; Sills and Stewart 1982, Touzi et al. 1982; De Mot et al. 1984, 1985a, b; De Mot and Verachtert 1985; de Moreas et al. 1995; Eksteen et al. 2003a, b).

The species *Schw. occidentalis* expresses the most significant debranching activity (Sills and Stewart 1982), which is a property of the glucoamylase of this yeast (Wilson and Ingledew 1982; Sills et al. 1984a, b). From a brewing perspective, both α-amylase and glucoamylase secreted by *Schw. occidentalis* also have the advantage of being inactivated during the pasteurization conditions usually employed (Sills et al. 1987). Whereas, the glucoamylase produced by some strains of *S. cerevisiae var. diastaticus* exhibits two negative brewing properties (Erratt and Stewart 1981b). Namely, it does not exhibit debranching ability, consequently, it does not hydrolyze starch (only dextrin), and its activity is not inactivated during pasteurization.

The synergistic action of α-amylase and glucoamylase secreted by *Schw. occidentalis* results in efficient and complete degradation of starch (Fuji and Kawamura 1985; Boze et al. 1989). Kinetic studies on the degradation of starch by the two amylases from *Schw. occidentalis* revealed that α-amylase is especially important to processes which employ complex substrates such as glycogen, potato and barley starch (Boze et al. 1989). Release of glucose from these substrates by glucoamylase is slow. However, the role of α-amylase is to generate free nonreducing ends, which subsequently provide the substrates for the exoenzyme glucoamylase. The expression of these enzymes and their biochemistry have been extensively studied (e.g. Calleja et al. 1980; Oteng-Gyang et al. 1981; Sills et al. 1984a; Clementi and Rossi 1986; Dowhanick et al. 1990).

Two different *Schw. occidentalis* α-amylase genes *AMY1*and *SWA2* have been cloned and sequenced (Wu et al. 1991; Claros et al. 1993). The two proteins display 66% identity and 77% similarity. Both proteins share highly conserved regions found in all α-amylases that have been analysed (Claros et al. 1993). They are about 54 kDa in size and contain two potential N-glycosylation sites.

A glucoamylase gene (*GAM1*) from *Schw. occidentalis* has been cloned and sequenced (Dohmen et al. 1990). It encodes a 958-amino acid protein (including the signal sequence) which, without further post-translational modification, has a calculated molecular weight of 106.5 kDa. A *Schw. occidentalis* strain, that was completely deleted for the sequence of the (*GAM1*) coding region, has been generated (Dohmen et al. 1990). Glucoamylase in the culture medium or on the surface of intact cells of such a *gam* strain was in the range of the detection limit (<1% of the activity found in cultures of *GAM1* strains). This phenomenon could be explained by either the presence of a second weakly expressed glucoamylase gene or by the ability of α-amylase together with a putative α-glucosidase to release a small amount of glucose from these substrates. The *GAM1* gene is clearly the main contributor to glucoamylase activity of the strains tested (Poinsot et al. 1987).

Expression of amylase genes is tightly regulated in *Schw. occidentalis* and related species. The α-amylase genes *AMY1* and *SWA2*, as well as the *GAM1*, are subject to glucose repression (Ingledew 1987; Dowhanick et al. 1990; Abarca et al. 1991). The expression of α-amylase and glucoamylase is induced (in the absence of glucose) by maltose, soluble starch and melizitose—a nonreducing trisaccharide sugar (Clementi and Rossi 1986) and is repressed at elevated temperatures (>37 °C) (Calleja et al. 1984). Both enzymes from this yeast are inactivated at pH values below 3.5 and temperatures above 55 °C.

Schw. occidentalis displays a Kluyver effect on maltose and starch media. Both compounds cannot be metabolized under anaerobic conditions due to repression of α-amylase and α-glucosidase expression (Calleja et al. 1982; De Mot et al. 1985a; Boze et al. 1989). Glucose, the product of glucoamylase activity, is at the same time a repressor of its expression. This feedback regulation of glucoamylase expression is apparently the reason why the highest expression of this enzyme is frequently observed in stationary growth phase (Clementi and Rossi 1986). When the concentration of a glucoamylase substrate is sufficiently low (usually in stationary growth phase) such that substrate hydrolysis does not generate repressing glucose concentrations, glucoamylase expression can proceed uninhibited.

Schw. occidentalis and related species have become a target for investigators involved with biotechnological processes because of their ability to completely degrade starch. It has already been discussed (Chap. 7) that starch is one of the most abundant renewable carbon sources on this planet (Hollenberg and Wilhelm 1987). The enzymes required for starch breakdown, α-amylase, glucoamylase and debranching enzymes (pullulanase) (Fig. 7.1) are amongst the biotechnologically most highly produced enzymes worldwide (Stewart 1987). Starch is used as a raw material in processes leading to high dextrose and high-fructose corn syrup (HFCS), the latter being used in the food and drink industries and in ethanol fermentation (https://en.wikipedia.org/wiki/High-fructose_corn_syrup).

In processes such as beer brewing or the production of malt whisky (Russell and Stewart 2014), saccharification by the amylolytic enzymes present in barley malt is incomplete and approximately 25% of starch material remains as unfermented dextrin (Tubb et al. 1986). These dextrins are, besides ethanol, the main contributors of calories in beer. In order to reduce the dextrin content of beer, a variety of efforts involving *Schw. occidentalis* have been made to introduce amylase activity, in particular glucoamylase with debranching activity into the fermentation step. These included the addition of amylases obtained from *Schw. occidentalis* cultures (Sills et al. 1983) and the introduction of amylase genes into *S. cerevisiae* (Dohmen et al. 1990; Dowhanick et al. 1990; Naim et al. 1991).

Another possible application (not to produce alcohol but will be discussed for completeness) for *Schw. occidentalis* strains is their direct utilization in single-cell protein (SCP) production from inexpensive starchy raw materials. This process still requires considerable yeast strain improvement (Ingledew 1987). The ability of *Schw. occidentalis* to grow in inexpensive media also makes it a choice as a host for the production of heterologous proteins.

The yeast *Brettanomyces bruxellensis* is the asexual reproductive stage (anamorph) of *Dekkera bruxellensis*. It is a yeast associated with, and named after, the Senne valley near Brussels, Belgium. It is one of several members of the genus *Brettanomyces*, which were first classified in the Carlsberg Laboratory. This research was conducted in 1904 by N. Hjelte Claussen, who was investigating this yeast as a cause of spoilage in English ales—hence the name. Despite its Latin species name, *B. bruxellensis*, it is found all over the globe. It often occurs on the skins of fruit (grapes, etc).

B. bruxellensis plays a key role in the production of typical Belgian beer styles such as Lambic, Flanders red ales, Gueuze, Kriek and Orval. It is naturally found in the brewery environment living within oak barrels that are used for the storage of beer during the secondary conditioning stage. Here it completes the long slow fermentation of super-attenuation of beer, often in symbiosis with the Gram-positive bacterium *Pediococcus* sp.

B. bruxellensis is increasingly being used by American craft brewers, especially in California and Colorado. For example, Port Brewing Company, Sierra Nevada Brewing Company, Russian Brewing Company, New Belgium Brewing Company and Rocket Brewing Company all have beers fermented with *B. bruxellensis*. These beers have a slightly sour, earthy character. Some tasters have described these beers as having a "barnyard" or "wet horse blanket" flavour.

B. bruxellensis is considered to be a spoilage yeast in the wine industry. It and other members of this genus are often referred to as *brett*. Its metabolic products can impact "sweaty saddle leather", "barnyard", "burnt plastic" or "band-aid" aromas to wine. Some winemakers in France (and elsewhere) consider it as a desirable addition to wine. New World vintners generally consider it to be a defect. Some authorities consider *brett* to be responsible for 90% of the spoilage problems in premium red wines. A defence against *brett* is to limit potential sources of contamination. It occurs more commonly in some vineyards than others. Consequently, producers can avoid purchasing grapes from such sources. Used wine barrels obtained from other vintners are another common source. Some producers sanitize used barrels with ozone. Others steam or soak them for many hours in very hot water or wash them with either citric acid or peroxycarbonate. If wine becomes contaminated by *brett*, some vintners sterile filter it, add SO_2, or treat it with dimethyl dicarbonate.

17.5 Pentose Fermenting Yeast Species

Despite the wealth of literature in the field of pentose (five-carbon) fermentation (Schneider et al. 1983), relatively little attention has been devoted to the performance of pentose-fermenting strains with industrial substrates and environments (Hahn-Hägerdal and Pamment 2004). Four industrial benchmarks have the greatest influence on the price of ethanol produced from a pentose containing substrate, particularly lignocellulosics (Wingren et al. 2003)—no order of priority:

- Ethanol yield.
- Specific ethanol productivity.
- Inhibitor tolerance.
- Process water economy.

It has already been discussed that *Saccharomyces* strains cannot metabolize pentoses. There are non-*Saccharomyces* species that can ferment pentoses, and the primary species in this category are discussed.

17.5.1 Pichia stipitis

Pichia stipitis (syn *Candida shehatae*) is distantly related to brewer's yeast. Found in other locations, such as the gut of passalid beetles, *P. stipitis* is capable of both aerobic and oxygen-limited fermentation and has the greatest known ability of any yeast to directly ferment xylose, converting it to ethanol. Xylose is a hemicellulosic sugar found in plants. As such, xylose constitutes the second most abundant carbohydrate moiety in nature (after glucose). Xylose can be produced from wood or agricultural residues through auto- or acid hydrolysis. Ethanol production from such lignocellulosic residues does not compete with food production through the consumption of grain. Given the abundance of xylose and its potential for the bioconversion of lignocellulosic materials to renewable fuels, *P. stipitis* has been extensively studied (Björling and Lindman 1989).

The complete sequence of this yeast's genome was published in 2007 (Jeffries et al. 2007). Native strains of *P. stipitis* have been shown to produce ~50 g/L ethanol in 48 h from pure xylose in a defined minimal medium with urea as the nitrogen source. *P. stipitis* is a predominantly haploid yeast. However, strains can be induced to mate with themselves or other strains of *P. stipitis* by cultivating cells on minimal media containing limiting amounts of carbon sources and nitrogen. Engineered strains of *P. stipitis* will produce 57 g/L ethanol from pure xylose in under 48 h (Spencer and Spencer 1997).

The natural ability of *P. stipitis* to ferment xylose to ethanol has inspired efforts to engineer this property into *S. cerevisiae*. *S. cerevisiae* prefers grain and cane sugar for ethanol production because it ferments hexose sugars rapidly and robustly. However, it does not naturally metabolize xylose. This limits the usefulness of *S. cerevisiae* in the production of fuels and chemicals from plant cell walls, which contain a large amount of xylose. Consequently, *S. cerevisiae* has been engineered to ferment xylose through the addition of the *P. stipitis*, *XYL1* and *XYL2*, as coding for xylose reductase and xylitol dehydrogenase, respectively. The combined action of these enzymes converts xylose to xylulose, which is naturally fermented by *S. cerevisiae*.

17.5.2 *Pachysolen tannophilus*

Pachysolen is a genus of yeast isolated from sulphite liquor (Spencer and Spencer 1997). This yeast differs from other pentose-metabolizing yeasts in that appreciable yields can be obtained in sealed vessels; it does not grow on D-xylose in the absence of oxygen (Maleszka and Schneider 1982; Slinger et al. 1982). Considerable efforts have been devoted to developing economic processes that employ this yeast to produce ethanol from both D-xylose and hexoses in phytomass hydrolyzates. However, to date large-scale applications are still pending (Schneider et al. 1983).

17.6 *Torulaspora delbrueckii*

Using an unconventional non-*Saccharomyces* yeast offers a unique opportunity for craft brewers to distinguish their beers from competitors (Basso et al. 2016). Research globally has started to focus on the possibilities of finding non-*Saccharomyces* yeast with potential brewing ability (Steensels and Verstrepen 2014). One example is *Torulaspora delbrueckii* (Canonico et al. 2016; Michel et al. 2016). This is a ubiquitous yeast species with both wild and anthropic (relating to human beings) habitats. *T. delbrueckii* is the most studied species of the *Torulaspora* genus, which currently consists of eight species in total. Although this species has been associated with winemaking for decades (Albertin et al. 2014), its application in brewing and wine production is still being investigated (Michel et al. 2016). The main fermentation control parameters that influence beer quality with *T. delbrueckii* are (1) wort oxygenation, (2) cell pitching rate and (3) fermentation temperature (Maximilian et al. 2017).

The fermentation conditions were chosen according to prior testing at varying fermentation temperatures between 15 and 25 °C and pitching rates, due to relatively small cell sizes of between 50×10^6 and 120×10^6 cells/mL. Further, the optimal pitching time has been investigated and the viability, vitality and cell count investigated. Fermentation at 20 °C and a pitching rate of 60×10^6 cells/mL led to the most acceptable beer with a blackcurrant and honey-like flavour. A flavour shift was found from honey-like at low fermentation temperatures (approximately 10 °C) to wine-like at higher temperatures (20 °C).

17.7 *Zymomonas mobilis*

Zymomonas mobilis is a Gram-negative, facultative anaerobic, non-sporulating, polarly flagellated, rod-shaped bacterium https://en.wikipedia.org/wike/Zymomonas_mobilis. It is the only species in the genus *Zymomonas*. It has notable bioethanol-producing capabilities, which compares yeast in some (not all) aspects (details in Chap. 9). It was originally isolated from alcoholic beverages such as African palm wine, Mexican pulque and also as a contaminant of cider and beer (cider and beer spoilage) in European countries. This problem has been known to cause British brewers nightmares (Paradh 2015)!

As an unwanted waterborne bacteria in beer, it causes an unpleasant estery-sulphury flavour due to the production of acetaldehyde and hydrogen sulphide. It is commonly found in cask-conditioned ales where priming sugar is used to carbonate (prime) the beer. This bacterium's optimum growth temperature is 25–30 °C.

Z. mobilis degrades sugars to pyruvate using the Entner-Doudoroff (pentose phosphate pathway) (Fig. 17.7). The pyruvate is then decarboxylated to produce ethanol and carbon dioxide (analogous to *S. cerevisiae* with respect to the production of bioethanol). The principal factors affecting this process are:

- Higher sugar uptake and ethanol yield (approx 2.5 times higher) (Rogers et al. 1982).
- Lower amounts of biomass produced (Chen et al. 2009).
- Enhanced ethanol tolerance [up to 16% (v/v)].
- Does not require controlled addition of oxygen during the fermentation (Agrawal et al. 2011).
- Amenability to genetic manipulation.

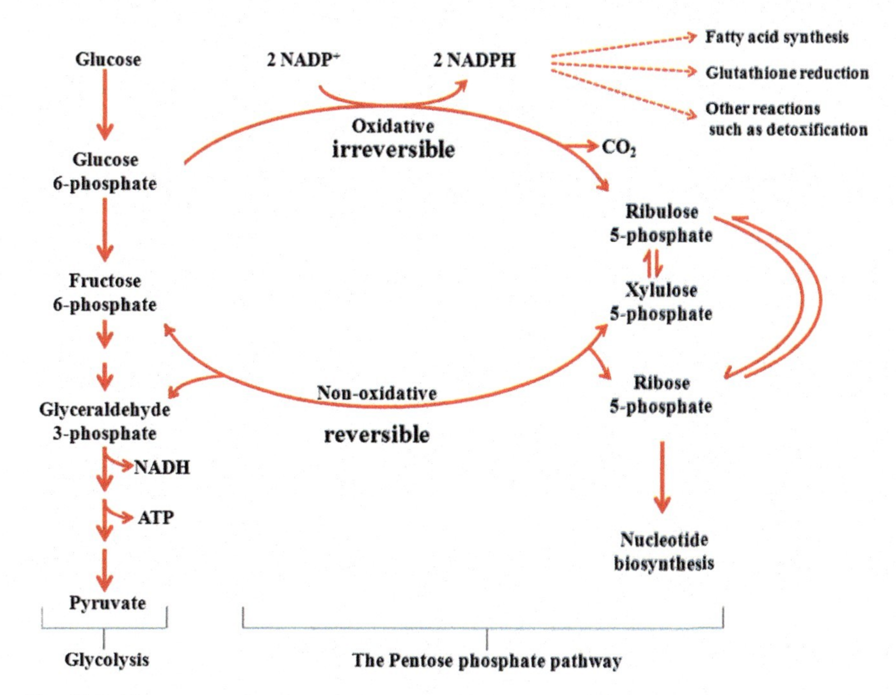

Fig. 17.7 The pentose phosphate pathway (Entner-Douderoff pathway)

In spite of these attractive advantages, several factors currently prevent the commercial use of *Z. mobilis* for ethanol production from cellulose. The most important hurdle is that its substrates are limited to glucose, fructose and sucrose; it cannot take up maltose and maltotriose. Wild-type *Z. mobilis* strains cannot ferment C5 sugars (xylose and arabinose) which are important components of lignocellulosic hydrolysates. Unlike *E. coli* and *S. cerevisiae*, *Z. mobilis* cannot tolerate toxic inhibitors present in lignocellulosic hydrolysates such as acetic acid and various phenolic compounds (Doran-Peterson et al. 2008). The concentration of acetic acid in lignocellulosic hydrolysates can be as high as 1.5% (w/v)—due to the tolerance threshold of *Z. mobilis*.

Several attempts have been made to engineer *Z. mobilis* to overcome the deficiencies already discussed. The substitute range has been expanded to include xylose and arabinose (Deanda et al. 1996; Yang et al. 2016). Also, acetic acid-resistant strains of *Z. mobilis* have been developed employing metabolic engineering and mutagenesis techniques (Joachimsthal and Rogers 2000). However, when these engineered strains metabolize mixed sugars in the presence of inhibitors, the yield and productivity are considerably lower and thus prevent their industrial application.

An interesting characteristic of *Z. mobilis* is that its plasma membrane contains compounds similar to eukaryotic sterols. This permits it to exhibit an environmental tolerance to ethanol of around 13% (v/v).

17.8 Summary

In excess of a thousand unique yeast species have been identified, and many of them have been characterized (to a lesser or greater extent). Ninety percent (and more) of the fermentation ethanol produced globally employs species of the genus *Saccharomyces* (predominantly *S. cerevisiae* and *S. pastorianus*). However, there are a number of non-*Saccharomyces* yeast species that can produce ethanol (also called nonconventional yeast species). Nonconventional yeasts are a large and barely exploited resource of yeast biodiversity. Many of these nonconventional yeast species exhibit industrially relevant traits such as an ability to utilize complex substrates, nutrient tolerance against stresses and fermentation inhibition. The evolution of most of these yeast species was independent of *Saccharomyces* sp. Many of them possess novel and unique mechanisms that are not present in *Saccharomyces* yeasts. Most of them have been characterized as spoilage yeasts because they have been isolated from contaminated foods and beverages. Yeast species that are included in this category are *Schizosaccharomyces pombe*, *Kluyveromyces marxianus*, *Schwanniomyces occidentalis*, *Brettanomyces bruxcellensis*, *Pichia stipitis*, *Pachysolen tannophilus* and *Torulaspora delbrueckii*.

References

Abarca D, Fernandez-Lobato M, Del Poso L, Jimenez A (1991) Isolation of a new gene (SWA2) encoding a new α-amylase *Swanniomyces occidentalis* and its expression in *Saccharomyces cerevisiae*. FEBS Lett 279:41–44

Abdel-Banat BMA, Hoshida H, Ano A, Nonklang S, Akada R (2010) High-temperature fermentation: how can processes for ethanol production at high temperatures become superior to the traditional process using mesophilic yeast? Appl Microbiol Biotechnol 85:861–867

Agrawal M, Mao Z, Chen RR (2011) Adaptation yields a highly efficient xylose-fermenting *Zymomonas mobilis* strain. Biotechnol Bioeng 108:777–785

Albertin W, Chasseriaud L, Comte G, Panfili A, Delcamp A, Salin F, Marullo P, Bely M (2014) Winemaking and bioprocesses strongly shaped the genetic diversity of the ubiquitous yeast *Torulaspora delbrueckii*. PLoS One:e94246

Barnett JA (1992) Some controls on oligosaccharide utilization by yeasts: the physiological basis of the Kluyver effect. FEMS Microbiol Lett 100:371–378

Baruffini E, Goffrini P, Donnini C, Lodi T (2006) Galactose transport in *Kluyveromyces lactis*: major role of the glucose permease Hgt1. FEMS Yeast Res 6:1235–1242

Basso RE, Alcarde AR, Portugal CB (2016) Could non-*Saccharomyces* yeasts contribute on innovative brewing fermentations? Food Res Int 86:112–120

Becerra M, Rodríguez-Belmonte E, Esperanza Cerdán M, González Siso MI (2004a) Engineered autolytic yeast strains secreting *Kluyveromyces lactis* beta-galactosidase for production of heterologous proteins in lactose media. J Biotechnol 8:131–137

Becerra M, Tarrío N, González-Siso MI, Cerdán ME (2004b) Genome-wide analysis of *Kluyveromyces lactis* in wild-type and rag2 mutant strains. Genome 47:970–978

Belloch C, Barrio E, García MD, Querol A (1998a) Inter- and intraspecific chromosome pattern variation in the yeast genus *Kluyveromyces*. Yeast 14:1341–1354

Belloch C, Barrio E, García MD, Querol A (1998b) Phylogenetic reconstruction of the yeast genus *Kluyveromyces*: restriction map analysis of the 5.8S rRNA gene and the two ribosomal internal transcribed spacers. Syst Appl Microbiol 21:266–273

Björling T, Lindman B (1989) Evaluation of xylose-fermenting yeasts for ethanol production from spent sulfite liquor. Enzyme Microb Technol 11(4):240–246

Boekhout T (2005) Biodiversity: gut feeling for yeasts. Nature 434(7032):449–451

Boze H, Moulin G, Galzy P (1987a) A comparison of growth yields obtained from *Schwanniomyces castellii* and an alcohol dehydrogenase mutant. Biotechnol Lett 9:461–466

Boze H, Moulin G, Galzy P (1987b) Influence of culture conditions on the yield and amylase biosynthesis in continuous culture by *Schwanniomyces castellii*. Arch Microbiol 148:162–166

Boze H, Guyot JB, Moulin G, Galzy B (1989) Isolation and characterization of a derepressed mutant of *Schwanniomyces castellii* for amylase production. Appl Microbiol Biotechnol 31:366

Breunig KD, Bolotin-Fukuhara M, Bianchi MM, Bourgarel D, Falcone C, Ferrero II, Frontali L, Goffrini P, Krijger JJ, Mazzoni C, Milkowski C, Steensma HY, Wésolowski-Louvel M, Zeeman AM (2000) Regulation of primary carbon metabolism in *Kluyveromyces lactis*. Enzym Microb Technol 26:771–780

Calleja GB, Zuker M, Johnson BF, Yoo BY (1980) Analyses of fission scars as permanent records of cell division in *Schizosaccharomyces pombe*. J Theor Biol 84:523–544

Calleja GB, Levy-Rick S, Lusena CV, Moranelli F, Nasim A (1982) Direct and quantitative conversion of starch to ethanol by the yeast *Schwanniomyces alluvius*. Biotechnol Lett 4:543–546

Calleja GB, Levy-Rick S, Moranelli F, Nasim A (1984) Thermosensitive export of amylases in the yeast *Schwanniomyces alluvius*. Plant Cell Physiol 25:757–761

Canonico C, Agarbati A, Comitini F, Ciani M (2016) *Torulaspora delbrueskii* in the brewing process: a new approach to enhance bioflavour and reduce ethanol content. Food Microbiol 56:45–51

Chen RR, Wang Y, Shin H-D, Agrawal M, Mao Z (2009) Strains of *Zymomonas mobilis* for fermentation of biomass. US Patent Appl. No. US20090269797

Clapp C, Portt L, Khoury C, Sheibani S, Norman G, Ebner P, Eid R, Vali H, Mandato CA, Madeo F, Greenwood MT (2012) 14-3-3 protects against stress-induced apoptosis. Cell Death Dis 3:e348

Claros MG, Abarca D, Fernández-Lobato M, Jiménez A (1993) Molecular structure of the SWA2 gene encoding an AMY1-related alpha-amylase from *Schwanniomyces occidentalis*. Curr Genet 24:75–83

Clementi F, Rossi J (1986) Alpha-amylase and glucoamylase production by *Schwanniomyces castellii*. Antonie Van Leeuwenhoek 52:343–352

Coenen TM, Bertens AM, de Hoog SC, Verspeek-Rip CM (2000) Safety evaluation of a lactase enzyme preparation derived from *Kluyveromyces lactis*. Food Chem Toxicol 38:671–677

Colussi PA, Taron CH (2005) *Kluyveromyces lactis* LAC4 promoter variants that lack function in bacteria but retain full function in K. lactis. Appl Environ Microbiol 71:7092–7098

Colussi PA, Specht CA, Taron CH (2005) Characterization of a nucleus-encoded chitinase from the yeast *Kluyveromyces lactis*. Appl Environ Microbiol 71:2862–2869

Crichton PG, Affourtit C, Moore AL (2007) Identification of a mitochondrial alcohol dehydrogenase in *Schizosaccharomyces pombe*: new insights into energy metabolism. Biochem J 401:459–464

De Moreas LMPS, Astolfi-Filhole S, Oliver SG (1995) Development of yeast strains for the efficient utilization of starch: evaluation of constructs that express amylase and glucoamylase separately or as bifunctional fusion proteins. Appl Microbiol Biotechnol 43:1067–1076

De Mot R, Verachtert H (1985) Purification and characterization of extracellular amylolytic enzymes from the yeast *Filobasidium capsuligenum*. Appl Environ Microbiol 50:1474–1482

De Mot R, Andries K, Verachtert H (1984) Production of extracellular debranching activity by amylolytic yeasts. Biotechnol Lett 6:581–586

De Mot R, Van Dijck K, Donkers A, Verachtert H (1985a) Potentialities and limitations of direct alcoholic fermentations of starchy material with amyloytic yeast. Appl Microbiol Biotechnol 22:222–226

De Mot R, Van Oudendijck E, Verachtert H (1985b) Purification and characterization of an extracellular glucoamylase from the yeast *Candida tsukubaensis* CBS 6389. Antonie Van Leeuwenhoek 51:275–287

Deanda K, Zhang M, Eddy C, Picataggio S (1996) Development of an arabinose-fermenting *Zymomonas mobilis* strain by metabolic pathway engineering. Appl Environ Microbiol 62:4465–4470

Dohmen RJ, Strasser AW, Dahlems UM, Hollenberg CP (1990) Cloning of the *Schwanniomyces occidentalis* glucoamylase gene (GAM1) and its expression in *Saccharomyces cerevisiae*. Gene 95:111–121

Domingues L, Guimarães PM, Oliveira C (2010) Metabolic engineering of *Saccharomyces cerevisiae* for lactose/whey fermentation. Bioeng Bugs 1:164–171

Doran-Peterson J, Cook DM, Brandon SK (2008) Microbial conversion of sugars from plant biomass to lactic acid or ethanol. Plant J 54:582–592

Dowhanick TM, Russell I, Scherer SW, Stewart GG, Seligy VL (1990) Expression and regulation of glucoamylase from the yeast *Schwanniomyces castellii*. J Bacteriol 172:2360–2366

Dujon B (2010) Yeast evolutionary genomics. Nat Rev Genet 11:512–524

Eksteen JM, Steyn AJ, van Rensburg P, Cordero Otero RR, Pretorius IS (2003a) Cloning and characterization of a second alpha-amylase gene (LKA2) from *Lipomyces kononenkoae* IGC4052B and its expression in *Saccharomyces cerevisiae*. Yeast 20:69–78

Eksteen JM, Van Rensburg P, Cordero Otero RR, Pretorius IS (2003b) Starch fermentation by recombinant *Saccharomyces cerevisiae* strains expressing the alpha-amylase and glucoamylase genes from *Lipomyces kononenkoae* and *Saccharomycopsis fibuligera*. Biotechnol Bioeng 84:639–646

Erratt JA, Stewart GG (1978) Genetic and biochemical studies on yeast strains able to utilize dextrins. J Am Soc Brew Chem 36:151–161

Erratt JA, Stewart GG (1981a) Fermentation studies using *Saccharomyces diastaticus* yeast strains. Dev Ind Microbiol 22:577–586

Erratt JA, Stewart GG (1981b) Genetic and biochemical studies on glucoamylase from *Saccharomyces diastaticus*. In: Advances in Biotechnol. Pergamon Press, Toronto, pp 177–183

Fantes PA, Nurse P (1977) Control of cell size at division in fission yeast by a growth-modulated size control over nuclear division. Exp Cell Res 107:377–386

Ferreira JD, Phaff HJ (1959) Life cycle and nuclear behaviour of a species of the yeast genus *Schwanniomyces*. J Bacteriol 78:352–361

Fonseca GG, Heinzle E, Wittmann C, Gombert AK (2008) The yeast *Kluyveromyces marxianus* and its biotechnological potential. Appl Microbiol Biotechnol 79:339–345

Forsburg SL, Rhind N (2006) Basic methods for fission yeast. Yeast 23:173–183

Fujii M, Kawamura Y (1985) Synergistic action of α-amylase and glucoamylase on hydrolysis of starch. Biotechnol Bioeng 27:260–265

Fukuhara H (2003) The Kluyver effect revisited. FEMS Yeast Res 3:327–331

Fuson GB, Presley HL, Phaff HJ (1987) Deoxyribonucleic acid base sequence relatedness among members of the yeast genus *Kluyveromyces*. Int J Syst Bacteriol 37:371–379

Goffeau A (1996) A vintage year for yeast. Yeast 12:1603–1605

Goffeau A, Barrell BG, Bussey H, Davis RW, Dujon B (1996) Life with 6000 genes. Science 274:546–547

Hagen I, Carr AM, Grallert A, Nurse P (2016) Fission yeast: a laboratory manual. Cold Spring Harbor Laboratory Press, Cold Spring Harbor, NY

Hahn-Hägerdal B, Pamment N (2004) Microbial pentose metabolism. Appl Biochem Biotechnol 113-116:1207–1209

Heinisch JJ, Buchwald U, Gottschlich A, Heppeler N, Rodicio R (2010) A tool kit for molecular genetics of *Kluyveromyces lactis* comprising a congenic strain series and a set of versatile vectors. FEMS Yeast Res 10:333–342

Hnatova M, Wesolowski-Louvel M, Dieppois G, Deffaud J, Lemaire M (2008) Characterization of KlGRR1 and SMS1 genes, two new elements of the glucose signaling pathway of *Kluyveromyces lactis*. Eukaryot Cell 7:1299–1308

Hollenberg CP, Wilhelm M (1987) New substrates for old organisms. In: Biotec I: microbiol genetic engineering and enzyme technology. Gustav Fisher, Stutgart, pp 21–31

Hoshida H, Murakami N, Suzuki A, Tamura R, Asakawa J, Abdel-Banat BM, Nonklang S, Nakamura M, Akada R (2014) Non-homologous end joining-mediated functional marker selection for DNA cloning in the yeast *Kluyveromyces marxianus*. Yeast 31:29–46

Ingledew WM (1987) *Schwanniomyces*: a potential superyeast? Crit Rev Biotechnol 5:159–176

Jeffries TW, Grigoriev IV, Grimwood J, Laplaza JM, Aerts A, Salamov A, Schmutz J, Lindquist E, Dehal P, Shapiro H, Jin YS, Passoth V, Richardson PM (2007) Genome sequence of the lignocellulose-bioconverting and xylose-fermenting yeast *Pichia stipitis*. Nat Biotechnol 25:319–326

Jeong H, Lee D-H, Kim SH, Kim HJ, Lee K, Song JY, Kim BK, Sung BH, Park JC, Sohn JH, Koo HM, Kim JF (2012) Genome sequence of the thermotolerant yeast *Kluyveromyces marxianus* var. *marxianus* KCTC 17555. Eukaryot Cell 11:1584–1585

Joachimsthal EL, Rogers PL (2000) Characterization of a high productivity recombinant strain of *Zymomonas mobilis* for ethanol production from glucose/xylose mixtures. Appl Biochem Biotechnol 84–86:343–356

Kegel A, Martinez P, Carter SD, Aström SU (2006) Genome wide distribution of illegitimate recombination events in *Kluyveromyces lactis*. Nucleic Acids Res 34:1633–1645

Krijger JJ, Baumann J, Wagner M, Schulze K, Reinsch C, Klose T, Onuma OF, Simon C, Behrens SE, Breunig KD (2012) A novel, lactase-based selection and strain improvement strategy for recombinant protein expression in *Kluyveromyces lactis*. Microb Cell Factories 11:112

Lane MM, Burke N, Karreman R, Wolfe KH, O'Byrne CP, Morrissey JP (2011) Physiological and metabolic diversity in the yeast *Kluyveromyces marxianus*. Antonie Van Leeuwenhoek 100:507–519

Lee K-S, Kim J-S, Heo P, Lee K-S, Kim J-S, Heo P, Yang T-J, Sung Y-J, Cheon Y, Koo HM, Yu BJ, Seo J-H, Jin Y-S, Park JC, Kweo D-H (2013) Characterization of *Saccharomyces cerevisiae* promoters for heterologous gene expression in *Kluyveromyces marxianus*. Appl Microbiol Biotechnol 97:2029–2041

Leupold U (1950) Die verebung van homothallie und homothallie und heterothallis bei *Saccharomyces pombe*. CR Trav Lab Carlsberg Ser Physiol 24:381–480

Leupold U (1993) The origin of *Schizosaccharomyces pombe* genetics. In: Hall MN, Linder P (eds) The early days of yeast genetics. Cold Spring Harbor Laboratory Press, Cold Spring Harbor, NY, pp 125–128

Limtong S, Sringiew C, Yongmanitchai W (2007) Production of fuel ethanol at high temperature from sugar cane juice by a newly isolated *Kluyveromyces marxianus*. Bioresour Technol 98:3367–3374

Lin Z, Li WH (2011) The evolution of aerobic fermentation in *Schizosaccharomyces pombe* was associated with regulatory reprogramming but not nucleosome reorganization. Mol Biol Evol 28:1407–1413

Llorente B, Malpertuy A, Blandin G, Artiguenave F, Wincker P, Dujon B (2000) Genomic exploration of the hemiascomycetous yeasts: 12. *Kluyveromyces marxianus* var. *marxianus*. FEBS Lett 487:71–75

Lodder J (ed) (1984) The yeasts: a taxonomic study, 3rd edn. North Holland Publishing, Amsterdam

Lomer M, Parkes G, Sanderson J (2008) Review article: lactose intolerance in clinical practice – myths and realities. Aliment Pharmacol Ther 27:93–103

Maleszka R, Schneider H (1982) Concurrent production and consumption of ethanol by cultures of *Pachysolen tannophilus* growing on D-xylose. Appl Environ Microbiol 44:909–912

Martinez LA, Naguibneva I, Lehrmann H, Vervisch A, Tchénio T, Lozano G, Harel-Bellan A (2002) Synthetic small inhibiting RNAs: efficient tools to inactivate oncogenic mutations and restore pathways. In: Vogt PK (ed) Proceedings of the National Academy of Sciences of the USA. The Scripps Research Institute, La Jolla, CA, p 53

Martini AV, Martini A (1987) Taxonomic revision of the yeast genus *Kluyveromyces* by nuclear deoxyribonucleic acid reassociation. Int J Syst Bacteriol 44:380–385

Martorell P, Stratford M, Steels H, Fernández-Espinar MT, Querol A (2007) Physiological characterization of spoilage strains of Zygosaccharomyces bailii and Zygosaccharomyces rouxii isolated from high sugar environments. Int J Food Microbiol 114:234–242

Maximilian M, Meier-Dörnberg T, Jacob F, Schneiderbanger H, Haselbeck K, Zarnkow M, Hutzler M (2017) Optimization of beer fermentation with a novel brewing strain of *Torulaspora delbrueckii* using response surface methodology. MBAA Tech Quart 54:23–33

Michel M, Kopecká J, Meier-Dörnberg T, Zarnkow M, Jacob F, Hutzler M (2016) Screening for new brewing yeasts in the non-*Saccharomyces* sector with *Torulaspora delbrueckii* as model. Yeast 33:129–144

Mitchison JM (1957) The growth of single cells. I. *Schizosaccharomyces pombe*. Exp Cell Res 13:244–262

Naim HY, Niermann T, Kleinhans U, Hollenberg CP, Strasser AW (1991) Striking structural and functional similarities suggest that intestinal sucrase-isomaltase, human lysosomal alpha-glucosidase and *Schwanniomyces occidentalis* glucoamylase are derived from a common ancestral gene. FEBS Lett 294:109–112

Nonklang S, Abdel-Banat BMA, Cha-aim K, Moonjai N, Hoshida H, Limtong S, Yamada M, Akada R (2008) High-temperature ethanol fermentation and transformation with linear DNA in the thermotolerant yeast *Kluyveromyces marxianus* DMKU3–1042. Appl Environ Microbiol 74:7514–7521

Novak B, Mitchison JM (1990a) CO_2 production after induction synchrony of the fission yeast *Schizosaccharomyces pombe*: the origin and nature of entrainment. J Cell Sci 96:79–91

Novak B, Mitchison JM (1990b) Change in the rate of oxygen consumption in synchronous cultures of the fission yeast *Schizosaccharomyces pombe*. J Cell Sci 96:429–433

Oteng-Gyang K, Moulin G, Galzy P (1981) A study of amylolytic system of *Schwanniomyces castelii*. J Basic Microbiol 21:537–544

Panesar PS, Kumari S, Panesar R (2010) Potential applications of immobilized β-galactosidase in food processing industries. Enzyme Res 2010:12–27

Paradh AD (2015) Gram-negative spoilage bacteria in brewing. In: Hill AE (ed) Brewing microbiology. Woodhead Publishing, Cambridge, pp 175–194

Pecota DC, Rajgarhia V, Da Silva NA (2007) Sequential gene integration for the engineering of *Kluyveromyces marxianus*. J Biotechnol 127:408–416

Phaff HJ (1970) Genus 20. *Schwanniomyces klocker*. In: Lodder J (ed) The yeasts, a taxonomic study, 2nd edn. North-Holland Publishing, Amsterdam, pp 756–766

Phaff HJ (1981) The species concept in yeasts: physiology, morphology, genetic and ecological parameters. In: Stewart GG, Russell I (eds) Current developments in yeast research. Pergamon Press, Toronto, pp 635–643

Phaff HJ, Miller MW, Mrak EM (1966) The life of yeasts. Harvard University Press, Cambridge, MA

Poinsot C, Moulin G, Claisse M, Galzy P (1987) Isolation and characterization of a mutant of *Schwanniomyces castellii* with altered respiration. Antonie Van Leeuwenhoek 53:65–70

Prazeres AR, Carvalho F, Rivas J (2012) Cheese whey management: a review. J Environ Manag 110:48–68

Price CW, Fuson GB, Phaff HJ (1978) Genome comparison in yeast systematics: delimitation of species within the genera *Schwanniomyces, Saccharomyces, Debaryomyces*, and *Pichia*. Am Soc Microbiol 42:161–193

Radecka D, Mukherjee V, Mateo RQ, Stojiljkovic M, Foulquié-Moreno MR, Thevelein JM (2015) Looking beyond *Saccharomyces*: the potential of non-conventional yeast species for desirable traits in bioethanol fermentation. FEMS Yeast Res 15 pii: fov053

Rigamonte TA, Silveira WB, Fietto LG, Castro IM, Breunig KD, Passos FM (2011) Restricted sugar uptake by sugar-induced internalization of the yeast lactose/galactose permease Lac12. FEMS Yeast Res 11:243–251

Rives J, Fernandez-Rodriguez I, Rieradevall J, Gabarrell X (2011) Environmental analysis of the production of natural cork stoppers in Southern Europe (Catalonia e Spain). J Clean Prod 19:259–271

Rogers P, Lee K, Skotnicki M, Tribe D (1982) Microbial reactions: ethanol Production by *Zymomonas mobilis*. Springer, New York, Berlin, pp 37–84

Rossi J, Clementi F (1985) Protein production by *Schwanniomyces castelli* on starchy substrates, in liquid and solid cultivation. J Food Technol 20:318–330

Rouwenhorst RJ, Visser LE, Baan AA, Scheffers WA, Van Dijken JP (1988) Production, distribution, and kinetic properties of inulinase in continuous cultures of *Kluyveromyces marxianus* CBS 6556. Appl Environ Microbiol 54:1131–1137

Russell I, Stewart GG (2014) Whisky: technology production and marketing, 2nd edn. Academic Press (Elsevier), Boston, MA

Saliola M, Shuster JR, Falcone C (1990) The alcohol dehydrogenase system in the yeast, *Kluyveromyces lactis*. Yeast 6:193–204

Schneider H, Maleszka R, Neirinck LG, Veliky IA, Wang PY, Chan YK (1983) Ethanol production from D-xylose and several other carbohydrates by *Pachysolen tannophilus* and other yeasts. In: Fiechter A (ed) Advances in biochemical engineering biotechnology. Springer, Berlin

Seyis I, and Aksoz N (2004) Production of lactase by *Trichoderma* sp. Food Tech. Biotechnol 42:121–124

Sicard D, Legras JL (2011) Bread, beer and wine: yeast domestication in the *Saccharomyces sensu stricto* complex. C R Biol 334:229–236

Sills AM, Stewart GG (1982) Production of amylolytic enzymes by several yeast species. J Inst Brew 88:313–316

Sills AM, Sauder ME, Stewart GG (1983) Amylase activity in certain yeasts and a fungal species (*Schwanniomyces castellii, Endomycopsis fibuligera, Aspergillus oryzae*). Dev Ind Microbiol 24:295–303

Sills AM, Zygora PSJ, Stewart GG (1984a) Characterization of *Schwanniomyces casteliii* mutants with increased productivity of amylases. Appl Microbiol Biotechnol 20:124–128

Sills AM, Sauder ME, Stewart GG (1984b) Isolation and characterization of the amylolytic system of *Schwanniomyces castellii*. J Inst Brew 90:311–314

Sills AM, Panchal CJ, Russell I, Stewart GG (1987) Production of amylolytic enzymes by yeasts and their utilization in brewing. Crit Rev Biotechnol 5:105–115

Slinger PJ, Bothast RJ, van Cauwenberge JE, Curtzman CP (1982) Conversion of D-xylose to ethanol by the yeast *Pachysolen tannophilus*. Biotechnol Bioeng 24:371–384

Snoek IS, Steensma HY (2006) Why does *Kluyveromyces lactis* not grow under anaerobic conditions? Comparison of essential anaerobic genes of *Saccharomyces cerevisiae* with the *Kluyveromyces lactis* genome. FEMS Yeast Res 6:393–403

Souciet JL, Dujon B, Gaillardin C, Johnston M, Baret PV, Cliften P, Sherman DJ, Weissenbach J, Westhof E, Wincker P, Jubin C, Poulain J, Barbe V, Ségurens B, Artiguenave F, Anthouard V, Vacherie B, Val ME, Fulton RS, Minx P, Wilson R, Durrens P, Jean G, Marck C, Martin T, Nikolski M, Rolland T, Seret ML, Casarégola S, Despons L, Fairhead C, Fischer G, Lafontaine I, Leh V, Lemaire M, de Montigny J, Neuvéglise C, Thierry A, Blanc-Lenfle I, Bleykasten C, Diffels J, Fritsch E, Frangeul L, Goëffon A, Jauniaux N, Kachouri-Lafond R, Payen C, Potier S, Pribylova L, Ozanne C, Richard GF, Sacerdot C, Straub ML, Talla E (2009) Comparative genomics of protoploid Saccharomycetaceae. Genome Res 19:1696–1709

Spencer JRT, Spencer DM (1997) Taxonomy: the names of the yeasts. In: Spencer JRT, Spencer DM (eds) Natural and artificial habitats. Springer, Berlin, pp 11–32

Spencer-Martins I, van Uden N (1979) Extracellular amylolytic system of the yeast *Lipomyces kononenkoae*. Eur J Appl Microbiol Biotechnol 6:241–250

Spencer-Martins I, van Uden N (1982) The temperature profile of growth, death and yield of the starch-converting yeast *Lipomyces kononenkoae*. J Basic Microbiol 22:503–505

Steensels J, Verstrepen KJ (2014) Taming wild yeast: potential of conventional and nonconventional yeasts in industrial fermentations. Annu Rev Microbiol 68:61–80

Stewart GG (1987) The biotechnological relevance of starch-degrading enzymes. CRC Crit Rev Biotechnol 5:89–94

Teoh AL, Heard G, Cox J (2004) Yeast ecology of Kombucha fermentation. Int J Food Microbiol 95:119–126

Touzi A, Prebois JP, Moulin G, Deschamps F, Galzy P (1982) Production of food yeast from starchy substrates. Eur J Appl Microbial Biotechnol 15:232–236

Tubb RS, Liljestrom PL, Torkkeli T, Korhola M (1986) In: Priest EG, Campbell I (eds) Proc 2nd Aviemore conf on malting, brewing and distilling. Institute of Brewing, London, pp 298–306

van den Berg JA, van den Laken KJ, van Ooyen AJ, Renniers TC, Rietveld K, Schaap A, Brake AJ, Bishop RJ, Schultz K, Moyer D (1990) *Kluyveromyces* as a host for heterologous gene expression: expression and secretion of prochymosin. Biotechnology (NY) 8:135–139

van Ooyen AJ, Dekker P, Huang M, Olsthoorn MM, Jacobs DI, Colussi PA, Taron CH (2006) Heterologous protein production in the yeast *Kluyveromyces lactis*. FEMS Yeast Res 6:381–392

Wilhelm BT, Marguerat S, Watt S, Schubert F, Wood V, Goodhead I, Penkett CJ, Rogers J, Bähler J (2008) Dynamic repertoire of a eukaryotic transcriptome surveyed at single-nucleotide resolution. Nature 453:1239–1243

Wilson JJ, Ingledew WM (1982) Isolation and characterization of *Schwanniomyces alluvius* amylolytic enzymes. Appl Environ Microbiol 44:301–307

Winderickx J, Delay C, De Vos A, Klinger H, Pellens K, Vanhelmont T, Van Leuven F, Zabrocki P (2008) Protein folding diseases and neurodegeneration: lessons learned from yeast. Biochim Biophys Acta 1783:1381–1395

Wingren A, Galbe M, Zacchi G (2003) Techno-economic evaluation of producing ethanol from softwood: comparison of SSF and SHF and identification of bottlenecks. Biotechnol Prog 19:1109–1117

Wolf K (1996) Nonconventional yeasts in biotechnology. Springer, Berlin

Wood V, Gwilliam R, Rajandream MA, Lyne M, Lyne R, Stewart A, Sgouros J, Peat N, Hayles J, Baker S, Basham D, Bowman S, Brooks K, Brown D, Brown S, Chillingworth T, Churcher C, Collins M, Connor R, Cronin A, Davis P, Feltwell T, Fraser A, Gentles S, Goble A, Hamlin N, Harris D, Hidalgo J, Hodgson G, Holroyd S, Hornsby T, Howarth S, Huckle EJ, Hunt S, Jagels K, James K, Jones L, Jones M, Leather S, McDonald S, McLean J, Mooney P, Moule S, Mungall K, Murphy L, Niblett D, Odell C, Oliver K, O'Neil S, Pearson D, Quail MA, Rabbinowitsch E, Rutherford K, Rutter S, Saunders D, Seeger K, Sharp S, Skelton J, Simmonds M, Squares R, Squares S, Stevens K, Taylor K, Taylor RG, Tivey A, Walsh S, Warren T, Whitehead S, Woodward J, Volckaert G, Aert R, Robben J, Grymonprez B, Weltjens I, Vanstreels E, Rieger M, Schäfer M, Müller-Auer S, Gabel C, Fuchs M, Düsterhöft A, Fritzc C, Holzer E, Moestl D, Hilbert H, Borzym K, Langer I, Beck A, Lehrach H, Reinhardt R, Pohl TM, Eger P, Zimmermann W, Wedler H, Wambutt R, Purnelle B, Goffeau A, Cadieu E, Dréano S, Gloux S, Lelaure V, Mottier S, Galibert F, Aves SJ, Xiang Z, Hunt C, Moore K, Hurst SM, Lucas M, Rochet M, Gaillardin C, Tallada VA, Garzon A, Thode G, Daga RR, Cruzado L, Jimenez J, Sánchez M, del Rey F, Benito J, Domínguez A, Revuelta JL, Moreno S, Armstrong J, Forsburg SL, Cerutti L, Lowe T, McCombie WR, Paulsen I, Potashkin J, Shpakovski GV, Ussery D, Barrell BG, Nurse P (2002) The genome sequence of *Schizosaccharomyces pombe*. Nature 415:871–880

Wu FM, Wang TT, Hsu WH (1991) The nucleotide sequence of *Schwanniomyces occidentalis* alpha-amylase gene. FEMS Microbiol Lett 66:313–318

Yang S, Fei Q, Zhang Y, Contreras LM, Utturkar SM, Brown SD, Himmel ME, Zhang M (2016) *Zymomonas mobilis* as a model system for production of biofuels and biochemicals. Microb Biotechnol 9:699–717

Yarimizu T, Nonklang S, Nakamura J, Tokuda S, Nakagawa T, Lorreungsil S, Sutthikhumpha S, Pukahuta C, Kitagawa T, Nakamura M, Cha-aim K, Limtong S, Hoshida H, Akada R (2013) Identification of auxotrophic mutants of the yeast *Kluyveromyces marxianus* by non-homologous end joining-mediated integrative transformation with genes from *Saccharomyces cerevisiae*. Yeast 30:485–500

Epilogue

This book has attempted to discuss research advances on yeast with a particular emphasis on the use of this microorganism to produce ethanol and related compounds in brewing and distilling. This discussion has been based on the yeast research my colleagues and I have conducted during the last five decades or so. Also, I have attempted to relate historical developments that have occurred and bring the reader's attention to notable scientists who conducted yeast research. Portraits of many relevant scientists have been employed to try and encourage a more personal appreciation of their achievements!

During my time as a brewing research scientist, I have been fortunate to develop many personal friendships with people all over the world. Some, unfortunately, are no longer with us. However, I am pleased that two of my closest friends, Charlie Bamforth and Ludwig Narziss, are still active and prominent members of our industry. In 2014, at a brewing conference in Chicago, I was proud and delighted to be photographed with them.

August 2017 Graham G. Stewart

G.G. Stewart, *Brewing and Distilling Yeasts*, The Yeast Handbook,
https://doi.org/10.1007/978-3-319-69126-8

Index

G.G. Stewart, *Brewing and Distilling Yeasts*, The Yeast Handbook,
https://doi.org/10.1007/978-3-319-69126-8

Z

Printed by Printforce, the Netherlands